W9-BVR-524

DIGITAL FILTERING
An Introduction

Edward P. Cunningham
Johns Hopkins University
Applied Physics Laboratory

HOUGHTON MIFFLIN COMPANY BOSTON TORONTO
Dallas Geneva, Illinois Palo Alto Princeton, New Jersey

Sponsoring Editor: Rodger Klas
Development Editor: Jean Andon
Senior Project Editor: Jean Andon
Assistant Design Manager: Pat Mahtani
Senior Production Coordinator: Renée Le Verrier
Senior Manufacturing Coordinator: Marie Barnes
Marketing Manager: Michael Ginley

Printed in the U.S.A.

Library of Congress Catalog Card Number: 91-71994

ISBN: 0-395-53989-7

ABCDEFGHIJ-H-954321

To my wife Marleen,
whose support and encouragement
made this book possible

Contents

CHAPTER 7 BASIC CONCEPTS OF PROBABILITY THEORY AND RANDOM PROCESSES 364

CHAPTER 8 QUANTIZATION EFFECTS IN DIGITAL FILTERS 407

Preface

Introduction

Digital filters are widely used today for processing signals in many different applications—for example, in communication, radar, and control systems, and in medical technology. For this reason it might be difficult for a newcomer to the field to realize that the subject of digital filtering is of comparatively recent origin. Its formal beginning, as an important discipline in its own right, might be traced to the 1966 survey chapter by J. F. Kaiser in the book *System Analysis by Digital Computer,* where he gave a unified framework to previous work on the subject done by himself and others. Since then, parallel advances in high-speed computers and large-scale-integration (LSI) technology have resulted in a proliferation of applications for digital filters.

In the past twenty years, several good books have been published on digital filters and on the more general subject of digital signal processing. This book differs from most of these in the following ways:

- The first four chapters place considerable emphasis on the analogies between digital and continuous filters. An entire chapter is devoted to the latter, rather than relegating them to an appendix. Signals from most physical phenomena are continuous even if they are measured at discrete intervals of time. Therefore, the student should be familiar with the concepts of sampling and the relationship between the spectra of the continuous and sampled signals. These concepts are developed in the context of the z-transform.
- Design of the discrete Kalman filter, which has long been ignored in books on digital filters, is treated in the last chapter of the book. The development is made in a logical and easy-to-understand fashion, with the necessary probabilistic concepts reviewed in an earlier chapter.

Prerequisites

This book is based on lecture notes for a first-year graduate course that I have taught in the Continuing Professional Programs of the Johns Hopkins University for the past fifteen years. The subject matter, however, is presented in such a manner that the book can be used either as a text for first-year graduate students or for senior undergraduates majoring in electrical or mechanical engineering or in physical science. No prior knowledge of digital signal processing techniques is assumed. Desirable prerequisites are a working knowledge of Fourier and Laplace transforms, and an introductory knowledge of complex variables and differential equations, all of which are usually treated in a first course on linear systems. Knowledge of a higher-level computer language such as FORTRAN would be useful for programming the filter algorithms. Chapter 8, on quantization effects, and Chapter 9, on estimation and the Kalman filter, require some familiarity with probabilistic concepts, which are reviewed in Chapter 7. Elementary matrix operations are used in deriving the Kalman filter.

Organization

This book deals with the analysis and design of digital filters. The design is taken to be completed when a transfer function or, equivalently, a difference equation has been obtained that can be converted into a computer algorithm. No attempt is made to treat dedicated hardware, which is sometimes used to implement a filter. Because of rapid advances in technology, the hardware is often outdated in a short time, whereas the fundamentals, up to and including the filter algorithms, remain relatively unchanged.

Most teachers have a preferred way to present a subject, and I am no exception. I have found by experience that the order in which the material is presented in this text enables the student to make an easy transition from continuous signals and systems to their discrete-time counterparts.

Chapter 1 provides a general introduction to terminology and a description of digital filters. Emphasis is placed on the time-domain description of discrete-time signals and systems and how the latter operate on the former. Chapter 2 is devoted to a review of continuous filter design, as it is advantageous to have some continuous filter design background because of the analogy to digital filter concepts. This review also provides a basis for a systematic design of recursive digital filters.

Chapter 3 deals with sampling theorem and the z-transform, which is the main mathematical tool used in the analysis and design of linear time-invariant digital filters. The z-transform provides a frequency-domain description of the filter that can be related to filter specifications, which are usually given in terms of desired frequency characteristics.

Chapter 4 treats the design of recursive digital filters by transforming a suitable continuous filter function obtained by the methods of Chapter 2. A computer-aided direct design method is also discussed. Finite impulse response (FIR) filters and nonrecursive filters are covered in Chapter 5. These do not have a useful continuous counterpart to aid the design and so are designed directly from the specified frequency response. The discrete Fourier transform (DFT) provides additional options for the design of FIR filters as discussed in Chapter 6. In particular, fast Fourier transform algorithms for efficient computation of the DFT facilitate the real-time processing of sequences using FIR filters.

Basic concepts in probability theory and random processes, needed for Chapters 8 and 9, are reviewed in Chapter 7. The operation of linear systems (filters) on random processes, modeling of correlated noise, and the Markov property are included. Up to this point, the effects of amplitude quantization or word length constraints on filter performance are ignored. The treatment of the subject in Chapter 8 indicates how these effects may be analyzed and taken into account in the filter structure.

The methods of design and analysis of the Kalman filter in Chapter 9 differ considerably from those given in the preceding chapters. The discrete Kalman filter is a very important example of a digital filter type and the extra effort required to understand the concepts involved are well worthwhile. Two appendices are included to review the required background. If desired, Chapter 9 can be omitted from a first reading of the text without loss of continuity.

Features

Each chapter contains worked analytical examples. Simplified diagrams illustrate the text where necessary. Problem sets are given at the end of each chapter to expand on the subject matter and to enable the reader to become more familiar with techniques.

A solutions manual containing the solutions to the problem sets is available to the instructor.

Software

A set of diskettes is available that form a useful adjunct to the book. The diskettes contain FORTRAN programs arranged for interactive operation. These will help the reader solve more difficult problems in filter design. The diskettes can be obtained free of charge from the publisher.

Acknowledgments

I am grateful to the Applied Physics Laboratory of Johns Hopkins University, which supported the publication of this work. This included a Janney fellowship to carry out necessary revisions. In particular, I wish to thank the staff of the Technical Publications Group at APL, especially Barbara Bankert, who typed most of the manuscript from a rough draft, and Robert Tharpe, who drew many of the figures and assisted considerably in manuscript preparation.

Jeffrey Patton wrote most of the software programs and, together with Vincent Neradka, Jeffrey Jordan, and James Christ, helped prepare the diskettes. I would also like to thank the following reviewers for their many valuable comments and suggestions:

Arthur R. Butz, Northwestern University
Antonio H. Costa, Southeastern Massachusetts University
P. E. Crouch, Arizona State University
Gary E. Ford, University of California, Davis
Lester Gerhardt, Rensselaer Polytechnic Institute
Monson H. Hayes, Georgia Institute of Technology
M. R. Ito, The University of British Columbia
Paul Milenkovic, University of Wisconsin, Madison
David C. Munson, Jr., University of Illinois, Urbana-Champaign
Robin N. Strickland, The University of Arizona
Keith A. Teague, Oklahoma State University
William H. Tranter, Univesity of Missouri, Rolla
Barry Van Veen, University of Wisconsin, Madison

Because the book was, for the most part, written during my off-duty hours, a special thanks is due to my wife Marleen for her help and understanding. Without her encouragement, this task could not have been completed.

Edward P. Cunningham

CHAPTER 0

INTRODUCTION

0.1 HISTORICAL BACKGROUND

Digital filters as we know them may be said to have evolved from simulation of analog filters on the early digital computers of the 1940s, though much of the underlying mathematics, such as numerical integration, had been known for centuries. The flexibility and accuracy of these simulations intrigued investigators, who wondered whether, instead of simply serving as an approximation to the analog filter, the digital algorithm could carry out some of the tasks previously reserved for analog filters. There were some formidable obstacles to this line of thought. The early computers were slow, bulky, and expensive. Consequently, no real-time signal processing was feasible. Why, then, would anyone switch from an analog filter that had proved adequate for the purpose? It happened that there were applications where the data were collected and stored for future processing—that is, non-real-time processing. One such case was in the field of seismic exploration [1], where the reflection seismic record, or seismogram, was stored on magnetic tape for later determination of whether the geological area sounded was suitable for oil exploration. These records produced primary reflections from the oil layers and secondary reflections, or noise, from other sources. In oil-rich areas, where the noise reflections were relatively small, one could inspect the seismograms and determine that drilling would be worthwhile. As the demand for domestic sources of oil grew in the early 1960s, it became necessary to explore less-promising areas such as offshore banks, where reflections from the water obscured the primary reflections from the oil layers. The seismologists found that analog signal-processing methods did not help them distinguish signal from noise. However, through discrete convolution and other noise-elimination techniques, they were able to process the seismograms digitally to yield a filtered form that was much easier to interpret, and thus they identified previously unsuspected sources of oil. This was one of the earliest applications of digital filtering. Today, the oil exploration industry records nearly all seismic data in digital form for digital processing.

In 1960, Kalman's paper [2] described a set of recursive algorithms for estimating states of a dynamic system from noisy discrete measurements. This recursive digital filter with time-varying gains is now known as the Kalman filter. Its publication was timely, and it soon found application in navigational problems because of the availability of faster computers.

At about the same time as attention was beginning to focus on the use of digital techniques in signal processing, researchers were considering the same techniques for automatic control. A probability generating function very similar to the z-transform had been used by mathematicians since the middle of the eighteenth century. In 1947 a monograph by Hurewicz [3] laid the foundation for using the z-transform in sampled-data control. In a classic paper [4] on sampled-data control in 1952, Ragazzini and Zadeh formalized the z-transform and the spectral concepts related to sampling. It was evident that this mathematical framework was equally applicable to digital signal processing. Digital filter design emerged in the mid-1960s as a discipline in its own right with some well-formulated techniques for analysis and design. The chapter by Kaiser in [5] discussed windowing techniques and the use of the bilinear transform for recursive-filter design. At about the same time, faster and less costly digital computers were being developed, and considerable strides were made in integrated circuitry. It became apparent that digital filters were becoming competitive with analog filters, even for real-time signal processing.

This was further demonstrated by Cooley and Tukey's 1965 paper [6] on the fast Fourier transform (FFT), an efficient means of computing the discrete Fourier transform (DFT). The FFT algorithms reduced the computation time for DFTs by orders of magnitude and enabled finite impulse response (FIR) digital filters to compete with infinite impulse response (IIR) digital filters for real-time signal processing. Both types challenged the analog filter in areas where the latter was traditionally employed. It even became feasible to sample a continuous signal, digitize the samples, process them in a digital computer, and reconvert the signal to a continuous form in about the same time as was taken for analog processing.

Analog filters typically consist of circuits containing resistors, capacitors, and inductors. The early digital filters were exclusively in software form; that is, they consisted of algorithms or difference equations with operations involving summing, multiplication, and storage on the associated digital computer. With the development of large-scale-integration (LSI) and very-large-scale-integration (VLSI) circuitry, it became feasible to implement the digital filter in hardware form as a special-purpose digital device.

This device still performed the basic operations of summing, multiplication, and storage in the process of iterating the difference equations as for a software design. Only the implementation was different. As the miniaturization of integrated circuitry progressed, it became clear that hardwired digital filters possessed several advantages over their analog counterparts:

1. Smaller size
2. Much lower component tolerances
3. Greater accuracy
4. Greater reliability
5. Multiplexing, or the ability of certain components to share multiple filtering tasks

The cost advantage that the analog filter once enjoyed has gradually been eroded. The one disadvantage of digital filters—the quantization error due to finite word lengths in the representation of signals and parameters—is negligible in software representations on general-purpose computers and can be controlled to manageable bounds by careful design of the filter in a special-purpose device.

In view of all these developments, it is not surprising that digital filters have replaced analog filters in many applications and have showed up in new applications as well. Some of these are discussed in the next section.

0.2 APPLICATIONS

Let us look at some areas where digital filters play a very important role.

The radar systems of World War II relied entirely on analog signal processing. By contrast, both the military and the civilian radar systems of today use digital signal-processing techniques where possible, with analog-to-digital (A/D) conversion at the receiver (Figure I.1). The primary reasons for the change were that (a) the digital processing can be made as accurate as desired by increasing word length and (b) the flexibility of both digital software and hardware components permits the implementation of very many different operations. In addition, digital processing has become more economical.

The main function of a typical radar system is to detect and track targets (usually), by emitting a sequence of pulses and detecting the return pulses reflected by the targets. Because of the noise on the return pulses, it is desirable to maximize the signal-to-noise ratio (SNR). This is done by means of a *matched filter*, a filter the impulse response of which is the time-inverse of the known pulse waveform. A *digital* matched filter is in the form of an FIR filter because the pulses are time-limited. If the impulse response of the filter is stored in a read-write memory, it can be changed quickly to match different signal shapes, thereby endowing the radar system's detec-

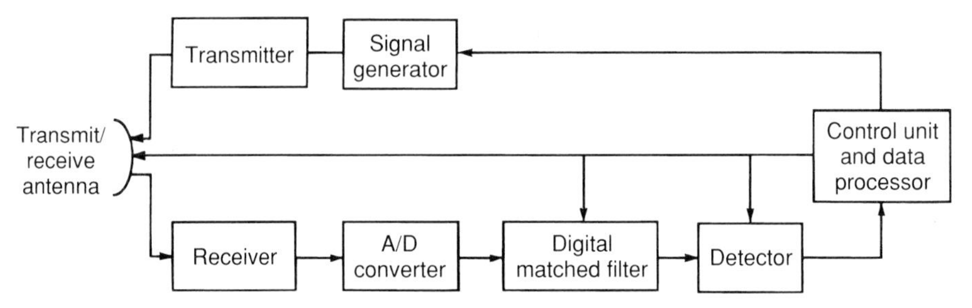

FIGURE I.1 Schematic layout of a modern radar system.

tion capability with considerable flexibility. Implementation of the filter can be carried out via high-speed convolution by using the FFT to compute the DFTs of signal and filter impulse response and to take the inverse DFT of the product (see Chapter 6).

When radar is applied to moving-target indication (MTI), the Doppler shift in frequency that is due to relative movement between target and radar can be used to distinguish between the target return and the noise or clutter from stationary, or near-stationary, objects such as the ground, clouds, or rain. If the clutter return is much greater than the target signal, it is necessary to design a clutter filter to remove much of the noise. Designing an FIR digital filter with stopbands to reject the clutter requires some knowledge of the frequency spectrum of the clutter [7].

As we have noted, one of the earliest applications of digital signal processing was in sampled-data control. Typically, the *plant* or dynamic system being controlled is continuous in nature and is modeled by differential equations. In order to ensure precise control of the output in some desired fashion, it is necessary to measure the output or some function of the output and to feed it back for comparison with the system input. The error formed by subtracting the feedback signal from the input is used to drive the system so that, ideally, its output corresponds to the input, and the error goes to zero. Because of the feedback, there is the possibility that the closed-loop system (Figure I.2) can go unstable. In addition, it is likely that some aspect of the system performance (such as speed of response or steady-state error) is not satisfactory. Hence it is usually necessary to put some compensation in the loop to change the frequency characteristics of the overall closed-loop system and thus ensure stability and good performance. For a sampled-data control system, this compensation takes the form of a recursive (IIR) digital filter that may be implemented as a computer algorithm or hardwired as a special-purpose device. The filter design is, of course, intimately linked to the modeling of the controlled dynamic system and the nature of the feedback signals. Often the latter are discrete or sampled measurements of the continuous output signals. In fact, because of the advantages already mentioned, the modern trend is toward complete digital control of the plant.

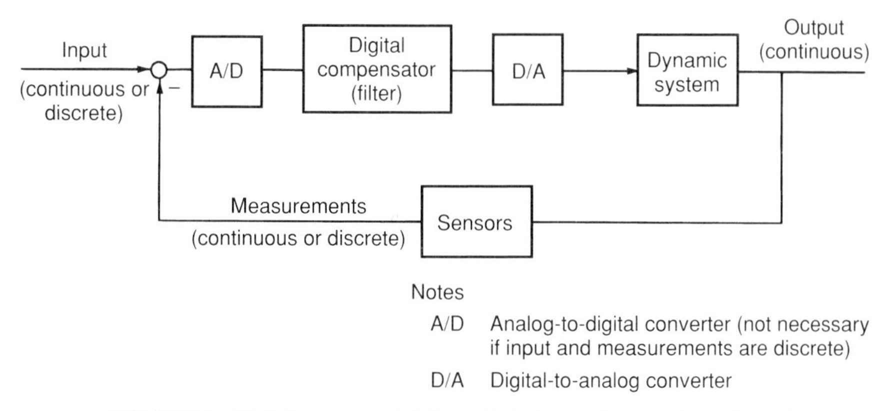

FIGURE I.2 Digital or sampled-data control of a continuous dynamic system.

In many control applications with multiple outputs or *states*, the optimal or best control law is some function of the states. Usually it is too expensive to measure all the states, especially when the system is of a high order. In such cases, a set of measurements of a smaller number of outputs can be made and fed as input to a discrete Kalman filter. This is a recursive digital filter with time-varying gains. The gains are functions of the assumed statistics of the random disturbances and measurement noise. The filter estimates the states of the system, and these estimates are then used for the control law. The estimates are optimal in the sense that the mean-square errors between them and the true values of the states are minimized (see Chapter 9).

The Kalman filter is used in many other areas besides automatic control. One of its first applications was in navigational systems. Periodic updates or measurements are made and passed through a Kalman filter to correct the navigational errors that tend to build up with time. An example is given in Chapter 9. Other applications of Kalman filtering include forecasting telephone loads and locating faults in power lines by estimating the steady-state voltages and currents in the line after the fault [8].

Digital signal processing has also found wide application in speech analysis and synthesis [7]. Speech is produced by the *vocal tract* (an acoustic tube) excited by quasi-periodic pulses of air flow caused by vibrations of the *vocal cords*. Unvoiced sounds are generated by constricting the vocal tract and pushing air through the constriction (*fricative sounds*) or by closing off the vocal tract completely and then suddenly releasing it (*plosive sounds*). The vocal tract has certain natural frequencies. It can be modeled, in general, by a linear time-varying system or, in the case of digital processing, by an IIR digital filter with time-varying coefficients and resonances arranged to occur at the natural frequencies of the vocal tract. For voiced sounds, the filter input is a quasi-periodic impulse train from a generator that adjusts the period of the impulses in accordance with the varying period of the vibrations of the vocal cords. For unvoiced sounds, the filter is excited with discrete white noise (a random sequence the samples of which are independent), so it has a flat frequency spectrum. The gain constant of the filter can be adjusted to reproduce speech *intensity*.

A bank of bandpass digital filters connected in parallel is used for bandpass analysis of speech signals. The passbands are adjacent to each other to cover the entire spectrum of the signal. Such a bank of filters is implemented in a vocoder (voice-coder), which is a processor for the analysis and synthesis of speech. For synthesis, the vocoder uses the speech model discussed above [7].

Another field in which digital technology is playing an increasing role is that of communications. In a communication system, a message or signal containing information is transmitted from one place to another through a communication channel that may be the atmosphere, a transmission line or cable, or a wave guide or optical fiber, for example. Generally, these channels have at least one thing in common: Because they are subject to noise, interference, and losses, the signal received differs from the signal transmitted. This problem can be alleviated by different filtering techniques, such as matched filtering and equalization. The digital form of an equalizer is a nonrecursive filter with coefficients that can be adjusted to suit the particular noisy channel (Figure I.3). A pulse train or some other test signal is sent through the channel, and the coefficients of the filter are adjusted to minimize the disturbance effects. These settings are retained for regular operation.

To use a communication channel more efficiently, signals or messages are *multiplexed* for transmission. In time-division multiplex (TDM), the continuous signals are sampled at a fixed rate. Suppose, for illustration purposes, that the sampling period is 1 second and that it is desired to multiplex ten signals. Samples of the first signal are transmitted at, say,

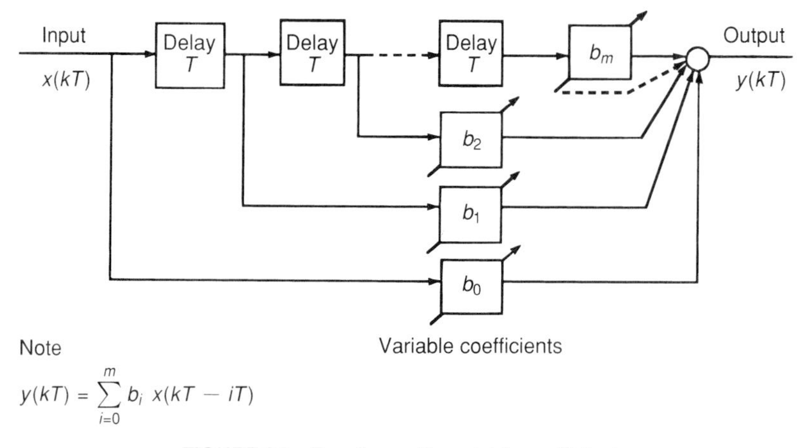

FIGURE I.3 Equalizer with variable coefficients.

$t = 0, 1.0, 2.0, \ldots$ samples of the second signal at $t = 0.1, 1.1, 2.1, \ldots,$ and so on through samples of the tenth signal at $t = 0.9, 1.9, 2.9, \ldots,$ seconds. In a pulse-code-modulation (PCM) multiplex, the signal samples are digitized—that is, represented by a finite number of bits. As shown in Figure I.4, digital filters are used at both the transmitter end and the receiver end to pass the desired band of frequencies. For example, for speech transmission, frequencies above, say, 3.5 kilohertz (kHz) are not of interest. The continuous signal might be sampled at a high rate such as 40 kHz, and the digital filter operates at the same rate to remove frequencies above 3.5 kHz. In addition, it is necessary to remove the 60-Hz "hum" due to the power supply. This can be done via a highpass digital filter with a cutoff frequency of about 100 Hz or, more precisely, via a comb filter (see Chapter 5) with a zero at 60 Hz. If it is desired to reduce the sampling rate to, say, 8 kHz after filtering, this can be done by discarding 4 out of every 5 samples (a process called decimation—see Chapter 5). In that case, it is necessary to filter the continous signal with a lowpass analog filter to reduce the possibility of aliasing at the 8-kHz sampling rate. A common digital filter of each type can be used with the multiplexed signals.

Another advantage of digital implementation is that no tolerance problems arise with digital filters. Although quantization effects (Chapter 8), cause some error, as we have noted, it can be kept within acceptable limits by careful design.

In digital telephony, detection of signaling tones is an important function. To detect the number called by a push-button telephone, the central office uses a digital-filter processor that has highpass filters to eliminate dial tone and 60-Hz hum, bandstop filters to reject certain groups of tones, and bandpass filters to effect tone selection [9].

Digital signal-processing techniques have been applied to great advantage in medicine. Measurements are made of various signals of clinical interest. Examples include electrocardiograms (EKG) to check heart function, electroencephalograms (EEG) for brain function, and readings of temperature and blood pressure. Often, noise on the measurements renders diagnosis difficult. (An EKG might have muscle noise superimposed on the voltage

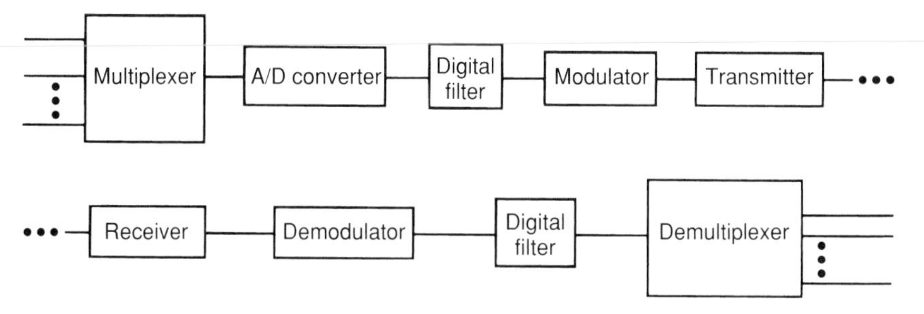

FIGURE I.4 Schematic layout of a digital communication channel.

that is due to heartbeat.) It is necessary to filter to reduce the noise, and the filtering should not distort the underlying signal. Hence a linear-phase filter with constant gain in the passband is desirable. Exact linear phase is virtually impossible to achieve with an analog filter but is readily attainable with an FIR digital filter (Chapter 5).

Digital processing of audio signals is no longer in the experimental stage and is now commercially available. Digital television receivers with requirements for multiple digital-filtering functions have also become a reality [10]. We could go on and on. But suffice it to say that so far, the horizon for applications of digital filtering is unbounded.

REFERENCES FOR INTRODUCTION

1. E. A. Robinson and S. Treitel, "Digital Signal Processing in Geophysics," Chap. 7 of *Applications of Digital Signal Processing*, ed. A. V. Oppenheim, (Englewood Cliffs, NJ: Prentice-Hall, 1978).
2. R. E. Kalman, "A New Approach to Linear Filtering and Prediction Problems," *Trans. of the ASME-Jour. of Basic Eng.* March 1960: 35–45.
3. W. Hurewicz, "Filters and Servo Systems with Pulsed Data," Chap. 5 of H. M. James, N. B. Nichols, and R. S. Phillips, *Theory of Servomechanisms* Vol. 25, M.I.T. Radiation Laboratory Series (New York: McGraw-Hill, 1947).
4. J. R. Ragazzini and L. A. Zadeh, "The Analysis of Sampled-Data Series Systems," *Trans. AIEE* 71, part 2 (1952): 225–234.
5. J. F. Kaiser, "Digital Filters," Chap. 7 of *System Analysis by Digital Computer*, ed. F. F. Kuo and J. F. Kaiser (New York: Wiley, 1966).
6. J. W. Cooley and J. W. Tukey, "An Algorithm for the Machine Calculation of Complex Fourier Series," *Math. Comp.* 19 (1965): 297–301.
7. V. Cappellini, A. G. Constantinides, and P. Emiliani, *Digital Filters and Their Applications* (London: Academic Press, 1978, Chap. 8).
8. R. G. Brown, *Introduction to Random Signal Analysis and Kalman Filtering* (New York: Wiley, 1983).
9. S. L. Freeny, J. F. Kaiser, and H. S. McDonald, "Some Applications of Digital Signal Processing in Telecommunications," Chap. 1 of *Applications of Digital Signal Processing*, ed. A. V. Oppenheim (Englewood Cliffs, NJ: Prentice-Hall, 1978).
10. E. J. Lerner, "Digital TV: Makers Bet on VLSI," *IEEE Spectrum* February 1983, 39–43.

CHAPTER 1

SIGNALS AND SYSTEMS

1.1 INTRODUCTION

A filter is a system that processes a signal in some desired fashion. Therefore, the concepts of signals and systems are fundamental in the design of digital filters. In this chapter we will discuss the way signals are classified and systems are characterized in the time domain. Thus equipped, we will be in a position to examine a category of systems that is mathematically tractable for design and yields a very broad class of filters suitable for a variety of applications. This is the class of linear time-invariant (LTI) filters. Digital LTI filters can be implemented either in a nonrecursive or a recursive form. The difference-equation representations for both forms are easy to understand and straightforward to program as computer algorithms. Much of this book is devoted to digital LTI filters. A notable exception is the Kalman filter (covered in Chapter 9), which, in its most general form, is an example of a linear time-varying system.

1.2 SIGNALS

A signal is a function of some independent variable or variables. Without loss of generality, we will assume that the independent variable is time, although it could be distance, area, or some other physical quantity. A continuous-time signal or continuous signal $x(t)$ is a function of the continuous variable t. It is sometimes called an analog signal. A discrete-time signal or discrete signal $x(kT)$ is defined only at discrete instances $t = kT$, where k is an integer and T is the uniform spacing or period between samples $x(kT)$. We will not consider discrete signals where T varies. For brevity, we will often write $x(k)$ or x_k for $x(kT)$, even if T is not unity. We note that $x(t)$ and $x(kT)$ can refer either to the signal or to its value at a particular instant. A discrete signal is often called a sequence because, like a continuous signal, its amplitudes are ordered in time.

A continuous signal such as the one shown in Figure 1.1(a), when sampled at regular intervals, is converted into a discrete signal consisting of a sequence of sample values of the original signal [Figure 1.1(b)]. Note that the amplitudes of the signals in Figure 1.1 are not restricted; in principle, any amplitude is permissible. However, if the sampled signal is to be processed in a digital computer, its values must be represented by a certain number of bits, so only a finite number of amplitude levels is possible. This results in amplitude quantization, as shown in Figure 1.2. The quantized discrete signal $x_q(kT)$ is called a *digital* signal. Typically, the arrangement could be as shown in Figure 1.3, where the filtering algorithm is in the computer. The processed digital signal can be converted to continuous form by means of the digital-to-analog (D/A) converter, which consists essentially of an interpolation or holding device. We should note that digital signals or sequences can also result from processes or measurements that are inherently discrete in both time and amplitude. Examples include the number of housing starts

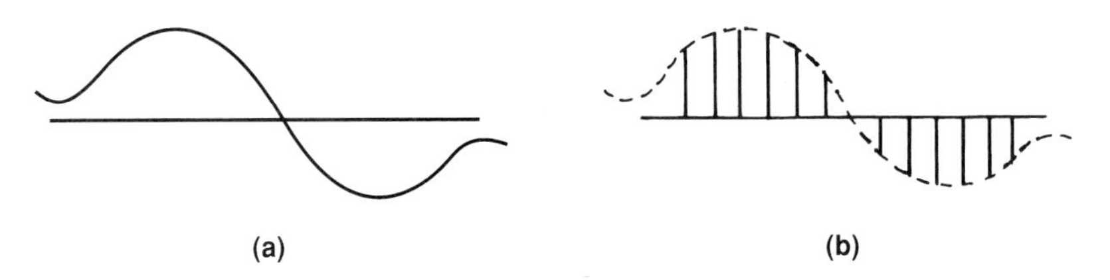

FIGURE 1.1 Sampling a continuous waveform. (a) Continuous signal. (b) Sampled signal.

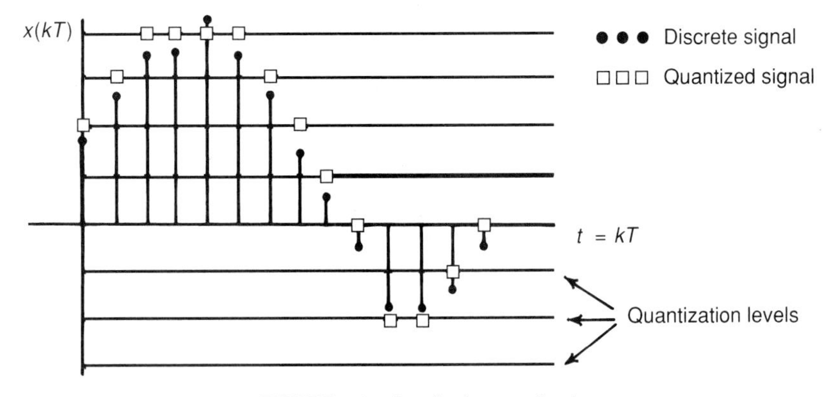

FIGURE 1.2 Amplitude quantization.

each month; the daily rainfall at a particular place to, say, the nearest tenth of an inch; and the number of cars passing a particular spot on a highway each hour.

In much of the literature, no distinction is made between the terms *discrete signal* and *digital signal*, it being assumed that the amplitude quantization steps are so fine that, for purposes of analysis and design, the signals are the same. Then, errors introduced by amplitude quantization are investigated separately. This approach is practical; it enables us to treat a system that processes digital signals as a discrete-time or a sampled-data system. If the signal values, uniformly spaced in time, are assumed to be processed by linear time-invariant digital filters or systems (to be defined soon), then the z-transform can be used for analysis and design of the filters. We will see that the z-transform (Chapter 3) has the same important role in the analysis of discrete linear systems that the Laplace transform has in the analysis of continuous linear systems.

So far we have considered three classes of signals: continuous, discrete, and digital. For various applications, it is convenient to divide signals further into different subclasses. For example, a signal may fall into one of these two categories:

a) Periodic signals
b) Nonperiodic, or transient, signals

In the three main classes, a periodic signal is defined as one for which the following relationships hold, respectively.

$$\begin{aligned} &1.\ x(t+T_p) = x(t) && \text{continuous signals} \\ &2.\ x[(k+N)T] = x(kT) && \text{discrete signals} \\ &3.\ x_q[(k+N)T] = x_q(kT) && \text{digital signals} \end{aligned} \tag{1.1}$$

where T_p and N (an integer) are arbitrary finite constants. A signal that does not satisfy (1.1) is called nonperiodic or transient. A sinusoid is a simple example of a periodic signal. The temperature of liquid in a container falls exponentially with time—that is, $x(t) = A\ e^{-ct}$—when the heat is turned off. Thus it is an example of a transient signal.

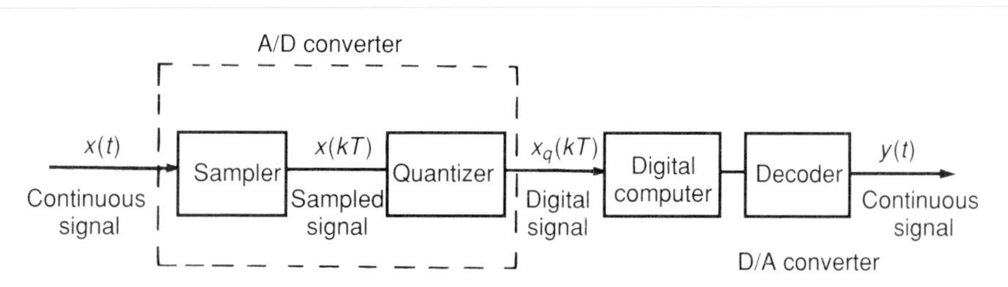

FIGURE 1.3 Processing a continuous signal by means of a digital computer.

Continuous and discrete signals are also classified in one of the following categories:

a) Deterministic signals
b) Random signals

A deterministic signal is one the amplitude of which at any time is predictable from the mathematical relationship $x(t)$ or $x(kT)$. The amplitude of a random signal is not predictable. For one thing, the noise or inaccuracies in a measurement may vary with both the measuring instrument and the person reading it. Thus it is impossible to predict a measurement's value before making the measurement. The amplitudes of random signals are described by probability distributions, and we deal with averages rather than instantaneous values. For most of this book, we will be concerned with deterministic signals, but for the later chapters on quantization effects and estimation, we will need some simple probabilistic concepts, which will be reviewed in Chapter 7.

One further classification of signals is of interest to us. They can be divided into the following categories:

a) Positive-time signals—that is, signals that are zero for t (or k) < 0. These are often called causal signals for a reason we will soon examine.
b) Negative-time signals—that is, signals that are zero for t (or k) > 0.
c) Two-sided signals that have some nonzero values for t (or k) $\lesseqgtr 0$

Note that a signal in category c) can be represented as the sum of a positive-time and a negative-time signal. For signals that start at a particular time, it is often convenient to choose the starting instant as the origin so that such signals are positive-time signals.

1.3 ELEMENTARY SIGNALS

There is an endless variety of signals that result from physical, economic, or other phenomena. However, a few elementary signal forms play a very important role in the classification, testing, evaluation, and design of the systems in which we are interested. These will be described in both their continuous and their discrete forms. Because this is a book on digital filters processing discrete signals, it may seem that we are devoting an undue amount of time and space to continuous functions. That time is well spent, however. Not only can continuous filters be used in a systematic manner to design a certain class of digital filters (Chapter 4), but the discrete signals themselves also have transforms that are *continuous* functions of frequency, so the elementary continuous functions play an important part in analysis and design.

1.3.1 Unit Impulse Function

The unit impulse or Dirac delta function $\delta(t)$ is a so-called generalized function [1]. It is defined by its integral property

$$\int_{-\infty}^{\infty} f(t)\delta(t)\,dt = f(0) \tag{1.2}$$

where $f(t)$ is any function, continuous at the origin. Loosely speaking, we can consider it as being zero everywhere except at $t = 0$, where it is infinite. That is,

$$\delta(t) = \begin{cases} \infty, & t = 0 \\ 0, & \text{elsewhere} \end{cases} \tag{1.3}$$

By a change of origin we can write (1.2) as

$$\int_{-\infty}^{\infty} f(t)\delta(t-a)\,dt = f(a) \tag{1.4}$$

where a is a constant. Thus the impulse function has the effect of sampling the function $f(t)$ when it is used under the integral sign. It follows from (1.4) that

$$\int_{-\infty}^{\infty} \delta(t-a)\,dt = \int_{a_-}^{a_+} \delta(t-a)\,dt = 1 \tag{1.5}$$

where a_- and a_+ denote values immediately below and above a, respectively. We can consider the impulse function $\delta(t-a)$ as the limiting value, as d approaches zero, of the function

$$g(t) = \begin{cases} 1/d, & a \le t \le a+d \\ 0, & \text{otherwise} \end{cases} \tag{1.6}$$

We see in Figure 1.4 that the *area* under the function remains unity as the limit is taken—hence the term *unit impulse*. Another limiting form [1] that will be useful when we discuss window functions (Chapter 5), is

$$\delta(t) = \lim_{b\to\infty} 2b\frac{\sin bt}{bt} \tag{1.7}$$

We note that the limits in (1.6) and (1.7) are defined in the distributional sense [1]. As pointed out by Papoulis [6], the foregoing definitions are meaningless unless the delta function is introduced as a function specified by its properties. Although the delta function seems like a mathematical abstraction, practical cases abound where approximations to it are generated. Examples include the force due to a sharp blow of a hammer and the voltage spike caused by rapidly switching a circuit on and off.

The impulse function as we have described it is generally used in conjunction with continuous signals. A similar, but somewhat simpler, elementary *discrete* signal can be defined. This is called the unit pulse, the unit

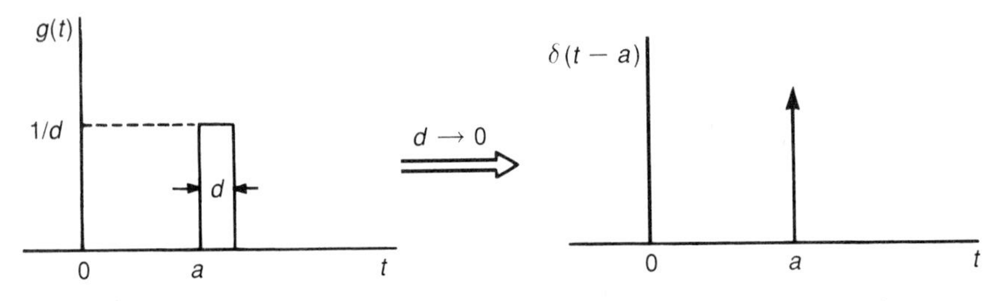

FIGURE 1.4 The unit impulse function as the limiting value of a rectangular pulse.

sample signal, or the Kronecker delta function. We will use the term *unit pulse*. The signal is defined as

$$\delta(k) = \begin{cases} 1, & k = 0 \\ 0, & k \neq 0 \end{cases} \tag{1.8}$$

Changing the origin yields the shifted unit pulse

$$\delta(k-n) = \begin{cases} 1, & k = n \\ 0, & k \neq n \end{cases} \tag{1.9}$$

where k and n are integers. The function is depicted in Figure 1.5. From the definition, we see that the unit pulse is finite and does not involve any limiting process. We note that for any discrete function $f(k)$,

$$\sum_{k=-\infty}^{\infty} f(k)\delta(k-n) = f(n) \tag{1.10}$$

so the unit pulse also has a sampling property.

1.3.2 Unit Step Function

The unit step function is defined as

$$u(t-a) = \begin{cases} 1, & t \geq a \\ 0, & t < a \end{cases} \tag{1.11}$$

where a is a finite constant that may be positive, negative, or zero. Thus the function is discontinuous at $t = a$, as shown in Figure 1.6(a). If a

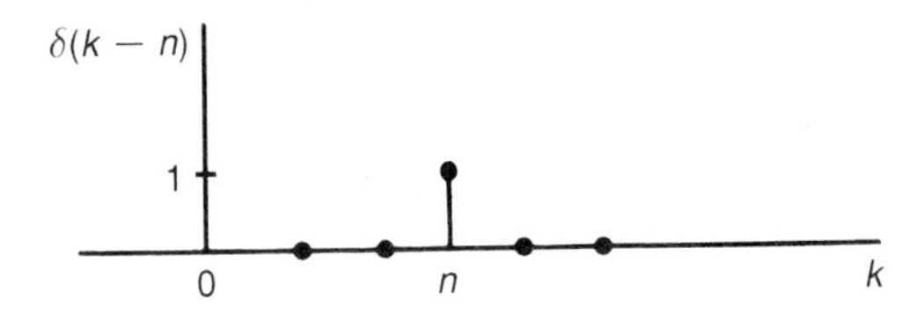

FIGURE 1.5 The unit pulse or unit sample function.

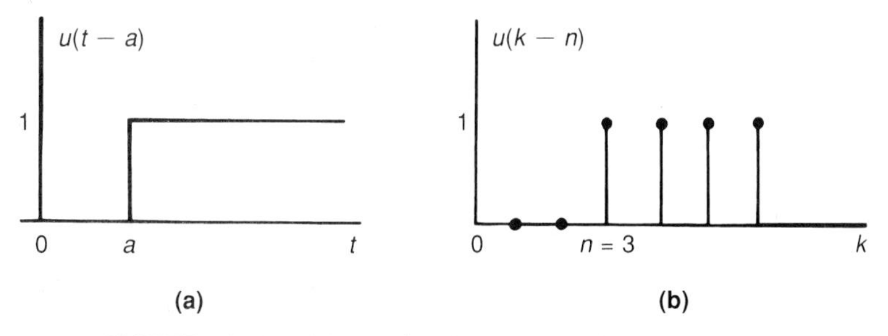

FIGURE 1.6 (a) Unit step function and (b) unit step sequence.

continuous function $f(t)$ is multiplied by $u(t-a)$, it is evident that the result is a function

$$g(t) = f(t)u(t-a) = \begin{cases} f(t), & t \geq a \\ 0, & t < a \end{cases} \tag{1.12}$$

The difference between two unit steps $u(t-a)$ and $u(t-b), b > a$, is a pulse of width $(b-a)$.

$$p(t) = u(t-a) - u(t-b) = \begin{cases} 1, & a \leq t < b \\ 0, & \text{otherwise} \end{cases} \tag{1.13}$$

If we multiply the continuous function $f(t)$ by $p(t)$, we get

$$g(t) = f(t)p(t) = \begin{cases} f(t), & a \leq t < b \\ 0, & \text{otherwise} \end{cases} \tag{1.14}$$

Hence multiplying $f(t)$ by $p(t)$ is equivalent to looking at $f(t)$ through a "window" of width $(b-a)$. Alternatively, a finite piece of data or signal of length $(b-a)$ can be considered as resulting from "windowing" an indefinitely long function. This facilitates the mathematical analysis of the data, as we will see in Chapter 5.

The unit step sequence, depicted in Figure 1.6(b), is defined as

$$u(k-n) = \begin{cases} 1, & k \geq n \\ 0, & k < n \end{cases} \tag{1.15}$$

where k and n are integers and n may be positive, zero, or negative.

It is evident from the definitions that the unit pulse and the unit step sequence are related by

$$\sum_{j=n}^{\infty} \delta(k-j) = u(k-n) \tag{1.16}$$

and

$$\delta(k-n) = u(k-n) - u(k-n-1) \tag{1.17}$$

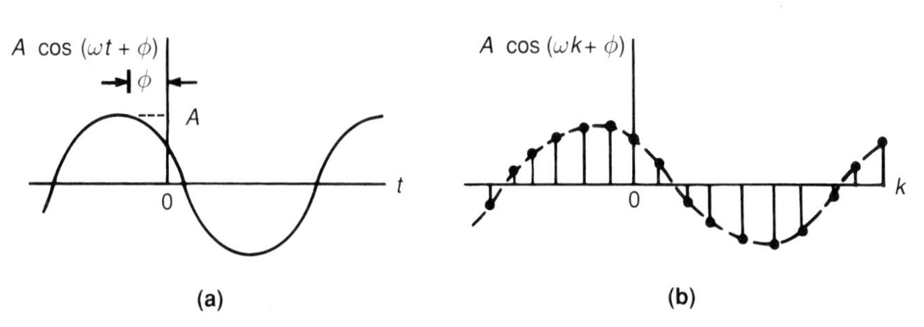

FIGURE 1.7 (a) Continuous and (b) discrete sinusoidal signals.

1.3.3 Sinusoidal Signals

The continuous cosinusoid [Figure 1.7(a)] is defined as

$$x(t) = A \cos (\omega t + \phi) \tag{1.18}$$

where A is the amplitude, ω is the frequency in radians per second, and ϕ is the phase in radians. Because

$$e^{j\omega t} = \cos \ \omega t + j \sin \ \omega t \tag{1.19}$$

where $j = \sqrt{-1}$, it is sometimes convenient for mathematical manipulations to deal with the complex sinusoid $e^{j\omega t}$ and use

$$x(t) = Re\{A \exp [j(\omega t + \phi)]\} \tag{1.20}$$

where Re denotes "the real part of." The frequency in cycles per second, or hertz, is $f = \omega/2\pi$, and the period $T_p = 1/f = 2\pi/\omega$.

The discrete cosinusoid [Figure 1.7(b)] is

$$x(k) = A \cos (\omega k + \phi) \tag{1.21}$$

or, to show the period T between samples more explicitly,

$$x(kT) = A \cos (\omega kT + \phi) \tag{1.22}$$

Here again, it is often convenient to use the complex sinusoid and write

$$x(kT) = Re \ \{A \exp [j(\omega kT + \phi)]\} \tag{1.23}$$

We will use these fundamental signal concepts to help us understand and define system characteristics.

1.4 SYSTEMS

If we operate on, or process, a signal, thereby producing another signal, the operation can be regarded as a system for which the *input* is the original signal and the *output* or response is the processed signal. Thus there is a cause-effect relationship between input and output. We can represent the system by the operator $\mathcal{H}$ so that

$$y(t) = \mathcal{H}\{x(t)\} \tag{1.24}$$

for continuous signals, and

$$y(k) = \mathcal{H}\{x(k)\} \tag{1.25}$$

for discrete signals. Systems (1.24) and (1.25) will be called continuous and discrete systems, respectively. In what follows, we will put more emphasis on discrete systems, though most of the classifications apply to continuous systems as well. The operator $\mathcal{H}$ might denote, for example, a gain constant,

$$y(k) = Kx(k) \tag{1.26}$$

a quadratic operation,

$$y(k) = Kx^2(k) \tag{1.27}$$

or a difference equation,

$$y(k) = B_0x(k) + B_1x(k-1) + B_2x(k-2) \tag{1.28}$$

where the coefficients B_i may be constant or time-varying (a function of k).

1.4.1 Linearity

Suppose that

$$y_1(k) = \mathcal{H}\{x_1(k)\} \tag{1.29}$$

and

$$y_2(k) = \mathcal{H}\{x_2(k)\} \tag{1.30}$$

where $x_1(k)$ and $x_2(k)$ are two different inputs. Then, the system is said to be *linear* if

$$\begin{aligned} y(k) &= \mathcal{H}\{\alpha_1x_1(k) + \alpha_2x_2(k)\} \\ &= \alpha_1y_1(k) + \alpha_2y_2(k) \end{aligned} \tag{1.31}$$

where α_1 and α_2 are constants. A similar definition holds for continuous systems. In other words, a linear system satisfies the *superposition principle*. If a system does not satisfy that principle, it is said to be *nonlinear*. We note that a system might be linear for signals of small or moderate amplitudes but might *saturate*, or become nonlinear, for large signals. For example, the input-output relationship for a voltage amplifier is linear for a certain range of input. If the input exceeds that range, however, the output no longer increases but reaches a limiting, or saturation, value. In this book, when we refer to an arbitrary signal being applied to a linear system, we are assuming that the bounds of linearity are not exceeded.

EXAMPLE 1.1

To illustrate the method for checking linearity, let us consider the system represented by the difference equation

$$y(k) = B_0 x(k) + B_1 x(k-1)$$

Solution

Application of input $x_1(k)$ gives

$$y_1(k) = B_0 x_1(k) + B_1 x_1(k-1)$$

whereas the response to input $x_2(k)$ is

$$y_2(k) = B_0 x_2(k) + B_1 x_2(k-1)$$

The response to input $\alpha_1 x_1(k) + \alpha_2 x_2(k)$ is

$$\begin{aligned} y(k) &= B_0[\alpha_1 x_1(k) + \alpha_2 x_2(k)] + B_1[\alpha_1 x_1(k-1) + \alpha_2 x_2(k-1)] \\ &= \alpha_1[B_0 x_1(k) + B_1 x_1(k-1)] + \alpha_2[B_0 x_2(k) + B_1 x_2(k-1)] \\ &= \alpha_1 y_1(k) + \alpha_2 y_2(k) \end{aligned}$$

Hence the system is linear whether or not the coefficients are time-varying.

Similarly, we can confirm that systems (1.26) and (1.28) are linear, whereas system (1.27) is not.

1.4.2 Time Invariance

We have seen that systems may be linear or nonlinear. They may also be classified in one of the following categories:

a) Time-invariant, or shift-invariant

b) Time-varying.

If the parameters or coefficients of the system do not vary with time, that system is said to be time-invariant. Otherwise, it is time-varying. For a time-varying system, if

$$y(k) = \mathcal{H}\{x(k)\} \tag{1.32}$$

then

$$y(k-n) = \mathcal{H}\{x(k-n)\} \tag{1.33}$$

If we let $\ell = k - n$ in (1.33), we see that the relationship between input and output does not vary with time. In a time-varying system, the relationship does vary with time. The system (1.26) is linear time-invariant (LTI), and the system (1.27) is nonlinear time-invariant if K is constant. The system (1.28) may be linear time-invariant or linear time-varying, depending on whether the coefficients B_i are constant or time-varying.

In system (1.26), the value of y at time k ($= kT$) depends on the value of x at the same time k, and not on any previous value $x(n), n < k$. Such a system is called a *memoryless* or *instantaneous* system. Of much more interest in our work will be the class of system exemplified by (1.28). There we see that the value of y at the current time k depends not only on the current input but also on two previous input values. In other words, the system "remembers" the previous input values. Such systems that have memory are called *dynamic* systems. System (1.28) has finite memory because it does not remember any inputs prior to two previous samples. A general finite-memory system of this type and of order m may be written

$$y(k) = \sum_{i=0}^{m} B_i x(k - i) \tag{1.34}$$

A finite-memory system is shown in block diagram form in Figure 1.8(a) for $m = 2$. It has what is called a *nonrecursive* form; the output does not depend on previous output values. For implementation in a special-purpose device, it requires three storage registers—one each for the coefficiente B_0, B_1 and B_2; two registers for previous output values $x(k-1)$ and $x(k-2)$; three multipliers; and one adder. If the system is designed so as to process its input in some desirable fashion, it is called a nonrecursive *filter* and is one of the two main types of digital filters treated in this book. (Chapter 5)

We can also have infinite-memory dynamic systems. A simple example is given by the difference equation

$$y(k) = A_1 y(k - 1) + B_0 x(k) + B_1 x(k - 1) \tag{1.35}$$

Here the current output $y(k)$ depends on the previous output $y(k - 1)$ as well as on the current input $x(k)$ and on the previous input $x(k - 1)$. The previous output $y(k-1)$ depends on $y(k-2)$, $x(k-1)$, and $x(k-2)$; and so on. It is evident that the current output depends on, or remembers, *all* the previous input values, no matter how long the system has been operating. It is said to have "infinite" memory. We can also confirm, as shown above, that this is a linear system.

A general system of type (1.35) and of order n can be written

$$y(k) = \sum_{i=1}^{n} A_i y(k - i) + \sum_{i=0}^{m} B_i x(k - i) \tag{1.36}$$

A block diagram of this system for $n = 1$ and $m = 1$ is shown in Figure 1.8(b). It has a recursive form, because the output depends on previous output values (as well as input values). Implementation in a special-purpose device requires separate storage registers for B_0, B_1, A_1, $x(k-1)$, and $y(k-1)$; three multipliers; and an adder. Again, if such a system is designed to process signals in some prescribed fashion, it is called a recursive *filter*. This

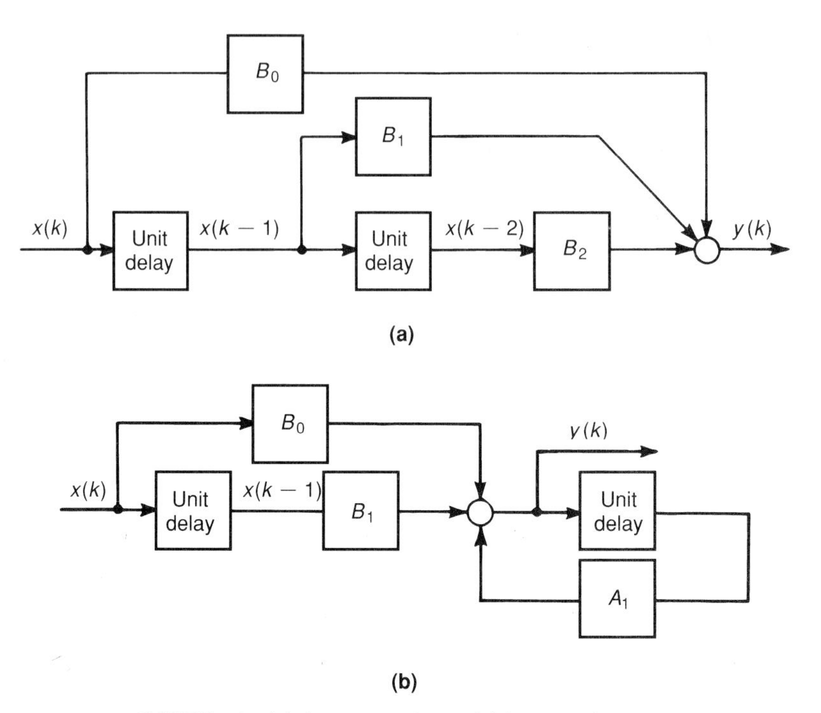

FIGURE 1.8 (a) A nonrecursive and (b) a recursive system.

is the second main type of filter we will discuss (Chapter 4). Because of the feedback due to dependence on previous outputs [Figure 1.8(b)], there is a possibility that the filter might go unstable, which means that the output would grow with time and become unbounded even with a bounded input. Means to check for this will be considered later.

The corresponding continuous linear system may be represented by the differential equation

$$\sum_{i=0}^{n} a_i y^{(i)}(t) = \sum_{i=0}^{m} b_i x^{(i)}(t) \tag{1.37}$$

where $y^{(i)}$ and $x^{(i)}$ denote, respectively, the ith derivative of y and x with respect to time.

In this book, we are concerned primarily with linear time-invariant dynamic discrete systems of the forms (1.34) and (1.36). Also, in Chapter 2, we will review the frequency characteristics of continuous LTI systems of the form (1.37) for the purpose of designing discrete equivalents to such systems in Chapter 4.

Let us proceed to develop the theory of discrete LTI systems with respect to their reaction with discrete signals.

1.5 INTERACTION OF SIGNALS AND SYSTEMS

A *system* may be characterized by a *signal*—that is, by its response to an elementary test signal. This is a very important concept. Suppose that a unit pulse $\delta(k)$ is applied as input to a discrete LTI system, and let the output response sequence be $h(k)$. This is called the unit pulse response or (sometimes) the unit impulse response. Simple though it may seem, the unit pulse response provides a complete description of the system, from which its response to *any* input sequence can be derived and the stability of the system investigated.

A system is said to be causal or *realizable* if the system output response cannot precede the input causing it. Because $h(k)$ is the response at time k to a unit pulse applied at time zero, it follows that, for a causal system, $h(k)$ must be zero for $k < 0$ [Figure 1.9(a)]. Otherwise the system is noncausal [Figure 1.9(b)]. Any positive-time sequence $y(k)$ is also called a causal sequence, because it could serve (at least in principle) as a unit pulse response for a causal system.

The system we are considering is assumed to be time-invariant, so a unit pulse applied at time i—that is, $\delta(k-i)$—results in an output sequence $h(k-i)$. Again, for a causal system, $h(k-i)$ equals zero for $k < i$ and for negative values of its argument $(k-i)$. We could regard i as the "date" of the pulse and $(k-i)$ as the "age" of the response at a particular time $k > i$, for such a system (Figure 1.10).

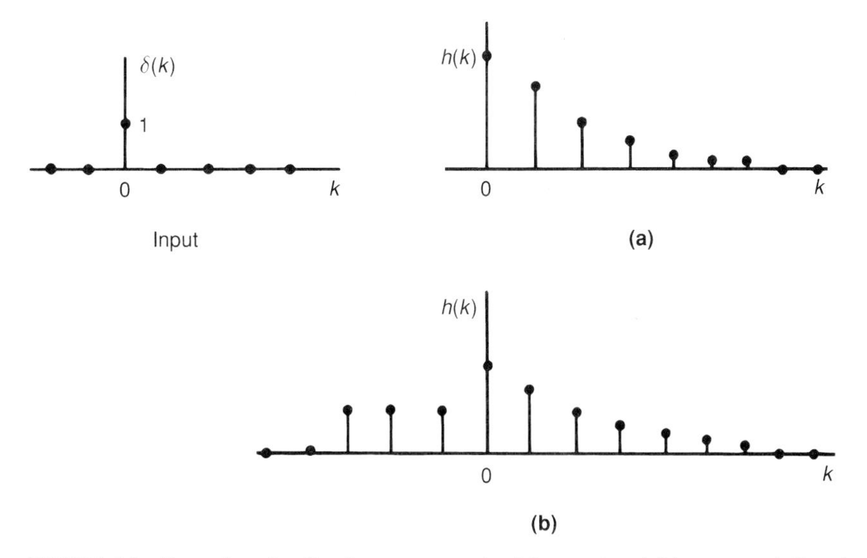

FIGURE 1.9 Examples of unit pulse responses for (a) causal and (b) noncausal discrete LTI systems.

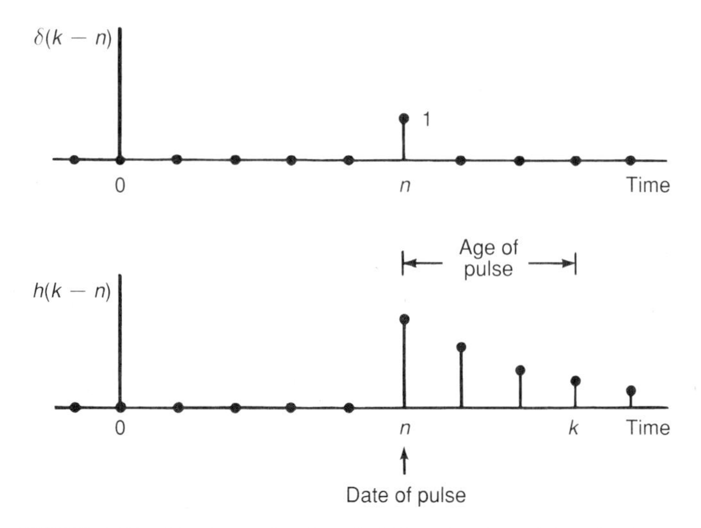

FIGURE 1.10 Response of a discrete LTI system to a delayed pulse.

Consider the arbitrary sequence $x(k)$ applied to the system. This produces an output sequence $y(k)$. Let us derive a relationship between the input x, the unit pulse response h, and the output y.

From the sampling property (1.10) of the unit pulse,

$$x(k) = \sum_{i=-\infty}^{\infty} x(i)\delta(k-i) \tag{1.38}$$

Each of the terms in the summation is a shifted unit pulse $\delta(k-i)$ multiplied, or *weighted*, by the input sample $x(i)$. Because the output response to $\delta(k-i)$ is $h(k-i)$, the response to a delayed pulse of magnitude $x(i)$ is $x(i)h(k-i)$, from the definition (1.31) of a linear system. The term $x(i)h(k-i)$ is but one component of the output $y(k)$. By the superposition principle, the total output $y(k)$ is obtained by adding together the responses $x(i)h(k-i)$ due to all the shifted weighted input samples, as illustrated in Figure 1.11. This yields

$$y(k) = \sum_{i=-\infty}^{\infty} x(i)h(k-i) \tag{1.39}$$

If we change the summation index i to m, where $m = k - i$, then (1.39) becomes

$$y(k) = \sum_{m=-\infty}^{\infty} x(k-m)h(m) = \sum_{i=-\infty}^{\infty} x(k-i)h(i) \tag{1.40}$$

where, in the last form, we have renamed the summation variable m as i. Either (1.39) or (1.40) is known as the *convolution summation* and, for

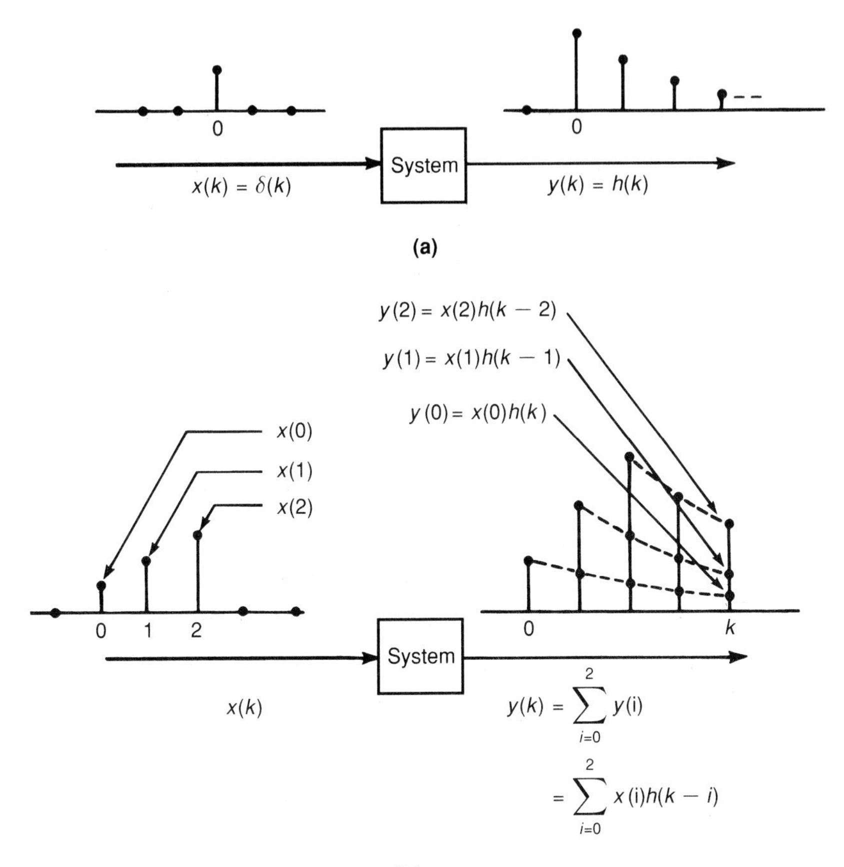

FIGURE 1.11 Derivation of the convolution summation. (a) Response to unit pulse. (b) Response to input $x(k)$.

brevity, is often written

$$y(k) = x(k) * h(k) \tag{1.41}$$

Here y is referred to as the *linear convolution* of x and h. Expressions (1.39) and (1.40) are valid for causal and noncausal discrete LTI systems. For causal systems, because h is zero for negative arguments,

$$y(k) = \sum_{i=-\infty}^{k} x(i)h(k-i) = \sum_{i=0}^{\infty} x(k-i)h(i) \tag{1.42}$$

If, in addition, the input $x(i)$ is a positive-time sequence, the summation may be written

$$y(k) = \sum_{i=0}^{k} x(i)h(k-i) = \sum_{i=0}^{k} x(k-i)h(i) \tag{1.43}$$

A unit pulse response can also be defined for a linear time-varying system. This is $h(k, i)$, which is a function of two variables, the date i of the pulse and the time k at which the response is observed. This is to be contrasted with the pulse response for an LTI system, which is a function of the single time variable $(k - i)$, the age of the pulse. The convolution summation for the time-varying system becomes

$$y(k) = \sum_{k=-\infty}^{\infty} x(i)h(k, i) \tag{1.44}$$

derived as for the time-invariant case. The two-dimensional character of $h(k, i)$ complicates the design in general. In particular, the corresponding frequency-domain representation will be a function of two frequency variables. This complicates frequency-domain techniques for the design of such systems. Accordingly, we will limit our attention to LTI systems until we deal in Chapter 9 with the Kalman filter, which is designed and analyzed in the time domain. The restriction to LTI systems is really not much of a handicap; such systems cover a very wide range of applications.

The unit pulse response is sometimes called the *weighting* sequence, because it weights the input samples $x(i)$ to form the output. We will use the terms interchangeably. The unit pulse response completely characterizes the system in the time domain—that is, as a function of time. Let us see how it is related to the difference-equation form, which is also a time-domain representation.

First, consider the finite-memory, or nonrecursive, system defined by (1.34), which is repeated here for convenience:

$$y(k) = \sum_{i=0}^{m} B_i x(k - i) \tag{1.45}$$

This is identical to the convolution summation

$$y(k) = \sum_{i=0}^{k} h(i)x(k - i), \tag{1.46}$$

if

$$h(i) = \begin{cases} B_i, & 0 \leq i \leq m \\ 0, & \text{otherwise} \end{cases} \tag{1.47}$$

Thus the samples of the pulse response are simply the coefficients of the difference equation in this case. Therefore the finite-memory, or nonrecursive, system has a finite pulse response. A filter with a finite pulse response is often called an FIR (finite impulse response) filter. It may seem from this that an FIR filter and a nonrecursive filter are the same thing, and in most cases they are. However, the term *nonrecursive* refers to how the filter is realized or implemented from the difference-equation representation (no previous output samples to form the current output). The term *FIR* refers to the length of the pulse response as given in the convolution summation representation and, strictly speaking, does not depend on how the filter is realized. In fact, an FIR filter can be implemented in a recursive manner if we ignore pole-zero cancellation (these terms will be explained in Chapter 2). Examples are given in Chapter 5.

The infinite-memory, or recursive, system of the form

$$y(k) = \sum_{i=1}^{n} A_i y(k-i) + \sum_{i=0}^{m} B_i x(k-i) \tag{1.48}$$

does not have a unit pulse response that can be obtained by inspection of the coefficients in the difference equation. Because of the infinite memory, the unit pulse response is infinite also. Such a filter is often called an IIR (infinite impulse response) filter. The terms *recursive filter* and *IIR filter* are almost synonymous, but the foregoing remarks about nonrecursive and FIR filters apply here as well.

We have defined a causal system with respect to the nature of its unit pulse response. We can also check on causality by examining the difference-equation representation. By inspection, we can tell that both systems (1.45) and (1.48) are causal for iteration in the forward direction. The current output is obtained from past outputs and/or current or past inputs. If the current output were formed from a future input, the system would be noncausal or nonrealizable.

1.5.1 Solving a Difference Equation

The system (1.48) is an nth-order difference equation. To solve for $y(k)$ requires n initial conditions, which correspond to the initial values in the registers storing $y(k-i), i = 1, \ldots, n$. Some or all of these values may be zero. Knowing the initial conditions and values for input x, one can always solve recursively for $y(k)$ by computer. A program to do this is given on the diskette. For some simple cases, it is possible to obtain an analytic solution for $y(k)$ by induction, as shown in Example 1.2. But before treating this example, we will prove the following lemma, which often arises in such manipulations.

Lemma

The sum to n terms of a geometric series (a series the terms of which have a common ratio r) is

$$S_n = \sum_{i=0}^{n-1} Cr^i = \frac{C(1-r^n)}{1-r} \tag{1.49}$$

Proof

$$S_n = C + Cr + Cr^2 + \ldots Cr^{n-1}$$

$$rS_n = Cr + Cr^2 + \ldots Cr^{n-1} + Cr^n$$

Subtract to get

$$(1 - r)S_n = C(1 - r^n)$$

Therefore

$$S_n = \frac{C(1 - r^n)}{1 - r}$$

Q.E.D.

We note that if $|r| < 1$, the sum of an infinite number of terms is

$$S_\infty = \sum_{i=0}^{\infty} Cr^i = \frac{C}{1 - r} \tag{1.50}$$

EXAMPLE 1.2

Solve the first-order difference equation

$$y(k) = Ay(k - 1) + x(k)$$

if y_{-1} is the initial value in the y register.

Solution

For $k = 0$:

$$y(0) = Ay_{-1} + x(0)$$

For $k = 1$:

$$\begin{aligned} y(1) &= Ay(0) + x(1) \\ &= A^2 y_{-1} + Ax(0) + x(1) \end{aligned}$$

For $k = 2$:

$$\begin{aligned} y(2) &= Ay(1) + x(2) \\ &= A^3 y_{-1} + A^2 x(0) + Ax(1) + x(2) \end{aligned}$$

By induction, we can write the solution as

$$y(k) = A^{k+1} y_{-1} + \sum_{i=0}^{k} A^i x(k - i)$$

Thus the output is a sum of two components: one due to the initial condition, the other due to the input. We note that if $x(k)$ is a unit step sequence, the solution becomes

$$y(k) = A^{k+1}y_{-1} + \sum_{i=0}^{k} A^i$$
$$= A^{k+1}y_{-1} + \frac{1 - A^{k+1}}{1 - A}$$

where the last term follows from the lemma proved above. If $|A| \geq 1$, $y(k)$ diverges as k increases. If $|A| < 1$, $y(k)$ converges as k increases. In that case, the steady-state solution is

$$y_{ss} = \lim_{k \to \infty} y(k) = \frac{1}{1 - A}$$

Furthermore, if $y_{-1} = 0$ and $x(k) = \delta(k)$, we have

$$y(k) = h(k) = \sum_{i=0}^{k} A^i \delta(k - i)$$
$$= A^k$$

In this simple case, we are able to solve for the unit pulse response, but for higher-order systems, this method is not very practical. After we have introduced the z-transform in Chapter 3, we will be in a position to derive the unit pulse response from the difference-equation representation.

Note that a system described by a recursive relationship is in a nonrecursive form when it is written as a convolution summation, because now the output is in terms of input values alone. This, however, is of little value for implementation purposes, because, in general, an infinite number of coefficients and previous input values would be required.

Many authors [2]–[5] divide digital filters into the two main classes, FIR and IIR, on the basis of the convolution representation

$$y(k) = \sum_{i=-\infty}^{\infty} h(k - i)x(i) \tag{1.51}$$

We prefer, however, to categorize them on the basis of their difference equation representations. We will refer to a filter represented by

$$y(k) = \sum_{i=0}^{m} B_i x(k - i) \tag{1.52}$$

as a *nonrecursive* filter of order m, where m is finite. A filter represented by

$$y(k) = \sum_{i=1}^{n} A_i y(k - i) + \sum_{i=0}^{m} B_i x(k - 1) \tag{1.53}$$

will be called a *recursive* filter of order n, where $n \geq m$. The rare cases wherein *FIR* is not synonymous with *nonrecursive* and *IIR* is not synonymous with *recursive* can be examined separately.

Next let us consider an example of the use of the convolution summation. The system involved is the same as that in Example 1.2, but now the output is obtained directly by evaluating the convolution summation, assuming that $h(k)$ is known, rather than by solving the difference equation.

EXAMPLE 1.3

A unit step is applied to a system the unit pulse response of which is

$$h(k) = \begin{cases} A^k, & k \geq 0, 0 < A < 1 \\ 0, & k < 0 \end{cases}$$

Find the resulting output $y(k)$.

Solution

The output is given by the convolution summation

$$y(k) = \sum_{i=-\infty}^{\infty} x(i)h(k-i)$$

For the signals involved, this becomes

$$y(k) = \sum_{i=0}^{k} A^{k-i} = A^k \sum_{i=0}^{k} A^{-i} = \frac{1 - A^{k+1}}{1 - A}$$

The steady-state response of this system to a unit step is

$$y_{ss} = \lim_{k \to \infty} y(k) = \frac{1}{1 - A}$$

For example, for $A = 0.8, y_{ss} = 5$.

It is also instructive to consider a *graphical* method for evaluating the convolution summation. The technique is illustrated for this example in Figure 1.12, which is drawn for $A = 0.8$. The functions $x(i)$ and $h(i)$ are plotted as shown in parts (a) and (b) of the figure. Reflect $h(i)$ about the time origin to get $h(-i)$, as shown in part (c). The value of the convolution for $k = 0$ is given by the product of the overlapping samples of $x(i)$ and $h(-i)$. Because they overlap at only one sample (which is at the origin), their product is 1 in this case. This is $y(0)$, and it is plotted as the first value for $y(k)$ in part (f). Next slide the weighting sequence h one sample to the right, as shown in part (d), obtaining $h(1 - i)$. Now there is a two-sample overlap of $x(i)$ and $h(1 - i)$, giving the convolution sum as $1 + 0.8 = 1.8$. Plot this value as $y(1)$ in part (f). Continue to slide h to the right, obtaining $h(2 - i)$ as in part (d), which yields a three-sample overlap with $x(i)$, resulting in the value $y(3) = 1 + 0.8 + 0.64 = 2.44$ to be plotted in part (f). Continue this

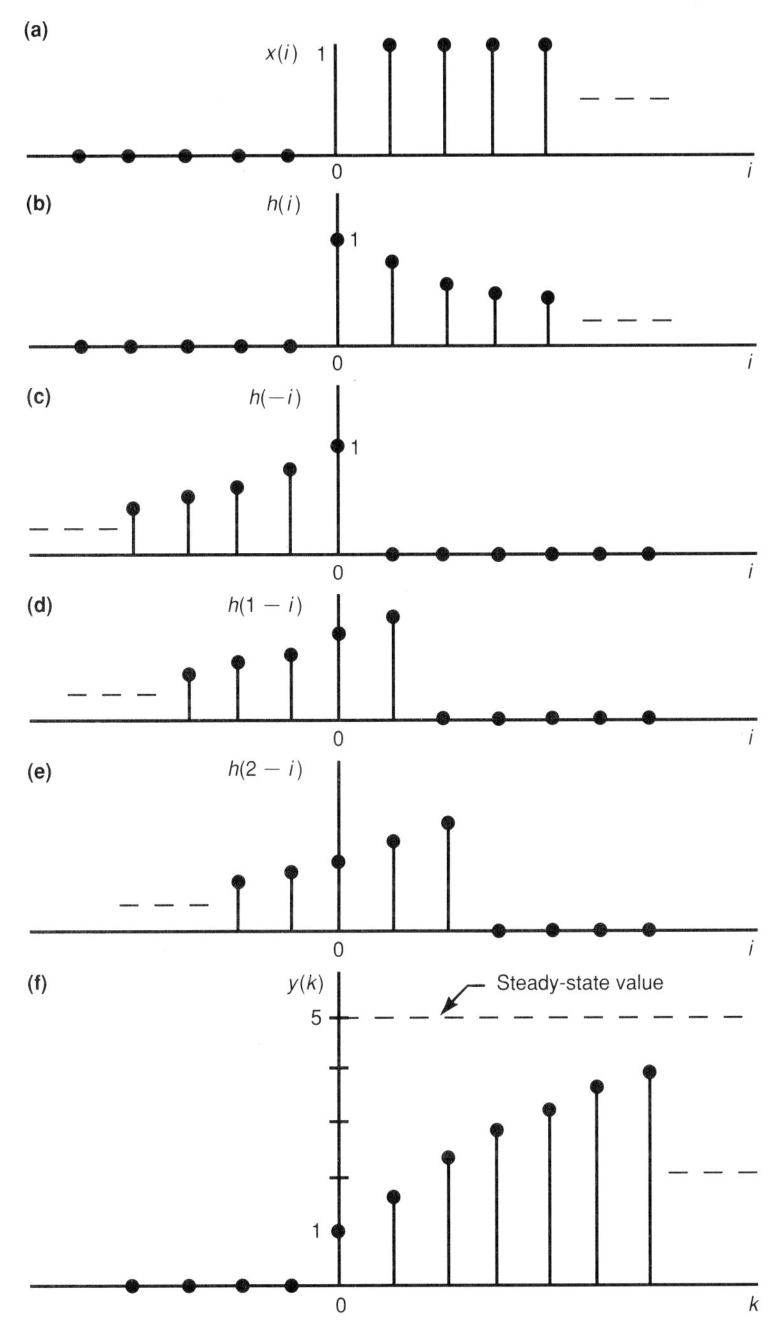

FIGURE 1.12 Graphical method for evaluating the convolution summation.

procedure to obtain as many values of $y(k)$ as are desired. These approach the steady-state value of 5, as indicated by the analytic solution.

For this simple example, it was possible to obtain an analytic solution for $y(k)$, and the evaluation by graphical means was not difficult. In many practical cases, however, a closed-form analytic result is difficult to obtain, and performing a graphical evaluation is tedious. It is desirable, therefore, to have a computer program that will take a given sequence $x(k)$ and convolve it with a particular weighting sequence $h(k)$, yielding the processed response $y(k)$. Such a program appears on the diskette.

1.5.2 Stability of a Linear System

A linear system is said to be stable if a bounded input produces a bounded output. We can use the weighting sequence $h(k)$ to check the stability of an LTI system according to the following theorem.

Theorem

A necessary and sufficient condition for stability of a discrete LTI system is that

$$\sum_{i=-\infty}^{\infty} |h(i)| < \infty \tag{1.54}$$

Proof

That this is a sufficient condition can be confirmed by setting the bounded input $|x(k)| \leq M$, where M is a positive constant. Then

$$|y(k)| \leq \sum_{i=-\infty}^{\infty} |h(i)||x(k-i)| \leq M \sum_{i=-\infty}^{\infty} |h(i)| \tag{1.55}$$

Hence the output is bounded if (1.54) holds. To check the necessity of condition (1.54), it suffices to find a bounded input that will result in an unbounded output if (1.54) is not satisfied. For example, consider the special bounded input $x(k)$ defined such that

$$x(k-i) = \begin{cases} 1 & \text{if } h(i) > 0 \\ -1 & \text{if } h(i) < 0 \end{cases} \tag{1.56}$$

where k is fixed at a particular value k_ϕ. Then

$$\begin{aligned} y(k_0) &= \sum_{i=-\infty}^{\infty} h(i)x(k_0 - i) \\ &= \sum_{i=-\infty}^{\infty} |h(i)| \end{aligned} \tag{1.57}$$

It follows that $y(k)$ is unbounded unless (1.54) is satisfied. This confirms the necessity of the condition.

This criterion simply states that if a sequence is to serve as a unit pulse response or weighting sequence for a stable LTI system, the sum of the absolute values of its samples must be finite. Figure 1.13 shows some examples of causal sequences that, when used as weighting sequences, would result in stable systems and some that would result in unstable systems. Included among the latter is the unit step.

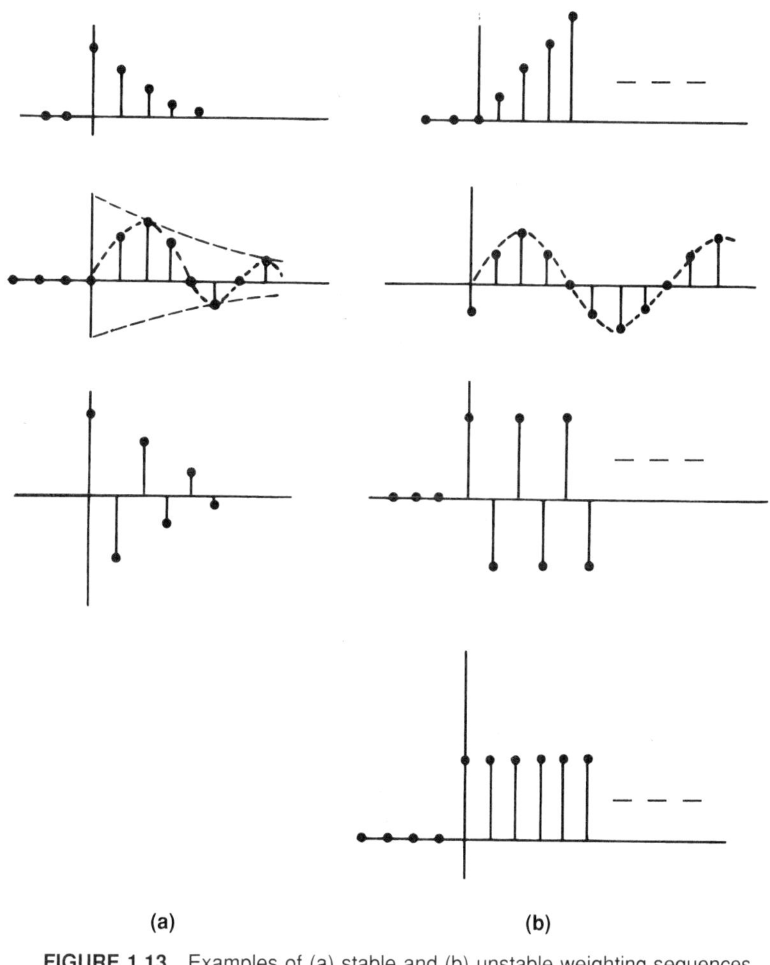

FIGURE 1.13 Examples of (a) stable and (b) unstable weighting sequences.

EXAMPLE 1.4

Use the stability criterion (1.54) to determine whether the first-order recursive filter defined by the following difference equation is stable.

$$y(k) = Ay(k-1) + Bx(k), \quad k \geq 0$$

Solution

To use criterion (1.54), we need the unit pulse response $h(k)$ of the filter. From Example 1.2, this is given by

$$h(k) = A^k$$

Hence

$$\sum_{i=-\infty}^{\infty} |h(i)| = \sum_{i=0}^{\infty} |A^i|$$

This is finite only when $|A| < 1$, which is a necessary and sufficient condition for stability of the filter.

A nonrecursive filter must be stable, because condition (1.54) is satisfied. As we have noted, a recursive filter can go unstable. It would be a simple matter to check stability if the weighting sequence $h(k)$ were available. However, obtaining $h(k)$ for higher-order recursive filters is mathematically complex. Fortunately, the frequency-domain representation of the filter (Chapter 3) enables us to check stability with far less effort.

1.6 ANALOGOUS FORMS FOR CONTINUOUS SYSTEMS

To close this chapter on signals and systems, we will list some attributes of continuous systems that parallel those of discrete systems. We will not include proofs, which are similar to those for the discrete case. Instead of the differential equation representation

$$\sum_{i=0}^{n} a_i y^{(i)}(t) = \sum_{i=0}^{m} b_i x^{(i)}(t), \tag{1.58}$$

we can write the output as a convolution integral

$$y(t) = \int_{-\infty}^{\infty} h(t, \tau) x(\tau) d\tau \tag{1.59}$$

where $h(t, \tau)$ is the weighting function or response at time t to a unit impulse applied at time τ, that is, it is the unit impulse response. For an LTI system, the coefficients a_i and b_i of the differential equation are constants, and the impulse response is a function of the age $(t - \tau)$ of the impulse. Thus

$$y(t) = \int_{-\infty}^{\infty} h(t - \tau) x(\tau) d\tau \tag{1.60}$$

For a causal system, $h(t-\tau)=0$ for $\tau > t$, so we can write

$$\begin{aligned} y(t) &= \int_{-\infty}^{t} h(t-\tau)x(\tau)d\tau \\ &= \int_{0}^{\infty} h(\tau)x(t-\tau)d\tau \end{aligned} \tag{1.61}$$

by a change of variable. Because of the feedback of previous output values, implied by the differential-equation representation, the system can go unstable. A necessary and sufficient condition for stability of a continuous LTI system is

$$\int_{-\infty}^{\infty} |h(t)|dt < \infty \tag{1.62}$$

which requires knowledge of the impulse response—often difficult to obtain for higher-order systems. Again, information on stability is generally easier to obtain from a frequency-domain representation of the system, which we will discuss in Chapter 2.

1.7 SOME GENERAL REMARKS ON DIGITAL FILTER DESIGN

When we process a signal $x(k)$ in a discrete system of the form (1.52) or (1.53), we get an output $y(k)$ that differs from $x(k)$. If $y(k)$ is to differ from $x(k)$ in some desired manner, we must tailor the system to meet this goal, and it is then called a digital filter. In its most basic form, digital-filter design involves determining the order of the filter and the values of the coefficients in the difference-equation representation such that the output meets certain specifications. In most cases, these specifications are related to frequency characteristics of the output.

For recursive digital filters, a systematic design method is available that draws on the wealth of knowledge available for *continuous*-filter design. This method involves finding a suitable frequency function for a continuous filter. This function is then transformed into a frequency function for a digital filter. The techniques are described in detail in Chapter 4. To prepare the reader for this approach, Chapter 2 includes a review of continuous filter design methods.

The nearest continuous counterpart to the finite-memory filter (1.52) is a tapped delay-line or transversal filter. The design of these continuous filters is not useful as a basis for nonrecursive filter design, but in many cases, the direct design of nonrecursive filters is somewhat easier than that of recursive filters. Suppose a desired frequency response is specified. An approximation to the response is obtained (see Chapter 5), and the corresponding pulse response is then known. Because the pulse response is finite and its samples are the coefficients of the difference equation [see (1.46) and (1.47)], the filter design is complete.

Another general method for nonrecursive filter design involves using the discrete Fourier transform (DFT). This is feasible because of the finite nature of the pulse response. The computationally efficient fast Fourier transform (FFT) algorithms facilitate the DFT evaluation. Design by this method is described in Chapter 6.

Filter specifications usually contain inequality terms such as *at least* and *not exceeding*. Thus a design based on such specifications is not unique. In addition, the variety of design methods further expands the number of filters that would meet the specifications. To avoid overdesign and to limit the number of choices, we take the lowest-order filter of each type that satisfies the specifications. To choose between filter types (for example, recursive or nonrecursive), we compare the amounts of computation (addition and multiplication) and storage required for the most efficient implementation. These considerations are directly related to cost. Design complexity may also be a factor in the decision. If the designer must devote an inordinate amount of time to come up with a lower-order filter, it may not be cost-effective. However, this factor is not so important now as it used to be, because the availability of computer programs takes most of the tedium out of the design of even complex filters.

1.8 SUMMARY

After discussing various types of signals, we considered the way systems are classified in terms of their effect on input signals. The class made up of discrete linear time-invariant dynamic systems is of particular interest for our work. These systems can be in either a nonrecursive form:

$$y(k) = \sum_{i=0}^{m} B_i x(k-i) \tag{1.63}$$

or a recursive form:

$$y(k) = \sum_{i=1}^{n} A_i y(k-i) + \sum_{i=0}^{m} B_i x(k-i) \tag{1.64}$$

If these systems are designed to process signals in some desired fashion, they are called nonrecursive and recursive digital filters, respectively. The design involves finding the order of the filter and values for the coefficients such that certain specifications for the output are met. These specifications are usually given in terms of the frequency characteristics of the output.

REFERENCES FOR CHAPTER 1

1. M. J. Lighthill, *An Introduction to Fourier Analysis and Generalized Functions* (New York: Cambridge University Press, 1959).
2. A. V. Oppenheim and R. W. Schafer, *Digital Signal Processing* (Englewood Cliffs, NJ: Prentice-Hall, 1975).
3. L. R. Rabiner and B. Gold, *Theory and Application of Digital Signal Processing* (Englewood Cliffs, NJ: Prentice-Hall, 1975).
4. N. K. Bose, *Digital Filters Theory and Application* (New York: Elsevier Science Publishing Co., 1985).
5. L. B. Jackson, *Digital Filters and Signal Processing* (Boston: Kluwer Academic Publishers, 1986).
6. A. Papoulis, *The Fourier Integral and Its Applications* (New York: McGraw-Hill, 1962).

EXERCISES FOR CHAPTER 1

1. List some continuous signals that you encounter in everyday life. Also list some familiar signals that are inherently discrete.

2. Determine whether the following represent continuous, discrete, or digital signals. Does their nature change if you measure them? Briefly explain.

a) Temperature in the shade at a particular spot
b) Temperature at a particular time, daily
c) Your body temperature
d) Your pulse rate
e) The path of a car along a highway
f) The speed of a car on a trip
g) Your daily outlay for food
h) The average price of a particular grade of oil on world markets
i) The Dow-Jones industrial average
j) Wind speed and direction at a particular place

3. Check whether the systems represented by the following equations are linear or nonlinear according to definition (1.31).

a) $y(k) = 0.5y(k-1) + x(k)$
b) $y(k) = x(k)y(k-1)$
c) $y(k) = x(k)x(k-1)$
d) $y(k) = ky(k-1) + x(k)$

4. The system

$$y(k) = \sum_{i=1}^{n} A_i y(k-i) + \sum_{i=0}^{m} B_i x(k-i)$$

is said to be time-invariant or shift-invariant if the coefficients A_i and B_i are constant. Show that this implies $y(k-p) = \mathcal{H}\{x(k-p)\}$, where p is any integer.

5. By *inspection*, check which of the following characteristics apply to each of the systems listed below: linear, nonlinear, time-invariant, time-varying, instantaneous, dynamic, recursive, nonrecursive.

a) $y(k) = kx(k)$
b) $y(k) = 2kx^2(k)$
c) $y(k) = 3x(k-2)$
d) $y(k) = x(k) + x^2(k-1)$
e) $y(k) = 1.8y(k-2) + 3ky(k-3) + x(k)$
f) $y(k) = (k-1)y(k-1) + (k-2)y(k-2) + 2x(k)$

6. Find the unit pulse response for each of the following filters. Are they causal or noncausal?

a) $y(k) = 2x(k) - 3x(k-1)$
b) $y(k) = x(k+1) + 3x(k) + 5x(k-1)$
c) $y(k) = u(k) + 2u(k-1)$
$u(k) = x(k-1) - 3x(k-2)$
(Here we have two filters in cascade. The pulse response relating y and x is required.)

7. Draw a block diagram for the following recursive filter.

$$y(k) = A_1 y(k-1) + A_2 y(k-2) + A_3 y(k-3) + B_0 x(k) + B_1 x(k-1)$$

where all coefficients are non-unity. Suppose that this filter is to be implemented in a special-purpose computer. From your diagram, determine how many adders, multipliers, and storage registers are required for the implementation.

8. Use graphical convolution to find the response of a discrete system with $h(k) = A^k u(k), 0 < A < 1$, to each of the following inputs. In each case, confirm your answer by finding a closed-form analytic solution.

a) $x(k) = \begin{cases} -1, & k = -1 \\ 0, & \text{otherwise} \end{cases}$

b) $x(k) = \begin{cases} 1/k, & 1 \le k \le 4 \\ 0, & \text{otherwise} \end{cases}$

c) $x(k) = \begin{cases} 2, & k = 1, 3, 5 \\ 0, & \text{otherwise} \end{cases}$

9. Repeat Exercise 8(a) for

$$h(k) = A^k[u(k) - u(k-4)], \quad 0 < A < 1$$

10. When you combine the finite input sequence with the finite weighting sequence in Exercise 9, you will note that the output is also a finite sequence. How is the length of the output sequence in each case related to the length of the input and weighting sequence? (This issue is discussed further in Chapter 6.)

11. Find the first eight terms of the unit step response of the following second-order recursive filter.

$$y(k) = 0.8y(k-1) - 0.64y(k-2) + x(k)$$

if the initial conditions $y_{-1} = y_{-2} = 0$. A program for the solution of a difference equation is given on the diskette. Use the program to verify the result you obtained.

12. Consider other inputs for the system of Exercise 11. Use the program to find the output response to the following inputs.

a) *Unit pulse:* $x(k) = \delta(k)$

b) *Unit ramp:* $x(k) = ku(k)$

c) *Sinusoid:* $x(k) = \sin \omega_0 kTu(k)$ where $\omega_0 T = \pi/6$, and $\pi/4$, and $T = 1$ second.

13. The integral $y(t) = \int_0^t x(t_1)dt_1$ may be considered a system with input $x(t)$ and output $y(t)$. Suppose we want to make a numerical approximation to $y(t)$—that is,

$$y(kT) = y(kT - T) + \Delta y$$

Find Δy if the area between $t = kT - T$ and $t = kT$ is approximated by a trapezoid. Then complete the difference equation. Does it represent a recursive or a nonrecursive system?

14. The compound interest formula

$$A_n = A_0(1 + i/100)^n$$

where

A_n = future value (value at the end of n time periods)

A_0 = present value (value at the beginning of the first time period)

n = number of time (compounding) periods

i = interest rate per time period (in percent)

is the solution of a difference equation. Write the difference equation. Does it represent a recursive or a nonrecursive system?

Diskette Program Examples

This appendix has examples of the use of two interactive programs, CONV and DIFF, that are on the diskette.

The CONV example is the linear convolution of a square wave with an exponential sequence. This is the default option. The user can input any other two sequences and obtain their convolution.

In DIFF an input sequence is filtered by a linear time-invariant difference equation (recursive or nonrecursive) the coefficients of which are selected by the user, who has three input options:

1. A square wave
2. A set of sine waves
3. A self-supplied sequence

The example deals with the unit step sequence response of a second-order recursive difference equation. The step is obtained from the third option. An example of the output response is shown in Figure 1A.1.

```
CONV
IF DATA FROM A USER SUPPLIED FILE IS USED IT SHOULD HAVE TWO
SEQUENCES X(K),
K = 1, N, AND H(K), K = 1, M, WITH N AND M SPECIFIED, M + N < = 200.
DATA
IN THE ORDER N, M, X(K) AND H(K) N AND M WITH FORMAT (I4), X(K) AND
H(K)
WITH FORMAT (5E16.9)
THE DEFAULT OPTION FOR THE LINEAR CONVOLUTION PROGRAM IS A SQUARE
WAVE
CONVOLVED WITH AN EXPONENTIAL SEQUENCE. IF THIS IS ACCEPTABLE, PRESS
RETURN.
OTHERWISE, ENTER A FILE NAME:
INPUT SEQUENCES AND OUTPUT SEQUENCE IN FILE CONV.DAT
Stop - Program terminated.
```

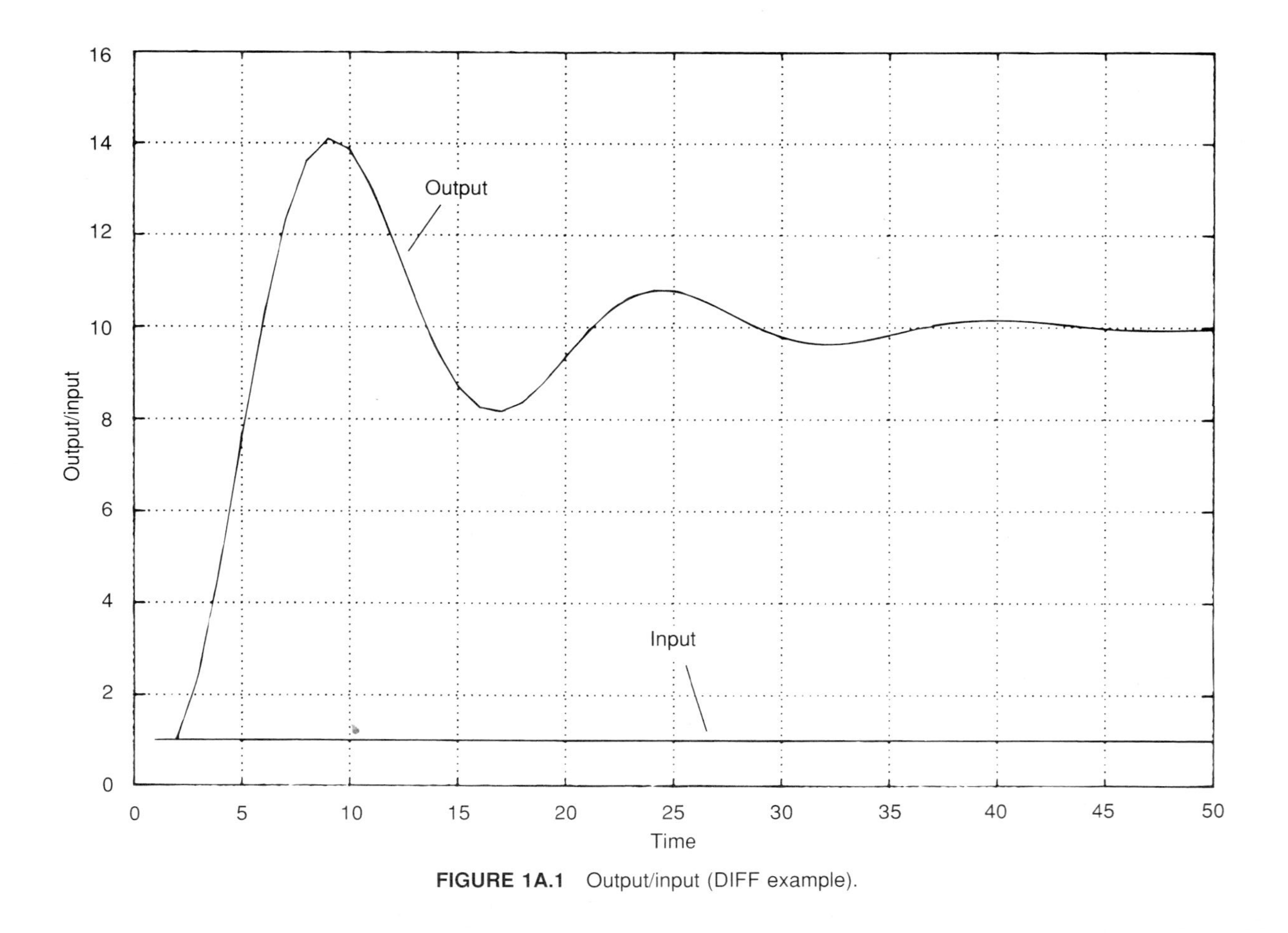

FIGURE 1A.1 Output/input (DIFF example).

```
INPUT SEQUENCE X(K)
 1.000  1.000  1.000  1.000  1.000  1.000  1.000  1.000  1.000  1.000
-1.000 -1.000 -1.000 -1.000 -1.000 -1.000 -1.000 -1.000 -1.000 -1.000
 1.000  1.000  1.000  1.000  1.000  1.000  1.000  1.000  1.000  1.000
-1.000 -1.000 -1.000 -1.000 -1.000 -1.000 -1.000 -1.000 -1.000 -1.000
 1.000  1.000  1.000  1.000  1.000  1.000  1.000  1.000  1.000  1.000
-1.000 -1.000 -1.000 -1.000 -1.000 -1.000 -1.000 -1.000 -1.000 -1.000
 1.000  1.000  1.000  1.000  1.000  1.000  1.000  1.000  1.000  1.000
-1.000 -1.000 -1.000 -1.000 -1.000 -1.000 -1.000 -1.000 -1.000 -1.000
 1.000  1.000  1.000  1.000  1.000  1.000  1.000  1.000  1.000  1.000
-1.000 -1.000 -1.000 -1.000 -1.000 -1.000 -1.000 -1.000 -1.000 -1.000


FILTER WEIGHTING SEQUENCE H(K)
 1.000   .950   .902   .857   .815   .774   .735   .698   .663   .630
  .599   .569   .540   .513   .488   .463   .440   .418   .397   .377
  .358   .341   .324   .307   .292   .277   .264   .250   .238   .226
  .215   .204   .194   .184   .175   .166   .158   .150   .142   .135
  .129   .122   .116   .110   .105   .099   .094   .090   .085   .081
  .077   .073   .069   .066   .063   .060   .057   .054   .051   .048
  .046   .044   .042   .039   .038   .036   .034   .032   .031   .029
  .028   .026   .025   .024   .022   .021   .020   .019   .018   .017
  .017   .016   .015   .014   .013   .013   .012   .012   .011   .010
  .010   .009   .009   .008   .008   .008   .007   .007   .007   .006


OUTPUT SEQUENCE Y(K)
 1.000  1.950  2.852  3.710  4.524  5.298  6.033  6.732  7.395  8.025
 6.624  5.293  4.028  2.827  1.685   .601  -.429 -1.407 -2.337 -3.220
-2.059  -.956   .092  1.087  2.033  2.931  3.784  4.595  5.365  6.097
 4.792  3.553  2.375  1.256   .194  -.816 -1.775 -2.687 -3.552 -4.375
-3.156 -1.998  -.898   .147  1.139  2.082  2.978  3.829  4.638  5.406
 4.136  2.929  1.782   .693  -.341 -1.324 -2.258 -3.145 -3.988 -4.788
-3.549 -2.372 -1.253  -.190   .819  1.778  2.689  3.555  4.377  5.158
 3.900  2.705  1.570   .492  -.533 -1.506 -2.431 -3.310 -4.144 -4.937
-3.690 -2.506 -1.380  -.311   .704  1.669  2.586  3.456  4.284  5.069
 3.816  2.625  1.494   .419  -.602 -1.572 -2.493 -3.368 -4.200 -4.990
-4.746 -4.515 -4.295 -4.086 -3.888 -3.699 -3.520 -3.350 -3.189 -3.035
-2.878 -2.728 -2.585 -2.450 -2.322 -2.200 -2.084 -1.974 -1.869 -1.770
-1.687 -1.609 -1.534 -1.463 -1.396 -1.332 -1.272 -1.214 -1.159 -1.107
-1.046  -.988  -.932  -.880  -.830  -.782  -.737  -.695  -.654  -.615
 -.591  -.567  -.544  -.523  -.503  -.484  -.465  -.448  -.432  -.416
```

-.389	-.364	-.340	-.317	-.295	-.274	-.255	-.236	-.218	-.202
-.197	-.193	-.190	-.186	-.183	-.180	-.176	-.174	-.171	-.168
-.154	-.140	-.127	-.115	-.103	-.092	-.082	-.072	-.062	-.053
-.056	-.060	-.062	-.065	-.068	-.070	-.073	-.075	-.077	-.079
-.069	-.060	-.051	-.043	-.035	-.027	-.020	-.013	-.006	.000

CHAPTER 2

BRIEF REVIEW OF CONTINUOUS LINEAR FILTERS

2.1 INTRODUCTION

The purpose of reviewing continuous-filter theory in this chapter is twofold.

1. Many of the techniques employed in studying digital filters are analogous to those used in studying continuous filters.
2. Considerable knowledge about, and experience with, the design of continuous filters have been amassed. Currently, one of the most systematic approaches to *recursive* digital filter design is based on obtaining a suitable continuous filter function and then transforming it into the discrete time domain.

In Chapter 1, we discussed the time-domain description of discrete and continuous systems. In this chapter, we will emphasize a *frequency*-domain description of continuous linear time-invariant systems. This approach both facilitates analysis and design and dovetails nicely with filter specifications, which are usually given in terms of the required frequency characteristics of the system. The Fourier and Laplace transforms are the most important mathematical tools for obtaining a frequency model for a continuous LTI system. We assume that the reader has a working knowledge of these transforms, and our treatment of them is brief. However, enough detail is included to refresh the reader's memory of the subject.

Designing a continuous filter for a desired magnitude response boils down to the task of designing a realizable approximation to an ideal *lowpass* prototype filter. Obtaining a suitable approximation is a classic problem that has occupied the attention of researchers since Butterworth [1]. Several techniques are available for designing the approximation, the most common of which are treated here. The resulting filter can then be turned into a practical design by an appropriate frequency transformation.

It is beyond the scope of this text to delve deeply into the mathematics underlying the various approximations. Books have been written with major sections devoted to the subject ([2] through [5]). For our purpose, it is enough to outline the important parameters and equations and to indicate how they can be used to obtain a practical design. Computer programs for each approximation method are given on the diskette accompanying the text so that the reader can design filters too complex to be handled analytically.

2.2 FOURIER SERIES AND INTEGRAL, LAPLACE TRANSFORM

2.2.1 Fourier Series Expansion

Trigonometric Form A periodic function $f(t)$ with period T_p, satisfying conditions given below, can be represented by a Fourier series expansion of sines and cosines [6] of the form

$$f(t) = a_0 + \sum_{n=1}^{\infty} a_n \cos n\omega_f t + \sum_{n=1}^{\infty} b_n \sin n\omega_f t \tag{2.1}$$

where $\omega_f = 2\pi/T_p$ is the fundamental frequency in radians per second if T_p is in seconds. This is called a *frequency-domain* representation, because the time function $f(t)$ is expressed in terms of the amplitudes or coefficients a_n, b_n of the sinusoidal components at the harmonic frequencies $n\omega_f, n = 0, 1, 2, \ldots$. It is also termed the harmonic content of $f(t)$. Some of the coefficients may be zero, depending on the nature of $f(t)$.

A nonperiodic function $f(t)$ that is defined in the interval $(0, T_p)$ can be regarded in that interval as being one period of a periodic function and, therefore, can also be expanded in a Fourier series of the form (2.1). This expansion represents $f(t)$ within the interval $(0, T_p)$ but not outside that interval. It follows from (2.1) that if we wish to examine the frequencies present (or the "frequency content") in a *finite* piece of continuous data of length T_p, we cannot distinguish between frequencies closer together than $\Delta f = 1/T_p = \omega_f/(2\pi)$, where f is in cycles per second, or hertz (Hz). The fundamental frequency, or frequency increment Δf in this case, is called the *frequency resolution* of the data.

If we select the fundamental period as that running from $-L$ to L, where $L = T_p/2$, noting that $\omega_f = \pi/L$, we can write (2.1) as

$$f(t) = a_0 + \sum_{n=1}^{\infty} a_n \cos \frac{n\pi t}{L} + \sum_{n=1}^{\infty} b_n \sin \frac{n\pi t}{L} \tag{2.2}$$

It is sometimes convenient to express the function to be expanded as $f(x), x = \omega_f t$, with fundamental period 2π, ranging from $-\pi$ to π. In this case, (2.1) takes the form

$$f(x) = a_0 + \sum_{n=1}^{\infty} a_n \cos nx + \sum_{n=1}^{\infty} b_n \sin nx \tag{2.3}$$

The coefficients in (2.1), (2.2), and (2.3) can be evaluated for a particular $f(t)$ or $f(x)$ from the orthogonal properties of sines and cosines. These are, for $m, n = 0, 1, 2, \ldots,$

$$\int_{-\pi}^{\pi} \cos mx \cos nx \, dx = \begin{cases} 0 & \text{if } m \neq n \\ \pi & \text{if } m = n \neq 0 \\ 2\pi & \text{if } m = n = 0 \end{cases}$$

$$\int_{-\pi}^{\pi} \sin mx \sin nx \, dx = \begin{cases} 0 & \text{if } m \neq n(2.4) \\ \pi & \text{if } m = n \end{cases} \tag{2.4}$$

$$\int_{-\pi}^{\pi} \sin mx \cos nx \, dx = 0 \text{ for all } m \text{ and } n$$

For example, multiply both sides of (2.3) by $\cos mx$ and by $\sin mx$, integrate with respect to x from $-\pi$ to π, and use the orthogonal relationships (2.4) to get, respectively,

$$a_0 = \frac{1}{2\pi} \int_{-\pi}^{\pi} f(x) \, dx$$

$$a_n = \frac{1}{\pi} \int_{-\pi}^{\pi} f(x) \cos nx \, dx, n = 1, 2, \ldots \tag{2.5}$$

$$b_n = \frac{1}{\pi} \int_{-\pi}^{\pi} f(x) \sin nx \, dx, n = 1, 2, \ldots$$

A similar process applied to (2.1) gives the coefficients

$$a_0 = \frac{1}{T_p} \int_0^{T_p} f(t) \, dt$$

$$a_n = \frac{2}{T_p} \int_0^{T_p} f(t) \cos n\omega_f t \, dt, n = 1, 2, \ldots \tag{2.6}$$

$$b_n = \frac{2}{T_p} \int_0^{T_p} f(t) \sin n\omega_f t \, dt, n = 1, 2, \ldots$$

The coefficients for (2.2) are

$$
\begin{aligned}
a_0 &= \frac{1}{2L}\int_{-L}^{L} f(t)\,dt \\
a_n &= \frac{1}{L}\int_{-L}^{L} f(t)\cos\frac{n\pi t}{L}dt, n = 1, 2, \ldots \\
b_n &= \frac{1}{L}\int_{-L}^{L} f(t)\sin\frac{n\pi t}{L}dt, n = 1, 2, \ldots
\end{aligned}
\tag{2.7}
$$

We note that the zero frequency coefficient a_0 is the average or mean value of $f(t)$ or $f(x)$ over the fundamental period.

A periodic function $f(t)$ can be expanded in a Fourier series in an interval (a, b) if it satisfies the following sufficient conditions, which are called the Dirichlet conditions.

1. It is single-valued; that is, there is only one value of $f(t)$ for each value of t.
2. It is finite everywhere, or, if an infinite value is present in $f(t)$, it is integrable (for example, a delta function).
3. It is absolutely integrable (or has finite energy) in the period. That is,

$$\int_0^T |f(t)|dt < \infty \tag{2.8}$$

4. There is a finite number of maxima and minima in one period.
5. There is a finite number of discontinuities in one period.

Most periodic signals of practical interest satisfy these conditions, so we will not concern ourselves further with the possible nonexistence of a Fourier series representation.

Our use of the =, or "equals," sign in (2.1) is not quite correct if $f(t)$ is discontinuous. At a discontinuity, the series representation converges to the mean of the value of the function immediately before the discontinuity and its value immediately after that discontinuity. In spite of this, we will continue to use the = sign. In this context, it will indicate that the function $f(t)$ can be represented by the expansion on the right-hand side of the symbol.

Fourier Cosine and Sine Series If $g(x) = g(-x)$, $g(x)$ is said to be an *even* function, and if $g(-x) = -g(x)$, it is said to be an *odd* function. By this definition, $\cos nx$ is even, $n = 0, 1, 2, \ldots$, and $\sin nx$ is odd, $n = 1, 2, \ldots$. To expand an even or an odd periodic function in a Fourier series, it is convenient to take the origin in the center of the fundamental period [see expressions (2.2) and (2.3)]. An even function $f(t)$ or $f(x)$ defined in the interval $(-L, L)$ or $(-\pi, \pi)$, respectively, can be expanded in a Fourier *cosine* series:

$$f(t) = a_0 + \sum_{n=0}^{\infty} a_n \cos \frac{n\pi t}{L} \tag{2.9}$$

with coefficients

$$\begin{aligned} a_0 &= \frac{1}{L} \int_0^L f(t)\, dt \\ a_n &= \frac{2}{L} \int_0^L f(t) \cos \frac{n\pi t}{L}\, dt, n = 1, 2, \ldots \end{aligned} \tag{2.10}$$

and

$$f(x) = a_0 + \sum_{n=0}^{\infty} a_n \cos nx \tag{2.11}$$

with coefficients

$$\begin{aligned} a_0 &= \frac{1}{\pi} \int_0^\pi f(x)\, dx \\ a_n &= \frac{2}{\pi} \int_0^\pi f(x) \cos nx\, dx, n = 1, 2, \ldots \end{aligned} \tag{2.12}$$

Similarly, an odd function $f(t)$ or $f(x)$ defined in $(-L, L)$ or $(-\pi, \pi)$, respectively, can be expanded in a Fourier *sine* series:

$$f(t) = \sum_{n=0}^{\infty} b_n \sin \frac{n\pi t}{L} \tag{2.13}$$

with coefficients

$$b_n = \frac{2}{L} \int_0^L f(t) \sin \frac{n\pi t}{L}, n = 1, 2, \ldots \tag{2.14}$$

and

$$f(x) = \sum_{n=0}^{\infty} b_n \sin nx \tag{2.15}$$

with coefficients

$$b_n = \frac{2}{\pi} \int_0^\pi f(x) \sin nx\, dx, n = 1, 2, \ldots \tag{2.16}$$

Polar Form Expressions (2.1), (2.2), and (2.3) are called the *trigonometric* form of the Fourier series. We can get another expression, called the *polar* form, by combining the sine and cosine terms in (2.1), for example. This gives

$$f(t) = \sum_{n=0}^{\infty} A_n \cos (n\omega_f t + \phi_n) \tag{2.17}$$

where the amplitude or magnitude is

$$A_n = \sqrt{a_n^2 + b_n^2} \tag{2.18}$$

and the phase is

$$\phi_n = \tan^{-1}\left(\frac{b_n}{a_n}\right) \tag{2.19}$$

From this, a frequency domain representation of $f(t)$ can be depicted on two diagrams: a plot of A_n versus $n\omega_f$ called the magnitude spectrum, and a plot of ϕ_n versus $n\omega_f$, called the phase spectrum. Both diagrams consist of a series of discrete *spectral* lines, one for each nonzero value of A_n and ϕ_n, respectively. Therefore, in general, the line spacing is the fundamental frequency ω_f. Examples of such line spectra are given in Figure 2.1.

Exponential Form A third form of the Fourier series expansion is obtained from (2.1) by using the Euler relationships:

$$\begin{aligned} \cos x &= \frac{1}{2}(e^{jx} + e^{-jx}) \\ \sin x &= \frac{1}{2j}(e^{jx} - e^{-jx}) \end{aligned} \tag{2.20}$$

This results in

$$f(t) = \sum_{n=-\infty}^{\infty} c_n e^{jn\omega_f t} \tag{2.21}$$

with coefficients given by

$$c_n = \frac{1}{T_p}\int_0^{T_p} f(t)e^{-jn\omega_f t}\,dt \tag{2.22}$$

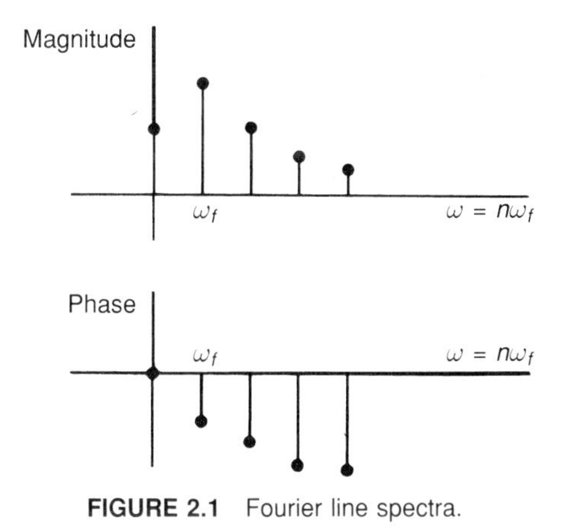

FIGURE 2.1 Fourier line spectra.

Expression (2.21) is called the *complex* or *exponential* form. If we write out the expansion, it appears that for negative values of n, we have negative frequencies. This is merely a convenient mathematical artifact. In practice, we are concerned solely with positive values of frequency.

We will now consider an example illustrating the use of the Fourier series expansion.

EXAMPLE 2.1

Find the Fourier series expansion of a square wave as shown in Figure 2.2.

Solution

It is evident from the figure that by making a suitable choice of the origin, we can treat the function as being either even or odd. In both cases, the fundamental period runs from $-L$ to L, where $L = T_p/2$. For illustration purposes, we will treat the function as odd and expand $f(t)$ in a Fourier sine series according to (2.13) and (2.14). Doing so yields

$$f(t) = \frac{2}{L} \sum_{n=1}^{\infty} \sin \frac{n\pi t}{L} \int_0^L f(t') \sin \frac{n\pi t'}{L} dt'$$

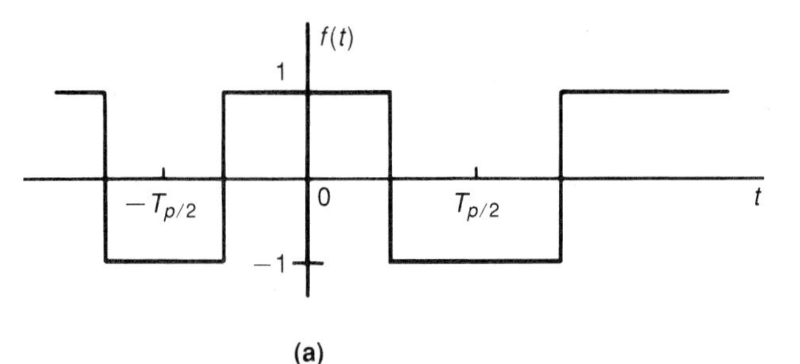

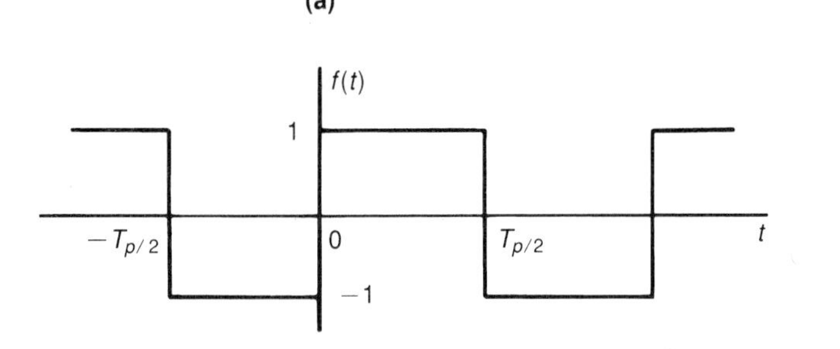

FIGURE 2.2 Square wave as (a) an even function and (b) an odd function.

When we evaluate the integral, we get the coefficients

$$b_n = \frac{2}{n\pi}(1 - \cos n\pi)$$
$$= \begin{cases} \frac{4}{n\pi} & \text{for odd } n \\ 0 & \text{for even } n \end{cases}$$

Therefore the required expansion is

$$f(t) = \frac{4}{\pi}\left(\sin\frac{\pi t}{L} + \frac{1}{3}\sin\frac{3\pi t}{L} + \frac{1}{5}\sin\frac{5\pi t}{L} + \ldots\right)$$

If the origin is selected other than as shown in Figure 2.2, the expansion will have both sines and cosines as in (2.2) or both magnitude and phase as in (2.17).

2.2.2 Fourier Transform

The Fourier series expansion gives the harmonic content of a periodic time function or of a function defined over a finite interval. It is natural to ask whether we can obtain similar useful frequency information for an aperiodic function defined over the entire time axis. The answer is yes. Such a signal $f(t)$ can be described in the frequency domain by the Fourier integral,

$$F(j\omega) = \int_{-\infty}^{\infty} f(t)e^{-j\omega t}dt \tag{2.23}$$

This is called the Fourier transform of $f(t)$ and can be written more briefly as

$$F(j\omega) = \mathfrak{F}\{f(t)\} \tag{2.24}$$

We can recover $f(t)$ from $F(j\omega)$ by the inverse Fourier transform,

$$f(t) = \frac{1}{2\pi}\int_{-\infty}^{\infty} F(j\omega)e^{j\omega t}d\omega = \mathcal{F}^{-1}\{F(j\omega)\} \tag{2.25}$$

One way to derive these expressions is to consider the Fourier integral as a limiting form of the Fourier series as the period approaches infinity. Recall that the spacing of the lines in the spectrum of a periodic function equals its fundamental frequency, which, in hertz, is the inverse of the fundamental period. Thus, as the period becomes larger and larger, the spectral lines move closer and closer together until, in the limit as $T_p \rightarrow \infty$ (that is, $f(t)$ is an aperiodic function), the spectral lines become dense, forming a continuous-frequency spectrum. To formalize this concept [7], consider the complex version of the Fourier series (2.21) with coefficients (2.22), written in the following form (recall that $T_p = 2\pi/\omega_f$).

$$f(t) = \sum_{n\omega_f=-\infty}^{\infty} \left(\frac{c_n}{\omega_f}\right) e^{jn\omega_f t}\delta(n\omega_f) \tag{2.26}$$

$$\left(\frac{c_n}{\omega_f}\right) = \frac{1}{2\pi}\int_{-\pi/\omega_f}^{\pi/\omega_f} f(t)e^{-jn\omega_f t}dt \tag{2.27}$$

Although as the spacing ω_f becomes smaller, the amplitudes c_n also shrink, the ratio c_n/ω_f approaches a limiting function. In (2.26), $\delta(n\omega_f)$ approaches ω_f, the increment in the line spacing. In the limit as period $T_p \to \infty$, we have $\omega_f \to 0, n \to \infty$, but the product $n\omega_f$ approaches ω, a continuous frequency variable. Then $\delta(n\omega_f) \to d\omega$, (2.26) becomes an integral, and $(c_n/\omega_f) \to F(\omega)$, or $F(j\omega)$, as it is usually written, to emphasize that F is a function of a complex variable. (We will omit the j in expressions where the notation becomes cumbersome.) Thus we obtain the Fourier transform pair (2.23) and (2.25). Although the proof just given is heuristic, it will suffice for our purposes. We note that $F(j\omega)$ itself is, in general, a complex variable for any fixed value of ω and hence has a real and an imaginary part,

$$F(j\omega) = R(\omega) + jI(\omega) \tag{2.28}$$

or, alternatively, can be expressed as a vector in the complex plane with magnitude M and phase ϕ,

$$F(j\omega) = M(\omega)\underline{/\phi(\omega)} = M(\omega)e^{j\phi(\omega)} \tag{2.29}$$

The conditions under which a function $f(t)$ possesses a Fourier transform are essentially the same as those required for a periodic function to have a Fourier series representation—the *Dirichlet conditions*—but now, because the period is infinite, condition (2.8) becomes

$$\int_{-\infty}^{\infty} |f(t)|dt < \infty \tag{2.30}$$

Properties of the Fourier Transform Let us list some important properties of the Fourier transform $\mathfrak{F}\{f(t)\} = F(j\omega)$.

1. *Linearity*

$$\mathfrak{F}\{\alpha_1 f_1(t) + \alpha_2 f_2(t)\} = \alpha_1 F_1(j\omega) + \alpha_2 F_2(j\omega) \tag{2.31}$$

where $F_1(j\omega) = \mathcal{F}\{f_1(t)\}$, $F_2(j\omega) = \mathcal{F}\{f_2(t)\}$, and α_1 and α_2 are constants.

2. *Time shift*

$$\mathcal{F}\{f(t \pm t_0)\} = e^{\pm j\omega t_0}F(j\omega) \tag{2.32}$$

3. *Frequency shift*

$$\mathcal{F}\left\{f(t)e^{\pm j\omega_0 t}\right\} = F[j(\omega \mp \omega_0)] \tag{2.33}$$

4. *Scale factor*

$$\mathcal{F}\{f(at)\} = \frac{1}{a}F(j\omega/a) \tag{2.34}$$

5. *Convolution theorem*

$$\mathcal{F}\left\{\int_{-\infty}^{\infty} f_1(\tau)f_2(t-\tau)d\tau\right\} = F_1(j\omega)F_2(j\omega) \tag{2.35}$$

6. *Complex convolution theorem*

$$\mathcal{F}\{f_1(t)f_2(t)\} = \frac{1}{2\pi}\int_{-\infty}^{\infty} F_1(p)F_2(\omega - p)dp \tag{2.36}$$

where p is a complex variable.

7. *Theorem for differentiation*

$$\mathcal{F}\left\{\frac{d^n f}{dt^n}\right\} = (j\omega)^n F(j\omega) \tag{2.37}$$

8. *Theorem for integration*

$$\mathcal{F}\left\{\int_{-\infty}^{t} f(t)\,dt\right\} = \frac{F(j\omega)}{j\omega} \tag{2.38}$$

This is a review, so we leave the proof of these properties as an exercise, with one exception. This is the convolution theorem (2.35), the proof of which follows.

$$\mathcal{F}\left\{\int_{-\infty}^{\infty} [f_1(\tau)f_2(t-\tau)\,d\tau]\right\} = \int_{-\infty}^{\infty} dt\left[\int_{-\infty}^{\infty} d\tau\, f_1(\tau)f_2(t-\tau)\right]e^{-j\omega t}$$

$$= \int_{-\infty}^{\infty} d\tau\, f_1(\tau)e^{-j\omega t}\int_{-\infty}^{\infty} dt\, f_2(t-\tau)e^{-j\omega(t-\tau)}$$

In the second integral, replace dt by $d(t-\tau)$ because τ is constant with respect to that integration. Then define a new variable $t_1 = (t-\tau)$ so that

$$\mathcal{F}\left\{\int_{-\infty}^{\infty} f_1(\tau)f_2(t-\tau)\,d\tau\right\} = \int_{-\infty}^{\infty} d\tau\, f_1(\tau)e^{j\omega t}\int_{-\infty}^{\infty} dt_1\, f_2(t_1)e^{-j\omega t_1}$$

$$= F_1(j\omega)F_2(j\omega) \tag{2.39}$$

Q.E.D.

This is a very useful theorem for system analysis. Recall from (1.61) that the output of a continuous linear system is the convolution of its input and its weighting function. It follows from the theorem that the Fourier transform of its output is the *product* of the individual transforms of the input and weighting function. That is,

$$Y(j\omega) = H(j\omega)X(j\omega) \tag{2.40}$$

Taking a product is certainly easier than evaluating a convolution. Consequently, the simplest way to find the output $y(t)$ if $x(t)$ and $h(t)$ are known is to multiply the transforms of $x(t)$ and $h(t)$ and then take the inverse Fourier transform of the product. We will see later that the same technique is applicable to the Laplace transform (Section 2.2.3) and to the z-transform (Chapter 3).

2.2.3 Laplace Transform

Many aperiodic functions of practical interest do not satisfy the Dirichlet conditions (2.30). For example, a unit step function has infinite area under the curve. So have a unit ramp $f(t) = tu(t)$ and an impulse train $f(t) = \sum_{k=-\infty}^{\infty} \delta(\tau - kT)$. In spite of this, we can often define a Fourier transform for such functions in the limit by making use of the generalized functions. The resulting transform includes delta functions. We will not consider further the Fourier transforms of such functions. Instead, for functions that start in time (positive-time functions, $f(t) = 0, t < 0$), a way around the difficulty of finding a Fourier transform is to multiply $f(t)$ by a convergence factor $e^{-\sigma t}$, where σ is a positive number, so that

$$\int_0^{\infty} |f(t)e^{-\sigma t}|dt < \infty \tag{2.41}$$

Take the Fourier transform of the composite function $f(t)e^{-\sigma t}$ to get

$$\mathcal{F}\left\{f(t)e^{-\sigma t}\right\} = \int_0^{\infty} f(t)e^{-\sigma t}e^{-j\omega t}\, dt = \int_0^{\infty} f(t)e^{-st}dt \tag{2.42}$$

where $s = \sigma + j\omega$. This is called the *Laplace transform* of $f(t)$ and is denoted by $F(s)$ or by $\mathcal{L}\{f(t)\}$. We can think of a complex s-plane with real axis σ and imaginary axis $j\omega$. For values of s along the $j\omega$ axis $(\sigma = 0), F(s) = F(j\omega)$ is the Fourier transform of the positive-time function. Thus, given a function of s, we can find its frequency characteristics by setting $s = j\omega$.

The inverse Laplace transform is obtained from (2.25) and results in

$$f(t) = \frac{1}{2\pi j}\int_{c-j\infty}^{c+j\infty} F(s)e^{st}ds \tag{2.43}$$

For brevity, this is often written

$$f(t) = \mathcal{L}^{-1}\{F(s)\} \tag{2.44}$$

Let us summarize now the conditions under which a function $f(t)$ is Laplace transformable.

1. $f(t)$ defined for $t \geq 0$
2. $f(t) = 0, \quad t < 0$
3. The composite function $f(t)e^{-\sigma t}$ satisfies the Dirichlet conditions. In particular,

$$\int_0^\infty |f(t)e^{-\sigma t}|dt < \infty \tag{2.45}$$

Suppose σ_{min} is the *minimum* value of the parameter σ that satisfies condition 3. Then $F(s)$ exists if the real part of s—that is σ—is greater than σ_{min}. The part of the s-plane where $\sigma > \sigma_{\text{min}}$ is called the *region of convergence* (ROC) for $F(s)$. The constant c in the limits of integration for the inverse transform (2.43) must lie in the ROC. Table 2.1 gives the Laplace transform of some simple time functions. These can be derived without much difficulty by evaluating the integral (2.42). In the case of the sinusoids, it is best to write them in exponential form (2.20) before performing the integration.

Properties of the Laplace Transform The properties of the Laplace transform $\mathcal{L}\{f(t)\} = F(s)$ are similar to those of the Fourier transform. Again, the proofs are left as an exercise.

TABLE 2.1 Some Basic Laplace Transform Pairs

$f(t), t \geq 0$	$F(s)$
$\delta(t - t_0), t_0 > 0$	e^{-st_0}
A	$\dfrac{A}{s}$
e^{-at}	$\dfrac{1}{s + a}$
$\sin \omega_0 t$	$\dfrac{\omega_0}{s^2 + \omega_0^2}$
$\cos \omega_0 t$	$\dfrac{s}{s^2 + \omega_0^2}$
$e^{-at} \sin \omega_0 t$	$\dfrac{\omega_0}{(s + a)^2 + \omega_0^2}$
At	$\dfrac{A}{s^2}$
At^{n-1}	$\dfrac{A(n-1)!}{s^n}$
$e^{-at}t^{n-1}$	$\dfrac{(n-1)!}{(s + a)^n}$

1. *Linearity*

$$\mathcal{L}\{\alpha_1 f_1(t) + \alpha_2 f_2(t)\} = \alpha_1 F_1(s) + \alpha_2 F_2(s) \tag{2.46}$$

where $F_1(s) = \mathcal{L}\{f_1(t)\}$, $F_2(s) = \mathcal{L}\{f_2(t)\}$, and α_1 and α_2 are constants.

2. *Time shift*

$$\mathcal{L}\{f(t \pm t_0)\} = e_0^{\pm st} F(s) \tag{2.47}$$

3. *Frequency shift*

$$\mathcal{L}\{f(t)e^{\pm bt}\} = F(s \mp b) \tag{2.48}$$

4. *Scale factor*

$$\mathcal{L}\{f(at)\} = \frac{1}{a} F(s/a) \tag{2.49}$$

5. *Convolution theorem*

$$\mathcal{L}\left\{\int_0^\infty f_1(\tau) f_2(t-\tau) d\tau\right\} = F_1(s) F_2(s) \tag{2.50}$$

6. *Complex convolution theorem*

$$\mathcal{L}\{f_1(t) f_2(t)\} = \frac{1}{2\pi j} \int_{-j\infty}^{j\infty} F_1(p) F_2(s-p) dp \tag{2.51}$$

where p is a complex variable and the imaginary axis lies in the ROC for $F(s)$.

7. *Theorem for differentiation*

a)

$$\mathcal{L}\left\{\frac{d\,f(t)}{dt}\right\} = sF(s) - f(0_+) \tag{2.52}$$

The argument 0_+ is the value of t immediately after $t = 0$. The term $f(0_+)$ is the value of $f(t)$ as the origin is approached from the right-hand side; that is, the positive value of time. This provides for the possibility that $f(t)$ is discontinuous at the origin—a step function, for example.

b)

$$\mathcal{L}\left\{\frac{d^2 f(t)}{dt^2}\right\} = s^2 F(s) - sf(0_+) - \frac{df}{dt}(0_+) \tag{2.53}$$

where $df/dt(0_+)$ is the initial value of the first derivative. In general,

c)

$$\mathcal{L}\left\{\frac{d^n f(t)}{dt^n}\right\} = s^n F(s) - s^{n-1} f(0_+) - s^{n-2}\frac{df}{dt}(0_+) \cdots - s\frac{d^{n-2}f}{dt^{n-2}}(0_+) - \frac{d^{n-1}f}{dt^{n-1}}(0_+) \tag{2.54}$$

8. *Theorem for integration*

If we define

$$f^{(-1)}(t) = \int f(t)dt = \int_0^t f(t)dt + f^{(-1)}(0_+)$$

then

$$\mathcal{L}\left\{\int f(t)dt\right\} = \frac{F(s)}{s} + \frac{f^{(-1)}(0_+)}{s} \tag{2.55}$$

9. *Final-value theorem*
   If the function $f(t)$ and its first derivative are Laplace transformable and there are no singularities of $sF(s)$ on the $j\omega$ axis or in the right half-plane ($\sigma > 0$), then

$$\lim_{t\to\infty} f(t) = \lim_{s\to 0} sF(s) \tag{2.56}$$

   This equation means that we can get the steady-state behavior of $f(t)$ from $F(s)$ without taking the inverse Laplace transform of $F(s)$.

10. *Initial-value theorem*
    If the function $f(t)$ and its first derivative are Laplace transformable and the limit of $sF(s)$ exists as s approaches infinity, then,

$$\lim_{t\to 0_+} f(t) = \lim_{s\to\infty} sF(s) \tag{2.57}$$

    This theorem is useful in determining the initial value of $f(t)$ at $t = 0_+$ (that is, immediately after switching on) without taking the inverse Laplace transform of $F(s)$.

System Transfer Function As noted in (1.58), a continuous LTI system can be represented by the differential equation

$$\sum_{i=0}^{n} a_i y^{(i)} = \sum_{i=0}^{m} b_i x^{(i)} \tag{2.58}$$

Using the theorem on differentiation and assuming zero initial conditions, we can transform (2.58) into an *algebraic* equation in s,

$$\left(\sum_{i=0}^{n} a_i s^i\right) Y(s) = \left(\sum_{i=0}^{m} b_i s^i\right) X(s) \tag{2.59}$$

where $X(s)$ and $Y(s)$ are the Laplace transforms of input $x(t)$ and output $y(t)$, respectively. From this, we can define the *transfer function* of the system as

$$H(s) = \frac{Y(s)}{X(s)} = \frac{\sum_{i=0}^{m} b_i s^i}{\sum_{i=0}^{n} a_i s^i} \tag{2.60}$$

In general, $H(s)$ is the ratio of two polynominals in s of order m and n (for numerator and denominator), respectively, with $m \le n$. The transfer function may be written in cascade form or parallel form.

$$H(s) = \frac{\prod_{i=1}^{m}(s-\beta_i)}{\prod_{i=1}^{n}(s-\alpha_i)} \quad \text{cascade form} \tag{2.61}$$

$$H(s) = \sum_{i=1}^{n} \frac{R_i}{(s-\alpha_i)} \quad \text{parallel form} \tag{2.62}$$

These forms are depicted in Figure 2.3 for $m < n$. Both forms are also useful in the implementation of digital filters. In (2.61) and (2.62), the α_i and β_i designate the poles and zeros, respectively, of $H(s)$. For each i, R_i is called the *residue of* $H(s)$ at the pole α_i.

If, instead of transforming the differential equation, we take the Laplace transform of the convolution representation,

$$y(t) = \int_0^{\infty} x(\tau)h(t-\tau)d\tau \tag{2.63}$$

where $h(t)$ is the impulse response or weighting function, we get, from the convolution theorem,

$$Y(s) = H(s)X(s) \tag{2.64}$$

so that

$$\frac{Y(s)}{X(s)} = H(s) \tag{2.65}$$

as before. However, in this case $H(s)$ is the Laplace transform of the weighting function. It follows that the transfer function is the Laplace transform of the weighting function.

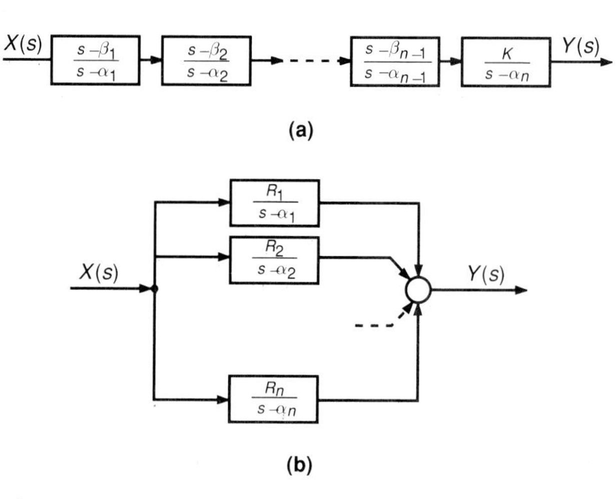

FIGURE 2.3 Linear system forms. (a) Cascade form ($m = n - 1$). (b) Parallel form ($m < n$).

Given an input signal $x(t)$ and the system weighting function $h(t)$ or transfer function $H(s)$, we can find the output $y(t)$ by multiplying the transfer function by the transform of $x(t)$ and taking the inverse transform of the product,

$$y(t) = \mathcal{L}^{-1}\{H(s)X(s)\} \tag{2.66}$$

In most cases, this is much easier than evaluating $y(t)$ by the convolution integral (2.63). At least three general methods of obtaining the inverse transform for a given $F(s)$ are available.

1. Refer to a table of transform pairs, such as Table 2.1. For more complex $F(s)$, this method is seldom productive.
2. Expand $F(s)$ in partial fractions, and take the inverse transforms of the simpler individual terms.
3. Use the residue method.

Methods 2 and 3 are similar. Consider the partial fraction expansion method. Let

$$F(s) = \frac{B(s)}{A(s)} \tag{2.67}$$

Factor the polynomial $A(s)$, obtaining

$$\begin{aligned} F(s) &= \frac{B(s)}{(s-\alpha_1)(s-\alpha_2)\ldots(s-\alpha_n)} \\ &= \frac{R_1}{s-\alpha_1} + \frac{R_2}{s-\alpha_2} + \ldots + \frac{R_i}{s-\alpha_i} + \ldots \frac{R_n}{s-\alpha_n} \end{aligned} \tag{2.68}$$

where the constants R_i, $i = 1, n$, are to be determined. The resemblance to the parallel form (2.62) is apparent. To evaluate a typical coefficient R_i, multiply both sides of (2.68) by $(s - \alpha_i)$, obtaining

$$(s-\alpha_i)F(s) = R_1\frac{s-\alpha_i}{s-\alpha_1} + \ldots + R_i + \ldots + R_n\frac{s-\alpha_i}{s-\alpha_n} \tag{2.69}$$

If we set $s = \alpha_i$, all the terms on the right-hand side of (2.57) are zero except the coefficient R_i, which was to be determined. Therefore,

$$R_i = (s-\alpha_i)F(s)\Big|_{s=\alpha_i} \tag{2.70}$$

The complete inverse transform is the sum of the inverse transforms from each term. That is,

$$\mathcal{L}^{-1}\{F(s)\} = \mathcal{L}^{-1}\left\{\sum_{i=1}^{n}\frac{R_i}{s-\alpha_i}\right\} = \sum_{i=1}^{n} R_i e^{\alpha_i t} \tag{2.71}$$

from the third entry in Table 2.1.

The partial-fraction expansion procedure must be modified if some of the roots of $A(s)$ in (2.68) are repeated. For example, suppose that there is a third-order pole, at $s = \alpha_k$.

$$F(s) = \frac{B(s)}{(s-\alpha_1)(s-\alpha_2)\ldots(s-\alpha_k)^3\ldots(s-\alpha_n)}$$

$$= \frac{R_1}{s-\alpha_1} + \ldots + \frac{R_{k1}}{(s-\alpha_k)^3} + \frac{R_{k2}}{(s-\alpha_k)^2}$$

$$+ \frac{R_{k3}}{s-\alpha_k} + \ldots + \frac{R_n}{s-\alpha_n} \tag{2.72}$$

We note that R_{k3} is the residue of $F(s)$ at the multiple pole α_k. Coefficients R_{k1} and R_{k2} are not residues. Multiply both sides by $(s-\alpha_k)^3$.

$$(s-\alpha_k)^3 F(s) = \frac{R_1(s-\alpha_k)^3}{s-\alpha_1} + \ldots + R_{k1} + R_{k2}(s-\alpha_k)$$

$$+ R_{k3}(s-\alpha_k)^2 + \ldots + \frac{R_n(s-\alpha_k)^3}{s-\alpha_n} \tag{2.73}$$

Set $s = \alpha_k$, obtaining

$$R_{k1} = (s-\alpha_k)^3 F(s)\big|_{s=\alpha_k} \tag{2.74}$$

The coefficients R_i, $i = 1, n$, and $i \neq k$, can be found from (2.70) as before. However, we still have to find R_{k2} and R_{k3}. To determine R_{k2}, differentiate (2.73) with respect to s, thereby eliminating the term $(s-\alpha_k)$ multiplying R_{k2}. Then set $s - \alpha_k$, obtaining

$$R_{k2} = \frac{d}{ds}[(s-\alpha_k)^3 F(s)]\big|_{s=\alpha_k} \tag{2.75}$$

Next differentiate (2.73) a second time and set $s = \alpha_k$ to get R_{k3}.

$$R_{k3} = \frac{1}{2}\frac{d^2}{ds^2}[(s-\alpha_k)^3 F(s)]\big|_{s=\alpha_k} \tag{2.76}$$

In general, for a pole of order m at $s = \alpha_k$,

$$R_{km} = \frac{1}{(m-1)!}\frac{d^{m-1}}{ds^{m-1}}[(s-\alpha_k)^m F(s)]\big|_{s=\alpha_k} \tag{2.77}$$

From the last entry in Table 2.1, the inverse transform of terms such as $R_{k1}/(s-\alpha_k)^m$ is obtained as follows:

$$\mathcal{L}^{-1}\left\{\frac{R_{k1}}{(s-\alpha_k)^m}\right\} = \frac{R_{k1}}{(m-1)!}t^{m-1}e^{\alpha_k t} \tag{2.78}$$

The residue method for evaluating the inverse Laplace transform is based on Cauchy's residue theorem in complex-variable theory [8]. Proving the theorem would cause us to digress too much. For our purpose here and in connection with the z-transform in Chapter 3, it will suffice to state the theorem and give an example of how it is used.

Cauchy's Integral Theorem Let $f(z)$ be analytic (that is, differentiable) in a simply connected domain D in the complex z-plane, except for a finite number of points α_i at which f may have isolated singularities (poles). Let C be a simple closed curve that lies in D and does not pass through any point α_i. Then,

$$\int_C f(z)dz = 2\pi j \sum_i \text{(residues of } f \text{ at } \alpha_i) \tag{2.79}$$

where the sum is extended over the α_i that are *inside* C.

We can apply this theorem to the inverse transform integral of (2.43) if we make the following adjustments. Identify z with s and $f(z)$ with $F(s)e^{st}$. Make (2.43) an integral around a closed curve in the s-plane by visualizing a circle of infinite radius going in a counterclockwise direction from $c + j\infty$ to $c - j\infty$ to close the curve. If $F(s)e^{st} \to 0$ as $s \to \infty$, this does not change the value of the integral. Then, the theorem states that the value of the integral is the sum of the residues of $F(s)e^{st}$ at the poles inside the closed contour.

To calculate the residue of $F(s)e^{st}$ at a pole α_k of order m, we use the formula

$$R_k = \frac{1}{(m-1)!}\frac{d^{m-1}}{ds^{m-1}}[(s-\alpha_k)^m F(s)e^{st}]\Big|_{s=\alpha_k} \tag{2.80}$$

If $m = 1$, this reduces to

$$F_k = (s-\alpha_k)F(s)e^{st}\Big|_{s=\alpha_k} \tag{2.81}$$

Expression (2.80) can be derived in the same manner as (2.77). We note, however, that in taking the residue with (2.80), we are processing $F(s)e^{st}$ instead of $F(s)$ alone, as in the partial-fraction expansion method. As a result, we get the inverse transform directly from (2.80) without the additional step implied by (2.78).

To compare the two methods for ease of use, we will consider an example where we want to find the response of a known LTI system to a unit ramp.

EXAMPLE 2.2

Find the unit ramp response of a system the transfer function of which is

$$H(s) = \frac{8(s+1)}{s(s+2)}$$

Solution

The output transform is

$$Y(s) = H(s)X(s) = \frac{8(s+1)}{s^3(s+2)}$$

where $X(s) = 1/s^2$. Expand in partial fractions.

$$Y(s) = \frac{R_1}{s+2} + \frac{R_{21}}{s^3} + \frac{R_{22}}{s^2} + \frac{R_{23}}{s}$$

We have

$$R_1 = (s-2)Y(s)\big|_{s=-2} = 1$$

$$R_{21} = s^3 Y(s)\big|_{s=0} = 4$$

$$R_{22} = \frac{d}{ds}[s^3 Y(s)]\big|_{s=0} = 2$$

$$R_{23} = \frac{1}{2}\frac{d^2}{ds^2}[s^3 Y(s)]\big|_{s=0} = -1$$

Therefore,

$$Y(s) = \frac{1}{s+2} + \frac{4}{s^3} + \frac{2}{s^2} - \frac{1}{s}$$

and the response to a unit ramp is

$$y(t) = e^{-2t} + 2t^2 + 2t - 1$$

from (2.78). Now, by the residue method,

$$y(t) = \sum_i \text{residues of } Y(s)e^{st} \text{ at poles } s = 0 \text{ (triple) and } s = -2$$

$$= \sum_i \frac{1}{(m-1)!}\frac{d^{m-1}}{ds^{m-1}}[(s-\alpha_i)^m Y(s)e^{st}]\big|_{s=\alpha_i}$$

where $\alpha_i = -2$ and 0.

Residue at $s = -2$:

$$R_1 = \frac{8(s+1)}{s^3}e^{st}\bigg|_{s=-2} = e^{-2t}$$

Residue at $s = 0$:

$$R_2 = \frac{1}{2}\frac{d^2}{ds^2}\left[\frac{8(s+1)}{s+2}e^{st}\right]\bigg|_{s=0}$$

$$= \left[4t^2 e^{st}\frac{s+1}{s+2} + \frac{8te^{st}}{(s+2)^2} - \frac{8e^{st}}{(s+2)^3}\right]\bigg|_{s=0}$$

$$= 2t^2 + 2t - 1$$

Therefore,

$$y(t) = e^{-2t} + 2t^2 + 2t - 1$$

as before. It is evident that the residue method is more concise.

When we examine the response $y(t)$ for this example, we see that it is the sum of four terms. One of these, e^{-2t}, due to the pole at $s = -2$, dies out as t increases. One term is constant, and the other two terms increase with time. We would expect a divergent response, because the input is a ramp that is unbounded as time goes on. If instead we apply a step function, which is a bounded input, the response is

$$y(t) = -2e^{-t} + 4t + 2$$

Again the response is unbounded, though in this case the input is bounded. By definition, the system is unstable. This instability is not the fault of the pole at $s = -2$ that gives the decaying part e^{-2t}. Clearly it is due to the integration represented by the pole at $s = 0$. When we integrate a constant (the unit step), we get a response that increases linearly with time.

A necessary and sufficient condition for stability of a continuous LTI system with transfer function $H(s) = B(s)/A(s)$ can be simply stated: *All the poles of $H(s)$ must be in the left half of the s-plane* ($\sigma < 0$). In general, the poles may be either real (such as $s = -a$) or complex (such as $s = -a \pm jb$). The foregoing requirement for stability follows from the consideration that only when the real part of each pole is negative (the pole lies in the left half-plane) is the component of the overall response due to that pole a decaying time function, as shown in Example 2.2.

If we have the denominator $A(s)$ in factored form,

$$H(s) = \frac{B(s)}{\prod_{i=1}^{n}(s - \alpha_i)} \tag{2.82}$$

we can check for stability by inspection. Examples of transfer functions for stable and unstable systems are given in the pole-zero plots in Figure 2.4.

In the next section, we will discuss the frequency response of a system, an important factor in filter design. This implies that steady-state conditions have been attained (all transients have died out). Clearly, this is possible only if the system is stable. Hence a first requirement for any filter, continuous or digital, is that it be a stable system.

2.3 FREQUENCY RESPONSE

As shown in the last section, we can use the Laplace transform to convert a time-domain description of a system (such as that given by a differential equation or convolution integral) into a system description in terms of the

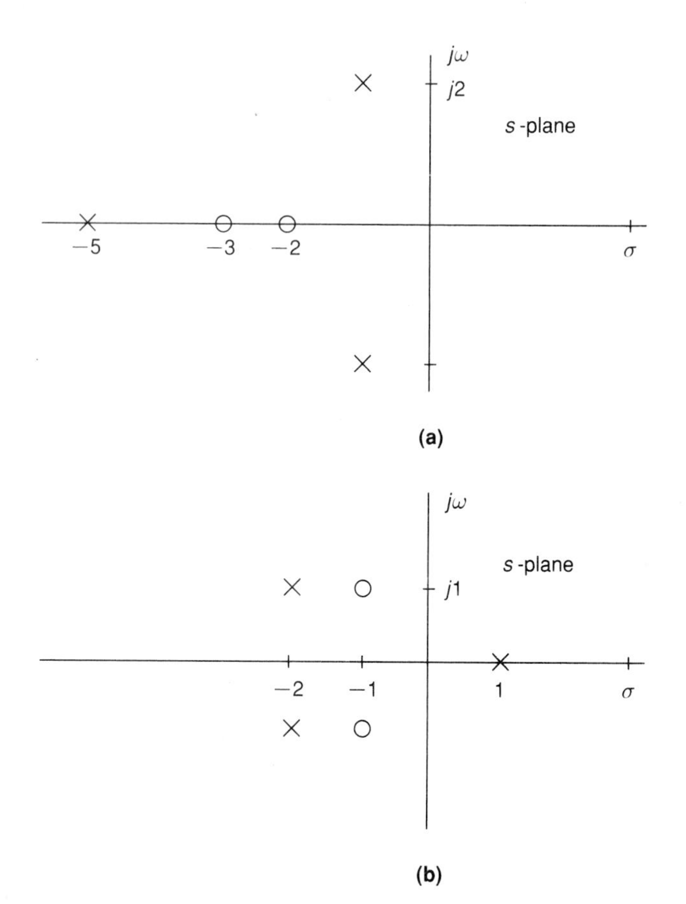

FIGURE 2.4 Pole-zero plots for a stable system and an unstable system. (a) Stable transfer function $H(s) = \frac{10(s+2)(s+3)}{(s+5)(s^2+2s+5)}$. (b) Unstable transfer function $H(s) = \frac{4(s^2+2s+3)}{(s-1)(s^2+4s+5)}$.

complex variable s. Let $H(s)$ be the transfer function. The system *frequency response* is obtained by setting $s = j\omega$. That is,

$$H(\omega) = A(\omega) + jB(\omega) = M(\omega)\exp\,[j\phi(\omega)] \tag{2.83}$$

where $M(\omega)$ is called the magnitude response and $\phi(\omega)$ is the phase response. At any frequency ω, $H(j\omega)$ is thus a complex number. As a simple example, the first-order system

$$H(s) = \frac{1}{T_1 s + 1} \tag{2.84}$$

has magnitude response

$$M(\omega) = |H(j\omega)| = \frac{1}{(\omega^2 T_1^2 + 1)^{1/2}} \tag{2.85}$$

and phase response

$$\phi(\omega) = \tan^{-1}(-\omega T_1) \tag{2.86}$$

We note that M is an even function and ϕ an odd function of ω.

In general, for higher-order systems the magnitude response and phase response for a given $H(j\omega)$ are tedious to calculate and can be obtained from a computer program. However, in order to get a qualitative feeling for the effect of the different poles and zeros of $H(s)$ on the frequency response, it is instructive to use a graphical method as illustrated in Figure 2.5. This involves expressing $H(s)$ in cascade form and plotting the poles and zeros in the s-plane. Vectors are drawn from the poles and zeros to a movable point $s = j\omega$ on the $j\omega$ axis. Then $M(\omega)$ is the product of the system gain constant K, the vector lengths from the zeros, and the reciprocals of the vector lengths from the poles. The phase response is the algebraic sum of the angles the vectors make with the positive σ direction, as shown. By varying s along the $j\omega$ axis, we can obtain as many values of $M(\omega)$ and $\phi(\omega)$ as desired. The proof that this is valid is left as an exercise. We will discuss later a similar technique that is useful for obtaining the frequency response of digital filters.

If a filter or system is defined in terms of its magnitude response, it generally falls into one of four categories: lowpass, highpass, bandpass, or bandstop. The *ideal* magnitude responses for these types of filters are depicted in Figure 2.6. For example, the ideal bandpass filter passes, without

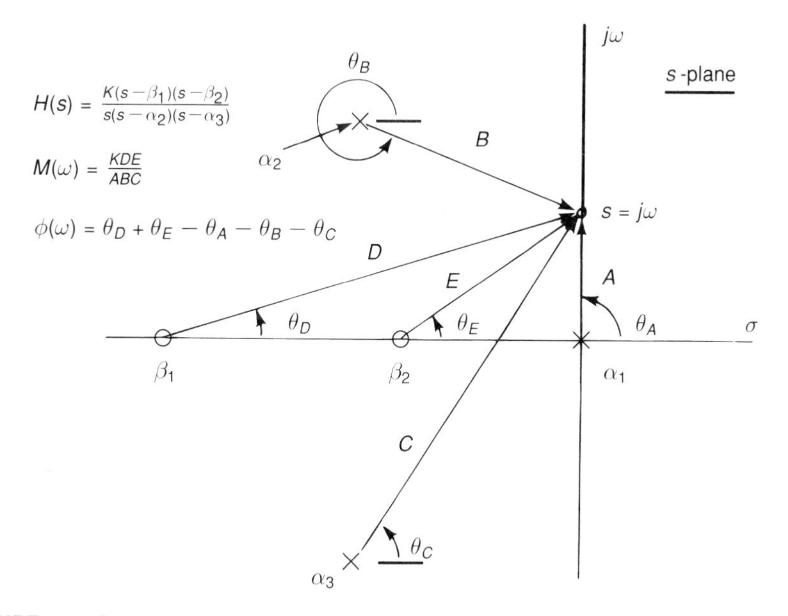

FIGURE 2.5 Graphical method for finding magnitude response and phase response.

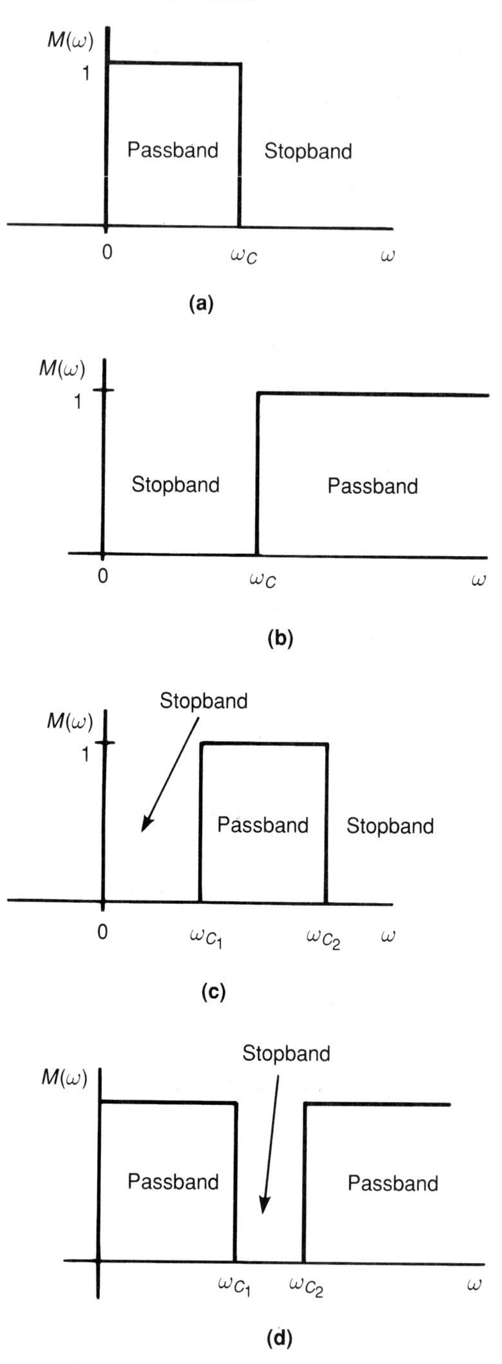

FIGURE 2.6 Ideal filters. (a) Lowpass. (b) Highpass. (c) Bandpass. (d) Bandstop.

attenuation, all frequency components of the input signal that lie in the passband stretching from the cutoff frequencies ω_{c_1} to ω_{c_2}. Any input frequency components outside that range do not appear in the output. For a lowpass or bandpass filter, the width of the passband is called the *bandwidth* of the filter.

In practice, for a reason to be discussed in the next section, we cannot achieve the infinitely sharp cutoff implied by the ideal filters. Instead we must compromise and accept a more gradual cutoff between passband and stopband, as shown in Figure 2.7. The magnitude units in this diagram are decibels (dB). If N is a number representing the actual magnitude or gain, its value in dB is $20\log_{10} N$. For example, a magnitude of unity is 0 dB, a magnitude of 10 is 20 dB, and so on. For a *practical* lowpass or bandpass design, the bandwidth is usually defined as the range of frequencies over which the magnitude response is not more than 3 dB below its maximum value in the passband. The *time constant* of a lowpass filter, defined as the time required for its output response to decay to 0.368 (that is, e^{-1}) of its initial value when left undisturbed, bears an inverse relationship to the bandwidth; the wider the bandwidth, the shorter the time constant.

2.4 APPROXIMATIONS TO THE IDEAL LOWPASS FILTER

2.4.1 Introduction

Although we cannot implement an ideal filter, we can find practical approximations to it. In designing a continuous filter for a desired magnitude response, it is convenient to standardize the procedure by concentrating on the design of approximations to an ideal *lowpass* prototype, as shown in Figure 2.8. This filter has cutoff frequency $\omega_c = 1$ and has a magnitude or gain of unity in the passband. The magnitude response is depicted as an even function of frequency in Figure 2.8(a). The phase response is not constrained when only the magnitude response is specified. Specifying the phase also would greatly increase the complexity of the design process. In Figure 2.8(c), a linear phase response is shown; that is, the phase decreases linearly with frequency. In that case, the frequency response of the ideal prototype is

$$H(j\omega) = \begin{cases} Ke^{-j\omega T} & \text{if } |\omega| \leq \omega_c \\ 0 & \text{otherwise} \end{cases} \tag{2.87}$$

where $-T$ is the slope of the phase response and gain constant $K = 1$. Taking the inverse Fourier transform of $H(j\omega)$ gives the weighting function

$$h(t) = \frac{K\omega_c}{\pi} \frac{\sin \omega_c(t-T)}{\omega_c(t-T)} \tag{2.88}$$

which is shown in Figure 2.8(b). The weighting function is nonzero for

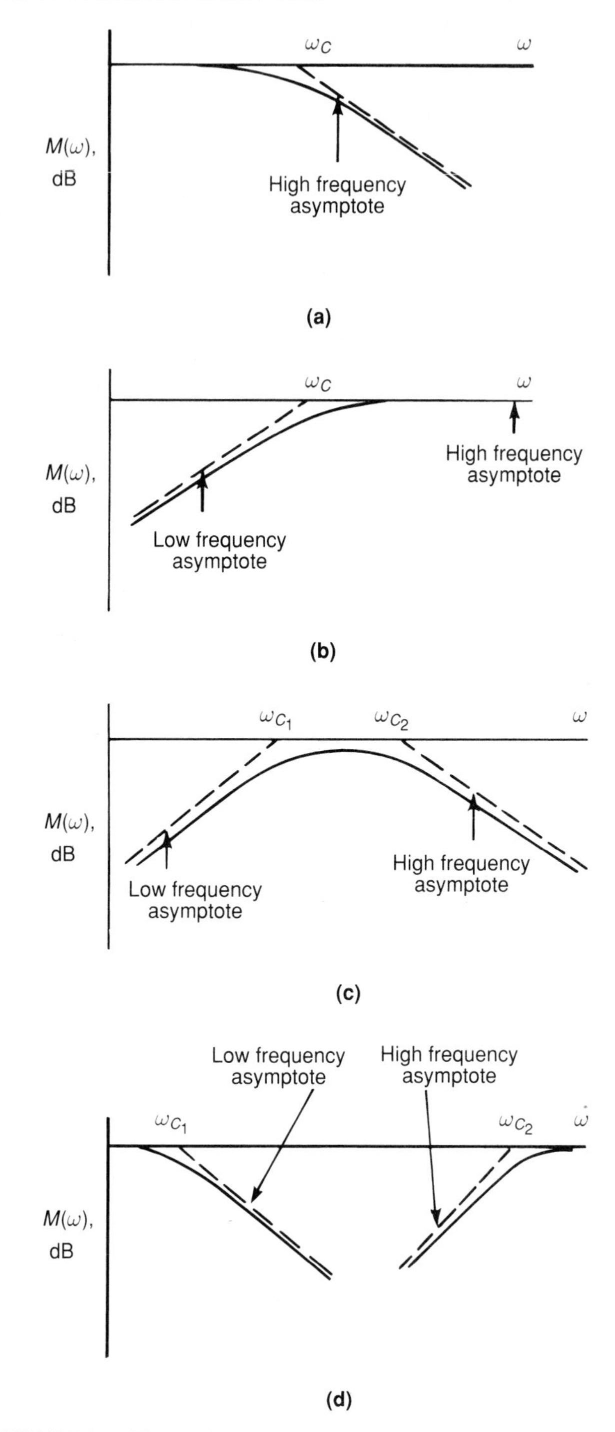

FIGURE 2.7 Types of filters. (a) Lowpass. (b) Highpass. (c) Bandpass. (d) Bandstop.

$t < 0$, so the ideal filter is not realizable. This is due to the infinitely sharp cutoff specified for the magnitude response. We note that the type of phase response does not affect this conclusion. For example, if $T = 0$ in (2.87), implying zero phase shift, we still get a sinc function for $h(t)$, but it is centered at the origin.

Because the magnitude response in Figure 2.8 implies an unrealizable filter, approximations to the ideal lowpass filter are necessary. Having designed a suitable approximation, we can obtain the actual filter we want

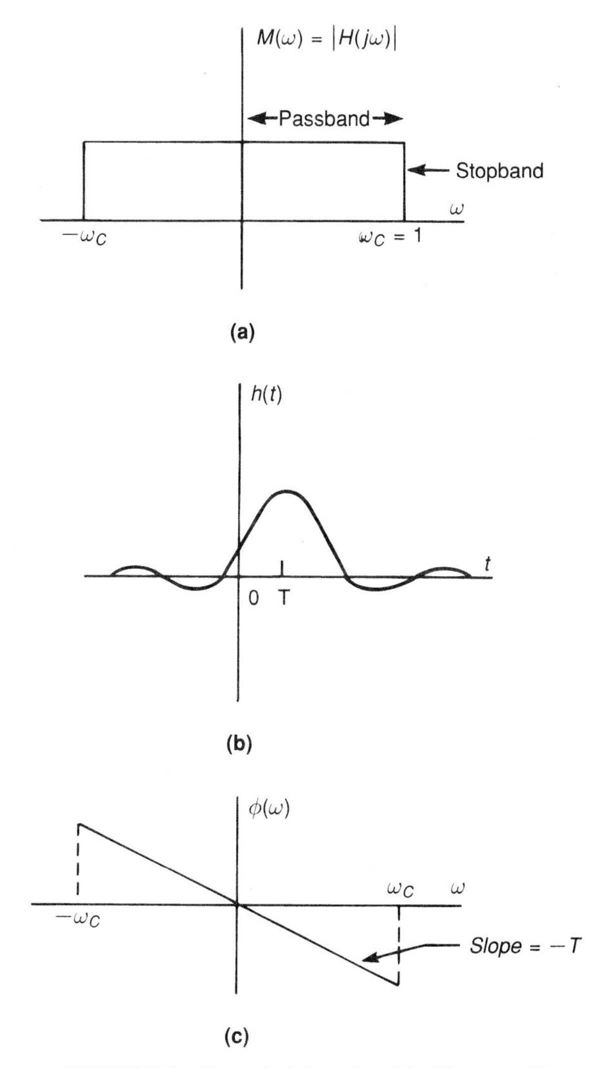

FIGURE 2.8 Characteristics of an ideal lowpass filter.

(such as bandpass or highpass) by means of an appropriate frequency transformation. The design is based on magnitude response specifications only, so the phase response of the filter that results from this procedure is not controlled. Whether this is important depends on the application. For example, in speech processing, a linear phase response is desirable to prevent distortion in transmission. No continuous filter has an exact linear phase response, but some (such as the Bessel filter) offer a reasonably good approximation to linear phase.

Approximations to the ideal lowpass prototype are obtained by first finding a polynomial approximation to the desired squared magnitude $|H(j\omega)|^2$ (using orthogonal polynomials such as Chebyshev and Legendre) and then converting this polynomial into a rational function via Padé approximants [3]. An error criterion is selected to measure how close the function obtained is to the desired function. On the basis of this error criterion, some orthogonal polynomial sets are found to be better than others for the approximation. We will omit the details of *finding* the approximating rational functions ([2] through [5]). Instead, we will assume that the approximations have been made and will examine their properties.

These approximations to the ideal prototype are discussed briefly in sections that follow:

1. Butterworth filter
2. Chebyshev filters type I and II
3. Elliptic or Cauer filter
4. Bessel filter

Some terms used in lowpass filter design are illustrated in Figure 2.9. Note that the magnitude scale is not calibrated in decibels. The figure is deliberately made general to show the various possibilities. The Butterworth and Bessel filters do not have frequency ripples. Filter specifications usually give information from which we can compute ω_r, the attenuation a, and (if appropriate) the passband ripple R. These parameters are defined in the notes under Figure 2.9.

For all-pole filters, which are filters of the form

$$H(s) = \prod_{i=1}^{n} \frac{1}{s - \alpha_i} \tag{2.89}$$

the following approximate relationship is useful as a first estimate of the order n of the required filter.

$$n \approx \frac{a}{20 \log_{10} \omega_r} \tag{2.90}$$

The proof is left as an exercise.

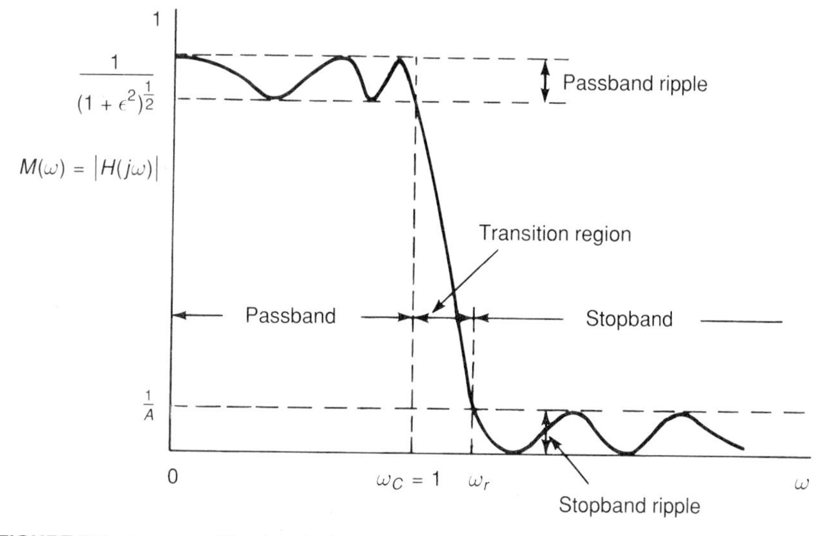

FIGURE 2.9 Lowpass-filter terminology.
Notes:
ω_C/ω_r = transition ratio of filter
ω_r = lowest frequency at which minimum attenuation in the stopband occurs
A = a measure of the minimum acceptable attenuation in the stopband; that is, $A = 10^{+a/20}$, where a is the minimum acceptable attenuation in dB [because $20 \log_{10} (1/A) = -a$]
ϵ = $(10^{R/10} - 1)^{1/2}$, where R is the maximum acceptable passband ripple in dB, [because $20 \log_{10} \frac{1}{(1+\epsilon^2)^{1/2}} = -R$]

2.4.2 Butterworth Filter

This is an all-pole approximation to the ideal filter. Its magnitude function is defined by

$$M(\omega) = |H(j\omega)| = \frac{1}{(1+\omega^{2n})^{1/2}} \tag{2.91}$$

where n is the order of the filter.

It is evident from (2.91) that

$$M(0) = 1$$

$$M(1) = 1/\sqrt{2} \text{ for all values of } n$$

For large ω,

$$M(\omega) \approx \frac{1}{\omega^n} \qquad 20 \log_{10} M(\omega) = -20n \log_{10} \omega$$

which implies that $M(\omega)$ falls off at $20n$ dB/decade ($6n$ dB/octave) for large values of ω (Figure 2.10). The frequencies ω_1 and ω_2 span an octave when $\omega_2 = 2\omega_1$; they span a decade when $\omega_2 = 10\omega_1$.

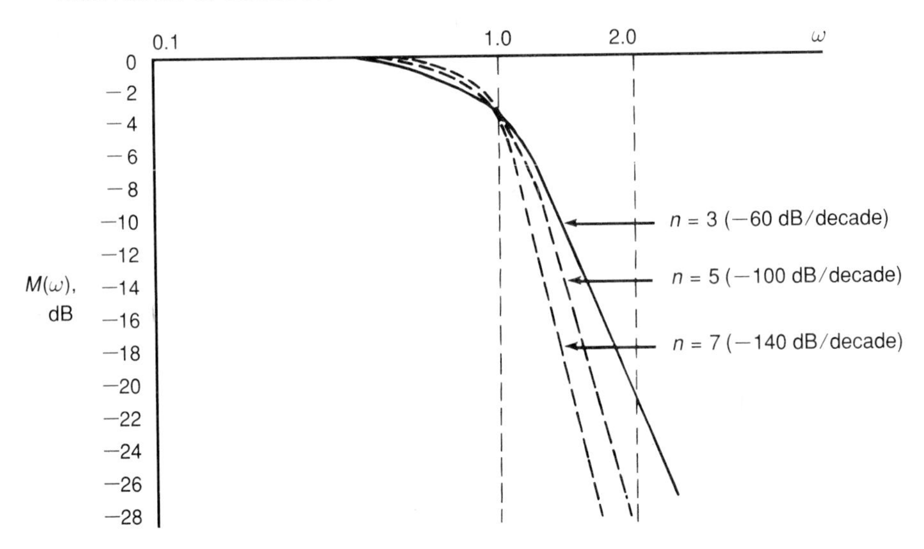

FIGURE 2.10 Magnitude response for a Butterworth filter.

Expand $M(\omega)$ in a power series.

$$M(\omega) = 1 - \frac{1}{2}\omega^{2n} + \frac{3}{8}\omega^{4n} - \frac{5}{16}\omega^{6n} + \frac{35}{128}\omega^{8n} \cdots \tag{2.92}$$

Note that the first $(2n - 1)$ derivatives of $M(\omega)$ vanish at $\omega = 0$. For this reason, the Butterworth filter is said to have a *maximally flat* amplitude characteristic at $\omega = 0$.

To obtain the transfer function $H(s)$ from the magnitude response, note that

$$M^2(\omega) = |H(j\omega)|^2 = H(j\omega)H(-j\omega) = \frac{1}{1 + (\omega^2)^n} \tag{2.93}$$

Because $s = j\omega$ for the frequency response, we have $s^2 = -\omega^2$.

$$H(s)H(-s) = \frac{1}{1 + (-s^2)^n} = \frac{1}{1 + (-1)^n s^{2n}} \tag{2.94}$$

The poles of this function are given by the roots of

$$(-1)^n s^{2n} = -1 = e^{j(2k-1)\pi}, \quad k = 1, 2, \ldots, 2n \tag{2.95}$$

so the $2n$ poles are

$$s_k = \begin{cases} e^{j[(2k-1)/2n]\pi} & n \text{ even}, k = 1, 2, \ldots, 2n \\ e^{j(k/n)\pi} & n \text{ odd}, k = 0, 1, 2, \ldots, 2n - 1 \end{cases} \tag{2.96}$$

To ensure stability, the left half-plane poles are identified with $H(s)$. The poles of $H(-s)$ are mirror images of these. Note that for any n, the poles of the normalized Butterworth filter lie on the unit circle in the s-plane.

EXAMPLE 2.3

Find the transfer function $H(s)$ that corresponds to the third-order ($n = 3$) Butterworth filter.

$$H(j\omega)H(-j\omega) = \frac{1}{1+\omega^6} = \frac{1}{1-(-\omega^2)^3}$$

$$H(s)H(-s) = \frac{1}{1-(s^2)^3} = \frac{1}{1-s^6}$$

Solution

The roots of the equation

$$s^6 = 1 = e^{j2k\pi}$$

are

$$s_k = e^{jk\pi/3}, \qquad k = 0, 1, 2, 3, 4, 5,$$

Therefore,

$$\begin{aligned} s_0 &= e^{j0} = 1 \\ s_1 &= e^{j\pi/3} = 1/2 + j\sqrt{3}/2 \\ s_2 &= e^{j2\pi/3} = -1/2 + j\sqrt{3}/2 \\ s_3 &= e^{j\pi} = -1 \\ s_4 &= e^{j4\pi/3} = -1/2 - j\sqrt{3}/2 \\ s_5 &= e^{j5\pi/3} = 1/2 - j\sqrt{3}/2 \end{aligned}$$

These are plotted in Figure 2.11. Identify the left half-plane poles with $H(s)$ to get the required transfer function,

$$H(s) = \frac{1}{(s+1)(s+1/2-j\sqrt{3}/2)(s+1/2+j\sqrt{3}/2)}$$

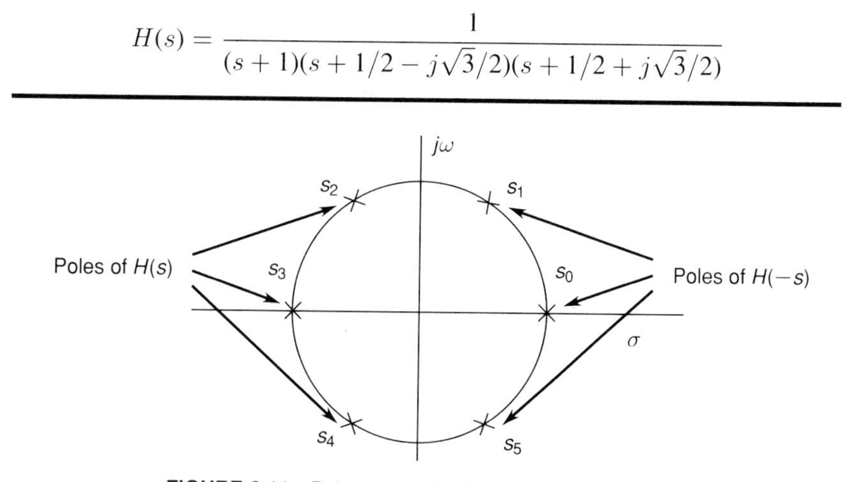

FIGURE 2.11 Poles of the third-order Butterworth filter.

Table 2.2 gives the denominator of $H(s)$ in factored form for $n = 1$ to $n = 8$.

Knowing $H(s)$, we can determine the phase characteristic of the Butterworth filter either analytically, by setting $s = j\omega$ in $H(s)$, or graphically, by plotting the poles of $H(s)$ on the s-plane and drawing vectors from the poles to a particular point in the $j\omega$ axis, as discussed above. For the third-order filter of Example 2.3, this process is illustrated in Figure 2.12. The phase response at a particular frequency ω_0 is

$$\begin{aligned}\phi(\omega_0) &= \sum_{i=1}^{m} \begin{array}{l}\text{angles of the vectors from the zeros}\\ \text{of } H(s) \text{ to the point } s = j\omega_0\end{array}\\ &\quad - \sum_{i=1}^{n} \begin{array}{l}\text{angles of the vectors from the poles}\\ \text{of } H(s) \text{ to the point } s = j\omega_0\end{array}\\ &= -(\theta_2 + \theta_3 + \theta_4) \end{aligned} \tag{2.97}$$

By selecting a number of points on the $j\omega$ axis and evaluating $\phi(\omega)$ for each, we can get the complete phase-response curve shown in Figure 2.13.

2.4.3 Chebyshev Approximations to the Ideal Filter

Consider the Chebyshev cosine polynomials [2] defined by

$$C_n(\omega) = \cos\,(n \cos^{-1} \omega) \quad |\omega| \le 1 \tag{2.98}$$

$$= \cosh\,(n \cosh^{-1} \omega) \quad |\omega| > 1 \tag{2.99}$$

Thus

$$C_0(\omega) = 1$$

$$C_1(\omega) = \omega$$

In general, we can obtain higher-order Chebyshev polynomials from the recursive relation

$$C_n(\omega) = 2\omega C_{n-1}(\omega) - C_{n-2}(\omega) \tag{2.100}$$

TABLE 2.2 Butterworth Filter Denominators in Factored Form

n	**Denominator of $H(s)$ for Butterworth Filter**
1	$s + 1$
2	$s^2 + \sqrt{2}s + 1$
3	$(s^2 + s + 1)(s + 1)$
4	$(s^2 + 0.765s + 1)(s^2 + 1,848s + 1)$
5	$(s + 1)(s^2 + 0.618s + 1)(s^2 + 1.618s + 1)$
6	$(s^2 + 0.517s + 1)(s^2 + \sqrt{2}s + 1)(s^2 + 1.932s + 1)$
7	$(s + 1)(s^2 + 0.445s + 1)(s^2 + 1.247s + 1)(s^2 + 1.802s + 1)$
8	$(s^2 + 0.390s + 1)(s^2 + 1.111s + 1)(s^2 + 1.663s + 1)(s^2 + 1.962s + 1)$

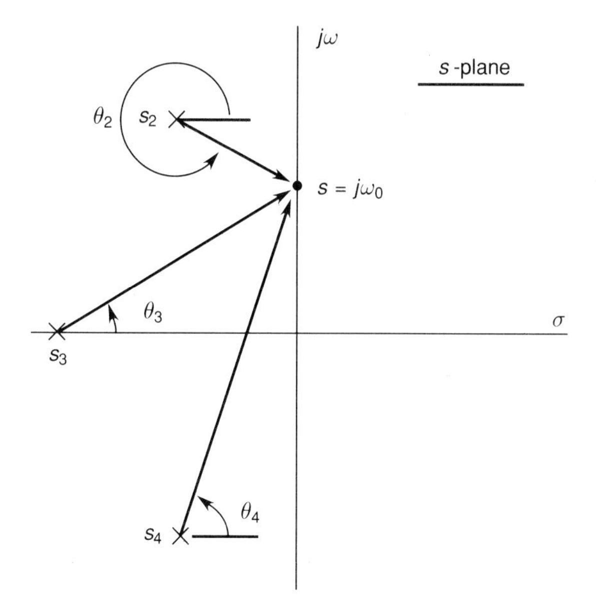

FIGURE 2.12 Graphical evaluation of Butterworth phase response ($n = 3$).

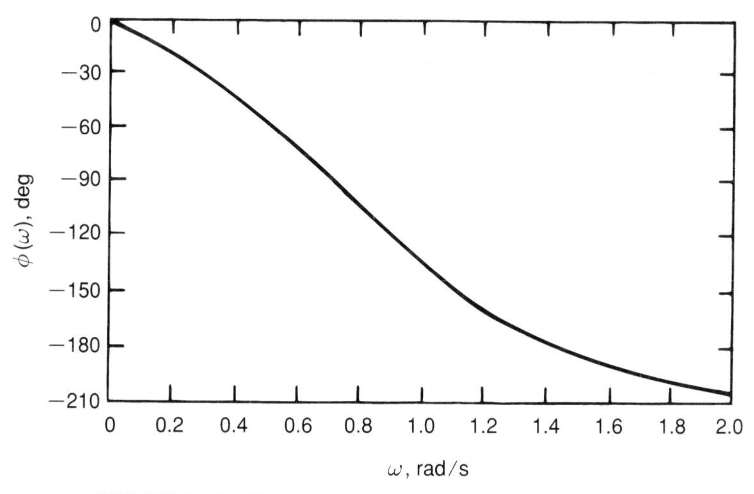

FIGURE 2.13 Phase response for Butterworth filter ($n = 3$).

Therefore, for $n = 2$,

$$C_2(\omega) = 2\omega C_1(\omega) - C_0(\omega) = 2\omega^2 - 1 \tag{2.101}$$

The first ten Chebyshev polynomials are listed in Table 2.3. The following are some of the properties of Chebyshev polynomials.

TABLE 2.3 The First Ten Chebyshev Polynomials

n	Chebyshev Polynomial $C_n(\omega)$
0	1
1	ω
2	$2\omega^2 - 1$
3	$4\omega^3 - 3\omega$
4	$8\omega^4 - 8\omega^2 + 1$
5	$16\omega^5 - 20\omega^3 + 5\omega$
6	$32\omega^6 - 48\omega^4 + 18\omega^2 - 1$
7	$64\omega^7 - 112\omega^5 + 56\omega^3 - 7\omega$
8	$128\omega^8 - 256\omega^6 + 160\omega^4 - 32\omega^2 + 1$
9	$256\omega^9 - 576\omega^7 + 432\omega^5 - 120\omega^3 + 9\omega$
10	$512\omega^{10} - 1280\omega^8 + 1120\omega^6 - 400\omega^4 + 50\omega^2 - 1$

1. $C_n(1) = 1$ for all n.
2. $|C_n(\omega)| \leq 1$ for $|\omega| \leq 1$.
3. The zeros of the polynomials $C_n(\omega)$ are located in the interval $|\omega| \leq 1$.
4. For $|\omega| > 1, |C_n(\omega)|$ increases rapidly with $|\omega|$.

EXAMPLE 2.4

Confirm these properties for $C_3(\omega)$ and sketch the function.

Solution

We have $C_3(\omega) = 4\omega^3 - 3\omega$.

1. $C_3(1) = 4 - 3 = 1$
2. $$\frac{dC_3(\omega)}{d\omega} = 12\omega^2 - 3 = 0 \text{ for an extremum}$$
   $$\omega^2 = \frac{1}{4}, \text{ or } \omega = \frac{1}{2}$$
   $$\frac{d^2C_3}{d\omega} = 24\omega > 0 \text{ for } \omega = \frac{1}{2} \Rightarrow \text{ min.}$$
   $$< 0 \text{ for } \omega = -\frac{1}{2} \Rightarrow \text{ max.}$$
   At $\omega = \frac{1}{2}$: $C_3(\frac{1}{2}) = 4(\frac{1}{8}) - \frac{3}{2} = -1$
   At $\omega = -\frac{1}{2}$: $C_3(-\frac{1}{2}) = 4(-\frac{1}{8}) + \frac{3}{2} = 1$
   Therefore, $|C_3(\omega)| \leq 1$
3. Zeros of $C_3(\omega)$ are roots of $4\omega^3 - 3\omega = 0$
   $\omega = 0$ or $4\omega^2 = 3 \Rightarrow \omega = \pm\sqrt{3}/2$
   The roots all lie in the interval $|\omega| \leq 1$.

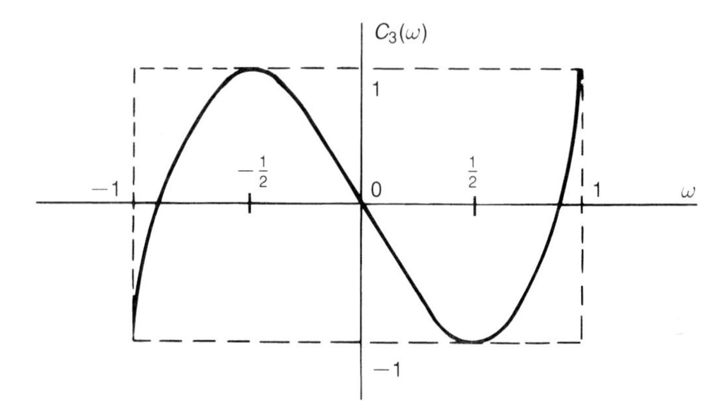

FIGURE 2.14 Third-order Chebyshev polynomial.

4. For $|\omega| > 1$, the ω^3 term dominates.
   The function $C_3(\omega)$ is sketched in Figure 2.14.

We get the lowpass filter approximation as follows. Choose a number $0 < \epsilon < 1$, and consider the function $\epsilon^2 C_n^2(\omega)$ in the interval $|\omega| < 1$.

$$0 \leq \epsilon^2 C_n^2(\omega) \leq \epsilon^2$$

Thus

$$1 \leq 1 + \epsilon^2 C_n^2(\omega) \leq 1 + \epsilon^2$$

Invert this function to get the square of the magnitude response of the nth-order Chebyshev filter.

$$|H(j\omega)|^2 = \frac{1}{1 + \epsilon^2 C_n^2(\omega)} \tag{2.102}$$

Therefore, for $|\omega| \leq 1, |H(j\omega)|^2$ oscillates between 1 and $1/(1+\epsilon^2)$. This oscillation is called a ripple. Outside this interval, as $|\omega|$ increases, $\epsilon^2 C_n^2(\omega)$ becomes very large so that $|H(j\omega)|^2$ approaches zero rapidly, as shown in Figure 2.15. The figure also shows the ripples of constant magnitude (equiripples) in the passband.

Note that $|H(j1)| = 1/(1+\epsilon^2)^{1/2}$ for all n, because $C_n(1) = 1$. If R is the maximum permissible ripple in decibels, then

$$20 \log_{10} \frac{1}{(1+\epsilon^2)^{1/2}} = -R \tag{2.103}$$

or

$$\epsilon = (10^{R/10} - 1)^{1/2} \tag{2.104}$$

This sets a bound on ϵ.

In the stopband, as $\epsilon^2 C_n^2(\omega) \gg 1$, we have

$$|H(j\omega)| \approx \frac{1}{\epsilon C_n(\omega)} \tag{2.105}$$

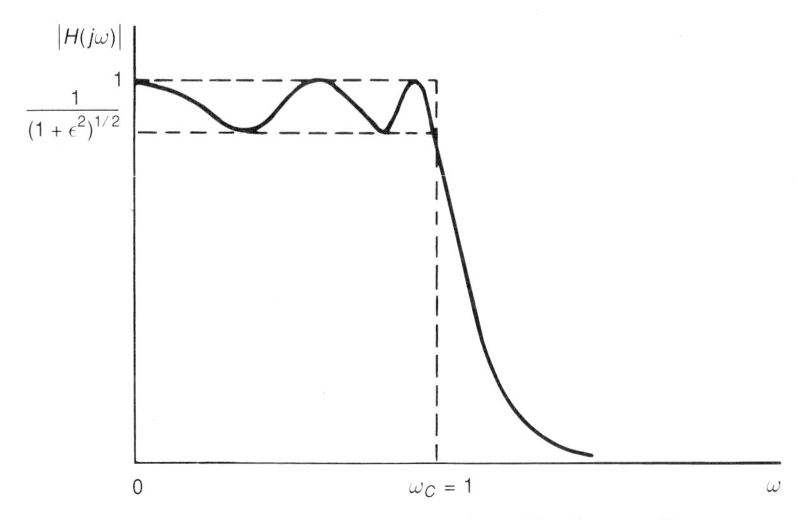

FIGURE 2.15 Chebyshev approximation to ideal lowpass filter.

so that the attenuation in decibels is

$$L = -20 \log_{10} H(j\omega) \tag{2.106}$$

$$= 20 \log_{10} \epsilon + 20 \log_{10} C_n(\omega) \tag{2.107}$$

For large $\omega, C_n(\omega) \approx 2^{n-1}\omega^n$ (see Table 2.3). Therefore,

$$L = 20 \log_{10} \epsilon + 20 \log_{10} 2^{n-1}\omega^n \tag{2.108}$$

$$\approx 20 \log_{10} \epsilon + 6(n-1) + 20n \log_{10} \omega \tag{2.109}$$

Because $0 < \epsilon < 1$, $20 \log \epsilon < 0$, so n must be selected sufficiently large to compensate for the decrease in attenuation.

From the foregoing, it is evident that the Chebyshev approximation depends on two variables ϵ and n, which can be determined directly from the filter specifications. The maximum permissible ripple puts a bound on ϵ. Once ϵ is determined, any desired value of attenuation in the stopband fixes n.

The transfer function of the Chebyshev filter is of the form

$$W(s) = \frac{C}{\prod_{k=1}^{n}(s - s_k)} \tag{2.110}$$

The poles $s_k = \sigma_k + j\omega_k$ are located on an ellipse in the s-plane (Figure 2.16) given by

$$\frac{\sigma_k^2}{\sinh^2 \gamma} + \frac{\omega_k^2}{\cosh^2 \gamma} = 1 \tag{2.111}$$

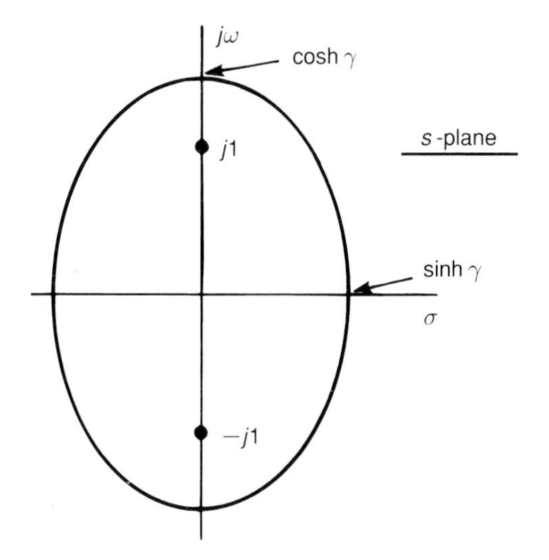

FIGURE 2.16 Locus of the poles of a Chebyshev filter.

where

$$\gamma = \frac{1}{n}\sinh^{-1}\frac{1}{\epsilon} = \frac{1}{n}\log_e\left(\frac{1}{\epsilon} + \sqrt{\frac{1}{\epsilon^2} + 1}\right) \tag{2.112}$$

The analytic expression for the poles is

$$s_k = \begin{cases} -\sinh\gamma\cos\frac{2k+1}{2n}\pi + j\cosh\gamma\sin\frac{2k+1}{2n}\pi, \\ \qquad n \text{ even}, \; k = \frac{-n}{2}, \frac{-n}{2}+1, \ldots, \frac{n}{2}-1 & (2.113) \\ -\sinh\gamma\cos\frac{k}{n}\pi + j\cosh\gamma\sin\frac{k}{n}\pi, \\ \qquad n \text{ odd}, \; k = 0, \pm 1, \pm 2, \ldots, \pm\frac{n-1}{2} & (2.114) \end{cases}$$

EXAMPLE 2.5

Find the transfer function $H(s)$ for a lowpass Chebyshev filter ($\omega_c = 1$ rad/s) with the following specifications:

a) Ripple not to exceed 1.1 dB in the passband

b) Attenuation of at least 20 dB at $\omega = 2$

Solution

First find ϵ from specification a).

$$20\log_{10}\frac{1}{(1+\epsilon^2)^{1/2}} = -1.1$$

from (2.103). Therefore,

$$\epsilon = 0.5369$$

Now find n from specification b).

$$L \approx 20 \log \epsilon + 6(n-1) + 20n \log \omega$$

$$20 = 20 \log 0.5369 + 6(n-1) + 20n \log 2$$

Therefore,

$$n = 2.61$$

To ensure that the specifications are met, we select the nearest higher integer, which is $n = 3$. Inserting these values for ϵ and n in (2.112) and (2.114) gives the poles

$$s_k = -0.4766, -0.2383 \pm j0.9593$$

On Figure 2.17 are plotted the magnitude response for the foregoing Chebyshev filter and that for a Butterworth filter of the same order. Note that for the same order n, the Chebyshev response falls off more rapidly in the transition region than does the Butterworth response. However, for large ω, the attenuation is the same: $-20n$ dB/decade.

The filter we have just discussed is often called a Chebyshev type I filter. It exhibits equiripple behavior in the passband and flat behavior in the stopband. The Chebyshev type II filter has the opposite characteristic: flat passband (near $\omega = 0$) and equiripple stopband. It is defined by the

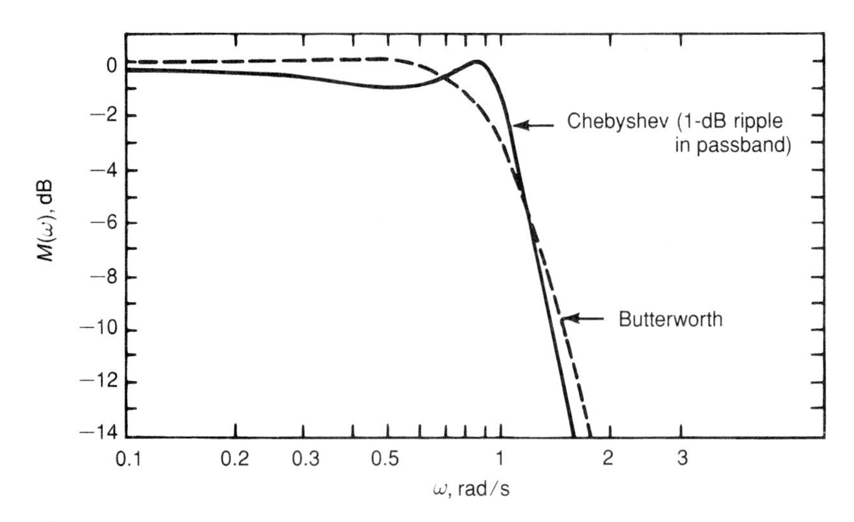

FIGURE 2.17 Comparison between magnitude responses for Chebyshev and Butterworth filters ($n = 3$).

relation [2]

$$|H(j\omega)|^2 = \frac{1}{1+\epsilon^2\left[\frac{C_n^2(\omega_r)}{C_n^2(\omega_r/\omega)}\right]} \tag{2.115}$$

$$= \frac{C_n^2(\omega_r/\omega)}{C_n^2(\omega_r/\omega)+\epsilon^2 C_n^2(\omega_r)} \tag{2.116}$$

We note the following:

1. This type of filter has both poles and zeros. From the numerator of the last expression, the zeros occur on the $j\omega$ axis for $\omega > \omega_r$—that is, in the stopband.
2. $|H(j1)| = (1+\epsilon^2)^{-1/2}$
3. $|H(j0)| = 1$
4. The function $|C_n(\omega_r/\omega)|^2$ for $\omega > \omega_r$ has a maximum value of 1. Hence the maximum value of $|H(j\omega)|$ in the stopband is

$$\frac{1}{A} = \frac{1}{[1+\epsilon^2 C_n^2(\omega_r)]^{1/2}} \tag{2.117}$$

5. The order of the filter is obtained from the design specifications for R, a, and ω_r (see Figure 2.9) by the relation

$$n = \frac{\cosh^{-1}(\sqrt{A^2-1}/\epsilon)}{\cosh^{-1}\omega_r} \approx \frac{1n(2A/\epsilon)}{(2\omega_r-1)^{1/2}} \tag{2.118}$$

where

$$\epsilon = (10^{R/10}-1)^{1/2} \tag{2.119}$$

$$A = 10^{a/20} \tag{2.120}$$

A typical magnitude response curve for a Chebyshev type II filter is shown in Figure 2.18.

2.4.4 Elliptic Filter

The elliptic filter has equiripple characteristics in both passband and stopband. Its transfer function has the form

$$H(s) = \frac{W_0}{D_0(s)}\prod_{i=1}^{r}\frac{s^2+c_i}{s^2+a_i s+b_i} \tag{2.121}$$

where

$$r = \begin{cases} \frac{n-1}{2} & \text{for odd } n \\ \frac{n}{2} & \text{for even } n \end{cases} \tag{2.122}$$

and

$$D_0(s) = \begin{cases} s+\sigma_0 & \text{for odd } n \\ 1 & \text{for even } n \end{cases} \tag{2.123}$$

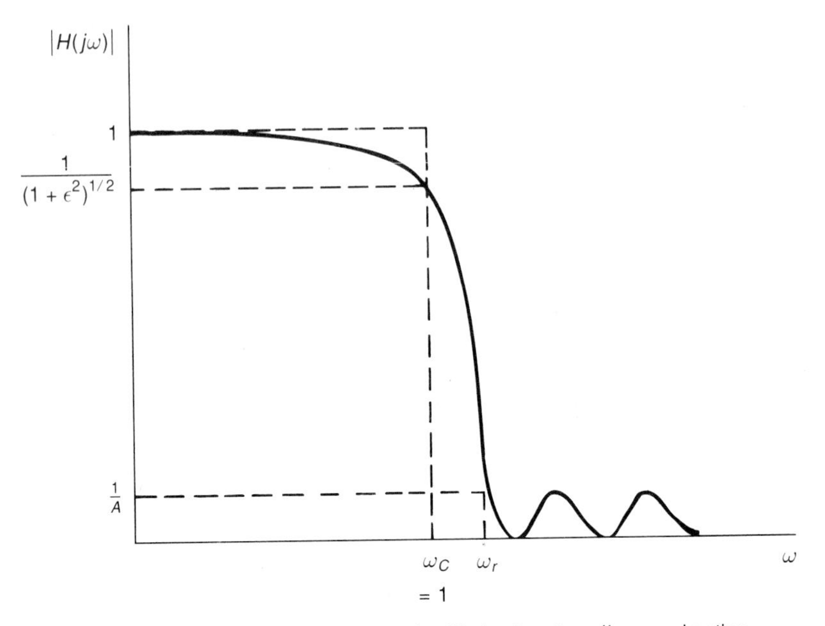

FIGURE 2.18 Magnitude response for Chebyshev type II approximation.

The magnitude-squared function is

$$|H_n(j\omega)|^2 = \frac{1}{1 + \epsilon^2 \psi_n^2(\omega)} \tag{2.124}$$

where $\psi_n(\omega)$ is called a Chebyshev rational function and can be defined in terms of Jacobi elliptic functions. A typical plot of $\psi_n(\omega)$ is given in Figure 2.19. It shows that $\psi_n^2(\omega)$ oscillates between 0 and 1 in the passband for $|\omega| \leq 1$, and between L^2 and ∞ for $|\omega| \geq \omega_r$. [9].

A detailed discussion of elliptic functions would take us too far afield. However, excellent treatments of the subject and its application to filter design are contained in [10], [11], and [12]. For our purposes, it will suffice to list a set of equations that, when solved in sequence, yield an elliptic filter design based on given specifications for a, R, and ω_r (see Figure 2.9). A derivation of these equations appears in [10].

First it is necessary to normalize ω_c and ω_r. For example, if ω_c' and ω_r' are the required values, $\omega_c' < \omega_r'$, we define

$$\omega_c = \frac{\omega_c'}{\omega_0} \tag{2.125}$$

$$\omega_r = \frac{\omega_r'}{\omega_0} \tag{2.126}$$

where the geometric mean

$$\omega_0 = \sqrt{\omega_c' \omega_r'} \tag{2.127}$$

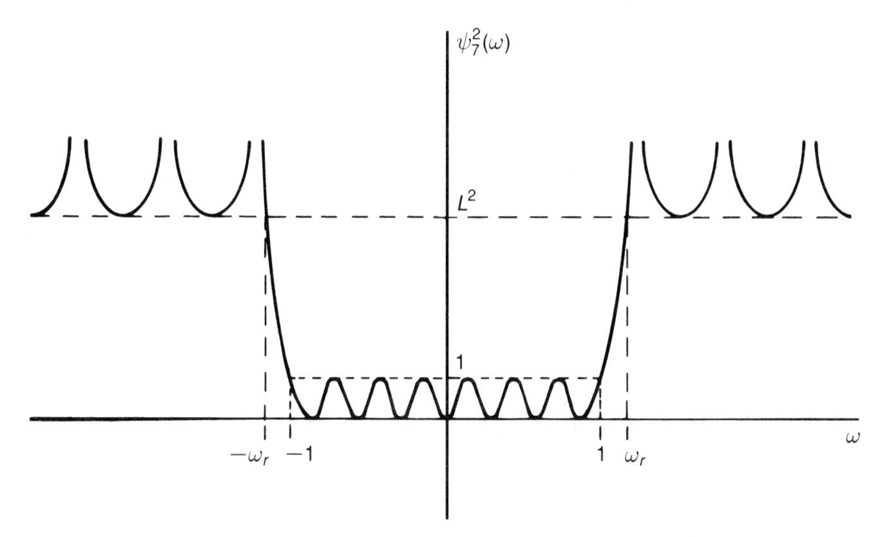

FIGURE 2.19 Chebyshev rational function for $n = 7$.

It follows that the normalized frequencies ω_c and ω_r are related by

$$\omega_r = 1/\omega_c \tag{2.128}$$

The remaining equations are

$$k = \omega_c^2 \tag{2.129}$$

$$k' = \sqrt{1 - k^2} \tag{2.130}$$

$$q_0 = \frac{1}{2}\frac{1 - \sqrt{k'}}{1 + \sqrt{k'}} \tag{2.131}$$

$$q = q_0 + 2q_0^5 + 15q_0^9 + 150q_0^{13} \tag{2.132}$$

$$A^2 = 10^{a/10} \tag{2.133}$$

$$\epsilon^2 = 10^{R/10} \tag{2.134}$$

$$D = \frac{A^2 - 1}{\epsilon^2 - 1} \tag{2.135}$$

$$n \geq \frac{\log_{10} 16D}{\log_{10} 1/q} \quad \text{(order of filter)} \tag{2.136}$$

$$\lambda = \frac{1}{2n} \ln\left(\frac{\epsilon + 1}{\epsilon - 1}\right) \tag{2.137}$$

$$\sigma_0 = \left| \frac{2q^{1/4} \sum_{m=0}^{\infty} (-1)^m q^{m(m+1)} \sinh [(2m+1)\lambda]}{1 + 2 \sum_{m=1}^{\infty} (-1)^m q^{m^2} \cosh 2m\lambda} \right| \tag{2.138}$$

$$W = \left[(1 + k\sigma_0^2) \left(1 + \frac{\sigma_0^2}{k} \right) \right]^{1/2} \tag{2.139}$$

$$\Omega_i = \frac{2q^{1/4} \sum_{m=0}^{\infty} (-1)^m q^{m(m+1)} \sin \dfrac{(2m+1)\pi\mu}{n}}{1 + 2 \sum_{m=1}^{\infty} (-1)^m q^{m^2} \cos \dfrac{2m\pi\mu}{n}} \tag{2.140}$$

where

$$\mu = \begin{cases} i, & \text{for odd } n \\ i - \frac{1}{2}, & \text{for even } n \end{cases} \tag{2.141}$$

and

$$i = 1, 2, \ldots, r$$

$$V_i = \left[(1 - k\Omega_i^2) \left(1 - \frac{\Omega_i^2}{k} \right) \right]^{1/2} \tag{2.142}$$

$$c_i = \frac{1}{\Omega_i^2} \tag{2.143}$$

$$b_i = \frac{(\sigma_0 V_i)^2 + (\Omega_i W)^2}{(1 + \sigma_0^2 \Omega_i^2)^2} \tag{2.144}$$

$$a_i = \frac{2\sigma_0 V_i}{1 + \sigma_0^2 \Omega_i^2} \tag{2.145}$$

$$W_0 = \begin{cases} \sigma_0 \prod_{i=1}^{r} \dfrac{b_i}{c_i} & \text{for odd } n \\ \frac{1}{\epsilon} \prod_{i=1}^{r} \dfrac{b_i}{c_i} & \text{for even } n \end{cases} \tag{2.146}$$

A typical elliptic-filter magnitude characteristic is shown in Figure 2.20.

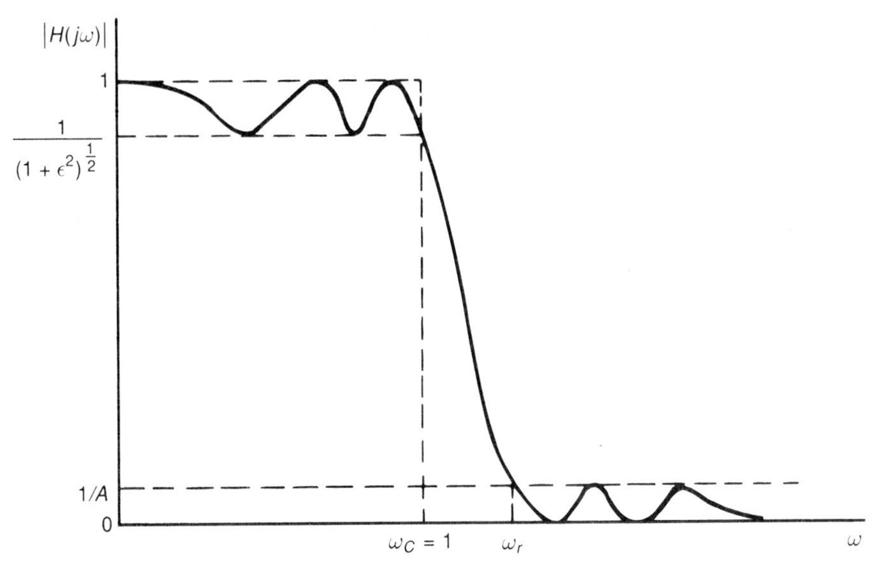

FIGURE 2.20 Magnitude response for an elliptic filter.

2.4.5 Comments

Let us take a moment to compare Butterworth, Chebyshev, and elliptic filters. All are approximations to a desired rectangular bandwidth. The Butterworth filter has a monotonic magnitude response. By allowing ripples in the passband for type I and in the stopband for type II, the Chebyshev filter can achieve sharper cutoff with the same number of poles. Elliptic filters have even sharper cutoff than the Chebyshev filters for the same network complexity, but they result in both passband and stopband ripple.

The designer of these filters strives to attain the ideal magnitude response without regard to the resulting phase response. With the Bessel filter outlined in the next section, however, the designer tries to achieve a linear phase as well.

2.4.6 Bessel Filter

The transfer function

$$H(s) = Ke^{-sT} \tag{2.147}$$

has amplitude response

$$M(\omega) = K \tag{2.148}$$

and linear phase response

$$\phi(\omega) = -\omega T \tag{2.149}$$

Thus an input signal $x(t)$ is simply multiplied by a constant and delayed by a time T, resulting in an output

$$y(t) = Kx(t-T)u(t-T) \tag{2.150}$$

Such a transfer characteristic is desirable for transmitting a series of pulses, for example. However, it is realizable only by a delay line or a lossless transmission line. We can approximate Ke^{-sT} with a rational function in s by means of the Bessel filter with

$$H(s) = \frac{K_0}{B_n(s)} \tag{2.151}$$

where $B_n(s)$ are Bessel polynomials defined by

$$\begin{aligned} B_0 &= 1 \\ B_1 &= s+1 \\ &\vdots \\ B_n &= (2n-1)B_{n-1} + s^2 B_{n-2} \end{aligned} \tag{2.152}$$

Note that all the zeros of the Bessel functions [poles of $H(s)$] lie in the left half-plane. Table 2.4 gives the roots of Bessel polynomials up to $n = 7$.

TABLE 2.4 Roots of Bessel Polynomials (up to $n = 7$)

	Non-normalized		Normalized	
n	$-a$	$\pm jb$	$-a$	$\pm jb$
1	1.000	0.000	1.000	0.000
2	1.500	0.867	0.866	0.500
3	2.322	0.000	0.942	0.000
	1.839	1.754	0.746	0.711
4	2.896	0.867	0.905	0.271
	2.104	2.657	0.657	0.830
5	3.647	0.000	0.926	0.000
	3.352	1.743	0.852	0.443
	2.325	3.571	0.591	0.907
6	4.248	0.868	0.909	0.186
	3.736	2.626	0.800	0.562
	2.516	4.493	0.539	0.962
7	4.972	0.000	0.919	0.000
	4.758	1.739	0.880	0.322
	4.070	3.517	0.753	0.650
	2.686	5.421	0.497	1.003

The roots of the polynomials must be normalized to ensure that $\omega_c = 1$ rad/s. This is done by scaling the roots of each nth-degree polynomial by $K_n^{1/n}$, where K_n is the product of the magnitudes of the non-normalized poles.

The magnitude responses and phase responses for Bessel and Butterworth filters ($n = 3$) are compared in Figure 2.21. Note that the Butterworth filter has a sharper initial cutoff in the magnitude response, whereas the phase response of the Bessel filter is approximately linear.

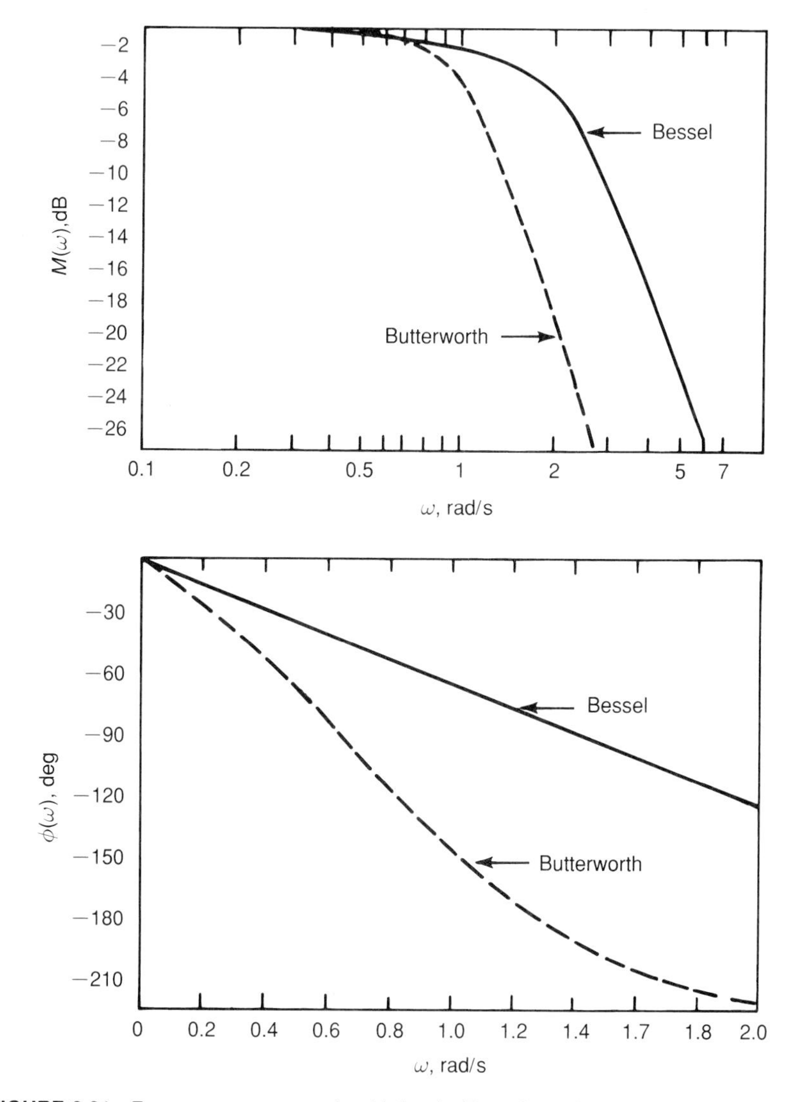

FIGURE 2.21 Frequency responses for third-order Bessel and Butterworth filters.

2.4.7 Transient Response of Low-Pass Filters

Although we have been concerned primarily with the frequency response of the lowpass filters we have discussed, the transient response is, of course, also very important in many applications. The following figures of merit are used to compare unit step function responses of filters. These definitions are illustrated in Figure 2.22.

1. *Rise time:* The time t_r required for the step response to rise from 10% to 90% of its final value.
2. *Overshoot:* The difference between the peak value and the final value of the step response, expressed as a percentage of the final value.
3. *Setting time:* The time t_s, measured from the first peak, beyond which the step response does not differ from the final value by more than ±2%.

In general, as the sharpness of the cutoff increases while the same cutoff frequency ($\omega_c = 1$) is maintained, the rise time, overshoot, and settling time also increase for the Butterworth filter, as shown in Figure 2.23. This figure also shows transient response curves for the Chebyshev and elliptic filters.

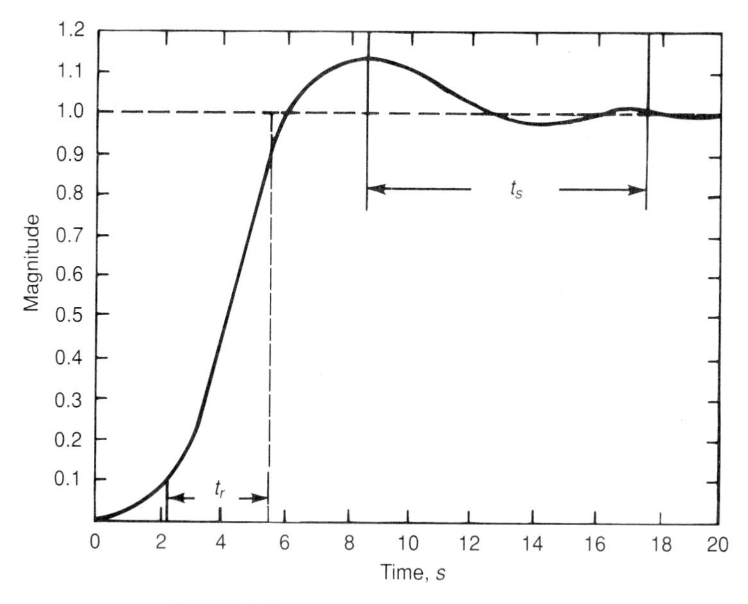

FIGURE 2.22 Measures of transient response.

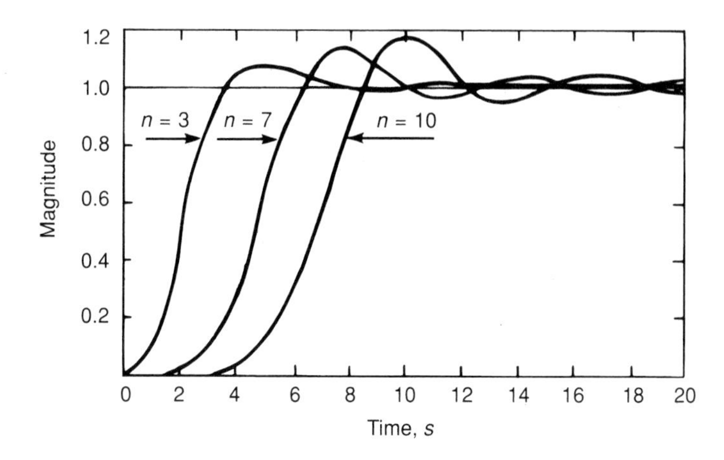

(a)

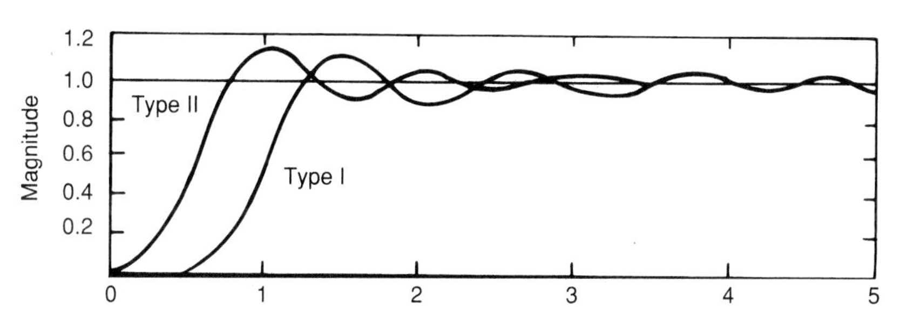

(b)

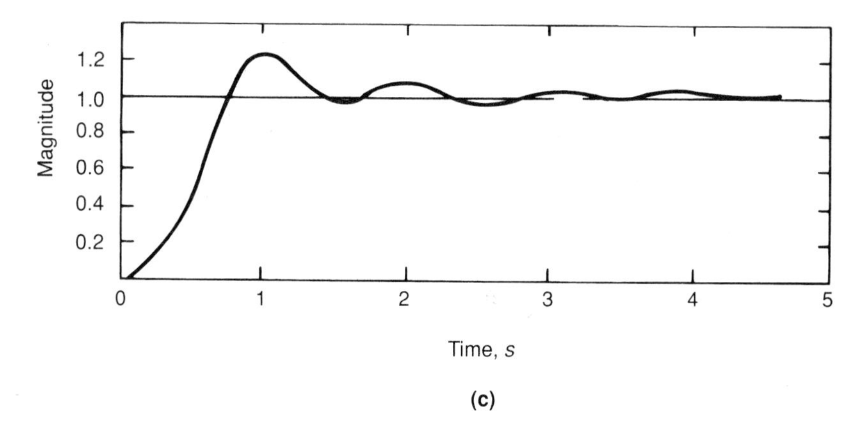

(c)

FIGURE 2.23 Unit step responses for different filters. (a) Butterworth filters. (b) Chebyshev filters ($n = 7$). (c) Elliptic filter ($n = 4$).

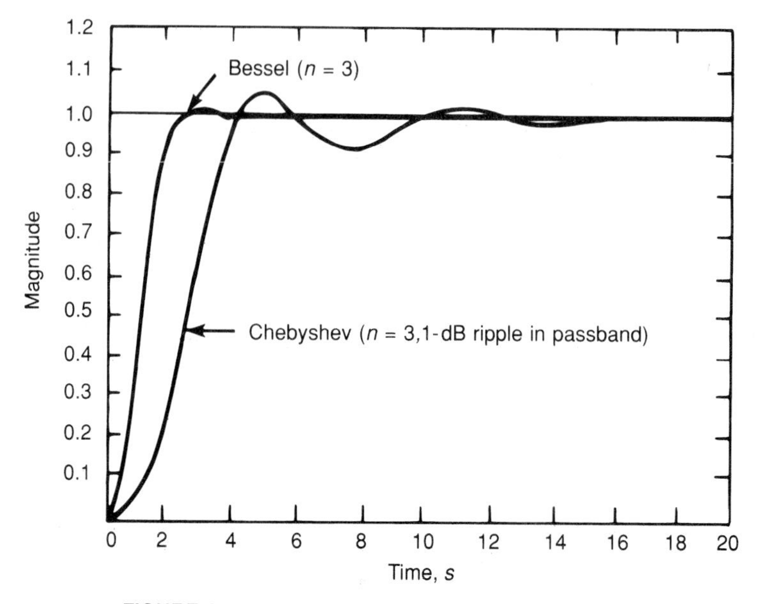

FIGURE 2.24 A comparison of filter transient responses.

Ripples in the magnitude response usually cause prolonged "ringing" in the step response. The best filter for transmitting a pulse train is the Bessel; because if t_r and t_s are small compared with the period of the pulses, it does not "smear" the output pulse train. A comparison between the step response for a Bessel filter and that for a Chebyshev type I filter is shown in Figure 2.24.

2.5 FREQUENCY TRANSFORMATIONS

We have discussed the design of prototype filters, lowpass filters with a normalized cutoff frequency $\omega_c = 1$ rad/s. From this prototype design, by means of a frequency transformation, we can obtain a lowpass, bandpass, bandstop, or highpass filter with specific cutoff frequencies.

Let $s_n = \sigma_n + j\omega_n$ be the complex frequency variable associated with the normalized lowpass design. Then, to obtain the desired filter, it is necessary to replace s_n in the prototype filter by the relevant expression. Each of these expressions is given, and its use illustrated, in one of the following sections.

2.5.1 Lowpass with Cutoff Frequency ω_u

$$\text{Transformation: } s_n = s/\omega_u \tag{2.153}$$

The segment $|\omega_n| \leq 1$ of the s_n-plane is mapped into the segment $|\omega_n| \leq \omega_u$ on the s-plane as shown in Figure 2.25. This scales the components of the prototype by the factor ω_u.

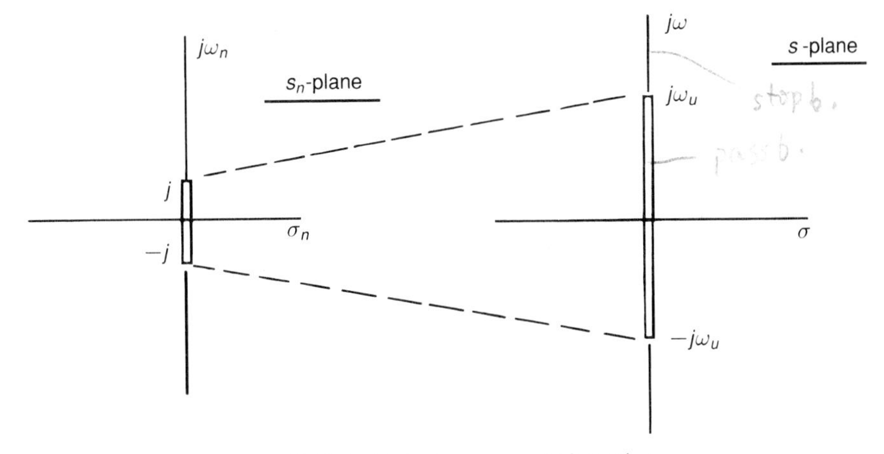

FIGURE 2.25 Lowpass transformation.

EXAMPLE 2.6

A fifth-order Butterworth filter (see Table 2.2),

$$H_n(s_n) = \frac{1}{(s_n + 1)(s_n^2 + 0.618s_n + 1)(s_n^2 + 1.618s_n + 1)}$$

was required to meet the specifications for a lowpass filter. If the actual cut-off frequency is 1000 Hz, what is the transfer function of the practical filter?

Solution

Because $\omega_u = 2000\pi$ rad/s, we replace s_n in $H_n(s_n)$ by $s/(2000\pi)$ to obtain the transfer function.

$$H(s) = \frac{9.7926 \times 10^{18}}{(s + 6283)(s^2 + 3883s + 3.9478 \times 10^7)(s^2 + 10166s + 3.9478 \times 10^7)}$$

2.5.2 Highpass with Cutoff Frequency ω_ℓ

$$\text{Transformation:} \quad s_n = \omega_\ell / s \tag{2.154}$$

The segment $|\omega_n| \leq 1$ is mapped into the segments $\omega_\ell \leq |\omega| \leq \infty$ as shown in Figure 2.26.

EXAMPLE 2.7

From the third-order Chebyshev type I approximation to the ideal lowpass prototype (Example 2.5), design a highpass filter to pass frequencies above 250 Hz. The maximum gain in the passband should be unity. In this case,

$$H_n(s_n) = \frac{K}{(s_n + 0.4942)(s_n^2 + 0.4942s_n + 0.9942)}$$

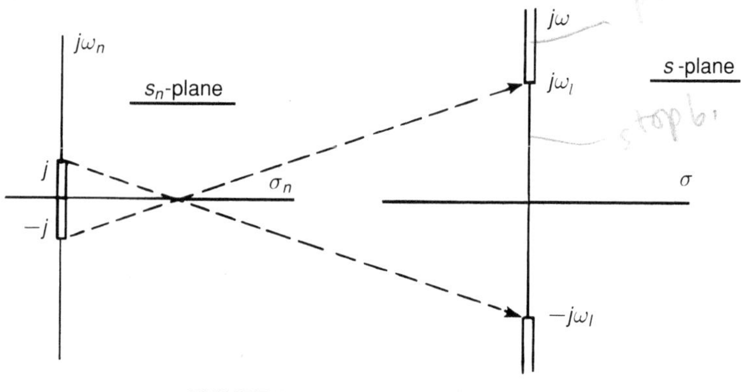

FIGURE 2.26 Highpass transformation.

Solution

Replace s_n by $500\pi/s$.

$$H(s) = \frac{s^3}{(s + 3179)(s^2 + 781s + 2.4819 \times 10^6)}$$

As $s \to \infty$ in the passband, $H(s) = 1$.

2.5.3 Bandpass with Cutoff Frequencies ω_ℓ and ω_u

Transformation:

$$s_n = \frac{s^2 + \omega_0^2}{Bs} \tag{2.155}$$

$$= \frac{\omega_0}{B}\left(\frac{s}{\omega_0} + \frac{\omega_0}{s}\right) \tag{2.156}$$

where

$$\omega_0 = \sqrt{\omega_u \omega_\ell} \quad \text{geometric mean} \tag{2.157}$$

$$B = \omega_u - \omega_\ell \quad \text{bandwidth} \tag{2.158}$$

The segment $|\omega_n| \leq 1$ in the s_n-plane is mapped into the segments $|\omega_\ell| \leq |\omega| \leq |\omega_u|$ in the s-plane as shown in Figure 2.27.

EXAMPLE 2.8

From the second-order normalized Bessel filter,

$$H_n(s_n) = \frac{K}{s_n^2 + 1.732s_n + 1}$$

design a filter to pass frequencies from 50 to 200 Hz.

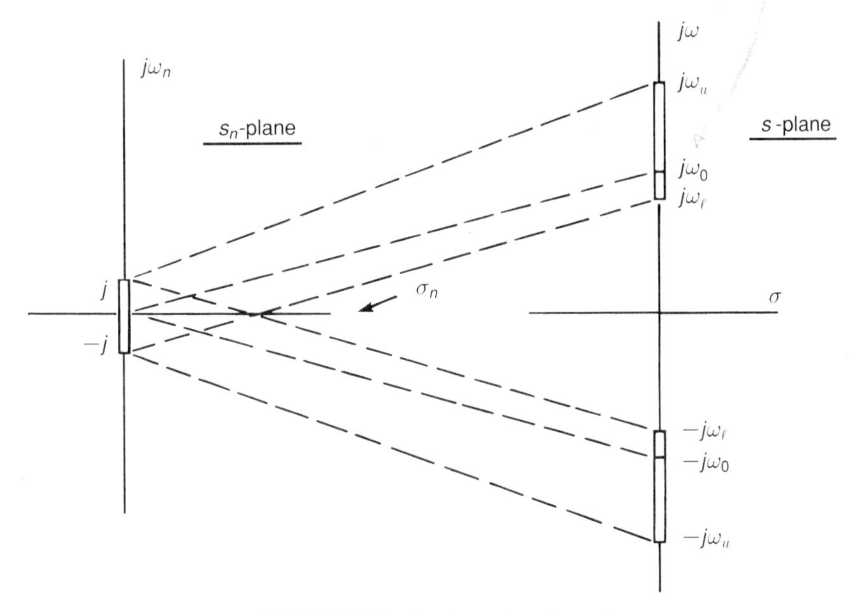

FIGURE 2.27 Bandpass transformation.

Solution

We have

$$\omega_0 = \sqrt{(100\pi)(400\pi)} = 200\pi \text{ rad/s}$$

and the bandwidth

$$B = 400\pi - 100\pi = 300\pi \text{ rad/s}$$

Make the substitution (2.155) for s_n in $H_n(s_n)$.

$$H(s) = \frac{888264Ks^2}{s^4 + 1.632s^3 + 1.6778 \times 10^6 s^2 + 6.4443 \times 10^8 s + 1.558 \times 10^{11}}$$

Note that the filter order is doubled as a result of the transformation.

2.5.4 Bandstop with Cutoff Frequencies ω_ℓ and ω_u

Transformation:

$$s_n = \frac{Bs}{s^2 + \omega_0^2} \tag{2.159}$$

$$= \frac{B}{\omega_0\left(\dfrac{s}{\omega_0} + \dfrac{\omega_0}{s}\right)} \tag{2.160}$$

The segment $|\omega_n| \leq 1$ in the s_n-plane is mapped into the segments $|\omega| \leq \omega_\ell$ and $|\omega| \geq \omega_u$ in the s-plane as shown in Figure 2.28.

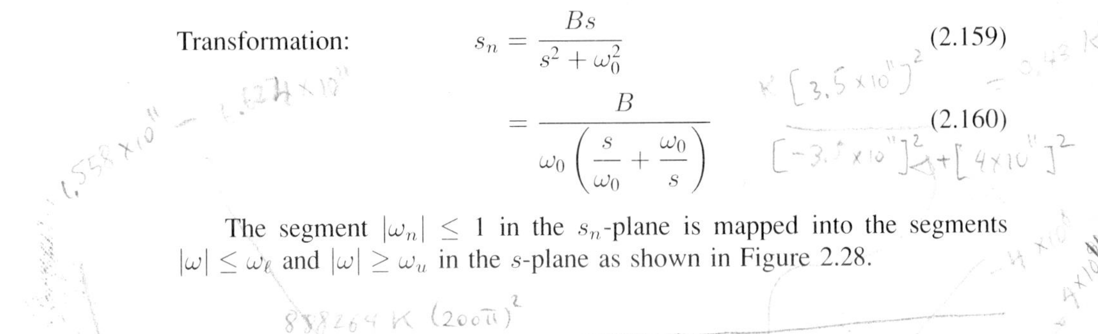

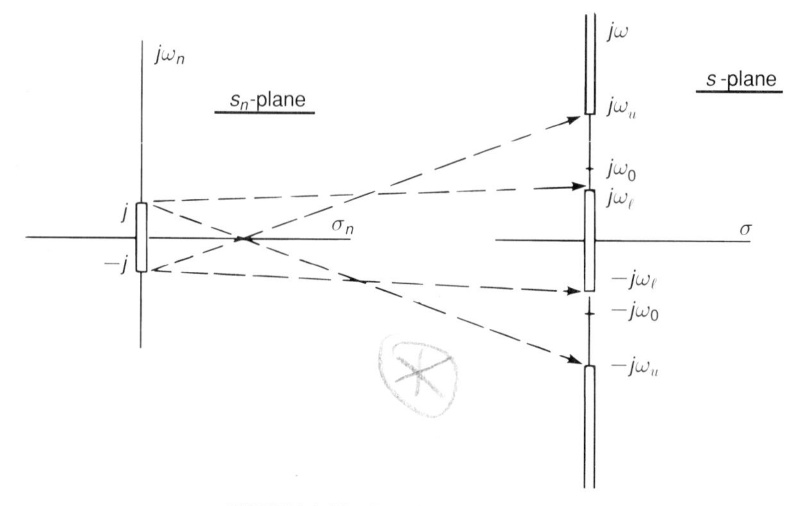

FIGURE 2.28 Bandstop transformation.

EXAMPLE 2.9

Use the first-order Butterworth filter

$$H_n(s_n) = \frac{1}{s_n + 1}$$

to find a bandstop filter with stopband from 50 to 70 Hz.

Solution

Make the substitution (2.159) for s_n in $H_n(s_n)$ to get

$$H(s) = \frac{s^2 + \omega_0^2}{s^2 + Bs + \omega_0^2}$$

where $B = 40\pi$ and $\omega_0^2 = 14000\pi^2$.

Therefore,

$$H(s) = \frac{s^2 + 138174}{s^2 + 126s + 138174}$$

2.6 TWO DESIGN EXAMPLES

To illustrate the foregoing design procedures for various types of filters, two examples are worked through in detail. The first example is for a lowpass filter, and the second example deals with a bandstop design. Using the programs on the diskette would eliminate much of the laborious work.

However, to get a feeling for the design methods, it is worthwhile to go through such analytic examples step-by-step at least once.

EXAMPLE 2.10

Design lowpass Butterworth and Chebyshev type I continuous filters to meet the following specifications:

a) The cutoff frequency is 10 kHz.
b) The attenuation at 200 kHz should be at least 130 dB.
c) The gain at $\omega = 0$ (dc gain) is such that the maximum magnitude in the passband is unity.
d) The maximum acceptable passband ripple (for the Chebyshev filter) is 1 dB.

Solution

1. *Butterworth filter*

From specifications a) and b) and (2.90), the normalized filter order is approximated by

$$n \approx \frac{a}{20\log_{10}(\omega_r/\omega_c)} = \frac{130}{20\log_{10} 20} \approx 5$$

Table 2.2 indicates that the transfer function of the normalized filter is

$$H_n(s_n) = \frac{K}{(s_n + 1)(s_n^2 + 0.618s_n + 1)(s_n^2 + 1.618s_n + 1)}$$

with $K = 1$ [specification c)].

To denormalize the filter, replace s_n in $H_n(s_n)$ by s/ω_c, where $\omega_c = 2\pi \times 10^4$ rad/s. This yields the Butterworth filter that satisfies the given specifications.

It is

$$H(s) = \frac{9.7926 \times 10^{23}}{(s + c)(s^2 + a_1 s + b_1)(s^2 + a_2 s + b_2)}$$

where

$$c = 6.2832 \times 10^4$$

$$a_1 = 3.8830 \times 10^4 \qquad b_1 = 3.9478 \times 10^9$$

$$a_2 = 1.0166 \times 10^5 \qquad b_2 = 3.9478 \times 10^9$$

2. *Chebyshev type I filter*

First find ϵ from specification d).

$$20\log_{10}\frac{1}{(1 + \epsilon^2)^{1/2}} = -1 \Rightarrow \epsilon = 0.509$$

Then find n from specifications a), b), and d).

$$L \approx 20\log_{10}\epsilon + 6(n - 1) + 20n\log_{10}(\omega_r/\omega_c)$$

or

$$130 \approx 20\log_{10}(0.509) + 6(n-1) + 20n\log_{10}20 \Rightarrow n = 4.43$$

Take $n = 5$, which agrees with the approximation given by (2.90).

Solve for γ according to (2.112), and insert values in (2.114) to obtain the normalized transfer function

$$H_n(s_n) = \frac{K'}{D_n(s_n)}$$

where

$$\begin{aligned} D_n(s_n) &= (s_n + .2895)(s_n^2 + .1789s_n + .9883)(s_n^2 + .4684s_n + .4293) \\ &= s_n^5 + \alpha_4 s_n^4 - \alpha_3 s_n^3 + \alpha_2 s_n^2 + \alpha_1 s_n + \alpha_0 \end{aligned}$$

with

$$\alpha_0 = 0.1228$$

Because of the passband ripple, calculation of dc gain depends on whether the filter order is even or odd.

From Figure 2.29, with dc gain $K = 1$, we must have

$$\begin{aligned} \frac{K'}{\alpha_0} &= \frac{1}{(1+\epsilon^2)^{1/2}} \longrightarrow K' = \frac{\alpha_0}{(1+\epsilon^2)^{1/2}} \qquad \text{for even } n \\ \frac{K'}{\alpha_0} &= 1 \longrightarrow K' = \alpha_0 \qquad \text{for odd } n \end{aligned}$$

Therefore, $K' = \alpha_0 = 0.1228$ in this case.

To denormalize the filter, replace s_n in $H_n(s_n)$ by s/ω_c. This yields the Chebyshev filter that satisfies the given specifications. It is

$$H'(s) = \frac{1.2028 \times 10^{23}}{(s+c)(s^2 + a_1 s + b_1)(s^2 + a_2 s + b_2)}$$

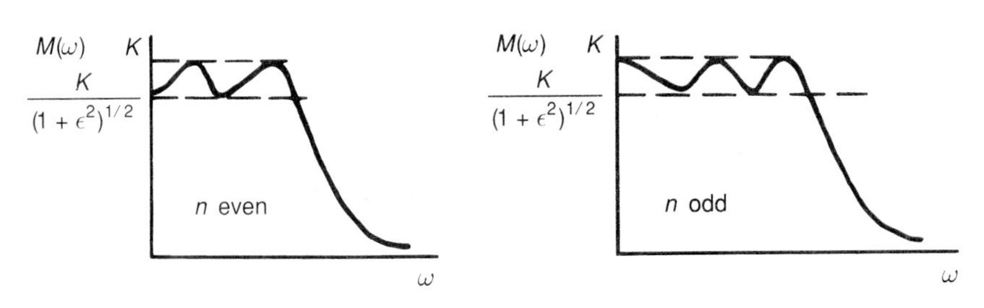

FIGURE 2.29 Calculation of dc gain for Chebyshev filter.

where

$$c = 1.8190 \times 10^4$$

$$a_1 = 1.1241 \times 10^4 \qquad b_1 = 3.9017 \times 10^9$$

$$a_2 = 2.9430 \times 10^4 \qquad b_2 = 1.6948 \times 10^9$$

Magnitude response and phase response for the two filters are shown in Figure 2.30. It is evident that the Chebyshev filter achieves the desired attenuation first. No restrictions were placed on the phase, and the phase responses are nonlinear.

3. *Elliptic filter*
Although the problem statement does not ask us to do so, it is interesting to design an elliptic filter for the same specifications. In order to use the elliptic filter equations in Section 2.44, we must first compute the geometric mean.

$$\omega_0 = 2\pi\sqrt{200 \times 10} \cdot 10^3$$

$$= 0.28099 \times 10^6 \text{rad/s}$$

Then we divide the specified cutoff frequency by ω_0 to obtain the normalized cutoff frequency,

$$\omega_c = 0.2236$$

Note that

$$\omega_r = 1/\omega_c = 4.4721$$

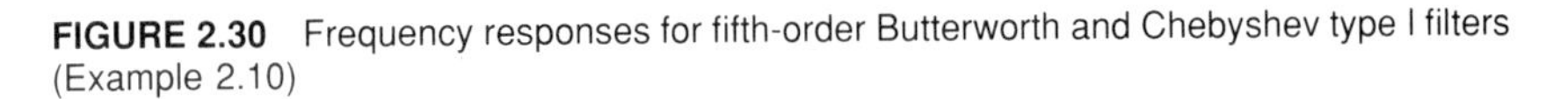

FIGURE 2.30 Frequency responses for fifth-order Butterworth and Chebyshev type I filters (Example 2.10)

A computer program (ELLIP) that appears on the diskette and is based on (2.121) through (2.123) and (2.128) through (2.146) yielded the following results.

A fourth-order elliptic filter will suffice. The transfer function of the normalized filter is

$$H_n(s_n) = \frac{0.1924 \times 10^{-6}(s_n^2 + 136.42)(s_n^2 + 23.427)}{(s_n^2 + 0.1507s_n + 0.0140)(s_n^2 + 0.0623s_n + 0.0493)}$$

To denormalize the filter, replace s_n by s/ω_0 to obtain

$$H(s) = \frac{K(s^2 + c_1)(s^2 + c_2)}{(s^2 + a_1 s + b_1)(s^2 + a_2 s + b_2)}$$

where

$K = 2.1594 \times 10^{-7}$

$a_1 = 4.2348 \times 10^4$ $\quad b_1 = 1.1043 \times 10^9$ $\quad c_1 = 1.07715 \times 10^{13}$

$a_2 = 1.7514 \times 10^4$ $\quad b_2 = 3.8949 \times 10^9$ $\quad c_2 = 1.84974 \times 10^{12}$

EXAMPLE 2.11

Design bandstop Butterworth and Chebyshev type I continuous filters to meet the following specifications:

a) Stopband 100 to 600 Hz.
b) The magnitude response should be down at least 20 dB at 200 and 400 Hz.
c) The gain at $\omega = 0$ is such that the maximum magnitude in the passband is unity.
d) The passband ripple (for the Chebyshev filter) is 1.1 dB.

Solution

First, it is necessary to relate the specified critical frequencies for the bandstop filter to those for the ideal lowpass filter. Let $f_1 = 100$, $f_2 = 200$, $f_3 = 400$, and $f_4 = 600$ Hz. Then the bandwidth is

$$B = 2\pi(f_4 - f_1) \text{ rad/s}$$

and the geometric mean is

$$\omega_0 = 2\pi(f_1 f_4)^{1/2} \text{ rad/s}$$

From (2.159), the normalized filter frequency ω_n and the bandstop filter frequency ω are related by

$$\omega_n = \frac{B\omega}{-\omega^2 + \omega_0^2}$$

It follows that

$$\omega_{n_1} = \frac{(500)(100)}{-(100)^2 + 60000} = 1 \text{ rad/s}$$

$$\omega_{n_2} = \frac{(500)(200)}{-(200)^2 + 60000} = 5 \text{ rad/s}$$

$$\omega_{n_3} = \frac{(500)(400)}{-(400)^2 + 60000} = -2 \text{ rad/s}$$

$$\omega_{n_4} = \frac{(500)(600)}{-(600)^2 + 60000} = -1 \text{ rad/s}$$

In these expressions the factors $(2\pi)^2$, which cancel, are omitted. Note that the stopband edge frequencies f_1 and f_4 map into the normalized frequencies ± 1 as they should (see Figure 2.28). The negative sign on ω_{n_3} and ω_{n_4} can be ignored, because the magnitude response is an even function of frequency.

For the normalized filter,

$$\omega_r = \min(\omega_{n_2}/\omega_{n_1}, \omega_{n_3}/\omega_{n_4}) = 2$$

in this case.

1. *Butterworth filter*

The normalized filter order is approximated by

$$n \geq \frac{a}{20 \log_{10}(\omega_r/\omega_c)} = \frac{20}{20 \log_{10} 2} = 3.32$$

Therefore, take $n = 4$. As a check, use (2.91) to determine that for a third-order filter, the attenuation at $\omega_n = 2$ rad/s is 18.1 dB, which does not meet specification b), whereas for $n = 4$ it is 24.1 dB.

From Table 2.2, the transfer function of the normalized filter is

$$H_n(s_n) = \frac{1}{(s_n^2 + 0.765s_n + 1)(s_n^2 + 1.848s_n + 1)}$$

To denormalize, replace s_n by $Bs/(s^2 + \omega_0^2)$, obtaining an *eighth-order* bandstop filter.

$$H(s) = \frac{(s^2 + \omega_0^2)^4}{(s^4 + a_1 s^3 + b_1 s^2 + c_1 s + d_1)(s^4 + a_2 s^3 + b_2 s^2 + c_2 s + d_2)}$$

where

$$\omega_0^2 = 2.3687 \times 10^6$$

$$a_1 = 2.4045 \times 10^3 \qquad a_2 = 5.8049 \times 10^3$$

$$b_1 = 1.4607 \times 10^7 \qquad b_2 = 1.4607 \times 10^7$$

$$c_1 = 5.6955 \times 10^9 \qquad c_2 = 1.3750 \times 10^{10}$$

$$d_1 = 5.6108 \times 10^{12} \qquad d_2 = 5.6108 \times 10^{12}$$

The filter poles in the s-plane are at

$$-180.0 \pm j620.2, \quad -1022.3 \pm j3522.6,$$
$$-667.3 \pm j511.7, \quad -2235.1 \pm j1714.0$$

The zeros are fourth-order and are located on the imaginary axis at $\pm j1539.1$. We note that $\omega_0 = 1539.1$ rad/s $= 245$ Hz is in the stopband.

2. *Chebyshev type I filter*
To find the order n of the normalized Chebyshev filter, first find ϵ from (2.104).

$$\epsilon = (10^{0.11} - 1)^{1/2} = 0.5369$$

The approximation (2.109) implies that

$$n \geq (L + 6 - 20\log_{10}\epsilon)/(6 + 20\log_{10}\omega_r) = 2.61$$

because $\omega_r = 2$ in this case. Therefore, take $n = 3$. It should be emphasized that (2.90) and (2.109) are merely useful approximations to the required filter order. It must be verified that the resulting normalized filter does in fact meet the attenuation specification. If, instead of taking $\omega_r = 2$ after making the appropriate transformation, we had set $\omega_r = f_4/f_3 = 1.5$ as is sometimes done, we would have had $n = 6$ for the Butterworth and $n = 4$ for the Chebyshev normalized filter in this case. It is easily verified that these values of n are unnecessarily high.

To return to the Chebyshev design, (2.114) gives the poles of a third-order normalized filter at

$$s = -0.4766, -0.2383 \pm j0.9593$$

The transfer function is

$$H_n(s_n) = \frac{0.4656}{(s_n + 0.4766)(s_n^2 + 0.4766s_n + 0.9771)}$$

The transformation $s_n = Bs/(s^2 + \omega_0^2)$ gives a sixth-order bandstop filter.

$$H(s) = \frac{(s^2 + \omega_0^2)^3}{(s^2 + es + f)(s^4 + a_1s^3 + b_1s^2 + c_1s + d_1)}$$

where

$$e = 6.5923 \times 10^3 \qquad f = \omega_0^2 = 2.3687 \times 10^6$$
$$a_1 = 1.5322 \times 10^3 \qquad b_1 = 1.4838 \times 10^7$$
$$c_1 = 3.6294 \times 10^9 \qquad d_1 = 5.6108 \times 10^{12}$$

The poles are at

$$-381.4, \quad -6210.9, \quad -109.8 \pm j620.0, \quad -656.3 \pm j3704.4$$

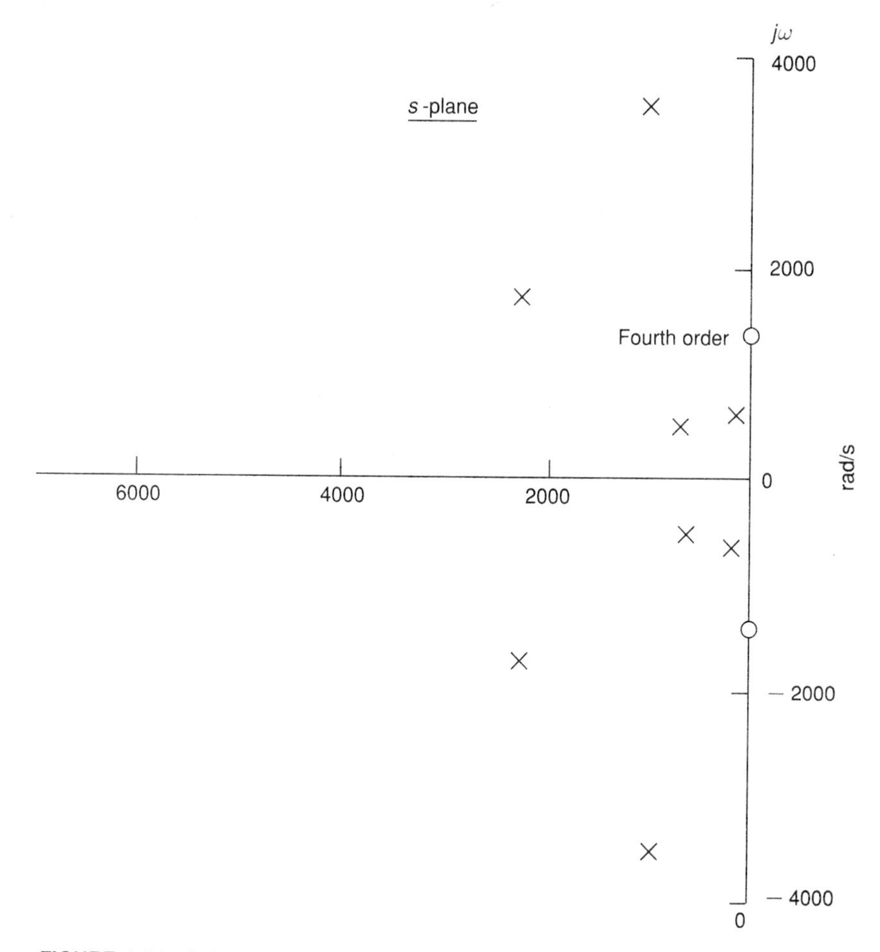

FIGURE 2.31 Pole-zero plot for an eighth-order continuous Butterworth bandstop filter (Example 2.11).

The zeros are third-order, and are located on the imaginary axis at $\pm j1539.1$.

Pole-zero plots for the Butterworth and Chebyshev filters are shown in Figures 2.31 and 2.32, respectively. A comparison of the magnitude responses is given in Figure 2.33. It is evident that both filters meet the attenuation specifications.

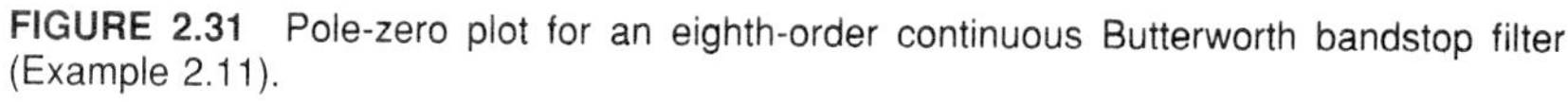

2.7 SUMMARY

Fourier series, Fourier transforms, and Laplace transforms are discussed as background for a review of continuous filter design. Sufficient detail is given to enable the reader who has a working knowledge of Laplace transforms to

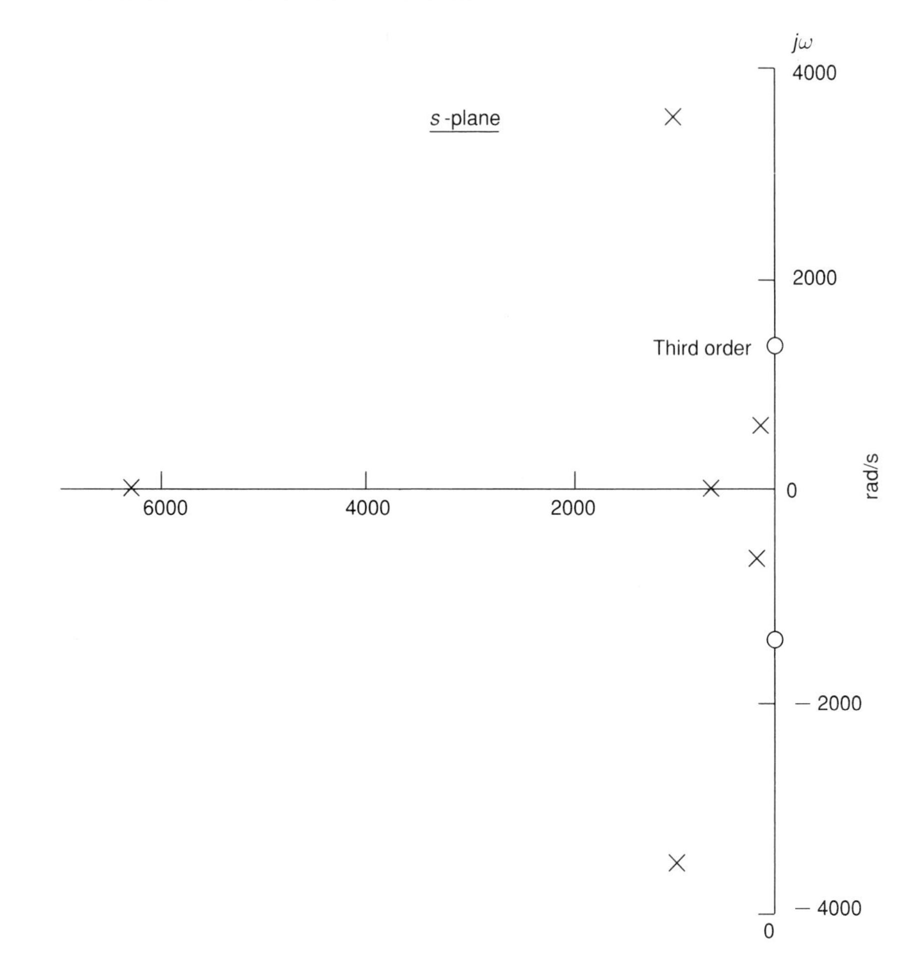

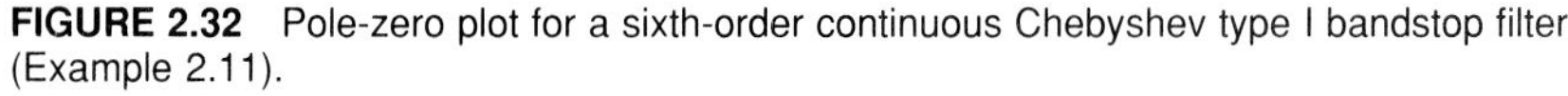

FIGURE 2.32 Pole-zero plot for a sixth-order continuous Chebyshev type I bandstop filter (Example 2.11).

design Butterworth, Chebyshev, elliptic, and Bessel filter approximations to the ideal lowpass prototype filter. By means of the frequency transformations given in Section 2.5, these approximations can be converted to practical lowpass, highpass, bandpass, or bandstop filters. Two worked examples are given. For higher-order filters, it is advisable to use the computer programs given on the diskette that accompanies the text.

The continuous filter can be converted to a recursive digital filter by means of a suitable transformation. Recursive filters are the subject of Chapter 4. First, however, in Chapter 3, we discuss the z-transform, which is a powerful tool for the design and analysis of digital filters.

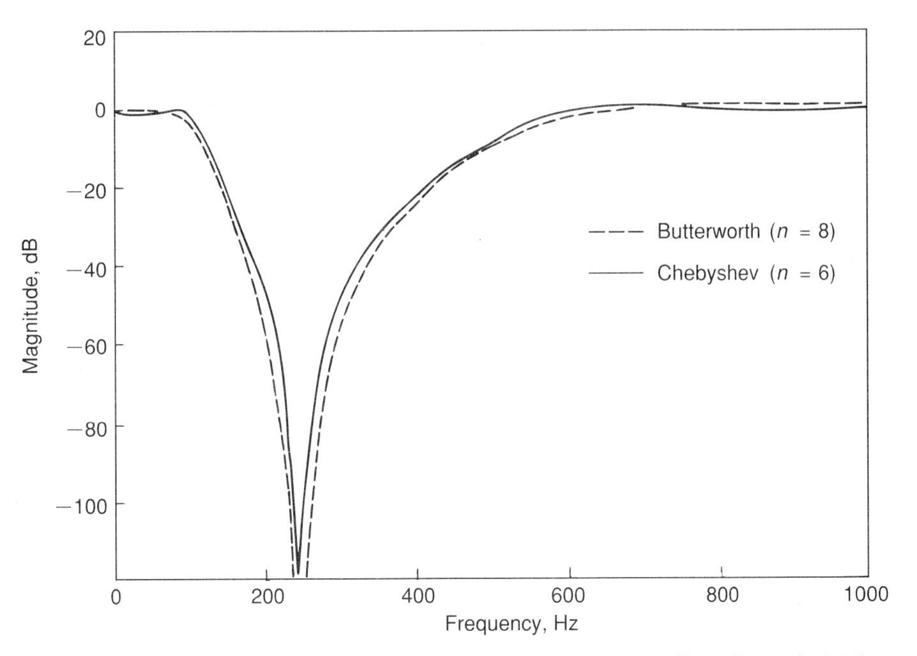

FIGURE 2.33 Magnitude responses for continuous bandstop filters (Example 2.11).

REFERENCES FOR CHAPTER 2

1. S. Butterworth, "On the Theory of Filter Amplifiers," *Wireless Engineer* 7 (October 1930): 536–541.
2. J. E. Storer, *Passive Network Synthesis* (New York: McGraw-Hill, 1957).
3. W. C. Yengst, *Procedures of Modern Network Synthesis* (New York: Macmillan, 1964).
4. M. E. Van Valkenburg, *Introduction to Modern Network Synthesis* (New York: Wiley, 1962).
5. R. W. Daniels, *Approximation Methods for the Design of Passive, Active and Digital Filters* (New York: McGraw-Hill, 1974).
6. R. V. Churchill, *Fourier Series and Boundary Value Problems* (New York: McGraw-Hill, 1941).
7. E. A. Guillemin, *The Mathematics of Circuit Analysis* (New York: M.I.T. Technology Press, Wiley, 1958).
8. N. Levinson and R. M. Redheffer, *Complex Variables* (Oakland, CA: Holden-Day, 1970), Chap. 4.
9. L. R. Rabiner and B. Gold, *Theory and Application of Digital Signal Processing* (Englewood Cliffs, NJ: Prentice-Hall, 1975).
10. A. Antoniou, *Digital Filters: Analysis and Design* (New York: McGraw-Hill, 1979), Chap. 5.
11. M. T. Jong, *Methods of Discrete Signal and System Analysis* (New York: McGraw-Hill, 1982), Chap. 8.

12. A. J. Grossman, "Synthesis of Tchebyscheff Parametric Symmetrical Filters," *Proc. IRE* 45 (April 1957): 454–473.

EXERCISES FOR CHAPTER 2

1. Expand the following periodic function in a Fourier series.

$$f(t) = \begin{cases} t, & \frac{-T_p}{4} < t < \frac{T_p}{4} \\ 0, & \frac{-T_p}{2} \le t \le \frac{-T_p}{4} \text{ and } \frac{T_p}{4} \le t \le \frac{T_p}{2} \end{cases}$$

where T_p is the period.

2. Expand the function

$$M(\omega T) = \begin{cases} 1, & 0 \le |\omega T| \le \frac{\pi}{2} \text{ and } \frac{2\pi}{3} \le |\omega T| \le \pi \\ 0, & \frac{\pi}{2} < |\omega T| < \frac{2\pi}{3} \end{cases}$$

in a Fourier series valid over the interval $[-\pi, \pi]$.

3. From the definitions of the Fourier transform pairs (2.23) and (2.25), prove the properties (2.31) through (2.34) and (2.36) through (2.38).

4. Find the Fourier transforms of the following functions.

a) $f(t) = \delta(t - t_0)$

b) $f(t) = \begin{cases} 1, & |t| \le t_1 \\ 0, & \text{otherwise} \end{cases}$

c) $f(t) = \begin{cases} 1 - t, & 0 \le t \le 1 \\ 1 + t, & -1 \le t \le 0 \\ 0, & \text{otherwise} \end{cases}$

By plotting these functions and their transforms, verify the principle of reciprocal spreading: As a signal spreads out in the time domain, it contracts in the frequency domain, and vice versa.

5. Prove the following form of *Parseval's theorem*, which relates signal energy to the integral of the squared spectrum.

$$\int_{-\infty}^{\infty} x^2(t)dt = \int_{-\infty}^{\infty} |X(f)|^2 df$$

where $X(f) = \mathcal{F}\{x(t)\}$ and $f = \omega/2\pi$.

6. From definition (2.42), verify the Laplace transforms given in Table 2.1.

7. From the definitions of the Laplace transform pair, (2.42) and (2.43), prove the properties (2.46) through (2.57).

8. Find the inverse Laplace transform of

$$F(s) = \frac{2(s+2)}{(s+1)^2(s+4)}$$

by a partial-fraction expansion and by the method of residues.

9. With reference to Figure 2.5, verify that the expressions for magnitude and phase in terms of vector lengths and angles are correct.

10. For the second-order filter

$$H(s) = \frac{16.25(s+1)}{s^2 + s + 16.25}$$

use a graphical method (Figure. 2.5) to derive the frequency response. Plot the magnitude response (in decibels) and the phase response.

11. Derive the approximate relationship (2.90) for the order n of an all-pole filter. What order normalized lowpass Butterworth filter will have an attenuation of at least 100 dB at 6 rad/s? Write the transfer function of the filter, and confirm that it does have the required attenuation.

12. Verify the frequency transformations given in Section 2.5.

13. A normalized lowpass filter has the transfer function

$$H_n(s_n) = \frac{K_n}{s_n^2 + A_1 s_n + A_2}$$

Suppose that transformation (2.155) is used to derive a bandpass filter $H(s)$ from $H_n(s_n)$. Prove that

$$|H(j\omega_1)| = |H_n(j\omega_{n_1})|$$

where

$$\omega_{n_1} = \frac{\omega_1^2 - \omega_0^2}{B\omega_1}$$

14. What order lowpass Butterworth filter will satisfy the following specifications?
   a) Cutoff frequency at 2 Hz
   b) At least 60 dB down at 8 Hz
   Plot the magnitude response and the phase response.

15. Find the transfer function of a highpass Butterworth filter to pass frequencies above 10 Hz. The stopband should be down at least 60 dB at 5 Hz. Plot the magnitude response and the phase response.

16. Repeat Exercise 14 for a Bessel filter.

17. Design a bandstop Butterworth filter with stopband from 100 to 125 Hz. The attenuation should be at least 20 dB at 105 and at 120 Hz. Plot the magnitude response and the phase response.

18. Design a bandpass filter with the following characteristics. Then plot the magnitude response and the phase response. Plot the poles on the s-plane.
   a) Passband 60 to 80 Hz
   b) Passband ripple not to exceed 0.5 dB
   c) Attenuation of at least 40 dB at 50 Hz and at 100 Hz
   d) Gain set so that the maximum magnitude in the passband is unity.

19. Confirm that your answers to Exercises 14 through 18 are correct by using the appropriate computer program on the diskette.

APPENDIX 2A

DISKETTE PROGRAM EXAMPLES

Examples of the use of three diskette programs are given. The programs, which are arranged for interactive operation, are

1. POLZER
2. CONFIL
3. ELLIP

In POLZER, the frequency response of a continuous or a digital filter is computed from the filter transfer function expressed in terms of its poles and zeros. The user is prompted for the poles and zeros. The magnitude response and the phase response are stored in a file POLZER.RSP created by the program. This file can be accessed to display or plot the results. The example treats a first-order continuous lead-lag filter.

The program CONFIL lets the user choose a Butterworth, Chebyshev, or Bessel continuous filter design. A lowpass, highpass, bandpass, or bandstop filter can be selected. Prompts are given for the passband and stopband specifications. The poles and zeros of the filter design are displayed. The magnitude response and the phase response are stored in a file CONFIL.RSP. In the example, a Butterworth bandstop filter is requested. An eighth-order filter is required to meet specifications.

An elliptic continuous-filter design is given by the program ELLIP. Interactive operation is similar to that for CONFIL. The frequency responses are stored in a file ELLIP.RSP. The same specifications as used for the CONFIL example resulted in a sixth-order elliptic bandstop design.

```
POLZER
FREQUENCY RESPONSE FROM POLES AND ZEROS
ARE POLES AND ZEROS IN S-PLANE OR Z-PLANE (ENTER S/Z):
S
ENTER NUMBER OF ZEROS:
1
ENTER REAL PART OF ZERO # 1:
-1.
```

```
ENTER IMAGINARY PART OF ZERO # 1:
0.
ENTER NUMBER OF POLES:
1
ENTER REAL PART OF POLE # 1:
-10.
ENTER IMAGINARY PART OF POLE # 1:
0.
ENTER HIGHEST FREQUENCY (RAD/SEC) FOR PLOT:
12.57
MAGNITUDE AND PHASE RESPONSES IN FILE POLZER.RSP
GAIN CONSTANT K FOR MAX. MAGNITUDE OF UNITY: .127382E+01
Stop - Program terminated.
```

```
CONFIL
ENTER CLASS OF FILTER (1: BUTTERWORTH, 2: CHEBYSHEV, 3: BESSEL):
1
BUTTERWORTH FILTER DESIGN
ENTER TYPE OF FILTER (LP,HP,BP,BS):
BS
ENTER LOWER CUTOFF FREQUENCY (HZ):
100.
ENTER UPPER CUTOFF FREQUENCY (HZ):
600.
ENTER MINIMUM ATTENUATION (DB) IN STOP BAND:
20.
ENTER LOWEST FREQUENCY (HZ) IN STOP BAND
WHERE MINIMUM ATTENUATION CAN OCCUR:
200.
ENTER HIGHEST FREQUENCY (HZ) IN STOP BAND
WHERE MINIMUM ATTENUATION CAN OCCUR:
400.
FILTER OF ORDER 8 REQUIRED TO MEET SPECIFICATIONS.
THERE ARE ZEROS OF ORDER 4 AT + AND - J .153906E+04
POLE LOCATIONS:
P( 1) =  -.102226E+04 + J  -.352265E+04
P( 2) =  -.179978E+03 + J   .620193E+03
P( 3) =  -.223511E+04 + J  -.171398E+04
P( 4) =  -.667341E+03 + J   .511746E+03
P( 5) =  -.223511E+04 + J   .171398E+04
P( 6) =  -.667341E+03 + J  -.511746E+03
P( 7) =  -.102226E+04 + J   .352265E+04
P( 8) =  -.179978E+03 + J  -.620193E+03
ENTER HIGHEST FREQUENCY (HZ) FOR PLOT:
1000.
MAGNITUDE AND PHASE RESPONSES IN FILE CONFIL.RSP
FILTER GAIN CONSTANT K FOR MAX.MAG OF UNITY =   .100000E+01
Stop - Program terminated.
```

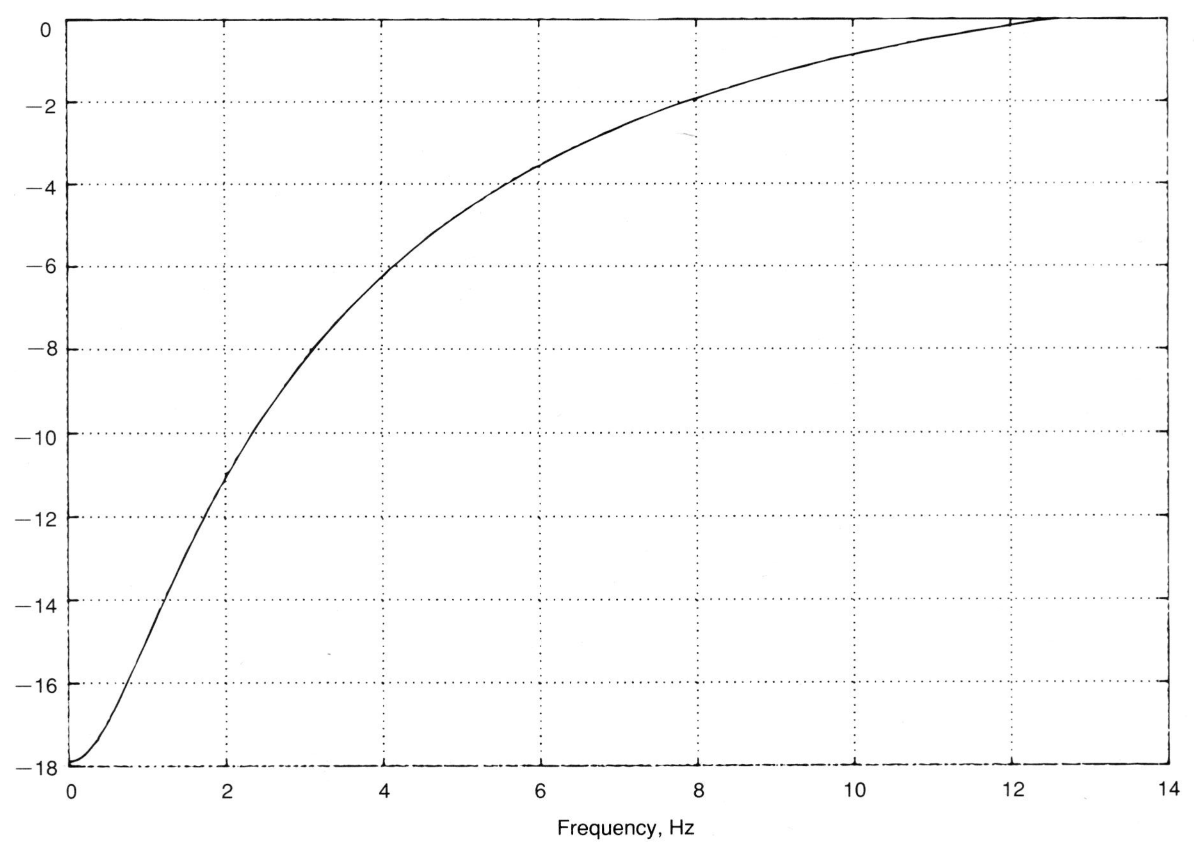

FIGURE 2A.1 Magnitude response (POLZER example).

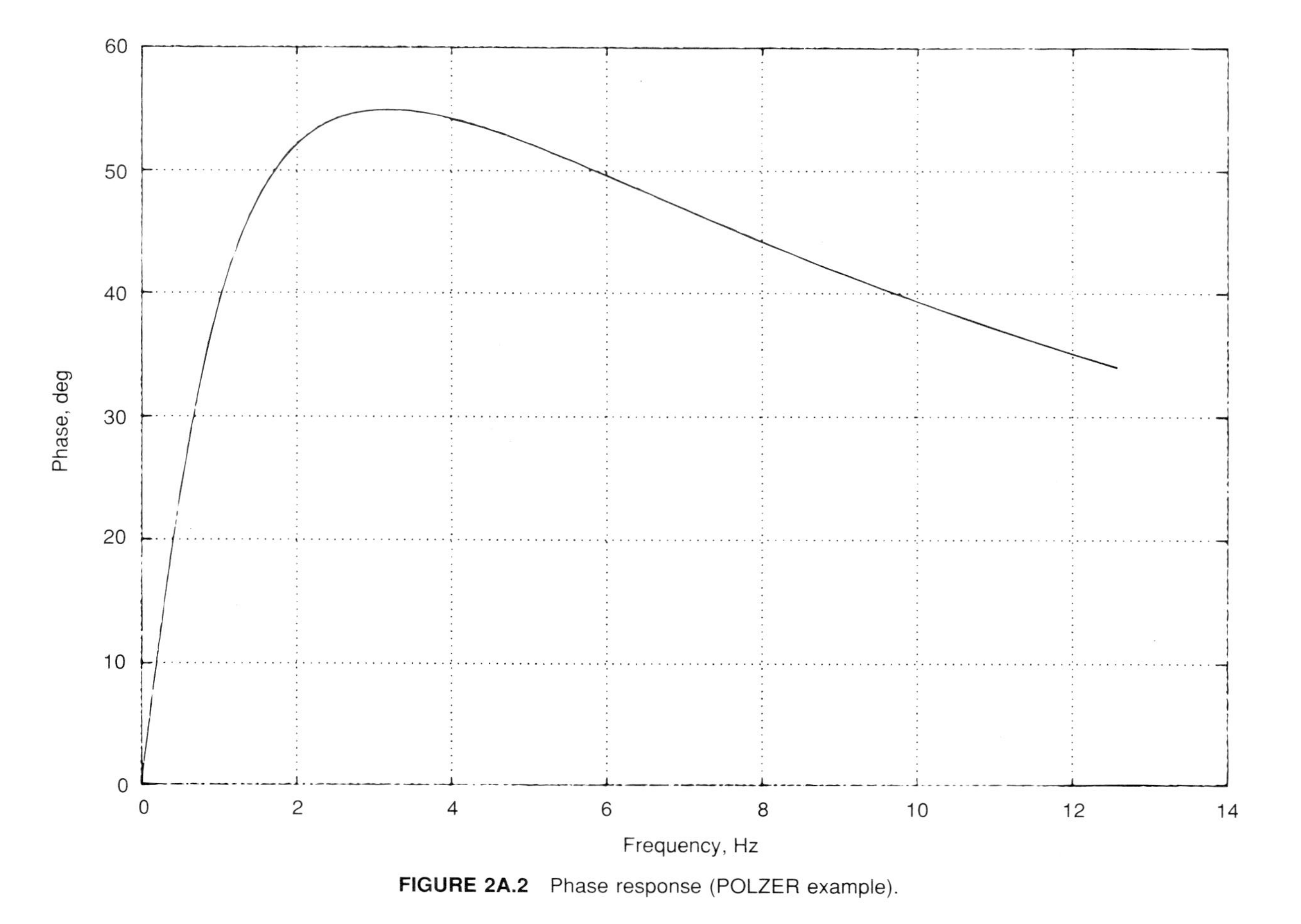

FIGURE 2A.2 Phase response (POLZER example).

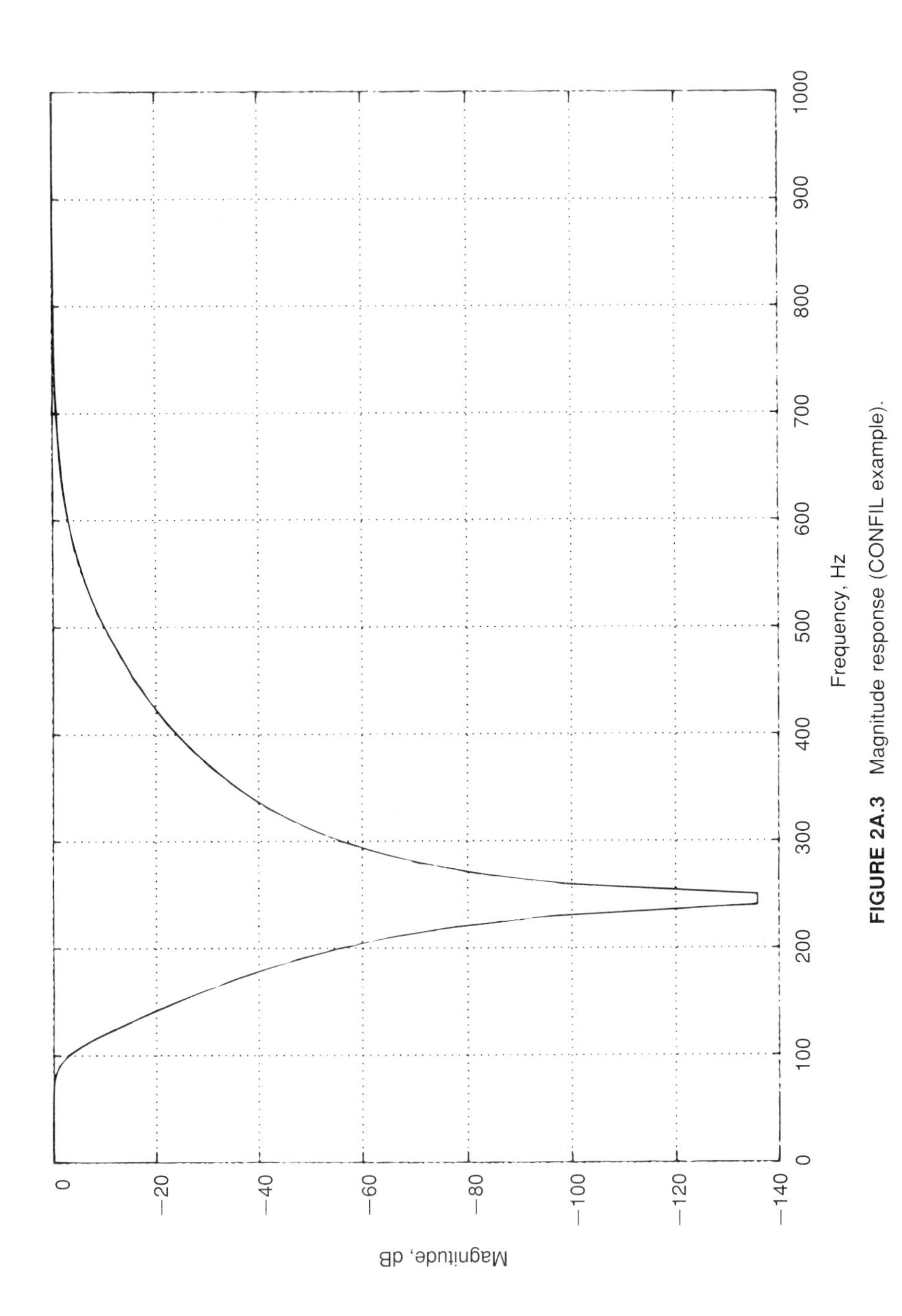

FIGURE 2A.3 Magnitude response (CONFIL example).

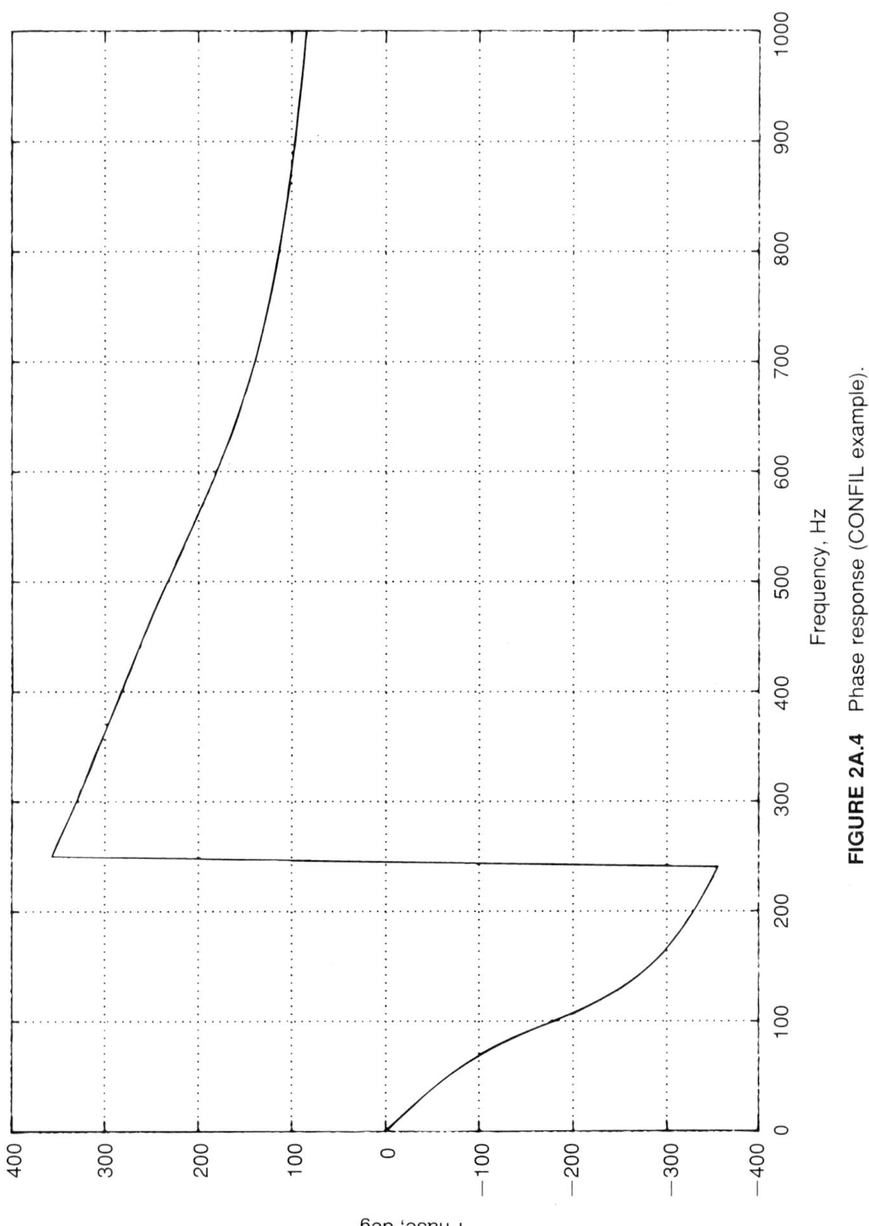

FIGURE 2A.4 Phase response (CONFIL example).

```
ELLIP
ELLIPTIC FILTER DESIGN
ENTER TYPE OF FILTER (LP,HP,BP,BS):
BS
ENTER LOWER CUTOFF FREQUENCY (HZ):
100.
ENTER UPPER CUTOFF FREQUENCY (HZ):
600.
ENTER MINIMUM ATTENUATION (DB) IN STOP BAND:
20.
ENTER PASSBAND RIPPLE (DB):
1.1
ENTER LOWEST FREQUENCY (HZ) IN STOP BAND
WHERE MINIMUM ATTENUATION CAN OCCUR:
200.
ENTER HIGHEST FREQUENCY (HZ) IN STOP BAND
WHERE MINIMUM ATTENUATION CAN OCCUR:
400.
ELLIPTIC FILTER OF ORDER 6 REQUIRED TO MEET SPECIFICATIONS.
ZERO LOCATIONS:
Z( 1) =   .000000E+00 + J   -.237942E+04
Z( 2) =   .000000E+00 + J    .237942E+04
Z( 3) =   .000000E+00 + J    .995498E+03
Z( 4) =   .000000E+00 + J   -.995498E+03
Z( 5) =   .000000E+00 + J   -.153906E+04
Z( 6) =   .000000E+00 + J    .153906E+04
POLE LOCATIONS:
P( 1) =  -.567264E+03 + J   -.370448E+04
P( 2) =  -.567264E+03 + J    .370448E+04
P( 3) =  -.825563E+04 + J    .000000E+00
P( 4) =  -.956699E+02 + J    .624766E+03
P( 5) =  -.956699E+02 + J   -.624766E+03
P( 6) =  -.286920E+03 + J    .000000E+00
ENTER HIGHEST FREQUENCY (HZ) FOR PLOT:
1000.
MAGNITUDE AND PHASE RESPONSES IN FILE ELLIP.RSP
FILTER GAIN CONSTANT K FOR MAX.MAG OF UNITY =    1.00000E+01
Stop - Program terminated.
```

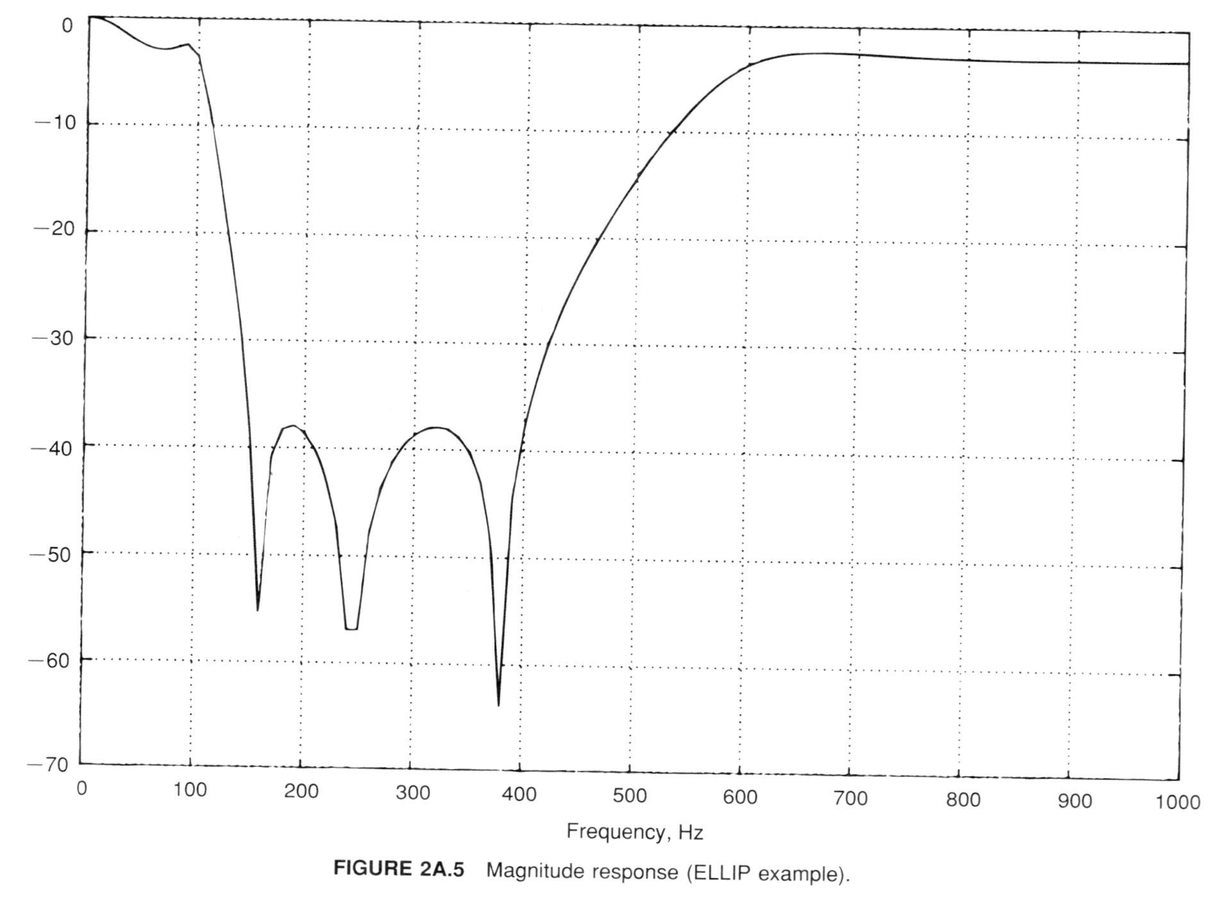

FIGURE 2A.5 Magnitude response (ELLIP example).

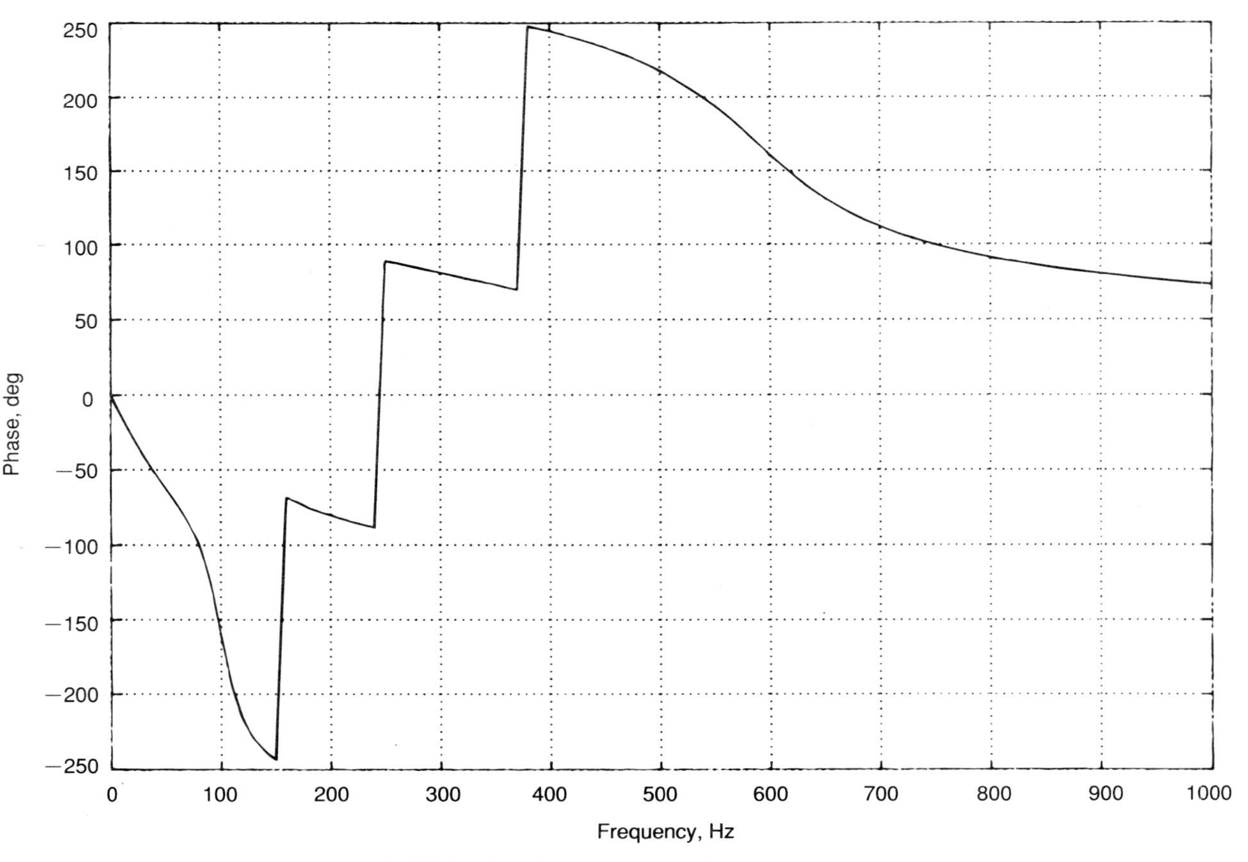

FIGURE 2A.6 Phase response (ELLIP example).

CHAPTER 3

SAMPLING AND THE z-TRANSFORM

3.1 INTRODUCTION

A discrete signal can be generated by a process or phenomenon that is inherently discrete, or it can result from sampling of a continuous signal. In either case, if the discrete signal is to be processed by a discrete LTI system or digital filter, the design and analysis of the filter are greatly facilitated by taking the z-transform of the discrete signals involved. In fact, the importance of the z-transform for such work can hardly be overemphasized. Although the concept of the z-transform dates back at least to 1730, when De Moivre introduced the generating function in probability theory [1], the technique really came into its own as a powerful analytic and design tool for discrete systems with the advent of the digital computer. Hurewicz [2] is generally credited with introducing the z-transform for discrete control application. The concept was developed further by a number of workers in sampled-data control, notably Barker [3] in England and Ragazzini and Zadeh in the United States [4].

In introducing the z-transform in this chapter, we had to choose between two approaches: (1) define the transform as an operation on a sequence of numbers or, more indirectly, (2) develop the transform by sampling a continuous signal and taking the Laplace transform of the sampled signal. The two approaches give the same final result. We chose the second because it provides much insight into the interaction between continuous and discrete signals through the sampling process and the related phenomenon of aliasing. Furthermore, the development that this approach entails indicates how we can transform directly from a continuous filter frequency function to a digital filter frequency function, a technique that we shall use in Chapter 4.

3.2 SAMPLING A CONTINUOUS SIGNAL

Many signals processed by digital filters are obtained by sampling a continuous signal at a uniform rate. We will call this rate f_s (samples per second). The sampling period is $T = 1/f_s$ (in seconds), and the angular frequency is $\omega_s = 2\pi f_s$ (in radians per second). Because the sampling cost goes up with sampling rate (or frequency), the following question arises: What is the minimum rate at which we should sample to avoid losing any of the information present in the continuous signal? Intuitively, we feel that a rapidly changing signal must be sampled at a higher rate than a slowly varying one. In other words, the answer to the question depends on the fluctuations of the signal—that is, on its frequency content. A bound on the minimum rate is given by the *sampling theorem.* First, however, we must define a *bandlimited signal.* This is a signal $f(t)$ that contains no frequency component above some finite frequency ω_h (or f_h); in other words, $F(j\omega) = 0,\ \omega > \omega_h$. The theorem follows.

Sampling Theorem

If a bandlimited continuous signal $f(t)$ is sampled at a rate ω_s greater than twice the highest frequency ω_h present, then the sample values contain all the information in the continuous signal, and the latter may be recovered from the samples $f(kT)$ by the interpolation formula.

$$f(t) = \sum_{k=-\infty}^{\infty} f(kT)\frac{\sin\left[\omega_s(t - kT)/2\right]}{\omega_s(t - kT)/2} \tag{3.1}$$

The minimum sampling rate $\omega_s = 2\omega_h$ is called the *Nyquist* rate for that signal. There are other sampling theorems [5], but this one fits our needs. Although it is often called the Shannon sampling theorem, the part dealing with minimum sampling frequency appears to have been enunciated first by H. Nyquist in 1928. A plausible argument [6] for the validity of that part of the theorem is as follows.

Suppose we have a very long record of a continuous signal containing no frequency components above f_h (in hertz). Let the record length be T_L. Strictly speaking, we cannot have a finite signal with finite spectrum [7], so we will visualize T_L as approaching infinity. If we expand $f(t)$ in a Fourier series valid over the interval $(0,\ T_L)$, the fundamental frequency is $f_1 = 1/T_L$. Therefore, the number of harmonics present is $f_h/f_1 = T_L f_h$. Each harmonic is completely specified by *two* constants, the magnitude and phase, or the coefficients of the sine and cosine terms, respectively. Thus, the entire record is uniquely described by $2T_L f_h$ Fourier coefficients. These can be computed from the amplitude of the record at $2T_L f_h$ sample times.

Therefore, the sampling rate should be

$$f_s > \frac{2T_L f_h}{T_L} = 2f_h \text{ samples per second} \tag{3.2}$$

For a more rigorous proof, we must wait until we model a periodic sampler in the next section.

What happens if we do not sample at a high enough rate, and an important related phenomenon, can be illustrated by a simple example. Suppose we sample a 1-Hz sinusoidal signal 4 times in 1 period as shown in Figure 3.1(a). In this case, $f_h = 1$ and $f_s = 4$, so the sampling theorem indicates that we can reconstruct the original signal as shown, without losing information. Then we sample a 5-Hz sinusoid at the same rate (4 times per second). This gives the result shown in Figure 3.1(b). We cannot distinguish the sampled signal for this case from that for the 1-Hz signal, so the reconstructed signal is not at all like the original signal. And according to the sampling theorem, it should not be; we are sampling a 5-Hz signal at a 4-Hz rate. What happens in this special case where the frequency of the two continuous signals differ by the sampling rate (5 Hz − 1 Hz = 4 Hz) is known as *aliasing* or *frequency folding*.

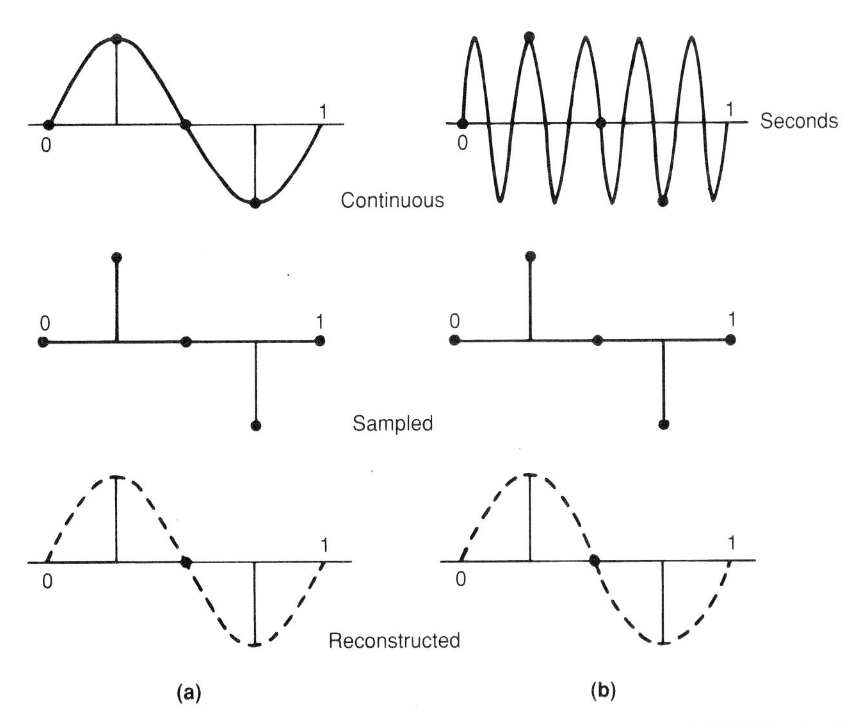

FIGURE 3.1 Illustration of the phenomenon of aliasing. (a) 1-Hz signal. (b) 5-Hz signal.

To generalize, let a continuous sinusoid

$$f(t) = \sin \omega_0 t \tag{3.3}$$

be sampled at a rate ω_s, resulting in samples

$$f(k) = \sin \omega_0 kT \tag{3.4}$$

where $T = 2\pi/\omega_s$. Now sample (at the same rate) a sinusoid the frequency of which differs from that of $f(t)$ by a multiple of the sampling frequency. That is,

$$g(t) = \sin [(\omega_0 + m\omega_s)t] \tag{3.5}$$

where m is an integer. This results in

$$g(k) = \sin [(\omega_0 + m\omega_s)kT] = \sin \omega_0 kT \tag{3.6}$$

because $\omega_s T = 2\pi$. Hence $f(k)$ and $g(k)$ are identical. When we sample two sinusoids the frequencies of which differ by a multiple of the sampling frequency, we cannot distinguish between the results. The higher-frequency signal is said to fold into the lower frequency. To avoid aliasing, we should sample the higher-frequency signal at its Nyquist rate, at least.

We are discussing bandlimited signals as though they were a regular occurrence. Yet many common signals have a spectrum that seems to have energy at all finite frequencies. For example, the exponentially decaying positive-time signal e^{-at} has a magnitude spectrum $1/\sqrt{\omega^2 + a^2}$ that goes to zero only when ω approaches infinity. Obviously, to sample at the Nyquist rate in this case would be impractical. Does this mean that we cannot sample such a signal and get meaningful results? Here theory and practice diverge somewhat. Even though the spectrum has energy at all frequencies, according to its mathematical model it falls off monotonically with frequency and has negligible values at the higher frequencies. For practical purposes, we can say that if the spectrum in this case is down by a factor of say, 100 (40 dB down) at a frequency ω_1 compared with its value at low frequency ($\omega \rightarrow 0$), we can take ω_1 as ω_h and sample at twice this frequency. Granted there will be an aliasing error due to the higher-frequency components, but this error will be negligible if we take ω_1 high enough. In practice, it is customary to sample at three to five times what appears to be the highest frequency in the signal.

Another important practical case is that of a decaying (or damped) cosinusoidal signal. Its spectrum peaks at a certain frequency ω_r, called the *resonant* frequency. This signal could be the impulse response of a system the transfer function of which has two complex conjugate poles at $s = -a \pm jb$. If these are the only poles, the transfer function could have the form

$$H(s) = \frac{1}{s^2 + 2as + a^2 + b^2} = \frac{1}{s^2 + 2\zeta\omega s + \omega_n^2} \tag{3.7}$$

where $\omega_n = \sqrt{a^2 + b^2}$ is called the undamped natural frequency, and $\zeta = a/\sqrt{a^2 + b^2}$ is called the damping ratio. The latter varies from zero to 1; it is zero when the poles lie on the $j\omega$ axis. The system is said to be underdamped if $\zeta < 0.3$. The lower the damping, the higher the resonant peak that occurs in the spectrum. The frequency response is given by

$$H(j\omega) = \frac{1}{\left(1 - \frac{\omega^2}{\omega_n^2}\right) + j2\zeta\frac{\omega}{\omega_n}} \tag{3.8}$$

This function is not bandlimited, but the magnitude response falls off at 40 dB per decade from its value of zero dB at $\omega = 0$. The resonant peak occurs at $\omega_r = \omega_n$. For this type of signal, it is customary to relate the required sampling frequency to the resonant frequency and to take ω_s as five to ten times ω_d, where $\omega_d = \sqrt{1 - \zeta^2}\omega_n$ is the *damped* natural frequency.

No physical device will pass unlimited high frequencies, no matter what the mathematical model says (the model simply is not very good in that frequency region). Therefore, all signals that are output from a physical device (that includes most, if not all, signals of interest) are bandlimited and can be sampled at a finite rate, as we have discussed, without much loss of information.

3.3 MODEL FOR A PERIODIC SAMPLER

Before a continuous signal can be processed in a digital computer, it must pass through an analog-to-digital (A/D) converter (see Figure 1.3). This device both samples the signal in time and quantizes its amplitude values at the sampling times, producing a sequence of numbers that the computer can handle. We will treat the amplitude quantization effects separately (Chapter 8). In this section, we discuss a model for the time-sampling operation. This model has a virtue that all mathematical models should possess: The results predicted by its use in analytic studies agree closely with experimental observations. In addition, the model leads naturally to a frequency-domain representation of a discrete signal—namely, the z-transform.

When a continuous signal is converted into a train of narrow pulses occurring at the sampling instances $0, T, 2T, \ldots, kT, \ldots$, it is convenient to represent the sampling operation by a *fictitious* switch (Figure 3.2). Here $f(t)$ is the continuous signal, T is the uniform sampling period, and $f^*(t)$ is the discrete signal resulting from sampling. It is assumed that the switch closes during the sampling for a duration τ. In practice, $\tau \ll T$. We seek a relationship between $f^*(t)$ and $f(t)$.

At this point, some confusion may arise. In Chapter 1 we stated that when we sample a continuous signal $f(t)$, we obtain a sequence of numbers that we called the discrete signal $f(k) = f(kT)$. There, we implicitly assumed that the sampling operation could be carried out *instantaneously*.

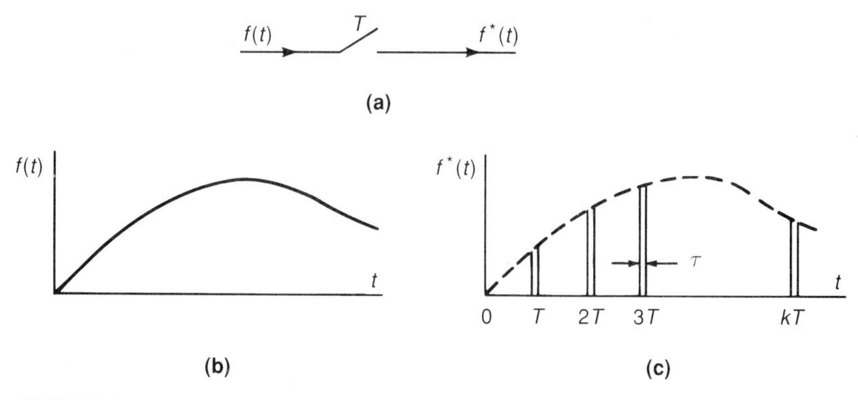

FIGURE 3.2 Sampling a continuous signal. (a) Sampling switch. (b) Continuous signal. (c) Sampled signal.

In fact, it has a duration τ that is non-zero, though it is very small. Thus the sampled signal $f^*(t)$ is a sequence of very narrow pulses, as shown in Figure 3.2, and is a function in the continuous time domain. The height or strength of each pulse is the value of $f(t)$ at the sampling instant. Because $\tau \ll T$, this value can be assumed to be constant over the sampling duration τ. Thus both the continuous function $f^*(t)$ and the discrete function $f(k)$ *contain the same information* about the sampled values of $f(t)$ at time $t = kT$. If τ is infinitesimal compared with T, and with the time constant of the system to which the sampled signal $f^*(t)$ is being applied, the latter signal can be considered to be a train of impulses, the strengths (the areas under the impulses) of which are equal to the value of the continuous function $f(t)$ at the corresponding sampling instances. This approach simplifies analysis and design and, in most cases, gives results that closely approximate those obtained in practice.

On the basis of the foregoing assumptions, an ideal sampling function can be defined,

$$\delta_T(t) = \sum_{k=-\infty}^{\infty} \delta(t - kT) \tag{3.9}$$

representing a train of unit impulses. Then

$$f^*(t) = f(t)\delta_T(t) \tag{3.10}$$

Thus, to borrow some terminology from communication theory, the sampler is equivalent to an *impulse modulator* with input $f(t)$ as the modulating signal and the ideal sampling function $\delta_T(t)$ as the carrier. This comparison is depicted in Figure 3.3.

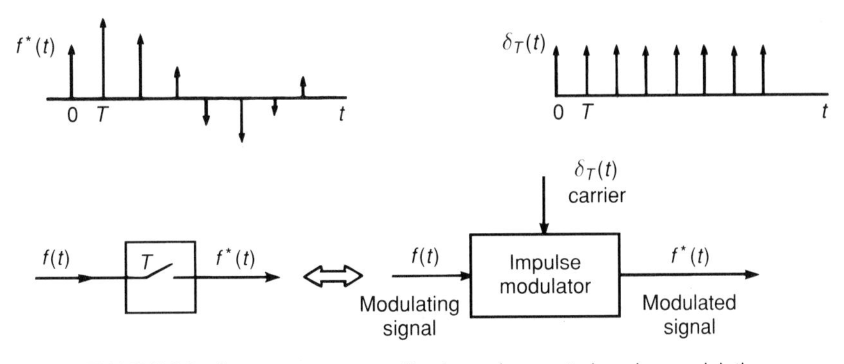

FIGURE 3.3 Instantaneous sampling is analogous to impulse modulation.

3.4 LAPLACE TRANSFORM OF THE SAMPLED FUNCTION

To obtain the frequency characteristics of the sampled function, we will consider three different ways to take the Laplace transform of $f^*(t)$ as it is given in (3.10). Assume that $f(t) = 0$ for $t < 0$.

3.4.1 Method 1

From (3.9) and (3.10),

$$f^*(t) = f(t) \sum_{k=-\infty}^{\infty} \delta(t - kT) = \sum_{k=0}^{\infty} f(kT)\delta(t - kT) \qquad (3.11)$$

because values of $f(t)$ between the sampling instances are lost. Taking the Laplace transform of (3.11) gives

$$F^*(s) = \sum_{k=0}^{\infty} f(kT)e^{-kTs} \qquad (3.12)$$

We will discuss the frequency interpretation of this result further after we have developed the other two methods.

3.4.2 Method 2

The impulse train $\delta_T(t)$ is periodic, so it can be expanded in a complex Fourier series.

$$\delta_T(t) = \sum_{n=-\infty}^{\infty} c_n e^{jn\omega_s t} \qquad (3.13)$$

where

$$c_n = \frac{1}{T}\int_{-T/2}^{T/2} \delta_T(t)e^{-jn\omega_s t}dt = \frac{1}{T} \qquad (3.14)$$

The integral is unity because only the delta function at the origin lies in the interval $[-T/2, T/2]$. Therefore,

$$\delta_T(t) = \frac{1}{T} \sum_{n=-\infty}^{\infty} e^{jn\omega_s t} \tag{3.15}$$

Combining (3.10) and (3.15) gives

$$f^*(t) = \frac{1}{T} \sum_{n=-\infty}^{\infty} f(t) e^{jn\omega_s t} \tag{3.16}$$

Taking the Laplace transformation of (3.16) term by term yields

$$F^*(s) = \frac{1}{T} \sum_{n=-\infty}^{\infty} F(s + jn\omega_s) \tag{3.17}$$

or, if $s = j\omega$, the frequency response is

$$F^*(j\omega) = \frac{1}{T} \sum_{n=-\infty}^{\infty} F[j(\omega + n\omega_s)] \tag{3.18}$$

Note that $F^*(j\omega)$ is a periodic function of frequency with fundamental period ω_s. This is a basic property of sampling. In general, periodic time functions have discrete frequency spectra, whereas discrete time functions resulting from periodic sampling have periodic frequency spectra.

If the highest-frequency ω_h in the unsampled signal is less than $\omega_s/2$ [Figure 3.4(a)], the periodic spectra of the sampled signal do not overlap [Figure 3.4(b)]. Then the original signal can be recovered by filtering the sample output through an ideal lowpass filter of bandwidth $\omega_s/2$ (or an approximation thereto). However, if $\omega_h > \omega_s/2$ [Figure 3.4(c)], then the periodic spectra overlap, creating the aliasing problem shown in Figure 3.4(d), and the original signal cannot be recovered by filtering. This behavior is in accordance with the sampling theorem stated earlier. The frequency range zero to $\omega_s/2$ defines the *operating range* of a discrete-time device, such as a digital filter, because if any frequencies greater than $\omega_s/2$ are processed, there will be aliasing, or folding into the lower frequencies.

The result (3.18) can be used to prove the sampling theorem with respect to recovery of the original time function from its samples by means of the interpolation formula (3.1). The proof follows.

From (3.18),

$$TF^*(j\omega) = \sum_{n=-\infty}^{\infty} F(j\omega + jn\omega_s) \tag{3.19}$$

This is a periodic extension of $F(j\omega)$ [see Figure 3.4(b)], and, as we have noted, if $\omega_s \geq 2\omega_h$, we can recover $F(j\omega)$ by passing $TF^*(j\omega)$ through an

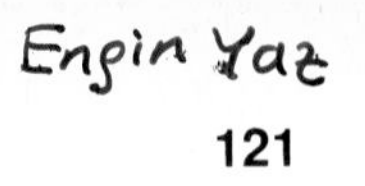

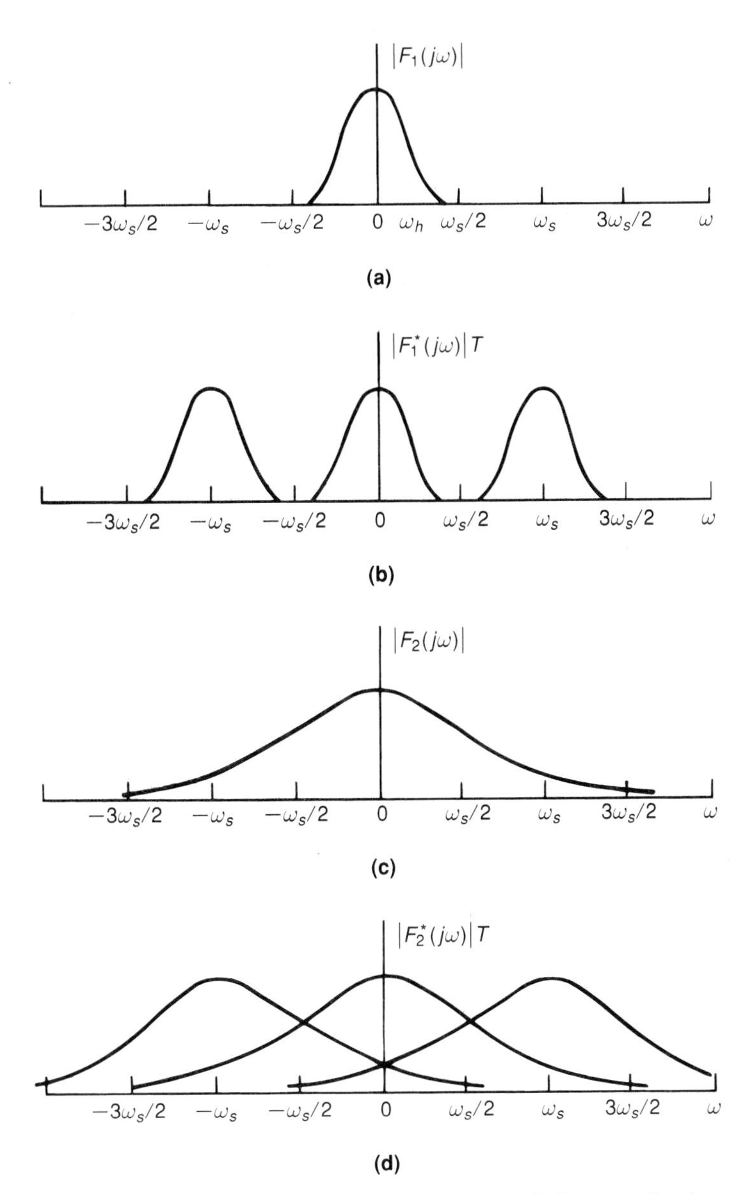

FIGURE 3.4 Effect of sampling on frequency response. (a) Before sampling ($\omega_h < \omega_S/2$). (b) After sampling. (c) Before sampling ($\omega_h > \omega_S/2$). (d) After sampling.

ideal lowpass filter.

$$F(j\omega) = H(j\omega)TF^*(j\omega) \tag{3.20}$$

where

$$H(j\omega) = \begin{cases} 1, & |\omega| \le \omega_s/2 \\ 0, & |\omega| > \omega_s/2 \end{cases} \tag{3.21}$$

Replace $F^*(j\omega)$ in (3.20) by its equivalent form (3.12), with $s = j\omega$ and the lower limit of the summation changed from 0 to $-\infty$. This does not change the value of the expression, because $f(kT) = 0$ for $k < 0$. Then (3.20) becomes

$$F(j\omega) = H(j\omega)T \sum_{k=-\infty}^{\infty} f(kT)e^{-jk\omega T} \tag{3.22}$$

Now

$$\begin{aligned} f(t) &= \mathcal{F}^{-1}\{F(j\omega)\} \\ &= T \sum_{k=-\infty}^{\infty} f(kT)\mathcal{F}^{-1}\{H(j\omega)e^{-jk\omega T}\} \end{aligned} \tag{3.23}$$

To evaluate the inverse Fourier transform, we note that

$$\begin{aligned} \mathfrak{F}^{-1}\{H(j\omega)\} &= \frac{1}{2\pi}\int_{-\infty}^{\infty} H(j\omega)e^{j\omega t}\,d\omega \\ &= \frac{1}{2\pi}\int_{-\omega_s/2}^{\omega_s/2} e^{j\omega t}\,d\omega \\ &= \frac{1}{\pi t}\sin\frac{\omega_s t}{2} \end{aligned} \tag{3.24}$$

Therefore, from the time-shift property (2.32) of the Fourier transform, we have

$$\begin{aligned} \mathcal{F}^{-1}\{H(j\omega)e^{-jk\omega T}\} &= \frac{\sin[\omega_s(t-kT)/2]}{\pi(t-kT)} \\ &= \frac{\sin[\omega_s(k-kT)/2]}{T\omega_s(t-kT)/2} \end{aligned} \tag{3.25}$$

Inserting (3.25) in (3.23) yields

$$f(t) = \sum_{k=-\infty}^{\infty} f(kT)\frac{\sin[\omega_s(t-kT)/2]}{\omega_s(t-kT)/2} \tag{3.26}$$

which is a reconstruction of the signal $f(t)$ from the samples $f(kT)$ of the sampled signal $f^*(t)$.

3.4.3 Method 3

Let

$$\mathcal{L}[f(t)] = F(s) \quad \text{and} \quad \mathcal{L}[\delta_T(t)] = \Delta_T(s)$$

where it is understood that we are taking the Laplace transform of the positive time function part of the ideal impulse train. This does not create a problem, because $f(t)$ is assumed to be zero for $t < 0$. Then, from (3.10) and the complex convolution theorem (2.51),

$$F^*(s) = \mathcal{L}\{f(t)\delta_T(t)\}$$

$$= \frac{1}{2\pi j}\int_{c-j\infty}^{c+j\infty} F(p)\Delta_T(s-p)dp \tag{3.27}$$

The next step is to find $\Delta_T(s)$.

$$\Delta_T(s) = \mathcal{L}\left\{\sum_{n=0}^{\infty}\delta(t-nT)\right\}$$

$$= \sum_{n=0}^{\infty} e^{-nTs}$$

$$= 1 + e^{-Ts} + e^{-2Ts} + \cdots$$

$$= \frac{1}{1-e^{-Ts}}, \quad \text{provided that } |e^{-Ts}| < 1 \tag{3.28}$$

Hence from (3.27),

$$F^*(s) = \frac{1}{2\pi j}\int_{c-j\infty}^{c+j\infty} F(p)\frac{1}{1-e^{-T(s-p)}}dp \tag{3.29}$$

To evaluate this as a contour integral in the p-plane, we need to find the zeros of $1 - e^{-T(s-p)}$.

$$1 - e^{-T(s-p)} = 0 \tag{3.30}$$

$$e^{-T(s-p)} = e^{j2\ell\pi}, \quad \ell = 0, \pm 1, \pm 2, \ldots \tag{3.31}$$

Therefore,

$$p = s + j\ell\omega_s \tag{3.32}$$

This gives the location of the poles of $1/(1 - e^{-T(s-p)})$ in the p-plane.

Suppose the poles of $F(p)$ are in the left half-plane with σ_2 the real part of the pole nearest the imaginary axis. Then by selecting $\sigma > \sigma_2$, we can draw a counterclockwise contour through c where $\sigma_2 < c < \sigma$ so that it encloses all the poles of $F(p)$ and excludes the poles of $1/(1 - e^{-T(s-p)})$, as shown in Figure 3.5. Thus to evaluate the integral (3.29), it suffices to evaluate the residues of the integrand *at the poles of* $F(p)$ *alone*.

$$F^*(s) = \sum \text{residues of } F(p)\frac{1}{1-e^{-T(s-p)}} \text{ at the poles of } F(p) \tag{3.33}$$

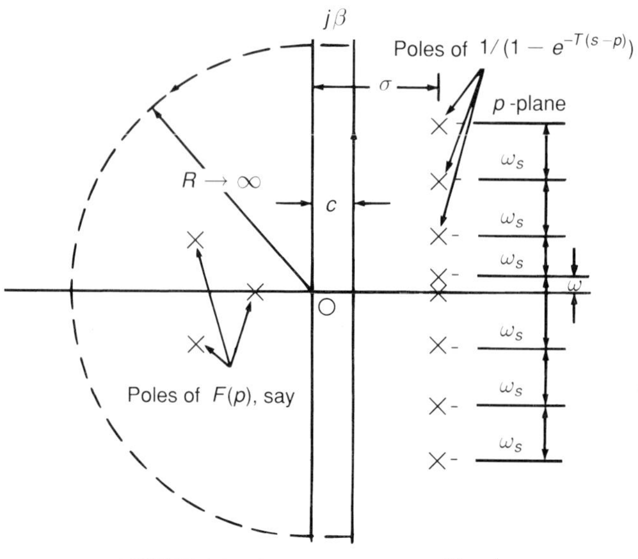

FIGURE 3.5 Poles of $F(p)/[1 - e^{T(s-p)}]$.

3.5 THE z-TRANSFORM

The three methods we examined in Section 3.4 gave three different expressions for the Laplace transform of the sampled signal $f^*(t)$. They are

1. $$F^*(s) = \sum_{k=0}^{\infty} f(kT)e^{-kTs} \tag{3.34}$$

2. $$F^*(s) = \frac{1}{T}\sum_{n=-\infty}^{\infty} F(s + jn\omega_s) \tag{3.35}$$

3. $F^*(s) = \sum$ residues of
$$F(p)\frac{1}{1 - e^{-T(s-p)}} \text{ at the poles of } F(p) \tag{3.36}$$

These expressions are all related and are useful in different circumstances, as we will see later.

In continuous linear systems theory, one of the advantages of the Laplace transform is that it reduces a differential equation to an algebraic equation. This advantage appears to be lost for discrete systems because of the transcendental term e^{Ts} in (3.34) and (3.36). Therefore, we make an additional transformation to put this term in a simpler form. Define a new complex variable z, where

$$z = e^{Ts} \quad \text{or} \quad s = \frac{1}{T} \ln z$$

If we substitute for e^{sT} in (3.34) and (3.36), we get two different expressions for what is called the *z-transform* of $f(t)$—that is, $F(z)$. The whole procedure can be summarized as follows:

$$\begin{array}{ll}
f(t) & \text{continuous function} \\
\downarrow & \text{(Sample.)} \\
f^*(t) & \text{sampled function} \\
\downarrow & \text{(Take the Laplace transform.)} \\
F^*(s) & \\
\downarrow & \text{(Let } z = e^{Ts}.\text{)} \\
F(z) & z\text{-transform of } f(kT)
\end{array}$$

Thus, from (3.34) and (3.36), respectively, the z-transform may be written either as

$$F(z) = \sum_{k=0}^{\infty} f(kT) z^{-k} \tag{3.37}$$

or as

$$F(z) = \sum \text{residues of } F(p) \frac{1}{1 - e^{Tp} z^{-1}} \text{ at the poles of } F(p) \tag{3.38}$$

Many writers *define* the $F(z)$ given in (3.37) as the z-transform of the sequence $f(kT)$, regardless of whether the latter is obtained by sampling a continuous function $f(t)$, as here, or is inherently discrete. It is also sometimes referred to as the "one-sided" z-transform to distinguish it from the "two-sided" z-transform defined by

$$F(z) = \sum_{k=-\infty}^{\infty} f(kT) z^{-k} \tag{3.39}$$

The two-sided form is more general in that the choice of the time origin, $k = 0$, is arbitrary. Thus the one-sided version is a special case of (3.39) if $f(kT) = 0$ for $k < 0$. In this text, the two forms are used interchangeably, depending on the nature of the discrete sequence being transformed. If $f(kT) = 0$ for $k < 0$, the one-sided form (3.37) is used; otherwise, the two-sided form (3.39) is understood.

Form (3.38) is useful when we wish to transform a frequency function $F(s)$ for a continuous system into a frequency function for a discrete system. In this context, $F(z)$ as given by (3.38) is understood to be the z-transform of the sequence $f(kT)$ resulting from sampling the continuous function $f(t)$

the Laplace transform of which is $F(s)$. For convenience, rather than selecting new symbols, we use $F(s)$ and $F(z)$ for the Laplace transform and z-transform of $f(t)$ and $f(k)$, respectively. It is understood that apart from having a different argument, $F(z)$ is an entirely different function from $F(s)$.

An example will illustrate the use of the two different expressions (3.37) and (3.38) for the z-transform.

EXAMPLE 3.1

Use the two different expressions (3.37) and (3.38) to get the z-transforms of the sequences obtained by sampling (with period T) the following continuous functions:

a) Unit ramp $\quad f(t) = tu(t)$

b) Exponential $\quad f(t) = e^{-at}u(t)$

Solution

a) For the unit ramp, replacing t by kT yields

$$f(kT) = kT, \qquad k = 0, 1, 2, \ldots$$

Therefore, by (3.37),

$$\begin{aligned} F(z) &= \sum_{k=0}^{\infty} kTz^{-k} \\ &= Tz^{-1}(1 + 2z^{-1} + 3z^{-2} + \cdots + kz^{-(k-1)} + \cdots) \\ &= \frac{Tz^{-1}}{(1 - z^{-1})^2} \end{aligned}$$

provided that $|z^{-1}| < 1$ or, if $z = re^{j\theta}$, $|z| = r > 1$.

To use (3.38), we must first find the Laplace transform of $f(t)$. In this case, $F(s) = 1/s^2$. Therefore, by (3.38),

$$\begin{aligned} F(z) &= \sum \text{residues of } \frac{1}{p^2}\frac{1}{1 - e^{TP}z^{-1}} \text{ at pole } p = 0 \text{ (double)} \\ &= \frac{d}{dp}\left\{p^2\left[\frac{1}{p^2}\frac{1}{1 - e^{TP}z^{-1}}\right]\right\}\Bigg|_{p=0} \\ &= \frac{Te^{TP}z^{-1}}{(1 - e^{TP}z^{-1})^2}\Bigg|_{p=0} \\ &= \frac{Tz^{-1}}{(1 - z^{-1})^2} \end{aligned}$$

As before, we note that with the first method, it is necessary to recognize the closed form of the infinite series; with the second method, the closed form is given directly.

If we write

$$F(z) = \frac{Tz}{(z-1)^2}$$

we see that it goes to zero at $z = 0$ and becomes infinite at $z = 1$. That is, it has a *zero* at $z = 0$ and a double *pole* at $z = 1$.

b) For the exponential,

$$f(kT) = e^{-akT}, \quad k = 0, 1, 2, \ldots$$

Therefore, by (3.37),

$$\begin{aligned} F(z) &= \sum_{k=0}^{\infty} e^{-akT} z^{-k} \\ &= 1 + e^{-aT} z^{-1} + (e^{-aT} z^{-1})^2 + (e^{-aT} z^{-1})^3 + \ldots \\ &= \frac{1}{1 - e^{-aT} z^{-1}} \end{aligned}$$

provided that $|e^{-aT} z^{-1}| < 1$—that is, provided that $|z| = r > e^{-aT}$.

For the second method, $F(s) = 1/(s+a)$. Then (3.38) gives

$$\begin{aligned} F(z) &= \sum \text{ residues of } \frac{1}{p+a} \frac{1}{1 - e^{Tp} z^{-1}} \text{ at pole } p = -a \\ &= \left. \frac{1}{1 - e^{Tp} z^{-1}} \right|_{p=-a} \\ &= \frac{1}{1 - e^{-aT} z^{-1}} \end{aligned}$$

as before. The function has a zero at $z = 0$ and a pole at $z = e^{-aT}$.

This example illustrates some important points. For a more complex time function $f(t)$ and sequence $f(kT)$, it would be difficult to recognize the closed form of the infinite sequence resulting from (3.37), whereas if we know $F(s)$ in factored form, we have no difficulty applying (3.38). For example, if

$$F(s) = \frac{s+2}{(s+1)^2} \tag{3.40}$$

then, by (3.38),

$$F(z) = \frac{1 - (1-T)e^{-T} z^{-1}}{(1 - e^{-T} z^{-1})^2} \tag{3.41}$$

To obtain this closed solution by applying (3.37) to the sequence resulting from $f(t) = \mathcal{L}^{-1}\{F(s)\}$ would be very difficult.

The second point to note in Example 3.1 is that the infinite series

$$F(z) = \sum_{k=0}^{\infty} f(kT)z^{-k} \tag{3.42}$$

does not converge for all values of z. In part a) it converged for $|z| = r > 1$, and in part b) it converged for $r > e^{-aT}$. The part of the z-plane for which the series converges is called the region of convergence (ROC) for $F(z)$. For values of z outside this region, the z-transform *does not exist.* The ROC depends on the particular sequence. For the sequences of Example 3.1, the ROCs are shown in Figure 3.6. We see that $F(z)$ for part a) has a zero at the origin of the z-plane and a double pole on the unit circle ($r = 1$) at $z = 1$. The ROC is the open region of the z-plane that lies outside the unit circle. The poles of $F(z)$ must lie outside the ROC, because they are the values of z for which $F(z)$ "blows up." Inside the unit circle, $F(z)$ does not exist. For $F(z)$ of part b), there is a zero at the origin and a pole at $z = e^{-aT}$. The ROC is the open region outside a circle of radius e^{-aT}.

These results are typical of positive-time or causal sequences $[f(kT) = 0, k < 0]$ that can be evaluated from the one-sided z-transform (3.37). The ROC is a region outside a circle the radius of which is the distance of the furthest pole of $F(z)$ from the origin.

Now let us investigate the ROC of the z-transform for a *negative-time* sequence ($f(k) = 0, k > 0$). For simplicity, we will consider the negative step sequence.

EXAMPLE 3.2

Find the z-transform and the region of convergence for

$$f(k) = u(-k) = \begin{cases} 1, & k \le 0 \\ 0, & k > 0 \end{cases}$$

Solution

In this case, we must use the two-sided z-transform (3.39).

$$\begin{aligned} F(z) &= \sum_{k=-\infty}^{\infty} f(k)z^{-k} \\ &= \sum_{k=-\infty}^{0} z^{-k} \end{aligned}$$

Let $\ell = -k$. Then,

$$\begin{aligned} F(z) &= \sum_{\ell=0}^{\infty} z^{\ell} \\ &= \frac{1}{1-z} \end{aligned}$$

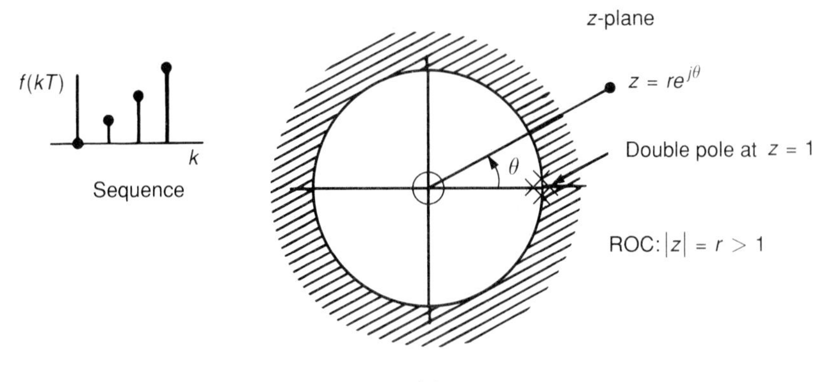

(a)

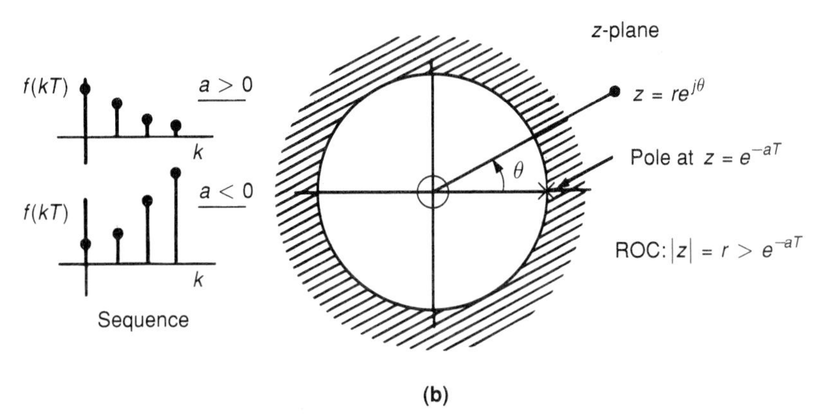

(b)

FIGURE 3.6 Regions of convergence for Example 3.1. (a) $F(z) = \frac{Tz}{(z-1)^2}$. (b) $F(z) = \frac{z}{z-e^{-aT}}$.

provided that $|z| = r < 1$. This function has a pole at $z = 1$, and the ROC is the *interior* of the unit circle (Figure 3.7). (By contrast, the z-transform for the unit step has a pole at $z = 1$, a zero at the origin, and a region of convergence that is the *exterior* of the unit circle.)

The result for this example is typical for negative-time sequences. The z-transform for such a sequence has an ROC that consists of the region of the z-plane that lies inside a circle the radius of which is the distance of the *nearest* pole of $F(z)$ from the origin. The nature of the ROC for the z-transform of a sequence that has non-zero values for both positive time and negative time is brought out in the following example.

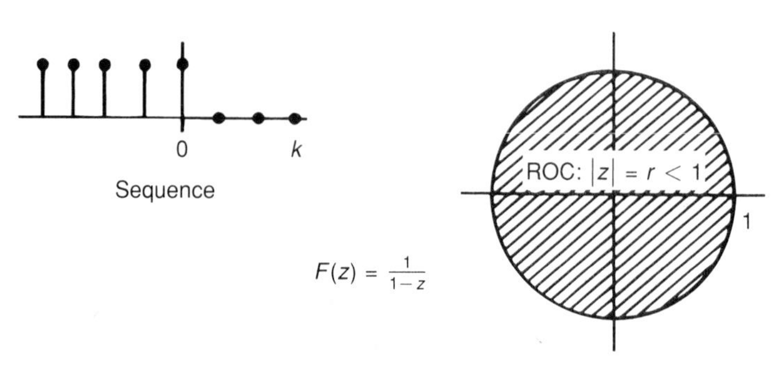

FIGURE 3.7 Region of convergence for negative step sequence.

EXAMPLE 3.3

As an example of a two-sided sequence, consider finding the z-transform and the region of convergence for

$$f(k) = a^{|k|}$$

where a is real and $|a| < 1$.

Solution

The z-transform of this sequence is

$$\begin{aligned} F(z) &= \sum_{k=-\infty}^{\infty} a^{|k|} z^{-k} \\ &= \sum_{k=-\infty}^{-1} a^{-k} z^{-k} + \sum_{k=0}^{\infty} a^k z^{-k} \\ &= \sum_{\ell=0}^{\infty} a^{\ell} z^{\ell} - 1 + \sum_{k=0}^{\infty} a^k z^{-k} \end{aligned}$$

The subsequence $\sum_{\ell=0}^{\infty} a^{\ell} z^{\ell}$ converges to $1/(1-az)$, provided that $|z| < 1/|a|$. The second subsequence $\sum_{k=0}^{\infty} a^k z^{-k}$ converges to $1/(1 - az^{-1})$ for $|z| > |a|$. Therefore,

$$F(z) = \frac{az}{1 - az} + \frac{z}{z - a} = \frac{z(1 - a^2)}{(1 - az)(z - a)}$$

with the region of convergence given by $|a| < |z| < 1/|a|$, the intersection of the ROCs for the subsequences, as shown in Figure 3.8. Note that if

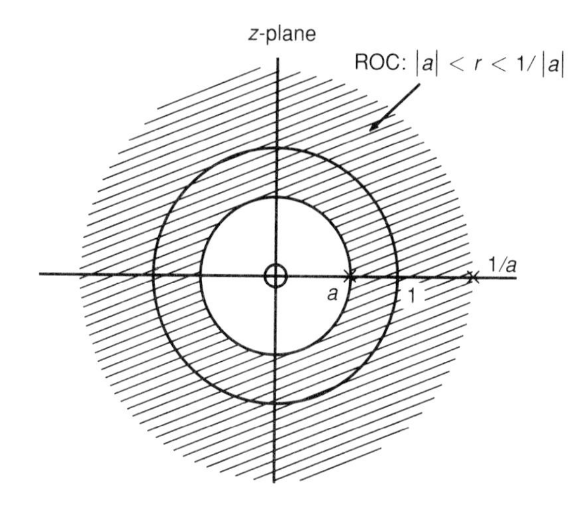

FIGURE 3.8 Region of convergence and singularities for the function $F(z) = z(1 - a^2)/[(1 - az)(z - a)]$.

$|a| > 1$ in this example, $F(z)$ does not converge anywhere in the z-plane; the z-transform does not exist.

The z-transform for a *finite* sequence where samples are finite-valued has no poles except at the origin and exists over the entire z-plane. In general, for an arbitrary sequence $f(k)$ the z-transform of which exists, the region of convergence is $R_- < |z| < R_+$, where R_+ may be infinite and R_- may be zero. A z-transform is not fully specified unless its associated region of convergence is given. In this book, when no ROC is specified, it is understood that we are dealing with the z-transform of a *positive-time* sequence.

3.6 PROPERTIES OF THE z-TRANSFORM

We will now discuss some important properties of the z-transform that make it a powerful tool for the analysis and design of discrete LTI systems. Many of these properties are analogous to those of the Laplace and Fourier transforms. In general, we will work with the two-sided z-transform (3.39), because it includes the one-sided transform (3.37) if $f(k) = 0,\ k < 0$, and we will write it as $F(z) = \mathcal{Z}\{f(k)\}$. Where a specific property applies only to the one-sided transform, we will write $F(z) = \mathcal{Z}_1\{f(k)\}$. Let the ROC for $F(z)$ be $R_- < |z| < R_+$.

1. *Linearity*
   If $F_1(z) = \mathcal{Z}\{f_1(k)\}$ and $F_2(z) = \mathcal{Z}\{f_2(k)\}$, then

$$\mathcal{Z}\{\alpha_1 f_1(k) + \alpha_2 f_2(k)\} = \alpha_1 F_1(z) + \alpha_2 F_2(z) \tag{3.43}$$

Proof

$$\mathcal{Z}\{\alpha_1 f_1(k) + \alpha_2 f_2(k)\} = \sum_{k=-\infty}^{\infty} [\alpha_1 f_1(k) + \alpha_2 f_2(k)] z^{-k}$$

$$= \alpha_1 \sum_{k=-\infty}^{\infty} f_1(k) z^{-k} + \alpha_2 \sum_{k=-\infty}^{\infty} f_2(k) z^{-k}$$

$$= \alpha_1 F_1(z) + \alpha_2 F_2(z)$$

Q.E.D.

2. *Time shift or translation*

a) If $F(z) = \mathcal{Z}\{f(k)\}$ and the initial conditions for $f(k)$ are zero, then

$$\mathcal{Z}\{f(k-m)\} = z^{-m} F(z) \tag{3.44}$$

where m is a positive or a negative integer.

Proof

$$\mathcal{Z}\{f(k-m)\} = \sum_{k=-\infty}^{\infty} f(k-m) z^{-k}$$

$$= z^{-m} \sum_{k=-\infty}^{\infty} f(k-m) z^{-(k-m)}$$

$$= z^{-m} \sum_{\ell=-\infty}^{\infty} f(l) z^{-\ell}, \quad \ell \triangleq k - m$$

$$= z^{-m} F(z)$$

Q.E.D.

b) If $F(z) = \mathcal{Z}_1\{f(k)\}$, then

i) $$\mathcal{Z}_1\{f(k-m)\} = z^{-m}\{F(z) + \sum_{p=1}^{m} f(-p) z^{p}\} \tag{3.45}$$

ii) $$\mathcal{Z}_1\{f(k+m)\} = z^{m}\{F(z) - \sum_{p=0}^{m-1} f(p) z^{-p}\} \tag{3.46}$$

where m is a positive integer in both cases.

Proof

$$\mathcal{Z}_1\{f(k-m)\} = \sum_{k=0}^{\infty} f(k-m)z^{-k}$$

$$= z^{-m}\sum_{k=0}^{\infty} f(k-m)z^{-(k-m)}$$

$$= z^{-m}\sum_{\ell=-m}^{\infty} f(\ell)z^{-\ell}, \quad \ell \triangleq (k-m)$$

$$= z^{-m}\left\{\sum_{\ell=0}^{\infty} f(\ell)z^{-\ell} + \sum_{\ell=-m}^{-1} f(\ell)z^{-\ell}\right\}$$

$$= z^{-m}\left\{F(z) + \sum_{p=1}^{m} f(-p)z^{p}\right\}, \quad p \triangleq -\ell$$

Q.E.D.

The proof of (3.46) is similar and is left as an exercise. Property 2b) is very useful for solving linear difference equations with constant coefficients, as we shall soon see.

3. *Multiplication by* a^k
   If $F(z) = \mathcal{Z}\{f(k)\}$, then

$$\mathcal{Z}\{a^k f(k)\} = F(a^{-1}z) \tag{3.47}$$

where a may be a real or a complex number.

Proof

$$\mathcal{Z}\{a^k f(k)\} = \sum_{k=-\infty}^{\infty} a^k F(k)z^{-k}$$

$$= \sum_{k=-\infty}^{\infty} f(k)(a^{-1}z)^{-k}$$

$$= F(a^{-1}z)$$

Q.E.D.

where the ROC is $|a|R_- < |z| < |a|R_+$.

4. *Time reversal*
   If $F(z) = \mathcal{Z}\{f(k)\}$, then

$$\mathcal{Z}\{f(-k)\} = F(z^{-1}) \tag{3.48}$$

Proof

$$
\begin{aligned}
\mathcal{Z}\{f(-k)\} &= \sum_{k=-\infty}^{\infty} f(-k)z^{-k} \\
&= \sum_{\ell=-\infty}^{\infty} f(\ell)z^{\ell} \qquad \ell \triangleq -k \\
&= \sum_{\ell=-\infty}^{\infty} f(\ell)(z^{-1})^{-\ell} \\
&= F(z^{-1}) \qquad \text{Q.E.D.}
\end{aligned}
$$

If the ROC for $F(z)$ is $R_- < |z| < R_+$, the ROC for $F(z^{-1})$ is $1/R_+ < |z| < 1/R_-$.

5. *Multiplication by time index*
   If $F(z) = \mathcal{Z}\{f(k)\}$, then

$$
\mathcal{Z}\{kf(k)\} = -z\frac{dF(z)}{dz} \tag{3.49}
$$

Proof

$$
\begin{aligned}
-z\frac{dF(z)}{dz} &= -z\frac{d}{dz}\left[\sum_{k=-\infty}^{\infty} f(k)z^{-k}\right] \\
&= -z\sum_{k=-\infty}^{\infty} f(k)\frac{d}{dz}(z^{-k}) \\
&= -z\sum_{k=-\infty}^{\infty} f(k)(-k)z^{-k-1} \\
&= \sum_{k=-\infty}^{\infty} kf(k)z^{-k} \\
&= \mathcal{Z}\{kf(k)\} \qquad \text{Q.E.D.}
\end{aligned}
$$

6. *Convolution theorem*
   If $F(z) = \mathcal{Z}\{f(k)\}$ and $G(z) = \mathcal{Z}\{g(k)\}$, then

$$
\mathcal{Z}\{f(k) * g(k)\} = F(z)G(z) \tag{3.50}
$$

where $f(k) * g(k)$ denotes the linear convolution of sequences f and g (see Chapter 1).

Proof

Let

$$y(k) = f(k) * g(k) = \sum_{n=-\infty}^{\infty} f(n)g(k-n)$$

Therefore,

$$Y(z) = \mathcal{Z}\{y(k)\} = \sum_{k=-\infty}^{\infty} \left[\sum_{n=-\infty}^{\infty} f(n)g(k-n) \right] z^{-k}$$

Interchange the order of the summations, assuming both summations converge.

$$Y(z) = \sum_{n=-\infty}^{\infty} f(n) \sum_{k=-\infty}^{\infty} g(k-n)z^{-k}$$

$$= \sum_{n=-\infty}^{\infty} f(n)z^{-n} \sum_{k=-\infty}^{\infty} g(k-n)z^{-(k-n)}$$

Replace $(k-n)$ by ℓ in the second summation. Then

$$Y(z) = \sum_{n=-\infty}^{\infty} f(n)z^{-n} \sum_{\ell=-\infty}^{\infty} g(\ell)z^{-\ell}$$

$$= F(z)G(z) \qquad \text{Q.E.D.}$$

The ROC for $Y(z)$ is the intersection of the ROCs for $F(z)$ and $G(z)$.

For a discrete LTI system, the output sequence $y(k)$ is related to the input $x(k)$ and the unit pulse response $h(k)$ through the convolution summation

$$y(k) = h(k) * x(k) \tag{3.51}$$

From the theorem, it follows that

$$Y(z) = H(z)X(z) \tag{3.52}$$

or

$$H(z) = Y(z)/X(z) \tag{3.53}$$

where $H(z) = \mathcal{Z}\{h(k)\}$ is the system transfer function (sometimes called the pulse transfer function).

7. *Complex convolution theorem*
The z-transform of the product of two sequences is related to the z-transforms of the individual sequences through the complex convolution theorem. This theorem states that if

$$x_3(n) = x_1(n)x_2(n) \tag{3.54}$$

then

$$X_3(z) = \frac{1}{2\pi j} \oint_C X_2(\nu) X_1\left(\frac{z}{\nu}\right) \nu^{-1} d\nu \tag{3.55}$$

The convergence region for $X_3(z)$ consists of all z such that if ν is in the region of convergence for $X_2(z)$, then z/ν is in the region of convergence for $X_1(z)$. The contour of integration C for (3.55) is a closed contour inside the intersection of the convergence regions for $X_2(\nu)$ and $X_1(z/\nu)$.

Proof

$$\begin{aligned} X_3(z) &= \sum_{n=-\infty}^{\infty} x_3(n) z^{-n} \\ &= \sum_{n=-\infty}^{\infty} x_1(n) x_2(n) z^{-n} \end{aligned}$$

Now

$$x_2(n) = \frac{1}{2\pi j} \oint_C X_2(\nu) \nu^{n-1} d\nu$$

We ask the reader to accept this with blind faith until we have covered the inverse z-transform in the next section. Therefore,

$$X_3(z) = \sum_{n=-\infty}^{\infty} x_1(n) z^{-n} \frac{1}{2\pi j} \oint_C X_2(\nu) \nu^{n-1} d\nu$$

Interchange the order of summation and integration.

$$\begin{aligned} X_3(z) &= \frac{1}{2\pi j} \oint_C X_2(\nu) \sum_{n=-\infty}^{\infty} x_1(n) (z/\nu)^{-n} \nu^{-1} d\nu \\ &= \frac{1}{2\pi j} \oint_C X_2(\nu) X_1(z/\nu) \nu^{-1} d\nu \end{aligned}$$

Q.E.D.

If C is the unit circle, we can write integral (3.55) so that it looks more like a convolution. Then $\nu = e^{j\theta}$ and $d\nu = je^{j\theta} d\theta$. Take $z = re^{j\phi}$. Then $z/\nu = re^{j(\phi-\theta)}$, and (3.55) becomes

$$\begin{aligned} X_3(z) &= \frac{1}{2\pi j} \oint_C X_2(e^{j\theta}) X_1(re^{j(\phi-\theta)}) e^{-j\theta} j e^{j\theta} d\theta \\ &= \frac{1}{2\pi} \int_{-\pi}^{\pi} X_2(e^{j\theta}) X_1(re^{j(\phi-\theta)}) d\theta \end{aligned} \tag{3.56}$$

Consider also the following *special case*. If $x_1(n) = x_2(n)$ and $z = 1$, then

$$X_3(1) = \sum_{n=-\infty}^{\infty} x_1^2(n)$$

$$= \frac{1}{2\pi j} \oint_C X_1(\nu) X_1(1/\nu) \nu^{-1} d\nu \tag{3.57}$$

Thus the sum of the squares, or the energy in the sequence, can be evaluated in terms of the z-transform of the sequence. This is *Parseval's relation* for a sequence.

8. *Initial-value theorem*
   If $F(z) = \mathcal{Z}_1\{f(k)\}$, then

$$f(0) = \lim_{z \to \infty} F(z) \tag{3.58}$$

Proof

$$F(z) = \sum_{k=0}^{\infty} f(k) z^{-k}$$

As $z \to \infty$, all the terms vanish except $f(0)$, which proves the theorem.

9. *Final-value theorem*
   If $F(z) = \mathcal{Z}_1\{f(k)\}$, where the ROC for $F(z)$ includes, but is not necessarily confined to, $|z| > 1$, and $(z-1)F(z)$ has no poles on or outside the unit circle, then

$$f(\infty) = \lim_{z \to 1} (z-1) F(z) \tag{3.59}$$

Proof

$$\mathcal{Z}_1\{f(k+1)\} - \mathcal{Z}_1\{f(k)\} = \lim_{n \to \infty} \sum_{k=0}^{n} [f(k+1) - f(k)] z^{-k}$$

From (3.46) with $m = 1$, we have

$$zF(z) - f(0) - F(z) = \lim_{n \to \infty} \sum_{k=0}^{n} [f(k+1) - f(k)] z^{-k}$$

Now let $z \to 1$.

$$\lim_{z \to 1} (z-1)F(z) - f(0) = \lim_{n \to \infty} \{[f(1) - f(0)] + [f(2) - f(1)] + \cdots + [f(n) - f(n-1)]\}$$

$$= f(\infty) - f(0)$$

Therefore,

$$f(\infty) = \lim_{z\to 1}(z-1)F(z) \qquad \text{Q.E.D.}$$

The conditions on $F(z)$ ensure that $f(k)$ is bounded and that the limit in (3.59) exists. Expression (3.59) can also be written

$$f(\infty) = \lim_{z\to 1}(1-z^{-1})F(z) \tag{3.60}$$

This theorem enables us to find the steady-state value of $f(k)$ without solving for the entire sequence. An example will illustrate this.

EXAMPLE 3.4

If

$$Y(z) = \frac{0.125(1-0.2z^{-1})}{(1-0.5z^{-1})(1-0.8z^{-1})(1-z^{-1})}$$

Find the steady-state value of $y(k)$ if it exists.

Solution

By inspection, we can see that the ROC for $Y(z)$ is $|z| > 1$ and that $(1-z^{-1})$. $Y(z)$ has no poles on or outside the unit circle, so the conditions of the final-value theorem are satisfied. Therefore, the steady-state value exists and is given by

$$\begin{aligned} y(\infty) &= \lim_{z\to 1}(1-z^{-1})Y(z) \\ &= \frac{0.125(1-0.2)}{(1-0.5)(1-0.8)} \\ &= 1 \end{aligned}$$

Several exercises on the above properties are given at the end of the chapter. In general, we will be working with causal or positive-time sequences, so we will be concerned mainly with the one-sided z-transform. Using (3.37) and the z-transform properties, we can generate a list of transforms for sequences that often arise. Table 3.1 is such a list.

3.7 THE INVERSE *z*-TRANSFORM

If the z-transform $F(z)$ is known, it is often desirable to find its inverse—that is, the sequence $f(k) = f(kT)$. For brevity, we write

$$f(k) = \mathcal{Z}^{-1}\{F(z)\} \tag{3.61}$$

where $f(k)$ is called the *inverse z-transform* of $F(z)$. Several methods are

TABLE 3.1 z-Transforms of Some Basic Sequences

Sequence	$f(kT)$, $k \geq 0$	$F(z)$
1. Unit pulse	$\delta(k)$	1
2. Unit step	$u(kT)$	$\frac{z}{z-1}$
3. Unit ramp	kT	$\frac{Tz}{(z-1)^2}$
4. Exponential	e^{-akT}	$\frac{z}{z-e^{-aT}}$
5. Power	a^k	$\frac{z}{z-a}$
6. Sinusoid	$\sin\omega_0 kT$	$\frac{z\sin\omega_0 T}{z^2-2z\cos\omega_0 T+1}$
7. Cosinusoid	$\cos\omega_0 kT$	$\frac{z(z-\cos\omega_0 T)}{z^2-2z\cos\omega_0 T+1}$
8. Damped sinusoid	$e^{-akT}\sin\omega_0 kT$	$\frac{ze^{-aT}\sin\omega_0 T}{z^2-2ze^{-aT}\cos\omega_0 T+e^{-2aT}}$
9. Damped cosinusoid	$e^{-akT}\cos\omega_0 kT$	$\frac{z^2-ze^{-aT}\cos\omega_0 T}{z^2-2ze^{-aT}\cos\omega_0 T+e^{-2aT}}$

available to find the inverse transform. We will discuss the three principal ones: long division, expansion in partial fractions, and the residue method.

In all cases, a region of convergence must be given, or understood, for $F(z)$. Recall from Section 3.5 that the ROC indicates the type of sequence, as shown in Table 3.2. If no ROC is explicitly specified, it is understood that a positive-time sequence is intended.

3.7.1 Long-Division Method

We express $F(z)$ as the ratio of two polynomials.

$$F(z) = \frac{B(z)}{A(z)} \tag{3.62}$$

TABLE 3.2 Region of Convergence for Different Types of Sequences

Sequence	ROC	Remarks
Positive-time	$\lvert z\rvert > r_{max}$	r_{max} is the distance of the furthest pole of $F(z)$ from the origin.
Negative-time	$\lvert z\rvert < r_{min}$	r_{min} is the distance of the nearest pole of $F(z)$ to the origin.
Both	$r_{min} < \lvert z\rvert < r_{max}$	r_{min} and r_{max} are as described above.

Divide $A(z)$ into $B(z)$ to obtain a series of negative powers of z if a positive-time sequence is indicated by the ROC. If a negative-time function is indicated, we express $F(z)$ as a series of positive powers of z. The method will not work for a sequence defined for both positive and negative time. When we have the series, we simply identify $f(k)$ with the coefficient of z^{-k} for a positive-time sequence and with the coefficient of z^k for a negative-time sequence. An example will illustrate the procedure.

EXAMPLE 3.5

a) Find the inverse z-transform of

$$F(z) = \frac{z + 0.2}{(z + 0.5)(z - 1)}, \quad |z| > 1$$

b) Find the inverse z-transform of

$$F(z) = \frac{z}{(z - 3)(z - 4)}, \quad |z| < 3$$

Solution

a) From the ROC, $f(k)$ is a positive-time sequence. Express $F(z)$ as

$$F(z) = \frac{z + 0.2}{z^2 - 0.5z - 0.5}$$

Divide denominator into numerator to get

$$\begin{aligned} F(z) &= z^{-1} + 0.7z^{-2} + 0.85z^{-3} + 0.775z^{-4} + \cdots \\ &= \sum_{k=0}^{\infty} f(k)z^{-k} \end{aligned}$$

Therefore,

$$f(0) = 0, f(1) = 1, f(2) = 0.7, f(3) = 0.85, \text{ and so on.}$$

b) From the ROC, $f(k)$ is a negative-time sequence. Write $F(z)$ as

$$\begin{aligned} F(z) &= \frac{z}{12 - 7z + z^2} \\ &= \frac{1}{12}z + \frac{7}{144}z^2 + \frac{37}{1728}z^3 + \cdots \\ &= \sum_{k=-\infty}^{0} f(k)z^{-k} \end{aligned}$$

Therefore,

$$f(0) = 0, f(-1) = \frac{1}{12}, f(-2) = \frac{7}{144}, f(-3) = \frac{37}{1728}, \text{ and so on.}$$

We note that it would be difficult, in general, to obtain a closed-form expression for the general term $f(k)$ via the long-division method. However, a computer program can be written to provide as many values of $f(k)$ as required. The next two methods do provide a closed-form solution for $f(k)$.

3.7.2 Expansion in Partial Fractions

In this case, we factor the denominator of $F(z)$, if it is not already in factored form, and expand $F(z)$ in partial fractions as described for the inverse Laplace transform in Section 2.2.3. We can then obtain from the tables the inverse transforms of the individual members of the partial-fraction expansion. Finally, we add the results to get $f(k)$. We will illustrate this with the functions of Example 3.5, in Examples 3.6 and 3.7, and with a two-sided sequence in Example 3.8.

EXAMPLE 3.6

Say

$$F(z) = \frac{z + 0.2}{(z + 0.5)(z - 1)}, \quad |z| > 1$$

Find $f(k)$ using a partial-fraction expansion.

Solution

Expand $F(z)$.

$$F(z) = \frac{R_1}{z + 0.5} + \frac{R_2}{z - 1}$$

where [see (2.70)]

$$R_1 = (z + 0.5)F(z)|_{z=-0.5} = 0.2$$

$$R_2 = (z - 1)F(z)|_{z=1} = 0.8$$

Therefore,

$$F(z) = \frac{0.2}{z + 0.5} + \frac{0.8}{z - 1}$$

If there were a z in the numerators of the terms on the right-hand side, we could read the inverse transforms of the terms directly from Table 3.1. To put them in this form, multiply through by z, obtaining

$$zF(z) = \frac{0.2z}{z + 0.5} + \frac{0.8z}{z - 1}$$

We can use property (3.46), with $m = 1$ and zero initial conditions, to get the inverse transforms of the left-hand side and the closed-form solution.

$$f(k+1) = \begin{cases} 0.2(-0.5)^k + 0.8, & k \geq 0 \\ 0, & k < 0 \end{cases}$$

Thus $f(0) = 0, f(1) = 1, f(2) = 0.7, f(3) = 0.85, f(4) = 0.775$, as before.

EXAMPLE 3.7

Find the inverse z-transform of

$$F(z) = \frac{z}{(z-3)(z-4)}, \quad |z| < 3$$

by partial fraction expansion.

Solution

Expand $F(z)$ in partial fractions.

$$F(z) = \frac{R_1}{z-3} + \frac{R_2}{z-4}$$

where

$$R_1 = (z-3)F(z)\big|_{z=3} = -3$$

$$R_2 = (z-4)F(z)\big|_{z=4} = 4$$

Therefore,

$$\begin{aligned} F(z) &= \frac{-3}{z-3} + \frac{4}{z-4} \\ &= \frac{1}{1 - 1/3z} - \frac{1}{1 - 1/4z} \\ &= \frac{z^{-1}}{z^{-1} - 1/3} - \frac{z^{-1}}{z^{-1} - 1/4} \end{aligned}$$

Because $\mathcal{Z}\{f(-k)\} = F(z^{-1})$ by the time reversal property (3.48), we have

$$\begin{aligned} f(k) &= \mathcal{Z}^{-1}\left\{\frac{z^{-1}}{z^{-1} - 1/3}\right\} - \mathcal{Z}^{-1}\left\{\frac{z^{-1}}{z^{-1} - 1/4}\right\} \\ &= \left(\frac{1}{3}\right)^{-k} - \left(\frac{1}{4}\right)^{-k} \\ &= 3^k - 4^k, \quad k \leq 0 \end{aligned}$$

Therefore, $f(0) = 0, f(-1) = 1/12, f(-2) = 7/144, f(-3) = 37/1728$, as before.

EXAMPLE 3.8

Find the inverse z-transform of

$$F(z) = \frac{z}{(z-0.5)(z-2)}, \quad 0.5 < |z| < 2$$

by partial-fraction expansion.

Solution

From the ROC, we see that $f(k)$ is non-zero for positive and negative k. Expand $F(z)$ in partial fractions.

$$F(z) = \frac{R_1}{z-0.5} + \frac{R_2}{z-2}$$

where

$$R_1 = \left.\frac{z}{z-2}\right|_{z=0.5} = -\frac{1}{3}$$

$$R_2 = \left.\frac{z}{z-0.5}\right|_{z=2} = \frac{4}{3}$$

Therefore,

$$F(z) = -\frac{1}{3(z-0.5)} + \frac{4}{3(z-2)}$$
$$= F_+(z) + F_-(z)$$

For the positive-time part of the sequence,

$$zF_+(z) = -\frac{z}{3(z-0.5)}$$

Therefore,

$$f(k+1) = -\frac{1}{3}(0.5)^k = -\frac{2}{3}(0.5)^{k+1}, \quad k \geq 0$$

or

$$f(k) = -\frac{2}{3}(0.5)^k, \quad k > 0$$

The first few terms are

$$f(1) = -\frac{1}{3}, f(2) = -\frac{1}{6}, f(3) = -\frac{1}{12}$$

For the negative-time part of the sequence,

$$F_-(z) = \frac{4}{3(z-2)} = -\frac{2z^{-1}}{3(z^{-1}-0.5)}$$

Therefore,

$$f(k) = -\frac{2}{3}(0.5)^{-k}$$
$$= -\frac{2}{3}2^k, \quad k \leq 0$$

If we combine the positive-time and negative-time parts, we get the sequence as

$$f(k) = \begin{cases} -\frac{2}{3}(0.5)^k, & k > 0 \\ -\frac{2}{3}2^k, & k \leq 0 \end{cases}$$

3.7.3 Residue Method

To *develop* the formula for this method requires a little more background in complex-variable theory than we needed before. However, to *use* the formula, the reader needs merely to know how to find the residues of a function as described in Chapter 2. With this in mind, we will derive the necessary expression.

The z-transform

$$F(z) = \sum_{n=-\infty}^{\infty} f(n)z^{-n} \tag{3.63}$$

is in the form of a Laurent series [1]. Hence both sides of (3.63) may be multiplied by z^{k-1} and integrated with respect to z about a closed contour C in the region of convergence of $F(z)$. Thus

$$\oint_C F(z)z^{k-1}dz = \oint_C \sum_{n=-\infty}^{\infty} f(n)z^{k-n-1}dz$$
$$= \sum_{n=-\infty}^{\infty} f(n) \oint_C z^{k-n-1}dz \tag{3.64}$$

It is permissible to interchange the order of summation and integration in (3.64), because C lies in the region of convergence of $F(z)$.

Assume that C encloses the origin of the z-plane. Then, by a theorem due to Cauchy,

$$\oint_C z^{k-n-1}dz = 2\pi j\delta_{kn} \tag{3.65}$$

where

$$\delta_{kn} = \begin{cases} 1, & \text{if } k = n \\ 0, & \text{if } k \neq n \end{cases} \tag{3.66}$$

Hence

$$\oint_C F(z)z^{k-1}dz = 2\pi j \sum_{n=-\infty}^{\infty} f(nT)\delta_{kn} \tag{3.67}$$

or

$$f(k) = \frac{1}{2\pi j}\oint_C F(z)z^{k-1}dz \tag{3.68}$$

$$= \sum \text{residues of } F(z)z^{k-1} \text{ at poles of } F(z)z^{k-1} \text{ within } C \tag{3.69}$$

by Cauchy's integral formula, provided that $F(z)$ has isolated singularities, which is generally true in practice.

EXAMPLE 3.9

a) Use the residue method to find the inverse z-transform of

$$F(z) = \frac{z+0.2}{(z+0.5)(z-1)}, \quad |z| > 1$$

b) Use the residue method to find the inverse z-transform of

$$F(z) = \frac{z}{(z-3)(z-4)}, \quad |z| < 3$$

c) Use the residue method to find the inverse z-transform of

$$F(z) = \frac{z}{(z-0.5)(z-2)}, \quad 0.5 < |z| < 2$$

Solution

a) From (3.69),

$$f(k) = \sum \text{residues of } \frac{z^{k-1}(z+0.2)}{(z+0.5)(z-1)} \text{ at poles of same within } C$$

The closed contour C, being in the ROC $|z| > 1$, encloses the poles at $z = -0.5$ and $z = 1$ and, for $k = 0$, the pole at $z = 0$. Therefore, for $k = 0$,

$$f(0) = \left.\frac{z+0.2}{(z+0.5)(z-1)}\right|_{z=0} + \left.\frac{z+0.2}{z(z-1)}\right|_{z=-0.5} + \left.\frac{z+0.2}{z(z+0.5)}\right|_{z=1}$$

$$= -0.4 - 0.4 + 0.8$$

$$= 0$$

For $k \geq 1$,

$$f(k) = \left.\frac{z^{k-1}(z+0.2)}{z-1}\right|_{z=-0.5} + \left.\frac{z^{k-1}(z+0.2)}{z+0.5}\right|_{z=1}$$

$$= 0.2(-0.5)^{k-1} + 0.8$$

which is the same result we obtained in Example 3.6.

b) Here there are two poles, $z = 3$ and $z = 4$, *outside* the ROC $|z| < 3$. Therefore, for $k \le 0$,

$$f(k) = -\sum \text{residues of } F(z)z^{k-1} \text{ at poles } z = 3 \text{ and } z = 4$$

$$= -\left.\frac{z^k}{z-4}\right|_{z=3} - \left.\frac{z^k}{z-3}\right|_{z=4}$$

$$= 3^k - 4^k$$

as in Example 3.7.

(c) The contour of integration C lies in the doughnut-shaped ROC, and the inverse transform is

$$f(k) = \begin{cases} -\sum \text{residue of } F(z)z^{k-1} \text{ at pole } z = 2, & k < 0 \\ \sum \text{residue of } F(z)z^{k-1} \text{ at pole } z = 0.5, & k \ge 0 \end{cases}$$

$$= \begin{cases} -\dfrac{2}{3}2^k, & k \le 0 \\ -\dfrac{2}{3}(0.5)^k, & k \ge 0 \end{cases}$$

which is the same result that the partial-fraction method yielded in Example 3.8.

Comparing the three methods reveals that we get a closed-form solution with partial fractions and with the residue method, but not with long division. The residue method is easiest to apply when the underlying sequence $f(k)$ is either positive-time or negative-time. For a two-sided sequence the residue method and the partial-fraction expansion method are easy ways to proceed.

3.8 MAPPING FROM THE *s*-PLANE TO THE *z*-PLANE

In Section 3.5, we obtained the z-transform from the Laplace transform of the sampled function by the substitution $z = e^{Ts}$. This change of variable is a mapping from the s-plane to the z-plane. Let $z = x + jy = re^{j\theta}$ and $s = \sigma + j\omega$. It follows that

$$re^{j\theta} = e^{T\sigma}e^{j\omega T} \tag{3.70}$$

or

$$r = e^{T\sigma}, \quad \theta = \omega T = 2\pi\frac{\omega}{\omega_s} \tag{3.71}$$

Thus, as shown in Figure 3.9, the left half of the s-plane maps into the interior of the unit circle, the right half of the s-plane maps into the exterior

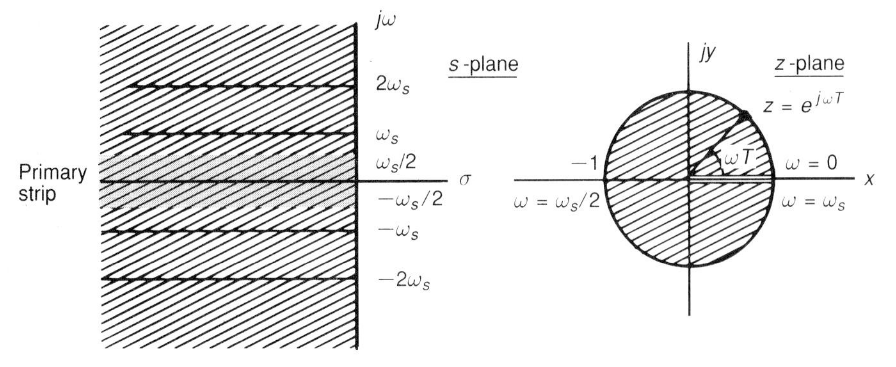

FIGURE 3.9 Mapping from the s-plane to the z-plane.

of the unit circle, and the imaginary axis ($\sigma = 0$) between is mapped onto the unit circle in the z-plane with $z = e^{j\omega T}$.

Note, however, that mapping a single horizontal strip of width ω_s of the left half s-plane (say, between $\omega = -\omega_s/2$ and $\omega = \omega_s/2$) completely fills the interior of the unit circle. It follows that the horizontal strips between $\ell\omega_s$ and $(\ell + 1)\omega_s, \ell = 0, \pm 1, \pm 2, \ldots$ are mapped on top of each other, or "aliased," in the interior of the unit circle. Thus we cannot distinguish between points $s = \sigma + j\omega$ and $\sigma + j(\omega + \ell\omega_s)$ when they are mapped into the z-plane. This is in agreement with our conclusions in Section 3.2, where we found that when we sample two sinusoids the frequencies of which differ by a multiple of the sampling frequency, we cannot distinguish between the results.

3.9 SOLVING DIFFERENCE EQUATIONS USING THE z-TRANSFORM

In Section 1.5, we used an inductive method to solve a first-order difference equation. It would be laborious to use such a method to solve second- and higher-order equations. Instead, we use the shift property of the z-transform to transform the difference equation so that we can solve for the transform of the output. Then the inverse transform yields the solution of the equation. This technique will be illustrated in two examples. For brevity in this section, we will write y_k for $y(k)$ or $y(kT)$.

EXAMPLE 3.10

Find the output y_k of the first-order recursive digital filter represented by the difference equation

$$y_k = Ay_{k-1} + x_k$$

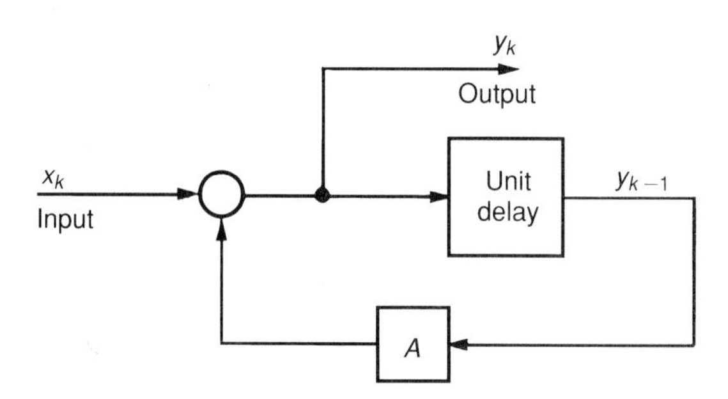

FIGURE 3.10 First-order recursive digital filter.

if x_k is a unit step sequence and y_{-1} is the initial condition on the storage register for y_k. A block diagram of the filter is given in Figure 3.10.

Solution

Multiply both sides of the equation by z^{-k} and sum over k from zero to infinity.

$$\sum_{k=0}^{\infty} y_k z^{-k} = A \sum_{k=0}^{\infty} y_{k-1} z^{-k} + \sum_{k=0}^{\infty} x_k z^{-k}$$

The first and last terms are the z-transforms of y_k and x_k, respectively. Also,

$$\begin{aligned} A \sum_{k=0}^{\infty} y_{k-1} z^{-k} &= A y_{-1} + A z^{-1} \sum_{k=1}^{\infty} y_{k-1} z^{-(k-1)} \\ &= A y_{-1} + A z^{-1} \sum_{\ell=0}^{\infty} y_\ell z^{-\ell}, \quad \ell \triangleq k-1 \\ &= A y_{-1} + A z^{-1} Y(z) \end{aligned}$$

Collecting terms gives

$$Y(z) = A y_{-1} + A z^{-1} Y(z) + X(z)$$

or

$$Y(z) = \frac{X(z)}{1 - A z^{-1}} + \frac{A y_{-1}}{1 - A z^{-1}}$$

The first part of the response is due to the input, the second part to the initial condition. If initial condition $y_{-1} = 0$, then

$$H(z) = \frac{Y(z)}{X(z)} = \frac{1}{1 - A z^{-1}}$$

where $H(z)$ is the *transfer function* (of the filter). In block diagram form, the relationship implied by the last expression can be represented as shown in Figure 3.11. Comparing Figures 3.10 and 3.11 indicates that the unit delay

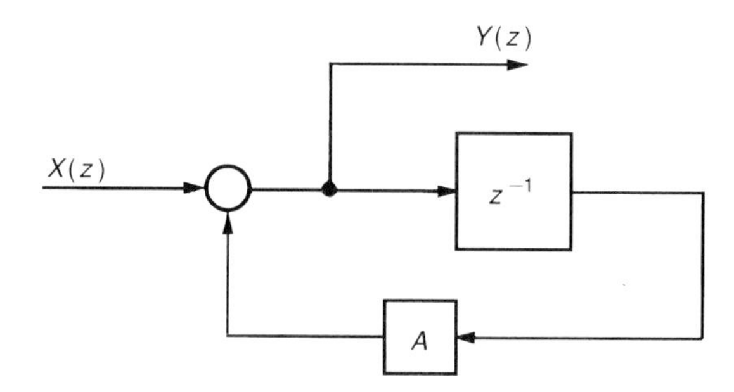

FIGURE 3.11 First-order digital filter in z-transform notation.

in the time domain is equivalent to multiplying by z^{-1} in the z-domain. We can write

$$H(z) = \frac{z}{z - A}$$

which implies that the filter has a zero at the origin and a pole at $z = A$.

Taking the inverse z-transform of $Y(z)$ yields the transient response y_k of the filter to a specific input x_k. If x_k is the unit step sequence $u(k)$, then

$$y_k = \frac{1 - A^{k+1}}{1 - A} + A^{k+1} y_{-1}$$

If $|A| < 1$, the steady-state response to a unit step is

$$\lim_{k \to \infty} y_k = \frac{1}{1 - A}$$

It is evident from the expression for y_k that if $|A| > 1$, the response diverges as k increases. In Figure 3.12, the unit step response is plotted as a function of k for $y_{-1} = 0, A = 0.5$, and $A = 0.8$. For convenience of representation, a continuous curve is drawn.

The rise time as a function of the coefficient A is plotted in Figure 3.13 for $0 < A < 1$. For this purpose, the rise time is arbitrarily defined as the time in samples for the unit step response to rise to 90% of its final value. Note that the rise time increases rapidly as A approaches 1. In other words, the system becomes more sluggish as the pole approaches the unit circle.

EXAMPLE 3.11

Derive a general expression for the output y_k of the second-order recursive digital filter represented by the difference equation

$$y_k = A_1 y_{k-1} + A_2 y_{k-2} + B_0 x_k + B_1 x_{k-1}$$

with initial conditions $x_{-1} = 0, y_{-1}$, and y_{-2}.

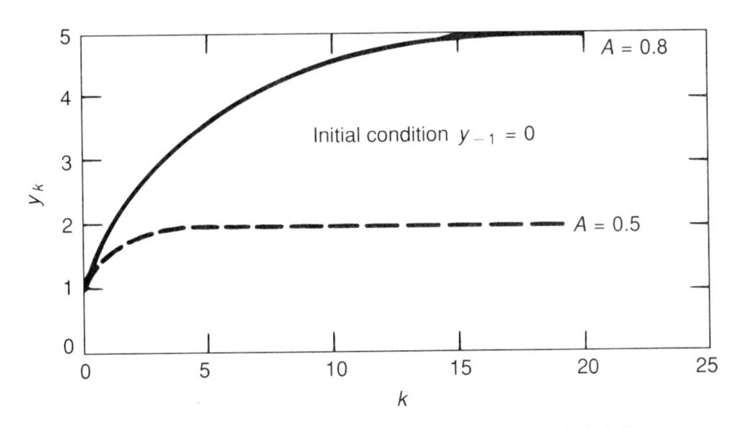

FIGURE 3.12 Unit step response of a first-order digital filter.

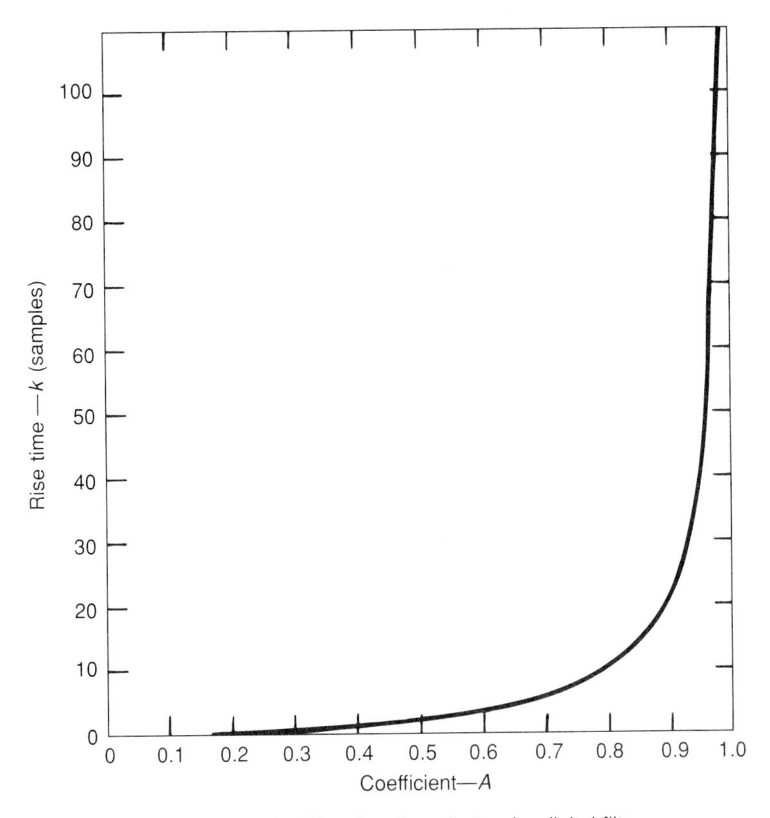

FIGURE 3.13 Rise time for a first-order digital filter.

Solution

Again, we multiply both sides of the equation by z^{-k} and sum over k from zero to infinity.

$$\sum_{k=0}^{\infty} y_k z^{-k} =$$

$$A_1 \sum_{k=0}^{\infty} y_{k-1} z^{-k} + A_2 \sum_{k=0}^{\infty} y_{k-2} z^{-k} + B_0 \sum_{k=0}^{\infty} x_k z^{-k} + B_1 \sum_{k=0}^{\infty} x_{k-1} z^{-k}$$

From Example 3.10, we already know how to find four of the terms. The remaining term with coefficient A_2 can be found from property (3.45) of the z-transform, which is repeated here for convenience.

$$\mathcal{Z}_1 \{f(k-m)\} = z^{-m} \left\{ F(z) + \sum_{p=1}^{m} f(-p) z^p \right\}$$

In this case $m = 2$, so

$$\begin{aligned}
\mathcal{Z}_1 \{y_{k-2}\} &= z^{-2} \left\{ Y(z) + \sum_{p=1}^{2} y_{-p} z^p \right\} \\
&= z^{-2} \{Y(z) + z y_{-1} + z^2 y_{-2}\} \\
&= z^{-2} Y(z) + z^{-1} y_{-1} + y_{-2}
\end{aligned}$$

When we collect terms, we get

$$Y(z) = \frac{(B_0 + B_1 z^{-1}) X(z)}{1 - A_1 z^{-1} - A_2 z^{-2}} + \frac{(A_1 + A_2 z^{-1}) y_{-1} + A_2 y_{-2}}{1 - A_1 z^{-1} - A_2 z^{-2}}$$

If we know the initial conditions y_{-1} and y_{-2} and the form of the input x_k, we can find y_k by taking the inverse z-transform. For zero initial conditions, the transfer function is

$$H(z) = \frac{Y(z)}{X(z)} = \frac{B_0 + B_1 z^{-1}}{1 - A_1 z^{-1} - A_2 z^{-2}} = \frac{z(B_0 z + B_1)}{z^2 - A_1 z - A_2}$$

This filter has zeros at the origin and at $z = -B_1/B_0$ and has two poles given by the roots of the *characteristic equation*

$$z^2 - A_1 z - A_2 = 0$$

which is obtained by setting the denominator polynomial equal to zero. We will return to the characteristic equation in the next section when we investigate the stability of digital filters.

These two simple examples are important for two reasons.

1. Because of quantization errors in implementing the coefficients of the difference equation (see Chapter 8), it is good design practice to build

higher-order recursive filters ($n > 4$, say) in first- and second-order blocks connected either in cascade or in parallel. Therefore, we should be familiar with the operation of these simple filters.

2. We see a pattern emerging for the transfer function. *By inspection*, and assuming zero initial conditions, we can write the transfer function for the digital filter given by the difference equation

$$y_k = \sum_{i=1}^{n} A_i y_{k-i} + \sum_{i=0}^{m} B_i x_{k-i} \tag{3.72}$$

as

$$H(z) = \frac{Y(z)}{X(z)} = \frac{\sum_{i=0}^{m} B_i z^{-i}}{1 - \sum_{i=1}^{n} A_i z^{-i}} \tag{3.73}$$

The reader who does not believe this should try a third- or fourth-order example to reinforce the concept. Furthermore, given the transfer function as the ratio of two polynominals in z^{-1} as in (3.73), it is straightforward to write the corresponding difference equation (3.72).

If $n \geq m$, (3.72) and (3.73) represent a recursive digital filter of order n. If $m > n$, then the filter can be divided up into a recursive filter of order n and a nonrecursive filter of order $m - n$, connected in cascade. Henceforth, when we refer to a recursive filter, we will assume that $n \geq m$. Of course, if $A_i = 0, i = 1, n$, in (3.72), we have a nonrecursive filter of order m, as discussed in Chapter 1. In that case, the transfer function is the numerator of $H(z)$ in (3.73).

Observe that we could divide the numerator of (3.73) by the denominator and, for $n > m$, obtain an infinite series in negative powers of z, which would imply that the IIR filter of (3.73) could be written in *nonrecursive* form. This fact has no practical significance, however, because an infinite number of previous values of input x would be required for implementation of the filter.

3.10 STABILITY OF DIGITAL FILTERS

In Section 1.5, we developed a time-domain criterion for stability of a digital filter in terms of its transient response to a unit pulse. The condition for stability is

$$\sum_{k=0}^{\infty} |h(k)| < \infty \tag{3.74}$$

The weighting sequence, or unit pulse response, $h(k)$, in most cases, is difficult to find for higher-order filters. It is much easier to examine the question of filter stability from its frequency-domain description, which is given by the transfer function $H(z)$.

Consider the transfer function for a general nth-order recursive digital filter in cascade form,

$$H(z) = \frac{B(z)}{A(z)} = \frac{K\prod_{i=1}^{m}(1 - \beta_i z^{-1})}{\prod_{i=1}^{n}(1 - \alpha_i z^{-1})} \tag{3.75}$$

where α_i and β_i are the poles and zeros, respectively. These may either be on the real axis of the z-plane or occur in complex conjugate pairs. We can expand $H(z)$ in partial fractions to get the parallel form

$$H(z) = \sum_{i=1}^{n} \frac{R_i}{1 - \alpha_i z^{-1}} = \sum_{i=1}^{n} H_i(z) \tag{3.76}$$

If we take the inverse transform of (3.76) term by term, we get the pulse response $h(k)$ as the sum of the components $h_i(k)$ due to the poles $\alpha_i, i = 1, n$. The criterion (3.74) implies that, for stability, $h(k)$ must converge to zero as k approaches infinity. Hence *all* the component responses $h_i(k)$ must do likewise. We need to examine the nature of $h_i(k)$ for different locations of the corresponding pole in the z-plane. It is necessary to consider only two cases: the response due to a pole on the real z-axis and the response due to a pair of complex conjugate poles.

1. *Response due to pole on real z-axis*

   Let us focus first on the component response $h_i(k)$ due to a particular pole $z = \alpha_i = A$. Then, from (3.76),

$$H_i(z) = \frac{R_i}{1 - Az^{-1}} \tag{3.77}$$

   Take the inverse transform of the term indicated to get the corresponding contribution to the overall pulse response as

$$h_i(k) = R_i A^k, \quad k \geq 0 \tag{3.78}$$

   This sequence diverges for $|A| > 1$, its absolute value is constant for $|A| = 1$, and it converges to zero for $|A| < 1$. Therefore, for $h_i(k)$ to be part of a stable response, we must have $|A| < 1$; that is, the real pole A must lie inside the unit circle. The forms of the response for various locations of the pole are shown in Figure 3.14. Note the oscillatory response (with period equal to twice the sampling period) that results from poles on the negative real axis. This is to be contrasted with real poles in the s-plane, which result in monotonic continuous time responses.

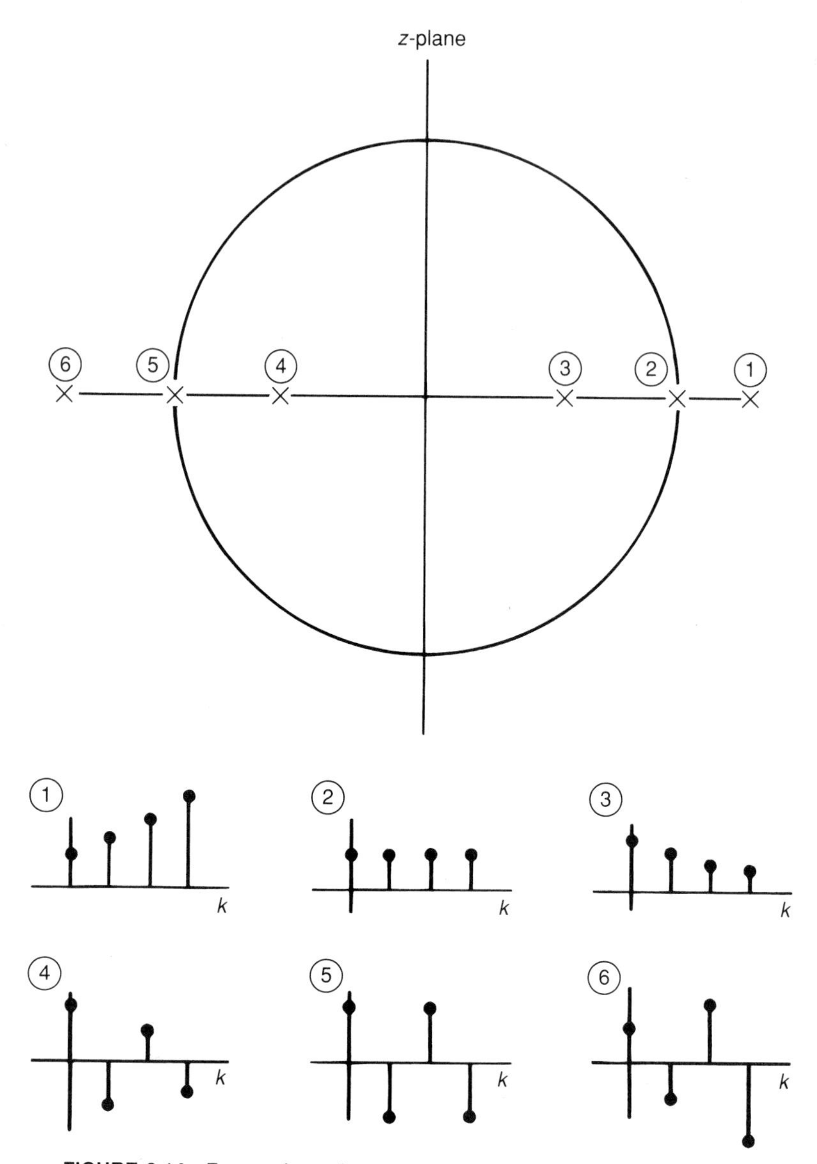

FIGURE 3.14 Forms of transient responses due to a pole on the real z-axis.

2. *Response due to a pair of complex conjugate poles*
   If $H(z)$ contains a pair of complex conjugate poles at $z = re^{\pm j\theta}$, then

$$H_i(z) = \frac{R_i}{1 - re^{j\theta}z^{-1}} + \frac{\overline{R}_i}{1 - re^{-j\theta}z^{-1}} \tag{3.79}$$

where $\overline{R}_i$ denotes the complex conjugate of $R_i = \rho e^{j\phi}$, say, and $\theta = \omega_0 T$. Then, taking the inverse z-transform of the two terms on the right-hand side in (3.79) and combining the results, we get

$$h_i(k) = 2\rho r^k \cos(k\theta + \phi), \quad k \geq 0 \tag{3.80}$$

The derivation of (3.80) is left as an exercise. This diverges for $r > 1$ and converges for $r < 1$. For $r = 1, h_i(k)$ is a constant-amplitude cosinusoid. Three different types of responses occur, depending on the location of the complex poles in the z-plane (Figure 3.15). In each case, the response is oscillatory. From (3.80), we see that the frequency of oscillation depends on $\theta = \omega_0 T$, and the phase on ϕ. The number of samples per cycle is $2\pi/\theta$, as may be verified from (3.80). The response diagrams in Figure 3.15 qualitatively illustrate the relationship between the sampling period and the frequency of oscillation of the waveform.

We see from the foregoing that for $h_i(k)$ to be part of a stable response $h(k)$, we must have $r < 1$. It follows from this and from the result for the real pole that the condition for stability of a discrete LTI system or digital filter may be simply stated in terms of its transfer function $H(z)$: *All the poles of the transfer function must lie within the unit circle in the z-plane.*

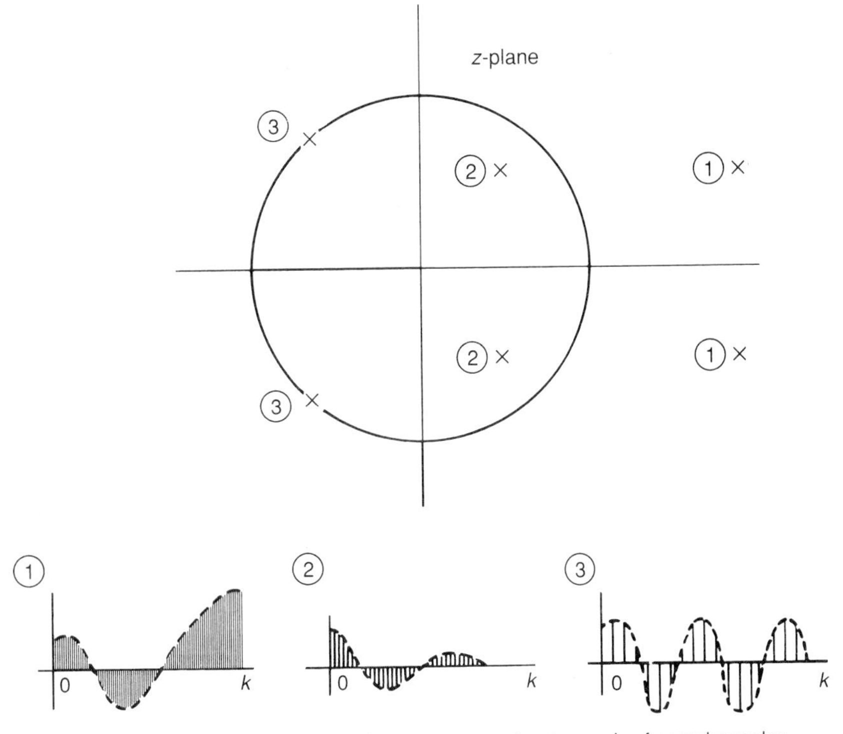

FIGURE 3.15 Forms of transient responses due to a pair of complex poles.

This is analogous to the stability condition for a *continuous* LTI system, which requires that all the poles of the transfer function $H(s)$ must lie in the left half of the s-plane. Multiple poles do not affect this result in either case.

The stability condition for the transfer function

$$H(z) = \frac{B(z)}{A(z)} \tag{3.81}$$

is often expressed as the requirement that the roots of the characteristic equation

$$A(z) = 0 \tag{3.82}$$

[which are the same as the poles of $H(z)$] must lie within the unit circle.

If the characteristic polynomial $A(z)$ is in factored form, stability can be checked by inspection. If not, there are tests available for checking stability without factoring $A(z)$. The easiest tests to use are the Schur–Cohn and the Jury criteria ([9], [8]). Such tests are more useful for digital control than for digital-filter design, and, because of the general availability of computer programs for factoring polynomials, we will not consider them further.

Filters the difference equations of which have the form

$$y_k = \sum_{i=0}^{m} B_i x_{k-i} \tag{3.83}$$

with transfer function

$$H(z) = \sum_{i=0}^{m} B_i z^{-i} \tag{3.84}$$

(that is, nonrecursive filters) cannot go unstable because they have no nonzero poles. This is merely a mathematical statement of the fact that there is no feedback of previous output values $y_{k-i}, i = 1, \ldots,$ to form the current output y_k.

3.11 FREQUENCY RESPONSE OF DIGITAL FILTERS

In Example 3.10, we found that the z-transform of the output of a first-order recursive filter to an arbitrary input x_k is

$$Y(z) = \frac{X(z)}{1 - Az^{-1}} + \frac{Ay_{-1}}{1 - Az^{-1}} \tag{3.85}$$

The response of the system to a sinusoidal input is of particular interest. Let $x_k = e^{jk\omega T}$, a one-sided complex sinusoid. Then

$$X(z) = \frac{1}{1 - e^{j\omega T} z^{-1}} \tag{3.86}$$

With this input, (3.85) becomes

$$Y(z) = \frac{1}{(1 - Az^{-1})(1 - e^{j\omega T}z^{-1})} + \frac{Ay_{-1}}{1 - Az^{-1}} \tag{3.87}$$

Taking the inverse z-transform gives

$$y_k = A^{k+1}y_{-1} + \frac{A^{k+1}}{A - e^{j\omega T}} + \frac{e^{j(k+1)\omega T}}{e^{j\omega T} - A} \tag{3.88}$$

The first two terms on the right-hand side decay for $A < 1$ and large k. Thus the *steady state* response for a sinusoidal input is

$$y_k = \frac{e^{j(k+1)\omega T}}{e^{j\omega T} - A} = \frac{x_k e^{j\omega T}}{e^{j\omega T} - A} \tag{3.89}$$

The term multiplying x_k can be regarded as a transfer function for the sinusoidal input and is called the *frequency response* of the digital filter. That is,

$$H(j\omega T) = \frac{e^{j\omega T}}{e^{j\omega T} - A} = M(\omega T)e^{j\phi(\omega T)} \tag{3.90}$$

where the *magnitude* response is

$$M(\omega T) = (1 + A^2 - 2A\cos\omega T)^{-1/2} \tag{3.91}$$

and the *phase* response is

$$\phi(\omega T) = \omega T - \tan^{-1}[\sin\omega T/(\cos\omega T - A)]. \tag{3.92}$$

Note that (3.90) could be obtained directly from (3.85) with zero initial condition by setting $z = e^{j\omega T}$, its value on the unit circle:

$$H(j\omega T) = \left.\frac{Y(z)}{X(z)}\right|_{z=e^{j\omega T}} = \left.\frac{1}{1 - Az^{-1}}\right|_{z=e^{j\omega T}} = \frac{e^{j\omega T}}{e^{j\omega T} - A} \tag{3.93}$$

For a filter to have a frequency response, the filter must be stable so that steady-state conditions can be attained. Note that if no sampling is involved, T is the computer time step.

In general, the frequency response may be obtained for this, and for higher-order examples, by writing a computer program to evaluate the magnitude and phase of $H(e^{j\omega T})$ for values of ω ranging from zero to $\omega_s/2$ (or of ωT ranging from 0 to π radians). Such a program may compute the magnitude and phase as it is done *graphically*. We will illustrate the graphical method with this simple example.

It is possible to obtain the frequency response graphically by plotting the pole and zero of the transfer function

$$H(z) = \frac{z}{z - A} \tag{3.94}$$

on the z-plane, as shown in Figure 3.16. Vectors are drawn from the singularities to a variable point $z = e^{j\omega T}$ on the unit circle. If $\vec{a}$ and $\vec{b}$ are the vectors from the pole and from the zero, respectively, it is clear that these will be functions of ω as the variable point z moves around the unit circle. Thus

$$H(j\omega T) = \frac{e^{j\omega T}}{e^{j\omega T} - A} = \frac{\vec{b}}{\vec{a}} \tag{3.95}$$

with magnitude response given by the ratio of the vector lengths,

$$M(\omega T) = \frac{|\vec{b}|}{|\vec{a}|} = \frac{1}{|\vec{a}|} \tag{3.96}$$

and phase response given by the difference between the angles made by the vectors with the positive z-axis,

$$\phi(\omega T) = \theta - \psi = \omega T - \psi \tag{3.97}$$

The magnitude and phase may be scaled directly from the pole-zero plot of Figure 3.16 for various values of z on the unit circle. Note that expressions (3.91) and (3.92) can be obtained directly from the geometry of Figure 3.16. We note also the periodic nature of the frequency response for a digital filter. It repeats, after the moving point reaches $\omega = \omega_s$ (or $\theta = \omega T = 2\pi$), as indicated by (3.18).

For comparison with the digital filter, we take the z-transform of samples of a continuous time function the Laplace transform of which is the first-order filter function

$$H(s) = \frac{1}{s - a} \tag{3.98}$$

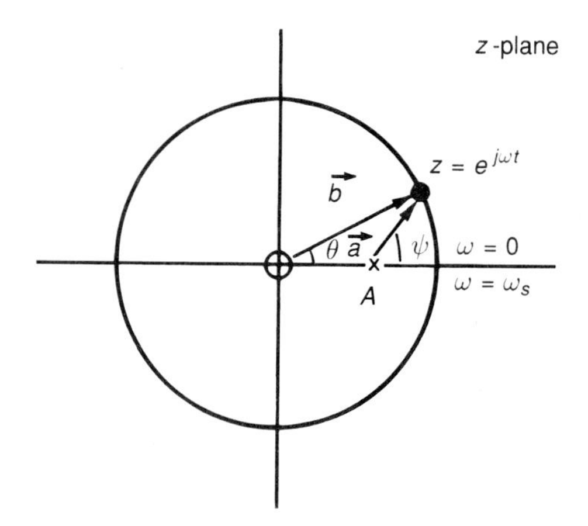

FIGURE 3.16 Graphical method to obtain the frequency response for a first-order digital filter.

Using (3.38), we get

$$H(z) = \frac{1}{1 - Az^{-1}} \tag{3.99}$$

where $A = e^{-aT}$. Figure 3.17 compares the magnitude responses of these two first-order filters for $a = 10\ \mathrm{s}^{-1}$ and $T = 10^{-2}$ s. The difference between the two responses in the range $0 \le \omega \le \omega_s/2$ is due to aliasing, as discussed in Section 3.2. The response of the continuous filter decreases monotonically, whereas that of the digital filter is periodic. The latter is of practical use up to $\omega = \omega_s/2$, in accordance with the sampling theorem.

The difference equation

$$y_k = A_1 y_{k-1} + A_2 y_{k-2} + x_k \tag{3.100}$$

for a second-order recursive filter can be transformed in the same manner as in Example 3.11. The result is

$$Y(z) = \frac{(A_1 + A_2 z^{-1})y_{-1} + A_2 y_{-2}}{1 - A_1 z^{-1} - A_2 z^{-2}} + \frac{X(z)}{1 - A_1 z^{-1} - A_2 z^{-2}} \tag{3.101}$$

If the initial conditions on the storage registers are zero, then the transfer function is

$$H(z) = \frac{Y(z)}{X(z)} = \frac{z^2}{z^2 - A_1 z - A_2} \tag{3.102}$$

A block diagram for this filter is shown in Figure 3.18. Setting $z = e^{j\omega T}$ in

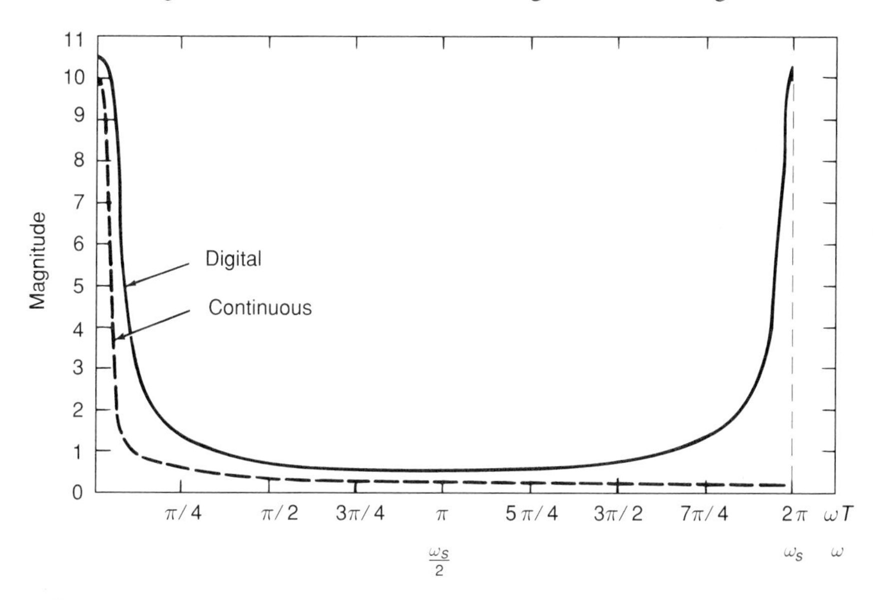

FIGURE 3.17 Comparison of the magnitude responses of first-order continuous and digital filters.

$H(z)$ gives the frequency response functions as

$$M(\omega T) = [(1 - A_1 \cos \omega T - A_2 \cos 2\omega T)^2 + (A_1 \sin \omega T + A_2 \sin 2\omega T)^2]^{-1/2} \tag{3.103}$$

$$\phi(\omega T) = -\tan^{-1} \frac{A_1 \sin \omega T + A_2 \sin 2\omega T}{1 - A_1 \cos \omega T - A_2 \cos 2\omega T} \tag{3.104}$$

Derivation of these expressions is left as an exercise.

The roots of the characteristic equation

$$z^2 - A_1 z - A_2 = 0 \tag{3.105}$$

which are the poles of the filter, may be either real or complex. For complex poles,

$$z_1, z_2 = r \exp(\pm j\theta) \tag{3.106}$$

where $r^2 = -A_2 < 1, \cos\theta = A_1/2r$, and $\theta = \omega_0 T$. The filter behaves as a *digital resonator* for r close to unity. This can be seen readily by means of the graphical approach, which is shown for this case in Figure 3.19.

The magnitude response is

$$M(\omega T) = \frac{1}{|\vec{a}_1|\,|\vec{a}_2|} \tag{3.107}$$

where $\vec{a}_1$ and $\vec{a}_2$ are the vectors from the poles to the variable point $z = e^{j\omega t}$ on the unit circle. It is evident that $|\vec{a}_1|$ becomes very small—and hence M very large—when z is on the radial line with pole z_1—that is, $\omega = \omega_0$. The magnitude has a peak at this resonant frequency, as shown in Figure 3.20. The closer r is to unity, the sharper the peak. The digital resonator is, in effect, an elementary bandpass filter with passband centered at the resonant frequency ω_0.

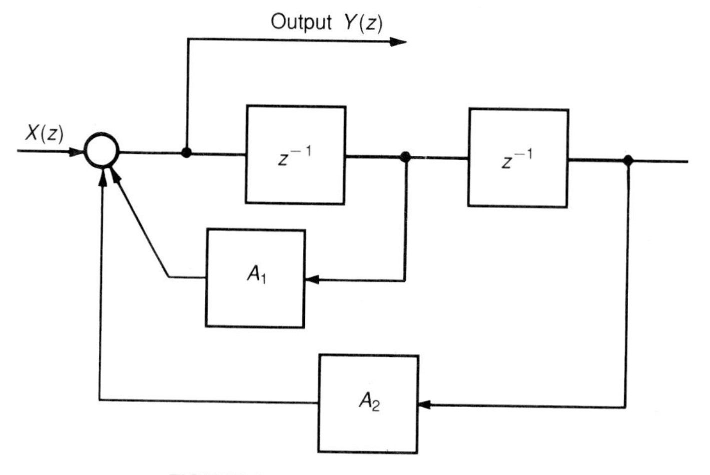

FIGURE 3.18 Second-order digital filter.

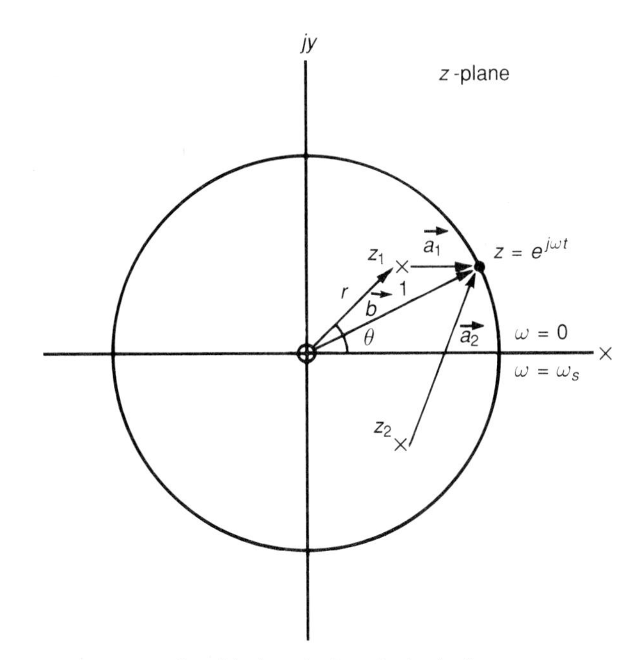

FIGURE 3.19 Graphical method to obtain the frequency response for a second-order digital filter.

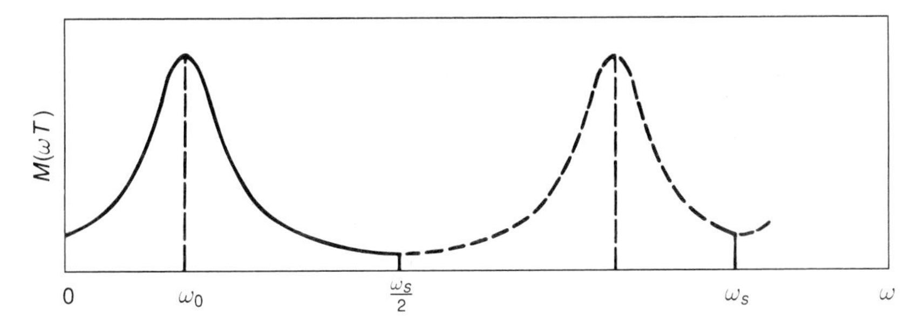

FIGURE 3.20 Magnitude response for a second-order digital filter with complex poles.

From the geometry of Figure 3.19, (3.107) may be written as

$$M(\omega T) = \{[1 + r^2 - 2r\cos((\omega_0 - \omega)T)][1 - r^2 - 2r\cos((\omega_0 + \omega)T)\}^{-1/2} \tag{3.108}$$

If a particular passband center frequency ω_0 is specified, it is often more convenient to use (3.108), with r close to unity, than to use (3.103).

In general, for an nth-order recursive filter described by the difference equation

$$y_k = \sum_{i=1}^{n} A_i y_{k-i} + \sum_{i=0}^{m} B_i x_{k-i} \tag{3.109}$$

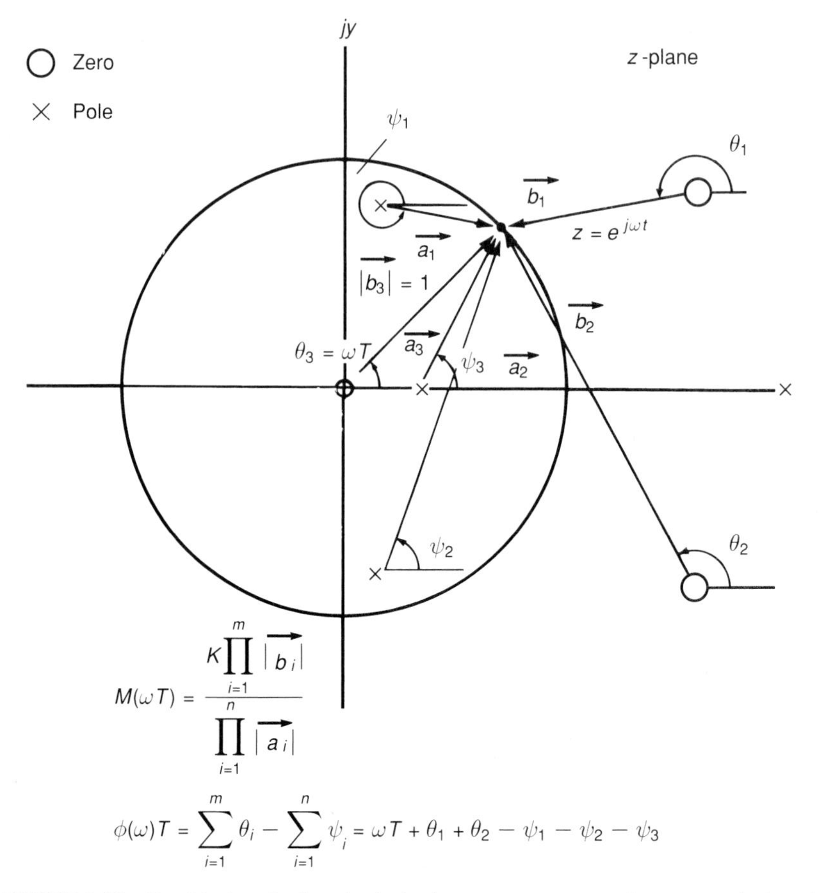

FIGURE 3.21 Graphical method to obtain the frequency responses for a general nth-order recursive filter.

the z-transform representation is

$$H(z) = \frac{Y(z)}{X(z)} = \frac{\displaystyle\sum_{i=0}^{m} B_i z^{-i}}{1 - \displaystyle\sum_{i=1}^{n} A_i z^{-i}} \tag{3.110}$$

The frequency response can be obtained by setting $z = e^{j\omega T}$ in (3.110).

$$H(e^{j\omega T}) = \frac{\displaystyle\sum_{i=0}^{m} B_i e^{-ji\omega T}}{1 - \displaystyle\sum_{i=1}^{n} A_i e^{-ji\omega T}} \tag{3.111}$$

The magnitude response and phase response of this function can then be computed for different values of ω. If the poles and zeros are known, the frequency responses can be obtained graphically for any higher-order filter in the same manner as for the simple first- and second-order filters we have examined. Figure 3.21 illustrates use of the graphical method for a third-order filter. We are not recommending that the graphical method be used for deriving *numerical* values for magnitude response and phase response of higher-order filters. The method is useful for providing *qualitative* information on how the individual poles and zeros affect the frequency response.

3.12 FORMS OF DIGITAL FILTERS

The recursive digital filter represented by the nth-order difference equation

$$y_k = \sum_{i=1}^{n} A_i y_{k-i} + \sum_{i=0}^{m} B_i x_{k-i} \tag{3.112}$$

can be realized in many different forms or structures, some of which have already been mentioned. For example, we could use n registers to store the n previous values of y and use m additional registers to store the m previous values of x. We shall see that this *direct form* is not very efficient. To illustrate with a simple example, consider the second-order filter

$$y_k = A_1 y_{k-1} + A_2 y_{k-2} + x_k + B x_{k-1} \tag{3.113}$$

with transfer function

$$H(z) = Y(z)/X(z) = \frac{1 + Bz^{-1}}{1 - A_1 z^{-1} - A_2 z^{-2}} \tag{3.114}$$

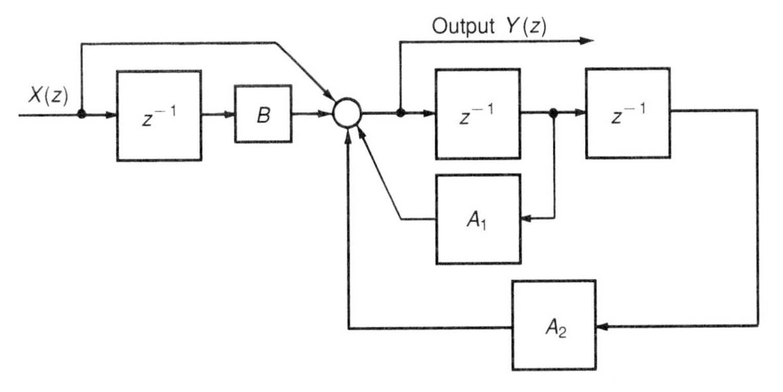

FIGURE 3.22 Direct form of a second-order filter with zero at $z = -B$.

In addition to the two poles, this filter has zeros at $z = 0$ and $z = -B$. If it is implemented directly according to (3.113), it takes the form shown in Figure 3.22, with $n + m = 3$ delay elements.

The direct form of Figure 3.22 is not efficient in that it uses one more delay element than is necessary. The *canonic form* or *standard form* of a recursive digital filter is defined as that filter arrangement that uses the minimum number of delay elements for the given transfer function. For a filter with m zeros and n poles, $n \geq m$, the minimum number of delays is equal to n, the number of poles. A canonic form for the filter of (3.113) is shown in Figure 3.23.

There are several different filter arrangements that will give a canonic form, but the one that follows will suffice for our treatment. In general, a canonic form can be obtained by taking the z-transform of (3.112),

$$Y(z) = \frac{\displaystyle\sum_{i=0}^{m} B_i z^{-i}}{1 - \displaystyle\sum_{i=1}^{n} A_i z^{-i}} X(z) \tag{3.115}$$

and writing it as

$$Y(z) = M(z) \sum_{i=0}^{m} B_i z^{-i} \tag{3.116}$$

where

$$M(z) = \frac{X(z)}{1 - \displaystyle\sum_{i=1}^{n} A_i z^{-i}} \tag{3.117}$$

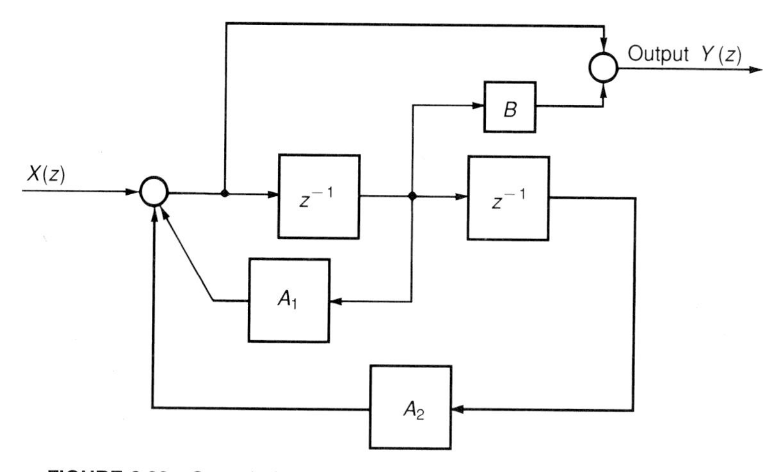

FIGURE 3.23 Canonic form of a second-order filter with zero at $z = -B$.

An auxiliary all-pole (zeros at $z = 0$) filter with output $M(z)$ is first arranged as shown in Figure 3.24, and then $Y(z)$ is formed according to (3.116). However, the remarks in Section 3.9 should be kept in mind; in general, because of quantization effects, it is better to realize a recursive filter with $n > 4$, say, by a cascade or parallel arrangement of first- and second-order filters.

The cascade form is obtained by factoring the numerator and denominator of transfer function $H(z)$,

$$H(z) = K\frac{\prod_{i=1}^{m} 1 - \beta_i z^{-1}}{\prod_{i=1}^{n} 1 - \alpha_i z^{-1}} \tag{3.118}$$

and the parallel form by expanding $H(z)$ in partial fractions,

$$H(z) = \sum_{i=1}^{n} \frac{R_i}{1 - \alpha_i z^{-1}} \tag{3.119}$$

for $m < n$. These two filter arrangements are shown in Figure 3.25. For second-order components, some of the blocks would be combined in pairs. For the cascade form, the output of one block is the input to the next block, and the computer algorithm, or hardware, is designed accordingly. To minimize the number of delays required, the first- and second-order components should themselves be in a canonic form.

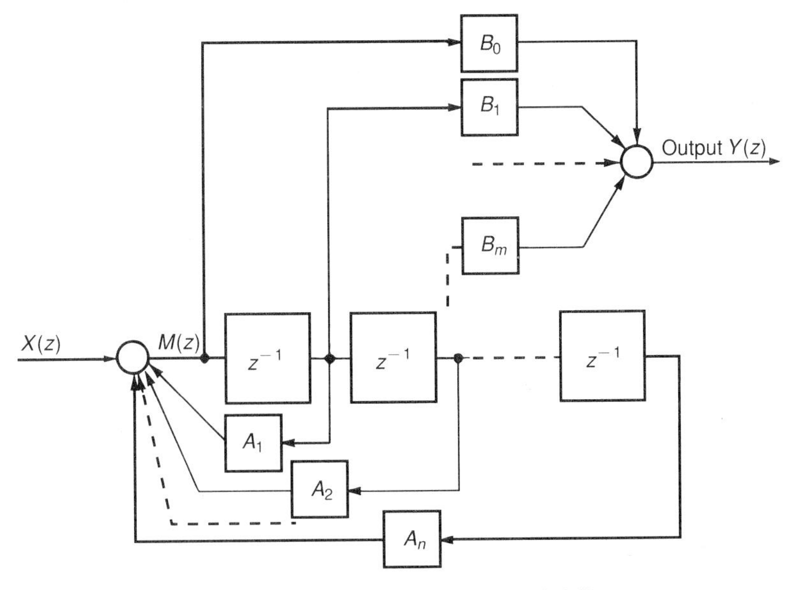

FIGURE 3.24 Canonic form of nth-order digital filter, $m \leq n$.

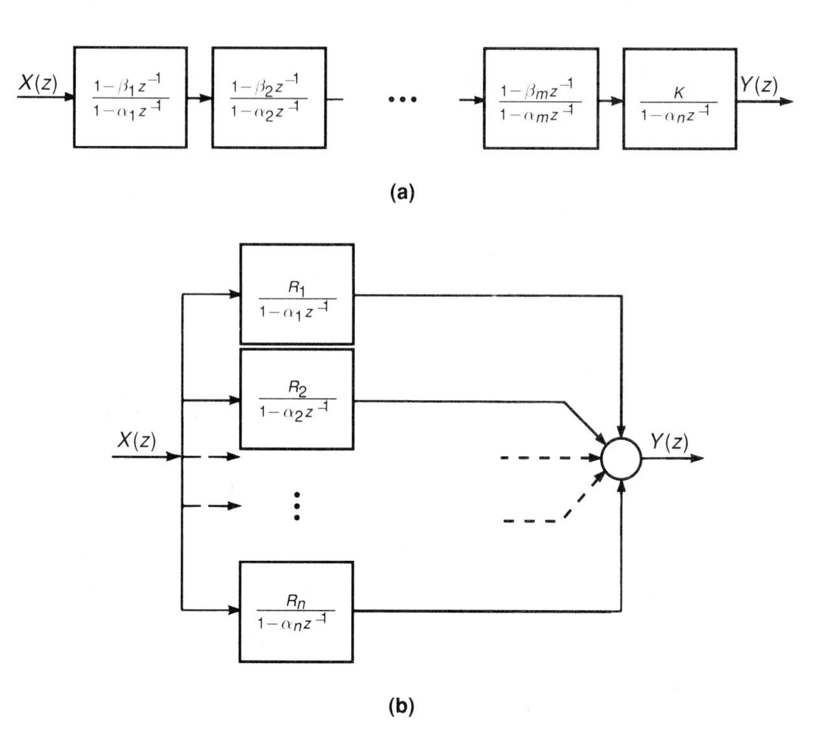

FIGURE 3.25 Cascade and parallel arrangements for digital filters. (a) Cascade form, $m = n - 1$. (b) Parallel form, $m < n$.

There are several other recursive-filter structures [10], but those we have mentioned are among the most commonly used in computer simulation and special-purpose hardware.

For a *nonrecursive* filter—a filter the difference equation of which has the form

$$y_k = \sum_{i=0}^{m} B_i x_{k-i} \tag{3.120}$$

with z-transform

$$Y(z) = \left(\sum_{i=0}^{m} B_i z^{-i} \right) X(z) \tag{3.121}$$

the minimum number of storage registers required is m. Therefore, the direct form shown in Figure 3.26 is also a canonic form. Another useful structure is a cascade arrangement of first- and second-order components,

$$H(z) = K \prod_{i=1}^{m} (1 - \beta_i z^{-1}) \tag{3.122}$$

where it is understood that if any of the zeros β_i are complex conjugates, they are combined to form second-order components. Because the nonrecursive filter has zeros only, it does not have a parallel structure.

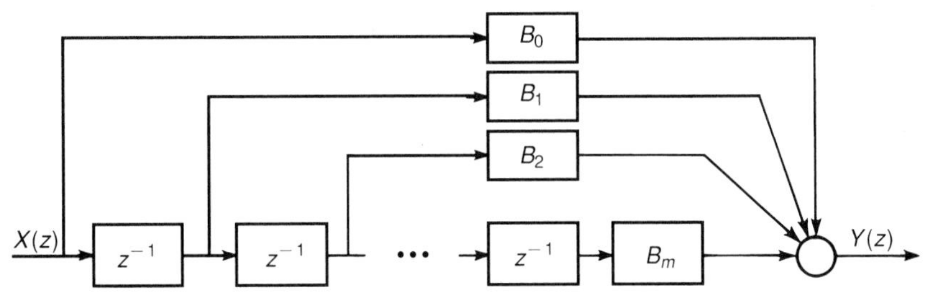

FIGURE 3.26 Direct and canonic forms of a nonrecursive filter.

3.13 SUMMARY

The implications of the sampling theorem, including the phenomenon of aliasing and the choice of sampling rate, are discussed. A model for a periodic sampler is obtained by considering the sampling process as a modulation of an ideal impulse train by the continuous signal being sampled. Three different ways to take the Laplace transform of the sampled function are described; they result in three different expressions, all equivalent. Two of these expressions give the z-transform by a change of variable $z = e^{Ts}$. Properties of the z-transform are examined, and different ways to derive the inverse z-transform are shown. The z-transform is applied to solve difference equations and to derive transfer functions for discrete systems. Requirements for stability of digital filters are discussed. By setting $z = e^{j\omega T}$ in the transfer functions, we can obtain the frequency response. Alternatively, by using a graphical method, we can derive the frequency response from the pole-zero plot in the z-plane. This method is very useful for providing a qualitative indication of the shape of the magnitude response curve. The chapter closes with a discussion of various forms that can be used in the realization of a digital filter.

REFERENCES FOR CHAPTER 3

1. A. DeMoivre, *Miscellanes Analytica de Seriebus et Quatratoris*, London, 1730.
2. W. Hurewicz, *Theory of Servomechanisms*. Radiation Laboratory Series, vol. 25 (New York: McGraw-Hill, 1947).
3. R. H. Baker, "The Pulse Transfer Function and Its Application to Sampling Servo Systems," *Proc. IEE* 99, pt. IV (December 1952), 302–317.
4. J. R. Ragazzini and L. A. Zadeh, "The Analysis of Sampled Data Systems," *Trans. AIEE* 71, pt. II (1952), 225–234.
5. D. Linden, "A Discussion of Sampling Theorems," *Proc. IRE* 47 (July 1949), 1219–1226.
6. W. W. Harman, *Principles of the Statistical Theory of Communication* (New York: McGraw-Hill, 1963).

7. R. N. Bracewell, *The Fourier Transform and Its Applications*, 2d ed. (New York: McGraw-Hill, 1986).
8. E. J. Jury, *Theory and Application of the z-Transform Method* (New York: Wiley, 1964).
9. B. C. Kuo, *Digital Control Systems* (Champaign, IL: SRL, 1977), Chap. 5.
10. L. R. Rabiner and B. Gold, *Theory and Application of Digital Signal Processing* (Englewood Cliffs, NJ: Prentice-Hall: 1975).

EXERCISES FOR CHAPTER 3

1. A certain continuous signal contains a dc component and additional sinusoidal components at frequencies of 2 kHz and 3 kHz, respectively.

a) The signal is sampled at 10 kHz. Make a rough sketch of its magnitude spectrum, showing all components between zero and 20 kHz.

b) Assume that this continuous signal is sampled at 5 kHz. By drawing a spectral diagram from zero to 10 kHz show that, with this sampling rate, it is impossible to recover the original signal by simple filtering.

2. The highest-frequency component in a certain continuous signal is known to be at about 5 kHz. The signal also has a component at 4.8 kHz. The signal is sampled for processing in a digital computer.

a) What is the *maximum* sampling period (the computer time step, in this case)?

b) If it is required that the spectral peaks at 4.8 and 5 kHz be distinguishable, what is the *minimum* length of continuous record that must be processed?

c) What is the least number of samples that must be taken from the finite record chosen in part b)?

3. The continuous signal $f(t) = e^{-1}u(t)$ has a frequency spectrum

$$F(j\omega) = \frac{1}{1 + j\omega}$$

that is not bandlimited. However, most of its energy lies in the range $0 \leq \omega \leq \omega_h$. The energy in the signal is computed from

$$E = \frac{1}{2\pi}\int_{-\infty}^{\infty} |F(j\omega)|^2 d\omega$$

Find the value of ω_h in this case if

$$E_1 = \frac{1}{2\pi}\int_{-\omega_h}^{\omega_h} |F(j\omega)|^2 d\omega$$

is 0.995 of the total energy E. Then find the minimum rate at which $f(t)$ must be sampled to ensure that most of the information in the signal is recovered.

4. With reference to Figure 3.5, a contour of integration could be closed about the poles of $1/(1 - e^{-T(s-p)})$ instead of about the poles of $F(p)$. Show that the sum of the residues in this case gives the same result as expression (3.33).

5. Find the z-transform of a discrete signal that is obtained from the continuous signal the Laplace transform of which is

$$F(s) = \frac{s+2}{(s+1)(s+4)}$$

6. Use expression (3.38) to get the z-transforms of the sequences obtained by sampling (with period T) the following continuous functions.

a) $f(t) = \sin(\omega_0 t)u(t)$

b) $f(t) = e^{-at}\cos(\omega_0 t)u(t)$

7. Find the z-transform, and its region of convergence, for the negative-time sequence

$$f(k) = \begin{cases} e^{akT}, & k \leq 0 \\ 0, & k > 0 \end{cases}$$

where $0 < a < 1$.

8. Find the z-transform and its ROC for the two-sided sequence

$$f(k) = \begin{cases} 1, & k < 0 \\ k, & k \geq 0 \end{cases}$$

9. Find the z-transform for the *finite* sequence

$$x(k) = \begin{cases} 1, & 0 \leq k \leq m-1 \\ 0, & \text{otherwise} \end{cases}$$

Locate the poles and zeros of the resulting function for $m = 6$.

10. Prove the property (3.46).

11. From the z-transform for the unit ramp sequence $f(k) = k, k \geq 0$, find the z-transform of

$$g(k) = a^k f(k)$$

12. From the z-transform for the power sequence $f(k) = a^k, k \geq 0$, find the z-transform of

$$g(k) = kf(k)$$

13. From the z-transform for the unit step sequence, find the z-transform for the negative unit step sequence.

14. Verify the entries in Table 3.1 that have not already been confirmed.

15. Given

$$F(z) = \frac{(1-e^{-T})z}{(z-1)(z-e^{-aT})}$$

find its inverse transform $f(kT)$, using the residue method, if $a > 0$ and the ROC is $|z| > 1$. Write an expression for the sampled function $f^*(t)$.

16. Find the inverse z-transform of

$$H(z) = \frac{z^3 - 0.5z^2 - 0.32z}{(z-0.5)(z-0.2)^2}, \quad |z| > 0.5$$

using the residue method. Check your result by long division.

17. Given

$$F(z) = \frac{z}{(z - 0.5)^2(z - 2)}, \quad 0.5 < |z| < 2$$

find $f(k)$ by the residue method.

18. A two-sided sequence is as follows:

$$h(k) = \begin{cases} a^k, & k < 0 \\ b^{-k}, & k \geq 0 \end{cases}$$

where a and b are real and $|a| > 1, |b| > 1$. Compute the z-transform $H(z)$. Indicate the region of convergence.

19. A certain causal digital system has input sequence $1, 1, 1, \ldots$ and output sequence $1, 0, 1, 0, 1, 0, \ldots$ Find the transfer function for the system, and indicate the region of convergence.

20. Find the output transform $Y(z)$ for the digital filter represented by the third-order difference equation

$$y_k = A_1 y_{k-1} + A_2 y_{k-2} + A_3 y_{k-3} + B_0 x_k + B_1 x_{k-1} + B_2 x_{k-2}$$

with initial conditions $y_{-1}, y_{-2}, y_{-3}, x_{-1}, x_{-2}$. Write the transfer function. Draw a block diagram for the filter in a) direct form and b) canonic form.

21. Plot the unit step response and the frequency responses (magnitude and phase) for the first-order digital filter

$$y_k = A y_{k-1} + x_k$$

if $A = -0.8$.

22. Derive expression (3.80).

23. Derive responses (3.91) and (3.92) from the geometry of Figure 3.16.

24. Derive responses (3.103) and (3.104).

25. Find the canonic form for the filter the transfer function of which is

$$H(z) = \frac{z(Cz + D)}{(z - A)(z^2 - A_1 z - A_2)}$$

Draw a block diagram of the resulting arrangement.

26. A block diagram for a digital filter is as shown in Figure P3.1. Write the transfer function $W(z) = Y(z)/X(z)$ and the difference equation. What is the order of the filter? Locate the poles and zeros.

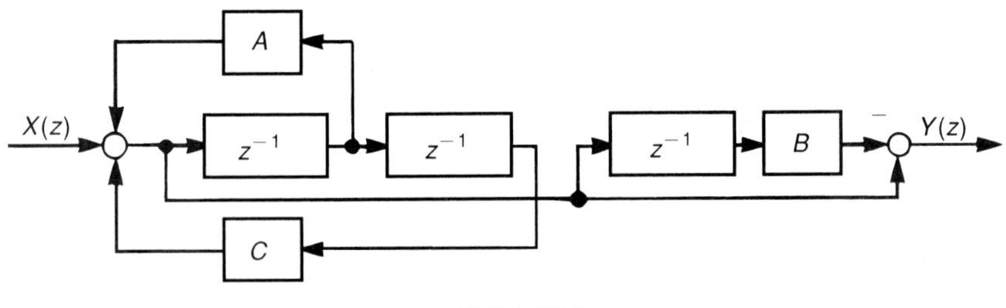

FIGURE P3.1

27. A pole-zero diagram of a digital filter is shown in Figure P3.2.
   a) Find the transfer function of the filter to within a gain constant K.
   b) Write the difference equation.
   c) Make a *rough* sketch of the magnitude response from $\omega = 0$ to $\omega = \omega_s$.

28. A pole-zero diagram of a digital filter is shown in Figure P3.3.
   a) Make a rough sketch of the magnitude response from $\omega = 0$ to $\omega = \omega_s$.
   b) Draw a block diagram for an arrangement of the filter with the minimum number of delays.

29. Magnitude response plots for various digital filters are given in Figure P3.4. In each case, write by inspection a transfer function that could give such a response.

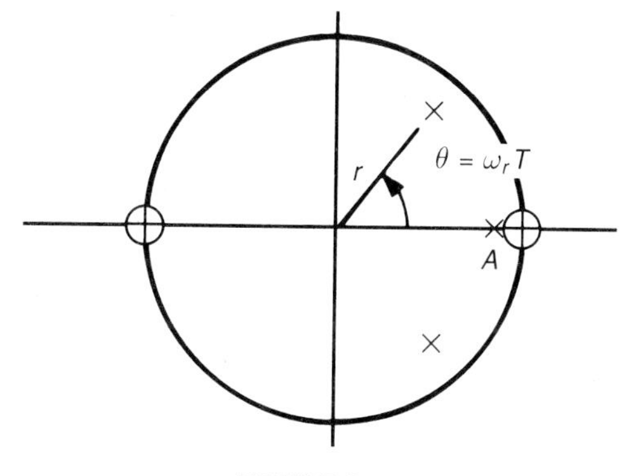

FIGURE P3.2

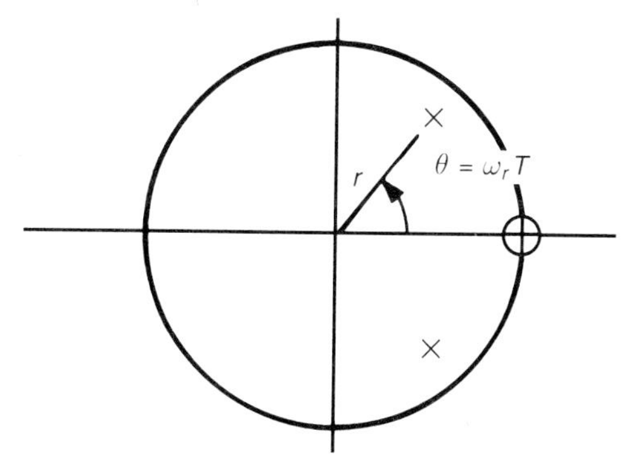

FIGURE P3.3

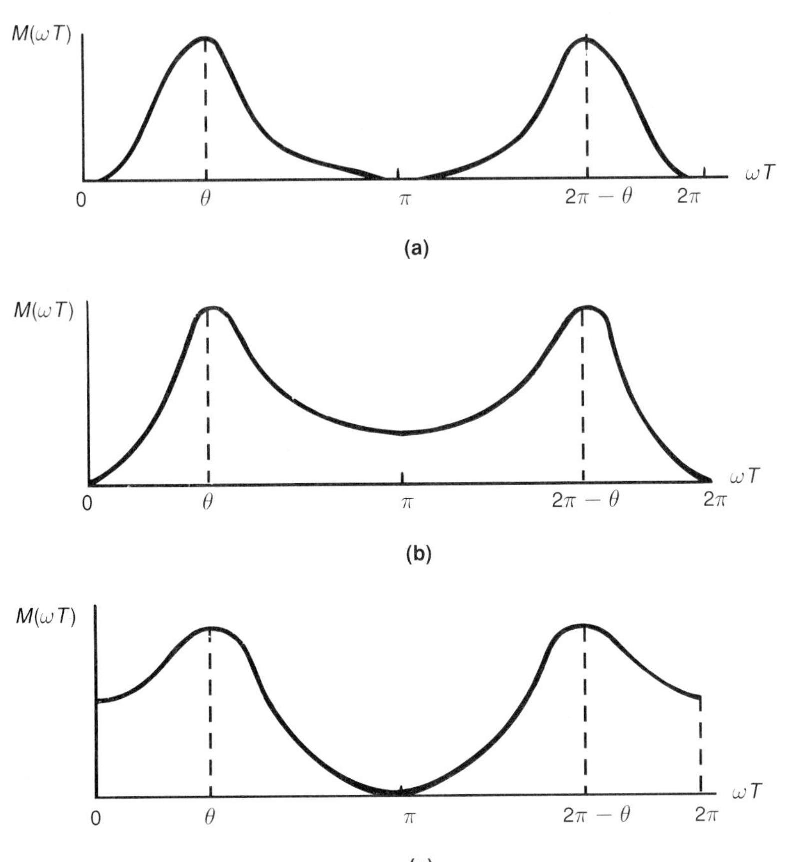

FIGURE P3.4

CHAPTER 4

RECURSIVE FILTER DESIGN

4.1 INTRODUCTION

Having reviewed the mathematical preliminaries, we are now in a position to design some digital filters. We will start with recursive filters. The reader will recall from Chapter 1 that these are filters wherein the current output is formed from previous values of the output as well as from the current input and previous input values, suitably weighted. A block diagram of the filter (see, for example, Figure 3.22) indicates feedback paths due to the previous output values and feedforward paths due to the previous input values. Because of the feedback, there is the possibility that the filter can be unstable. An unstable filter is of no practical use as a signal-processing device, so, for a start, designers must ensure that their designs are stable. Fortunately, as we saw in the last chapter, it is simple to check stability when the filter transfer function is known. The next step is to verify that the filter specifications are met. There are many filters that will satisfy a given set of specifications. An efficient design, however, ensures that the specifications are met by the lowest-order filter possible. In general, the cost of implementation goes up as the order increases.

To avoid having to take a trial-and-error approach to the problem of designing a digital filter to meet given specifications, we need a systematic method of design. For lowpass, highpass, bandpass, and bandstop filters (with frequency-domain specifications), which make up most of the practical applications, transforming a suitable *continuous* filter function into a digital filter affords us such a systematic procedure. Because of the large amount of experience and literature available on the design of continuous filters, this seems to be a fortuitous approach to the problem. It will also justify our review of continuous filter design in Chapter 2. However, some questions need to be answered.

1. What transformation should we use?
2. If the continuous filter is stable, will the transformation ensure that the digital filter is stable?
3. Does the continuous filter's meeting the specifications mean that the digital filter will also meet them? If not, how do we modify the procedure to ensure that it does?

Several transformations are available for converting from a continuous filter function to a discrete filter function. The more useful of these are described in this chapter, together with a few that are not so good—the latter to illustrate some pitfalls to avoid. By "useful" we mean that the transformation should yield a good design and should be easy to apply. We shall examine the theory behind each transformation and the advantages and disadvantages of each.

With regard to stability of the resulting digital filter, it is more efficient to check this from the transformation itself rather than to wait until the design is completed. As we noted in Chapter 2, a stable continuous filter must have all its poles in the left half of the s-plane. A stable digital filter must have all its poles within the unit circle in the z-plane, as shown in Chapter 3. Hence, using a transformation that maps the left half of the s-plane into the interior of the unit circle in the z-plane ensures that a stable digital filter will result from a stable continuous filter.

The third question—whether the fact that the continuous filter satisfies the specifications ensures that the digital filter does also, is not so easy to answer in advance as the stability question. Nevertheless, by using the bilinear transform and the regular z-transform (see Chapter 3) with reasonable care, we can ensure that the resulting digital filter meets the specifications, or at least comes close to meeting them.

We will examine the foregoing questions in more detail for the different transformations. To assist the designer with more complex designs than the analytic examples treated in our discussion, computer programs are given on the diskette.

Recursive filters can also be designed directly from filter specifications. Several methods for doing this are discussed by Rabiner and Gold [1]. One such method—that of Steiglitz [2]—is described here.

4.2 DIGITAL FILTER DESIGN BY MEANS OF THE *z*-TRANSFORM (IMPULSE-INVARIANT METHOD)

In light of our discussion in Chapter 3, it seems that a logical way to design a digital filter is to obtain a z-transform by applying expression (3.38) to a suitable continuous filter function $H(s)$. This method has both desirable and undesirable features. We will consider the good points first.

1. The method is easy to apply. If $H(s)$ is in factored form, we obtain $H(z)$, the transfer function for the digital filter, by using expression (3.38), which is repeated here for convenience.

$$H(z) = \sum \text{ residues of } H(p)\frac{1}{1 - e^{Tp}z^{-1}} \text{ at poles of } H(p) \tag{4.1}$$

2. The resulting digital filter is *impulse-invariant* with the continuous filter. Let us define what impulse-invariant means.

Definition

A continuous filter $G(s)$ and a digital filter $H(z)$ are said to be impulse-invariant if the pulse response of $H(z)$ is the same as the sampled impulse response of $G(s)$.

We will now prove that if $H(z)$ is the z-transform of the samples $g(kT)$ of a continuous function $g(t)$ the Laplace transform of which is $G(s)$, then $H(z)$ and $G(s)$ are impulse-invariant. Note that here we are using different symbols—$H(z)$ and $G(s)$ instead of the usual $H(z)$ and $H(s)$—to avoid confusion in the proof. This usage is temporary.

Proof

Expand $G(s)$ in partial fractions.

$$G(s) = \sum_{i=1}^{n} \frac{R_i}{s - s_i} \tag{4.2}$$

where $s_i, i = 1, n$, are the poles of $G(s)$. Take the inverse Laplace transform of (4.2) term by term to get the impulse response

$$g(t) = \sum_{i=1}^{n} R_i e^{s_i t} \tag{4.3}$$

The values of the samples of the impulse response are obtained by setting t equal to kT.

$$g(kT) = \sum_{i=1}^{n} R_i e^{s_i kT} \tag{4.4}$$

Now apply (4.1) to $G(s)$ term by term (we can do this because the z-transform is a linear operation). This gives

$$H(z) = \sum_{i=1}^{n} \frac{R_i}{1 - e^{Ts_i}z^{-1}} \tag{4.5}$$

The unit pulse response is obtained by taking the inverse z-transform.

$$h(k) = \sum \text{residues of } H(z)z^{k-1} \text{ at poles of same inside ROC for } H(z)$$

$$= \sum_{i=1}^{n} R_i e^{s_i kT} \tag{4.6}$$

which is the same as (4.4). Q.E.D.

3. The frequency variable ω for the digital filter bears a linear relationship to that for the continuous filter, within the operating range of the digital filter. This means that when ω varies from 0 to $\omega_s/2$ (and ωT from 0 to π) around the circumference of the unit circle in the z-plane, ω varies from 0 to $\omega_s/2$ along the $j\omega$-axis in the s-plane (see Figure 3.9). Thus critical frequencies such as cutoff and bandwidth frequencies specified for the digital filter can be used directly in selection of the continuous filter.
4. The next point to consider entails both an advantage and a serious disadvantage of the z-transform method. Taking the z-transform of a continuous filter function is a satisfactory procedure when applied to all pole filters such as Butterworth, Chebyshev (type I), and Bessel, provided that the filters we are designing are *bandlimited.* In a bandlimited filter, the magnitude response of the continuous filter is negligibly small at frequencies exceeding half the sampling frequency, in order to reduce the aliasing effect. Thus we must have

$$|H(j\omega)| \to 0 \quad \text{for } \omega \geq \omega_s/2 \tag{4.7}$$

It is clear that (4.7) can hold for lowpass and bandpass filters but not for highpass and bandstop filters. This is illustrated for ideal bandpass and highpass functions in Figure 4.1. The periodic spectra represent the magnitude response for the digital filters that result from taking the z-transform of $h(k)$ [samples of $h(t)$ the Laplace transform of which is $H(s)$]. The operating range $-\omega_s/2$ to $\omega_s/2$ is of interest for the digital filter. There is no overlap of the passbands in the periodic extension of the bandpass filter in Figure 4.1(b), provided that the upper limit of the passband of $H(s)$ is less than $\omega_s/2$. However, with the highpass filter, because the passband of $H(s)$ is unlimited, the periodic extension in Figure 4.1(d) indicates that the periodic lobes of the passband tend to "fill up" the stopband as a result of aliasing in the operating range $-\omega_s/2$ to $\omega_s/2$. Consequently, the digital filter is useless as a highpass filter. There is a way to get around this difficulty by modifying the procedure, as we shall see.

The procedure used for designing, for example, a bandpass filter, can be summarized as follows:

1. Select a suitable continuous, lowpass prototype to satisfy the sharpness-of-cutoff specification (see Chapter 2).

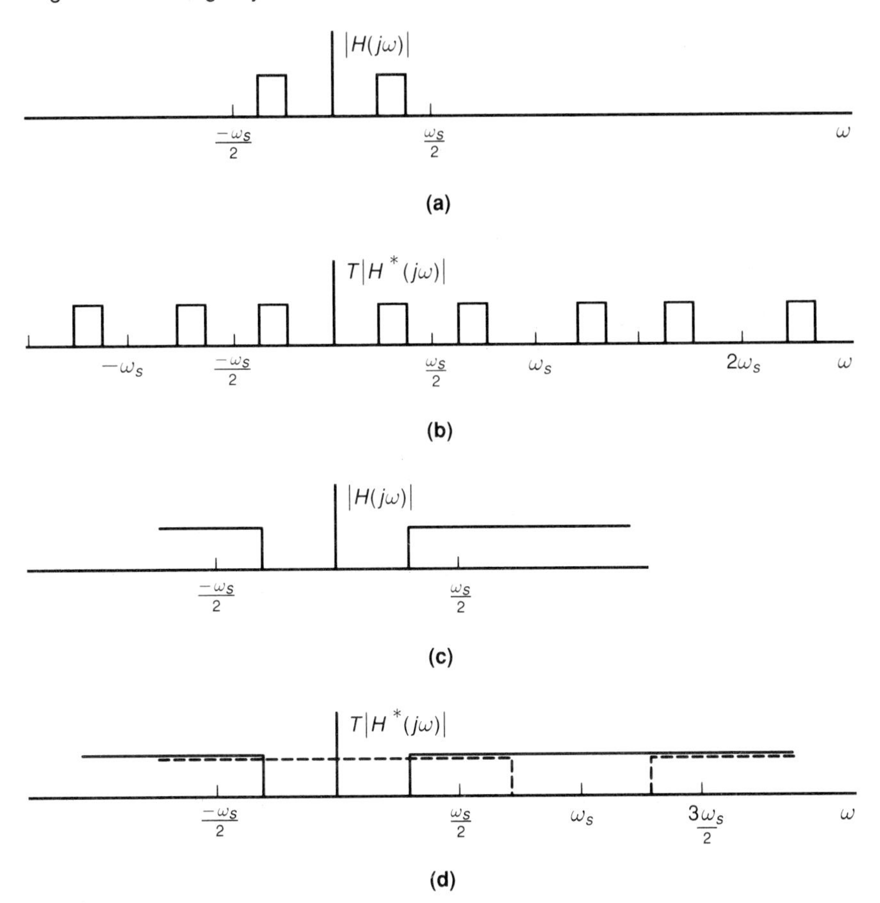

FIGURE 4.1 Periodic extension for bandlimited and for non-bandlimited functions. (a) Bandlimited function. (b) Periodic extension of bandlimited function. (c) Non-bandlimited function. (d) Periodic extension of non-bandlimited function.

2. Make a lowpass-to-bandpass frequency transformation for the specified passband (see Chapter 2).
3. Apply transformation (4.1) to the resulting continuous filter function.
4. Use the transfer function obtained to express the digital filter in difference-equation form for implementation as a computer algorithm.

An example of the application of this design procedure follows. The example is made as simple as possible to avoid obscuring the principles with algebraic complexity.

EXAMPLE 4.1

Use the impulse-invariant method to design a bandpass digital filter that is monotonic in passband and stopband, has a passband from 500 to 550 Hz, and is as low in order as practical. Sampling rates of 2 kHz and 10 kHz should be examined.

Solution

Because of the simplicity requirement, a second-order filter will suffice. It is convenient to carry out the design in symbolic form rather than with the numerical values, which may be inserted later. In general, in this book, we do not number the equations in an example unless it is necessary to refer to them. This, however, is such a case.

Because a second-order bandpass filter is required, the design procedure may begin with a first-order, continuous, lowpass prototype of the Butterworth type.

$$H_n(s_n) = \frac{1}{s_n + 1} \tag{4.8}$$

Applying the bandpass transformation (2.156) to $H_n(s_n)$ gives

$$H(s) = \frac{1}{\omega_0(s/\omega_0 + \omega_0/s)/B + 1} = \frac{Bs}{s^2 + Bs + \omega_0^2} \tag{4.9}$$

where

$$B = \omega_u - \omega_\ell, \omega_0^2 = \omega_u \omega_\ell \tag{4.10}$$

Equation (4.9) can be written

$$H(s) = \frac{Bs}{(s+a)^2 + b^2} \tag{4.11}$$

where

$$\begin{aligned} a &= \frac{B}{2} \\ b &= (\omega_0^2 - B^2/4)^{1/2} \end{aligned} \tag{4.12}$$

The magnitude response is

$$M(\omega) = B\omega/\{[a^2 + (\omega - b)^2][a^2 - (\omega + b)^2]\}^{1/2} \tag{4.13}$$

To obtain a digital filter that is impulse-invariant with $H(s)$, apply transformation (4.1) to $H(s)$.

$$\begin{aligned} H(z) &= \sum \text{residues of } H(p)\frac{1}{1 - e^{Tp}z^{-1}} \text{ at poles } p = -a \pm jb \\ &= \frac{B[1 - (a/b \sin bT + \cos bT)e^{-aT}z^{-1}]}{1 - 2e^{-aT}\cos bTz^{-1} + e^{-2aT}z^{-2}} \end{aligned} \tag{4.14}$$

This is the transfer function of the required digital bandpass filter. It may be written in the form

$$H(z) = \frac{Bz(z-q)}{z^2 - 2r(\cos bT)z + r^2} \tag{4.15}$$

where

$$r = e^{-aT}$$
$$q = \left(\frac{a}{b}\sin bT + \cos bT\right)e^{-aT} \tag{4.16}$$

The pole-zero diagram for the filter is shown in Figure 4.2. Note that the design is essentially complete when the transfer function $H(z)$ has been derived. From $H(z)$ we can obtain the frequency response and the difference equation for implementing the filter algorithm.

From the geometry of Figure 4.2, the magnitude response is

$$M(\omega T) = B\left\{\frac{1 + q^2 - 2q\cos\omega T}{[1 + r^2 - 2r\cos((\omega - b)T)][1 + r^2 - 2r\cos((\omega + b)T)]}\right\}^{1/2} \tag{4.17}$$

Figure 4.3 compares the magnitude responses for the continuous and digital filters, the latter with two different sampling rates. Note that it is necessary to multiply $M(\omega T)$, given by (4.17), by the scale factor T (0.0005 s or 0.0001 s in this example) in order to compare it with $M(\omega)$, given by (4.13). This follows from the relationship between the Fourier transform of the sampled function and the continuous function as given in (3.35). Note that there is negligible difference between the responses in the vicinity of the passband. However, in accordance with the sampling theorem, the digital filter with 2-kHz sampling frequency would be useless for processing signals above 1000 Hz. With the higher sampling frequency

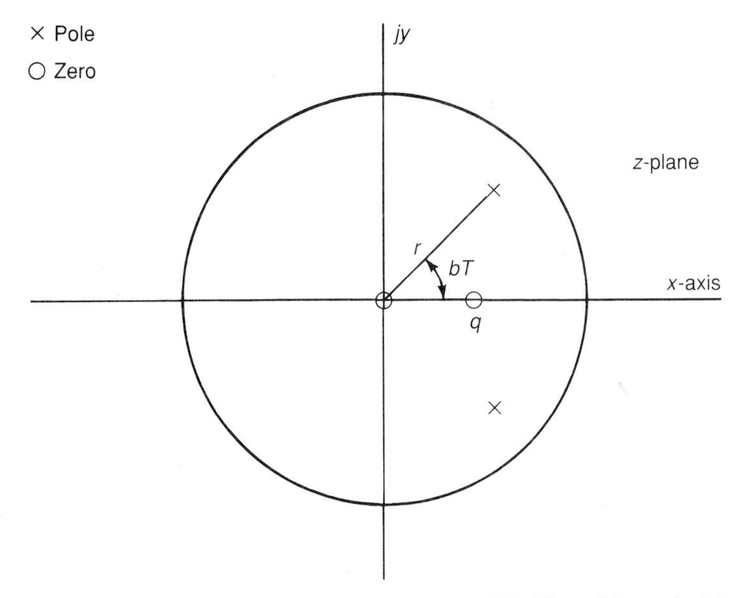

FIGURE 4.2 Pole-zero plot for the bandpass digital filter of Example 4.1.

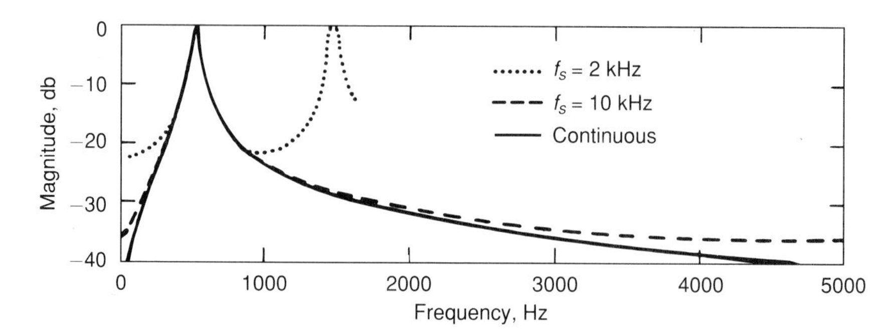

FIGURE 4.3 Comparison between the magnitude response of the second-order digital and continuous bandpass filters of Example 4.1.

of 10 kHz, there is a wider and deeper stopband—and a better approximation to the continuous filter response.

From the transfer function (4.14), the difference equation can be written by inspection. It is

$$y_k = A_1 y_{k-1} + A_2 y_{k-2} + B_0 x_k + B_1 x_{k-1} \tag{4.18}$$

where

$$A_1 = 2e^{-aT} \cos bT$$

$$A_2 = -e^{-2aT}$$

$$B_0 = B$$

$$B_1 = -B\left(\frac{a}{b} \sin bT + \cos bT\right) e^{-aT} \tag{4.19}$$

with a, b, and B replaced by their values in terms of the specified ω_u and ω_ℓ according to (4.10) and (4.12). The block diagram for the canonic implementation of this filter is shown in Figure 4.4.

Higher-order lowpass and bandpass filters may be designed in a similar manner. They should be implemented as a series or parallel arrangement

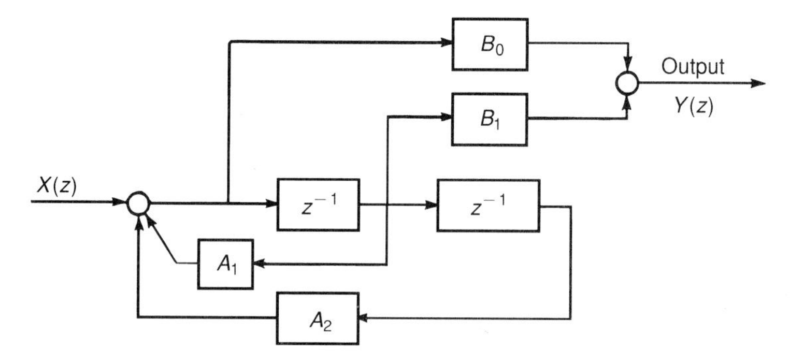

FIGURE 4.4 Implementation of the second-order bandpass filter of Example 4.1.

of first- and second-order filters. For example, take the z-transform of the third-order Butterworth filter [3],

$$H(s) = \frac{\omega_c^3}{(s+\omega_c)(s^2+\omega_c s+\omega_c^2)} \tag{4.20}$$

with cutoff frequency ω_c, after first expanding $H(s)$ in partial fractions. The result is

$$H(z) = \omega_c \left[\frac{1}{1-e^{-\omega_c T}z^{-1}} + \frac{-1+\eta z^{-1}}{1-\beta z^{-1}+e^{-\omega_c T}z^{-2}} \right] \tag{4.21}$$

where

$$\beta = 2e^{-\omega_c T/2} \cos \frac{\sqrt{3}\omega_c T}{2} \tag{4.22}$$

$$\eta = e^{-\omega_c T/2} \left[\cos \frac{\sqrt{3}\omega_c T}{2} + \sin\left(\frac{\sqrt{3}\omega_c T}{2}\right) / \sqrt{3} \right] \tag{4.23}$$

This gives a digital filter consisting of a first-order and a second-order filter in parallel.

Because of aliasing, we cannot design a bandstop or highpass digital filter by taking the z-transform without first bandlimiting the function $H(s)$ to be transformed. This involves multiplying $H(s)$ by a so-called guardband filter $G(s)$, which is a lowpass filter arranged such that

$$|G(j\omega)H(j\omega)| \to 0 \quad \text{for } \omega \geq \frac{\omega_s}{2} \tag{4.24}$$

Then operation (4.1) is applied to the product $G(s)\,H(s)$. The disadvantage of this method is that it yields digital filters of higher order than the original continuous design $H(s)$. It does, however, provide a means of obtaining useful z-transform functions of filters that are inherently not bandlimited.

Suppose, for example, a bandstop filter is required with stopband cutoff frequencies f_ℓ and f_u, and with f_h being the highest frequency in the signal to be processed. First, design a suitable continuous-filter $H(s)$ of order n_1 that has the required stopband with the specified sharpness of cutoff. Then design a lowpass filter $G(s)$ with cutoff frequency f_h and with attenuation a dB at frequency f_r, where

$$f_r = f_s - f_u \tag{4.25}$$

f_s being the sampling frequency. The order of the guardband filter is given by

$$n_2 = \frac{a}{20\log_{10}\left(\frac{f_r}{f_h}\right)} \tag{4.26}$$

The attenuation a must be sufficient to ensure negligible aliasing into the periodic response at f_r. Applying operation (4.1) to the product $G(s)\,H(s)$

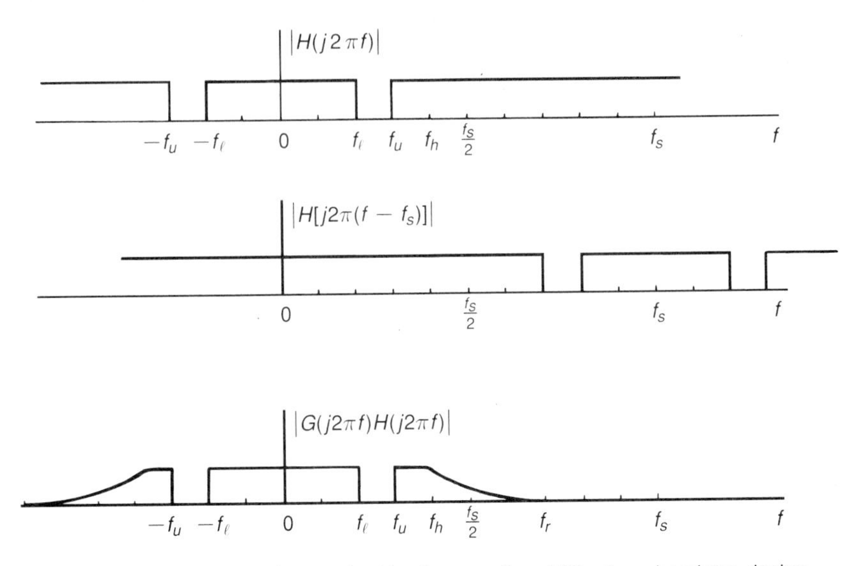

FIGURE 4.5 The attenuation required by the guardband filter for a bandstop design.

yields a digital filter of order $n = n_1 + n_2$. The procedure for obtaining the order of the guardband filter may be understood with reference to the idealized magnitude response shown in Figure 4.5. Note that if the sampling frequency is increased, the cutoff requirement—and hence the order of the guardband filter—may be reduced.

4.3 DESIGNING DIGITAL FILTERS BY NUMERICAL APPROXIMATION TO AN INTEGRATION

4.3.1 Introduction

We have seen that in designing a digital filter by means of the z-transform, we use the mapping $z = e^{sT}$ to go from points in the s-plane to points in the z-plane. As mentioned earlier, there are several different ways of going from the s-plane to the z-plane. The transformations we will now discuss are based on making a numerical approximation to an integration. The following simple example illustrates the general principles. Let the continuous system be

$$H(s) = \frac{Y(s)}{X(s)} = \frac{a}{s + a} \tag{4.27}$$

which represents the first-order differential equation

$$\frac{dy}{dt} = -ay + ax \tag{4.28}$$

Integrate both sides of (4.28) with respect to t to get

$$y(t) = a \int_0^t [-y(\tau) + x(\tau)]d\tau \tag{4.29}$$

which is the area under the curve $-ay(\tau)+ax(\tau)$ from 0 to t. If $t = (k-1)T$, we can write

$$y(k-1) = a \int_0^{(k-1)T} [-y(\tau) + x(\tau)]d\tau \tag{4.30}$$

Similarly, if $t = kT$, we have

$$\begin{aligned} y(k) &= a \int_0^{kT} [-y(\tau) + x(\tau)]d\tau \\ &= a \int_0^{(k-1)T} [-y(\tau) + x(\tau)]d\tau + a \int_{(k-1)T}^{kT} [-y(\tau) + x(\tau)]d\tau \\ &= y(k-1) + \Delta y \end{aligned} \tag{4.31}$$

This is in a recursive form. The representation for $y(k)$ is exact if we determine the incremental area Δy by evaluating the integral between $(k-1)T$ and kT. The situation is depicted in Figure 4.6. If we choose to *approximate* the incremental area Δy instead, we can develop several different rules for going from a continuous system to a discrete system. We will consider three different ways to approximate Δy:

1. Forward rectangular approximation
2. Backward rectangular approximation
3. Trapezoidal approximation

We shall see that each of these numerical approximations indicates how to convert a continuous filter to a digital filter by means of an appropriate replacement for s. Note that there is no actual sampling involved in going from a continuous filter function $H(s)$ to a digital-filter function $H(z)$ via any of the following transformations. However, the notation T is retained to denote the time step of the digital computer or filter. The "sampling" frequency ω_s is the frequency that corresponds to time step T. It is *defined* as

$$\omega_s = \frac{2\pi}{T}$$

4.3.2 Forward Rectangular Approximation

With this rule, we approximate Δy by a rectangle the height of which is the value of the integrand at the *start* of the incremental area and the width of which is the sampling period T (see Figure 4.6). We note in advance that

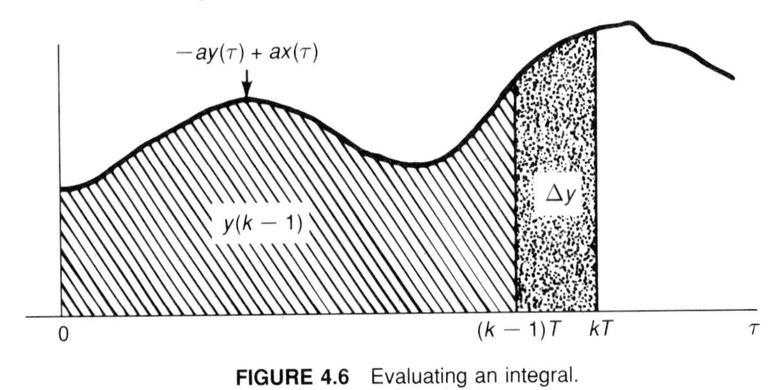

FIGURE 4.6 Evaluating an integral.

this could be a reasonable approximation if T were small and the integrand did not vary much, but otherwise it might not be very good. Thus

$$\Delta y \approx [-ay(k-1) + ax(k-1)]T \tag{4.32}$$

Insert this value of Δy into (4.31) to get the difference equation

$$\begin{aligned} y(k) &= y(k-1) - aTy(k-1) + aTx(k-1) \\ &= (1-aT)y(k-1) + aTx(k-1) \end{aligned} \tag{4.33}$$

Take the z-transform of (4.33) to get

$$H(z) = \frac{Y(z)}{X(z)} = \frac{aTz^{-1}}{1-(1-aT)z^{-1}} = \frac{a}{(z-1)/T + a} \tag{4.34}$$

When we compare this with

$$H(s) = \frac{a}{s+a} \tag{4.35}$$

from which it was derived, it is apparent that with this approximation, we can get a digital filter from a continuous filter by replacing s in the latter by $(z-1)/T$.

Let us check this transformation to see whether it gives a stable digital filter from a stable continuous filter. The mapping

$$s = \frac{z-1}{T} \tag{4.36}$$

implies that

$$z = 1 + Ts \tag{4.37}$$

Let $z = u + jv$ (rather than $x + jy$, because we are using x and y as signals in this context). Then (4.37) becomes

$$u + jv = 1 + T(\sigma + j\omega) \tag{4.38}$$

or

$$u = 1 + T\sigma \tag{4.39}$$

and

$$v = \omega T \tag{4.40}$$

From (4.39),

$$\begin{aligned} \sigma = 0 &\Rightarrow u = 1 \\ \sigma < 0 &\Rightarrow u < 1 \\ \sigma > 0 &\Rightarrow u > 1 \end{aligned} \tag{4.41}$$

This mapping is depicted in Figure 4.7. The $j\omega$-axis in the s-plane is mapped into the line $u = 1$ in the s-plane. The right half of the s-plane maps into the region to the right of the line $u = 1$ in the s-plane, and the left half of the s-plane maps into the region to the left of the line $u = 1$ in the z-plane. This last-mentioned region contains the unit circle, the interior of which is the stable zone of the z-plane. However, it also contains a much larger area of unstable points. It follows that a stable continuous filter (one the poles of which are in the left half of the s-plane) could produce an unstable digital filter by the forward rectangular approximation. For example, from (4.34), $H(z)$ has a pole at $z = 1 - aT$. Hence if $aT > 2$, or $T > 2/a$, the pole lies outside the unit circle, and $H(z)$ is unstable. We need examine this transformation no further; it is *not* suitable for our purposes.

4.3.3 Backward Rectangular Approximation

Let us see whether we can do any better with the backward rectangular approximation. Here we approximate Δy in Equation 4.31 by a rectangle the height of which is the value of $-ay + ax$ at the end of the incremental area.

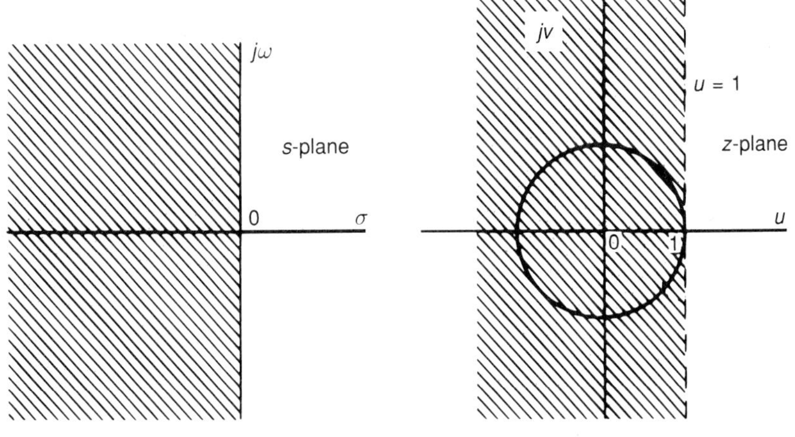

FIGURE 4.7 Mapping for the forward rectangular rule, $s = (z - 1)/T$.

Then

$$\Delta y \approx [-ay(k) + ax(k)]T \tag{4.42}$$

If we insert this value of Δy into (4.31), we get the difference equation

$$y(k) = y(k-1) + [-ay(k) + ax(k)]T \tag{4.43}$$

or

$$(1 + aT)y(k) = y(k-1) + aTx(k) \tag{4.44}$$

Take the z-transform to get

$$H(z) = \frac{Y(z)}{X(z)} = \frac{aT}{1 + aT - z^{-1}} = \frac{a}{(z-1)/Tz - a} \tag{4.45}$$

Comparing this with $H(s) = a/(s+a)$, we see that we could get $H(z)$ from $H(s)$ by replacing s by $(z-1)/Tz$. In the same way, we can transform higher-order continuous filters by replacing s by $(z-1)/Tz$ in each case.

We can write the mapping

$$s = \frac{z-1}{Tz} \tag{4.46}$$

as

$$\begin{aligned} z &= \frac{1}{1-Ts} \\ &= \frac{1}{2} + \left[\frac{1}{1-Ts} - \frac{1}{2}\right] \\ &= \frac{1}{2} + \frac{1}{2}\left[\frac{1+Ts}{1-Ts}\right] \end{aligned}$$

Therefore,

$$\begin{aligned} z - \frac{1}{2} &= \frac{1}{2}\left[\frac{1+Ts}{1-Ts}\right] \\ &= \frac{1}{2}\left[\frac{1+\sigma T + j\omega T}{1-\sigma T - j\omega T}\right] \\ &= \frac{1}{2}\frac{1 - \sigma^2T^2 - \omega^2T^2 + j2\omega T}{(1-\sigma T)^2 + \omega^2T^2} \end{aligned} \tag{4.47}$$

If $\sigma = 0$ (that is, $s = j\omega$), then

$$z - \frac{1}{2} = \frac{1}{2}\frac{1 - \omega^2T^2 + j2\omega T}{1 + \omega^2T^2} \tag{4.48}$$

Take the magnitude squared of both sides.

$$\left|z - \frac{1}{2}\right|^2 = \frac{1}{4}\frac{(1-\omega^2T^2)^2 + 4\omega^2T^2}{(1+\omega^2T^2)^2} = \frac{1}{4} \tag{4.49}$$

Therefore,

$$\left| z - \frac{1}{2} \right| = \frac{1}{2} \tag{4.50}$$

It follows that the $j\omega$-axis in the s-plane is mapped into the circumference of a circle that has radius $\frac{1}{2}$ and is centered at $z = \frac{1}{2}$. It is not difficult to show that the entire left half of the s-plane maps into the interior of this circle (see the shaded area in Figure 4.8). The right half of the s-plane maps into the exterior of the circle. The proof is left as an exercise.

Because the interior of the circle $|z - \frac{1}{2}| = \frac{1}{2}$ lies within the unit circle in the z-plane, a stable continuous filter produces a stable digital filter by this transformation. However, Figure 4.8 shows that a large part of the interior of the unit circle can be mapped only from *unstable* continuous filters. Thus this transformation severely restricts the pole locations and attainable frequencies in the digital filter. For this reason, it is not a recommended means of designing a digital filter from a continuous filter.

4.3.4 Trapezoidal Approximation (Bilinear Transform)

So far, we have not been doing very well in our quest for a useful transformation based on numerical approximation to an integration. We will try once more. Approximate Δy by the area of a trapezoid with parallel sides equal to $-ay + ax$, evaluated at $(k-1)T$ and kT, respectively. Then

$$\Delta y \approx \{-a[y(k-1) + y(k)] + a[x(k-1) + x(k)]\}T/2 \tag{4.51}$$

and the difference equation for this case is

$$y(k) = y(k-1) - aT[y(k-1) + y(k)]/2 + aT[x(k-1) + x(k)]/2 \tag{4.52}$$

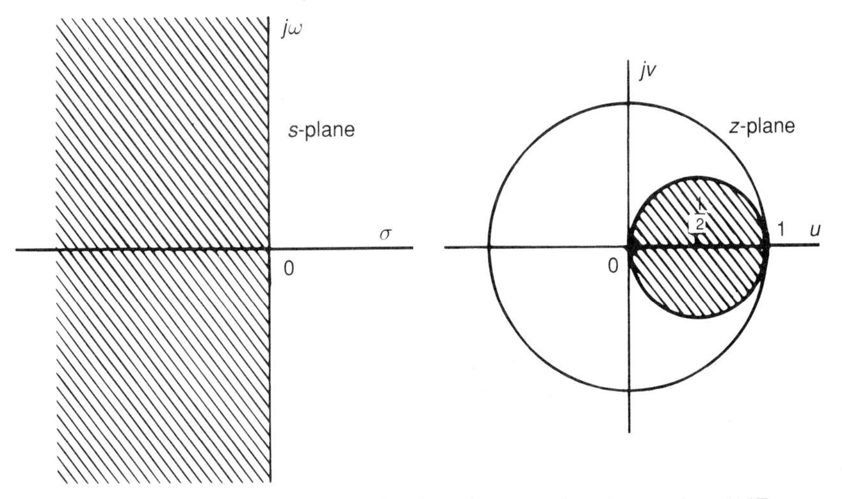

FIGURE 4.8 Mapping for the backward rectangular rule, $s = (z - 1)/Tz$.

or

$$(1 + aT/2)y(k) = (1 - aT/2)y(k-1) + aT[x(k-1) + x(k)]/2 \quad (4.53)$$

From this, the transfer function is

$$H(z) = \frac{aT(1+z^{-1})/2}{1 + aT/2 - (1 - aT/2)z^{-1}} = \frac{a}{\frac{2}{T}\frac{z-1}{z+1} + a} \quad (4.54)$$

Hence we can obtain $H(z)$ from $H(s) = a/(s+a)$ by the substitution

$$s = \frac{2}{T}\frac{z-1}{z+1} \quad (4.55)$$

This is called the *bilinear transform* because of the linear functions of z in the numerator and denominator of (4.55). It is often written

$$s = c\frac{z-1}{z+1} \quad (4.56)$$

where the constant c depends on the application. It may be $2/T$ as above. Alternatively, some writers [3] simply take $c = 1$, and others [4] make c a function of the desired cutoff frequency, as we will see later.

Consider the transformation

$$s = c\frac{z-1}{z+1} \quad (4.57)$$

from the s-plane to the z-plane. If $s = \sigma + j\omega$ and $z = re^{j\theta} = u + jv$, then equating the real and imaginary parts of (4.57) gives

$$\sigma = c\frac{u^2 + v^2 - 1}{(u+1)^2 + v^2} \quad (4.58)$$

$$\omega = c\frac{2v}{(u+1)^2 + v^2} \quad (4.59)$$

Because $r^2 = u^2 + v^2$, it follows from (4.58) that in this case, as for the standard z-transform, the $j\omega$-axis of the s-plane ($\sigma = 0$) maps onto the circumference of the unit circle in the z-plane, and the left and right halves of the s-plane map, respectively, into the interior and exterior of the unit circle. Therefore, a stable continuous filter gives a stable digital filter with this transformation. However, in contrast to the z-transform, with the bilinear transform the s-plane is not mapped in overlapping strips onto the z-plane; instead each point in the s-plane is uniquely mapped onto the z-plane. We can see how this comes about by considering the relationship between s-plane and z-plane frequencies.

Because the $j\omega$-axis maps onto the circumference of the unit circle (Figure 4.9), there is a direct relationship between the s-plane frequency ω and the z-plane frequency ω_D, where the subscript D stands for "digital." Be-

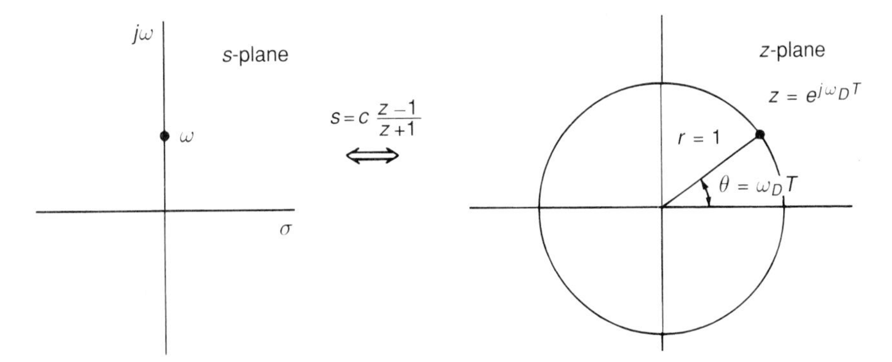

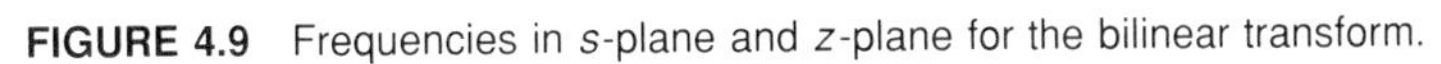
FIGURE 4.9 Frequencies in s-plane and z-plane for the bilinear transform.

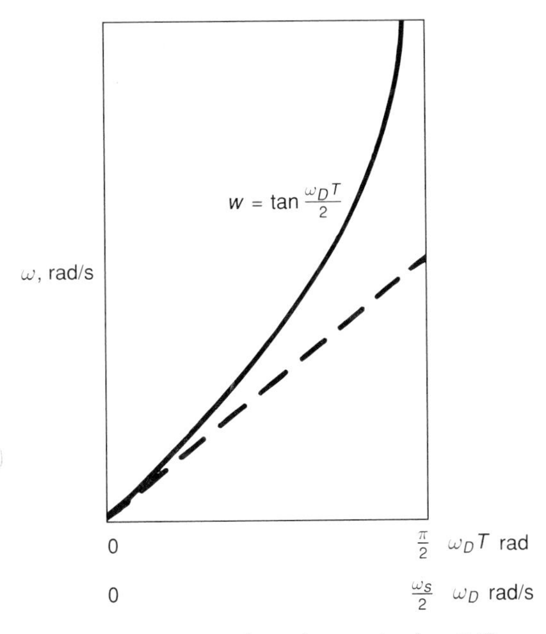

FIGURE 4.10 Frequency transformation $\omega = \tan(\omega_D\ T/2)$.

cause $z \neq \exp(sT)$, it follows that $\exp(j\omega_D T) \neq \exp(j\omega T)$ and therefore $\omega_D \neq \omega$. From (4.57), with $s = j\omega$ and $z = \exp(j\omega_D T)$,

$$j\omega = c\frac{\exp(j\omega_D T) - 1}{\exp(j\omega_D T) + 1} = jc\tan\frac{\omega_D T}{2} \tag{4.60}$$

Thus the relationship between frequencies with respect to the continuous time domain and the discrete time domain is given by

$$\omega = c\tan\frac{\omega_D T}{2} \tag{4.61}$$

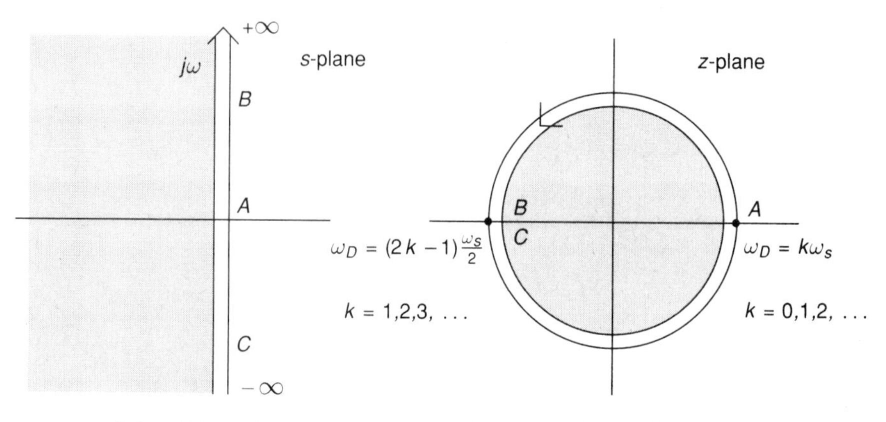

FIGURE 4.11 Mapping from s-plane to z-plane with the bilinear transform.

This nonlinear frequency transformation (4.61) is plotted in Figure 4.10 for $c = 1$. It is evident that frequency distortion, or warping, according to (4.61) is the price paid for having no aliasing. That is, each point of the infinite $j\omega$-axis maps uniquely into the circumference of the unit circle. Figure 4.11 shows the mapping from s-plane to z-plane with the bilinear transform.

Alternatively, the transformation could be carried out in two steps—from s-plane to p-plane via

$$s = c \tan h \frac{pT}{2} \tag{4.62}$$

and from p-plane to z-plane via

$$z = e^{-pT} \tag{4.63}$$

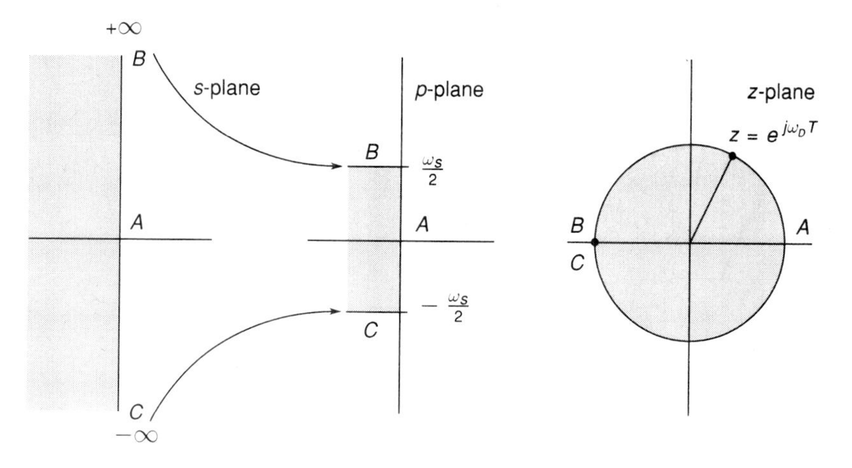

FIGURE 4.12 Bilinear transformation in two stages.

That these steps are together equivalent to (4.57) can be seen by combining (4.62) and (4.63). The mappings (4.62) and (4.63) are indicated in Figure 4.12.

Figure 4.13 shows how the magnitude response $|H(j\omega)|$ of a suitable continuous lowpass filter function is transformed into the magnitude response $|H(j\omega_D T)|$ of a digital filter by the bilinear transform. Note that the magnitude response of the digital filter *terminates* at a frequency $\omega_s/2$. Furthermore, the passband and stopband do not undergo much distortion, because the response is reasonably flat in those ranges. However, where the response is rapidly changing, as in the transition region, there is considerable distortion of the response, compared with that of the continuous filter. Thus, although this method of designing digital filters is suitable for filters that have piecewise-constant magnitude responses (such as lowpass, bandpass, highpass, and bandstop filters), it would not be suitable for designing, say, a differentiator, wherein the magnitude response is continuously changing. Because of the frequency distortion, the phase response of a digital filter that results from the bilinear transform of, say, a continuous filter with approximately linear phase (such as a Bessel filter) is decidedly nonlinear.

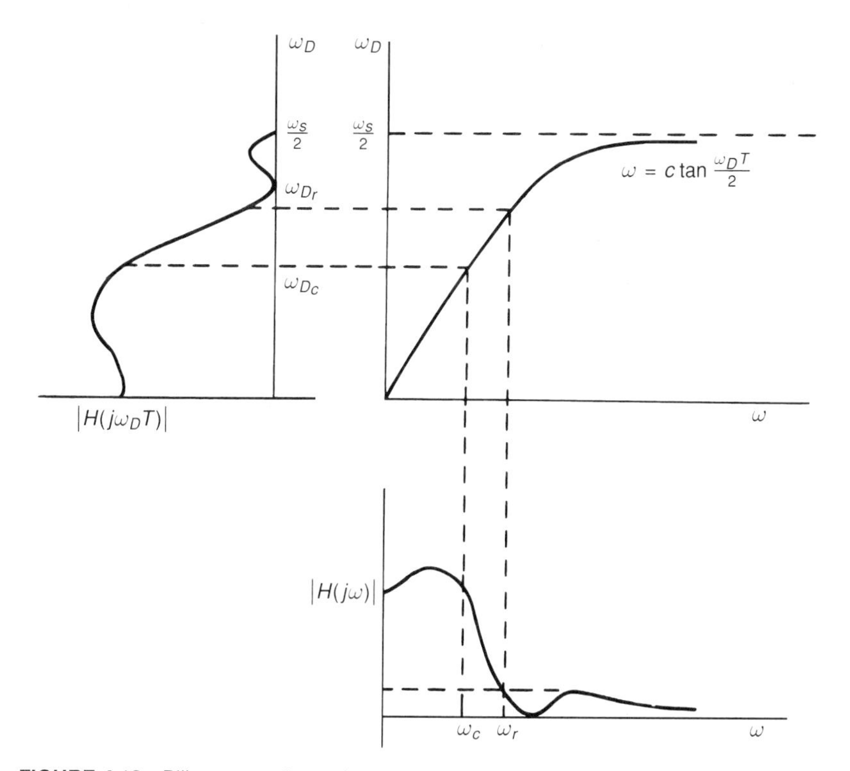

FIGURE 4.13 Bilinear transform of the magnitude response of a continuous lowpass filter.

The bilinear transform of a continuous filter function

$$H(s) = \frac{\prod_{i=1}^{m}(s - \beta_i)}{\prod_{i=1}^{n}(s - \alpha_i)} \tag{4.64}$$

is obtained by simply replacing s in (4.64) by $c(z-1)/(z+1)$. Thus

$$H(z) = \frac{\prod_{i=1}^{m}\left(c\frac{z-1}{z+1} - \beta_i\right)}{\prod_{i=1}^{n}c\left(\frac{z-1}{z+1} - \alpha_i\right)} \tag{4.65}$$

It seems that taking a bilinear transform of a continuous filter function is much easier than taking a standard z-transform via (4.1). There is a catch, however. Granted that it is easy to go from (4.64) to (4.65), how do we get (4.64) in the first place? The filter specification will be in terms of the critical frequencies ω_{D_i} (such as cutoff frequencies and bandwidth frequencies) of the *digital* filter. If we were to use the *same* critical frequencies for the continuous design and then apply the bilinear transform, the digital-filter frequencies would be in error because of the frequency warping (4.61). Therefore, we have to *pre-warp* the critical frequencies of the continuous filter.

The procedure for using the bilinear transform to design a digital filter to meet certain specifications is as follows:

1. Convert the critical frequencies ω_{D_i} specified for the digital filter to the corresponding frequencies in the s-domain by relationship (4.61),

$$\omega_i = c\tan\frac{\omega_{D_i}T}{2} \tag{4.66}$$

2. Design a continuous filter $H(s)$ that has the desired properties of the digital filter at the frequencies ω_i. The continuous filter does not have to be implemented.
3. Replace s by $c(z-1)/(z+1)$ in $H(s)$ to obtain $H(z)$, which is the required digital filter.

Two examples, a highpass and a bandstop filter design, will serve to illustrate the procedure.

EXAMPLE 4.2

Using the bilinear transform, design a highpass digital filter, monotonic in passband and stopband, with cutoff frequency of 1000 Hz and down 10 dB at 350 Hz. The sampling frequency is 5000 Hz. The gain should be unity in the passband.

Solution

The computer time step is $T = 1/f_s = 2 \times 10^{-4}$ s. We must pre-warp the critical frequencies as shown in Table 4.1.

The frequency ω_r for the normalized approximation to the continuous lowpass prototype is

$$\omega_r = \frac{\omega_2}{\omega_1} = \frac{7265}{2235} = 3.2504$$

Because of the monotonic passband and stopband, a Butterworth filter will suffice. The order of the filter is given by the approximation

$$n \approx \frac{a}{20 \log_{10} \omega_r} = \frac{10}{20 \log_{10} 3.2504} = 0.9767$$

Therefore, take $n = 1$ (the parameters in the specifications were deliberately chosen to give the simplest possible filter). The first-order Butterworth approximation to the ideal lowpass prototype is

$$H_n(s_n) = \frac{1}{s_n + 1}$$

By (2.154), we transform this into a continuous highpass filter with cutoff frequency ω_ℓ via the transformation

$$s_n = \frac{\omega_\ell}{s}$$

This gives

$$H(s) = \frac{1}{\omega_\ell / s + 1} = \frac{s}{s + \omega_\ell}$$

Now we transform $H(s)$ into a digital filter by replacing s with $c(z-1)/(z+1)$, where $c = 2/T$, obtaining the digital filter transfer function

$$H(z) = \frac{c}{c + \omega_\ell} \frac{z - 1}{z - (c - \omega_\ell)/(c + \omega_\ell)}$$

TABLE 4.1 Critical Frequencies for Digital and Continuous Highpass Filters

$\omega_{D_i}, i = 1, 2$ (rad/s)	$\omega_i = (2/T) \tan(\omega_{D_i} T/2), i = 1, 2$ (rad/s)
700π	2235
2000π	7265

With $c = 10^4$ and $\omega_\ell = 7265$, we get

$$H(z) = 0.5792\frac{z - 1}{z - 0.1584}$$

The corresponding difference equation is obtained by writing

$$H(z) = 0.5792\frac{1 - z^{-1}}{1 - 0.1584z^{-1}}$$

Hence, by inspection, the difference equation is

$$y_k = 0.1584y_{k-1} + 0.5792(x_k - x_{k-1})$$

A pole-zero plot and a rough magnitude response obtained graphically from the pole-zero plot are shown in Figure 4.14. The operating range of the filter ends at $\omega T = \pi$ ($\omega = \omega_s/2$), which is the point $z = -1$. At that point the maximum gain in the passband is attained. In this case,

$$H(z)|_{z=-1} = 0.5792\frac{-1 - 1}{-1 - 0.1584} = 1$$

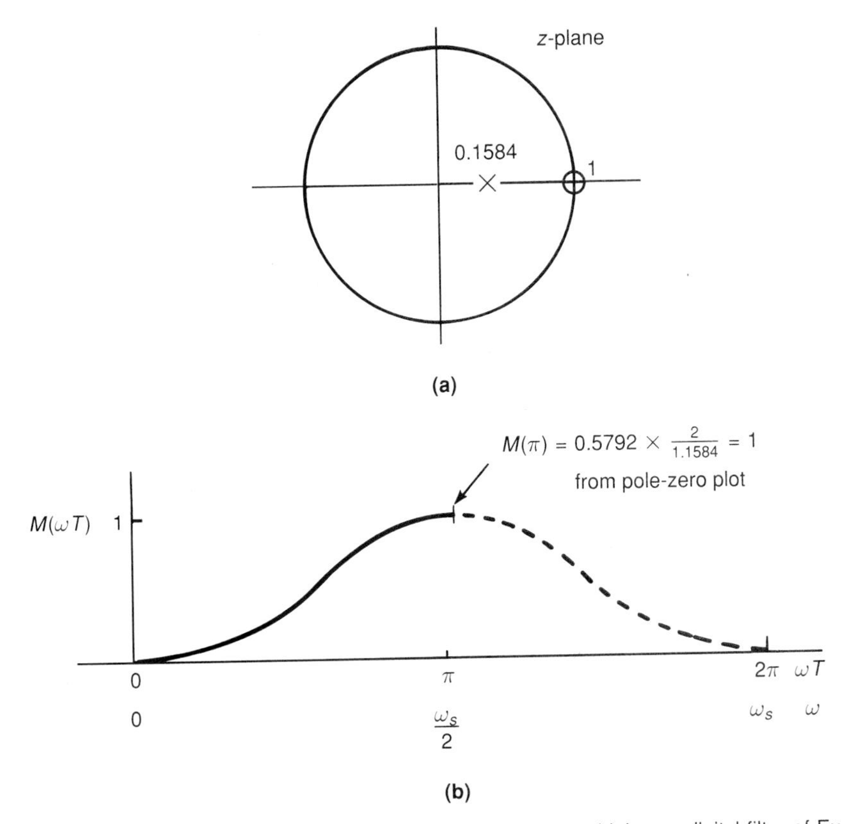

FIGURE 4.14 Pole-zero plot and magnitude response for the highpass digital filter of Example 4.2. (a) Pole-zero plot. (b) Magnitude response (approximate).

as specified. We got the specified gain in the passband because the lowpass prototype had unity gain at $s = 0$, and we used the value $2/T$ for c. If we had used a different value for c, it would have been necessary to adjust the digital filter gain accordingly.

In the next example, we will take $c = 1$. This will not affect the filter design except for the gain adjustment mentioned.

EXAMPLE 4.3

Using the bilinear transform, design a bandstop digital filter with stopband 100 to 600 Hz. The magnitude response should be monotonic in the stopband and should have a ripple of about 1.1 dB in the passband. The response should be down at least 20 dB at 200 Hz and 400 Hz. Use a 2000-Hz sampling rate.

Solution

The critical frequencies for the digital filter and the corresponding frequencies for the continuous filter are given in Table 4.2.

From (2.159), the normalized lowpass filter frequencies ω_{n_i} and the continuous bandstop filter frequencies ω_i are related by

$$\omega_{n_i} = \frac{B\omega_i}{-\omega_i^2 + \omega_0^2} \tag{4.67}$$

where $B = \omega_4 - \omega_1$, $\omega_0^2 = \omega_1\omega_4$. This gives $\omega_{n_1} = 1$, $\omega_{n_2} = 3.54$, $\omega_{n_3} = -2.85$, and $\omega_{n_4} = -1$ rad/s (see also Example 2.11). For the normalized filter, $\omega_r = \min(\omega_{n_2}/\omega_{n_1}, \ \omega_{n_3}/\omega_{n_4}) = 2.85$, in this case. Some designers simply use the minimum of the ratios ω_2/ω_1 and ω_4/ω_3 shown in Table 4.2 for ω_r. This approach happens to give the same filter order in this example, but is not recommended in general.

A ripple in the passband implies that the lowpass prototype approximation will have a ripple in the passband, that is, a Chebyshev type I filter will suffice in this case (see Chapter 2). The parameter ϵ and the order n are

TABLE 4.2 Critical Frequencies for Digital and Continuous Bandstop Filters

$\omega_{D_i}, i = 1, 4$ (rad/s)	$\omega_i = \tan \omega_{D_i}T/2, i = 1, 4$ (rad/s)	
200π	0.1584	$\omega_2/\omega_1 = 2.0511$
400π	0.3249	
800π	0.7265	$\omega_4/\omega_3 = 1.8946$
1200π	1.3764	

needed to define the filter. First find ϵ from

$$\epsilon = (10^{R/10} - 1)^{1/2} \tag{4.68}$$

which gives $\epsilon = 0.5369$ for $R = 1.1$.

The order of the filter is obtained by requiring that (2.109) be satisfied, which implies that

$$n \geq (L + 6 - 20 \log_{10} \epsilon)/(6 + 20 \log_{10} \omega_r) = 2.1 \tag{4.69}$$

where, in this case, $L = 20$ dB. The nearest integer value that satisfies the specifications is $n = 3$.

For a third-order Chebyshev type I filter, (2.114) gives the poles as

$$s = -0.4766, -0.2383 \pm j0.9593 \tag{4.70}$$

The transfer function for the lowpass prototype approximation is

$$H_n(s_n) = \frac{K_1}{(s_n + s_0)(s_n^2 + 2as_n + a^2 + b^2)} \tag{4.71}$$

where, from (4.70), $s_0 = 0.4766$, $a = 0.2383$, and $b = 0.9593$.

Use the frequency transformation

$$s_n = \frac{Bs}{s^2 + \omega_0^2} \tag{4.72}$$

to obtain the sixth-order continuous bandstop filter

$$H(s) = \frac{K_2(s^2 + \omega_0^2)^3}{(s^2 + Bs/s_0 + \omega_0^2)(s^4 + c_3 s^3 + c_2 s^2 + c_1 s + c_0)} \tag{4.73}$$

where

$$\begin{aligned}
K_2 &= K_1/[s_0(a^2 + b^2)] \\
B &= \omega_4 - \omega_1 \\
\omega_0^2 &= \omega_1 \omega_4 \\
c_3 &= 2aB/(a^2 + b^2) \\
c_2 &= B^2/(a^2 + b^2) + 2\omega_0^2 \\
c_1 &= 2aB\omega_0^2/(a^2 + b^2) \\
c_0 &= \omega_0^4
\end{aligned} \tag{4.74}$$

Applying the bilinear transform ($c = 1$)

$$s = \frac{z - 1}{z + 1} \tag{4.75}$$

gives the required digital filter

$$H(z) = \frac{K[z^2 - 2(1 - \omega_0^2)z/(1 + \omega_0^2) + 1]^3}{(z^2 - 2d_1 z + d_0)(z^4 + e_3 z^3 + e_2 z^2 + e_1 z + e_0)} \tag{4.76}$$

where

$$
\begin{aligned}
K &= (1-\omega_0^2)^3 K_2/[(1+\frac{B}{s_0}+\omega_0^2)(1+c_3+c_2+c_1+c_0)] \\
d_1 &= (1-\omega_0^2)/(1+B/s_0+\omega_0^2) \\
d_0 &= (1-B/s_0+\omega_0^2)/(1+B/s_0+\omega_0^2) \\
e_3 &= -2(2+c_3-c_1-2c_0)/(1+c_3+c_2+c_1+c_0) \\
e_2 &= 2(3-c_2+3c_0)/(1+c_3+c_2+c_1+c_0) \\
e_1 &= -2(2-c_3+c_1-2c_0)/(1+c_3+c_2+c_1+c_0) \\
e_0 &= (1-c_3+c_2-c_1+c_0)/(1+c_3+c_2+c_1+c_0)
\end{aligned}
\tag{4.77}
$$

When the numerical values for this example are substituted into (4.76), the result is

$$
H(z) = \frac{0.1285(z^2-1.2841z+1)^3}{(z-0.8376)(z+0.4232)(z^2-1.7950z+0.8872)(z^2+0.5229z+0.6893)} \tag{4.78}
$$

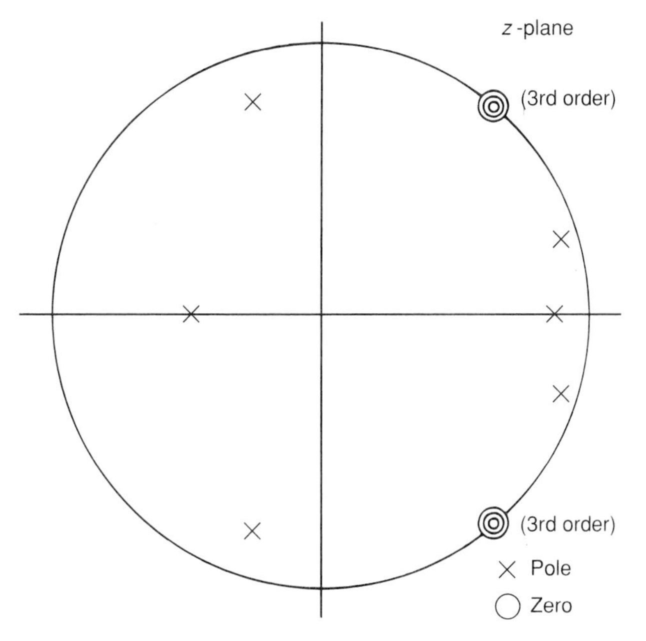

FIGURE 4.15 Pole-zero plot for a sixth-order digital bandstop filter of Example 4.3.

The pole-zero plot for the bandstop filter is shown in Figure 4.15. Note that the six complex zeros $[z = \exp(\pm j1747.40T)]$ lie on the unit circle. This means that the magnitude response goes to zero at frequency 1747.40 rad/s = 278.11 Hz, which, of course, is in the stopband. The magnitude response and the phase response can be evaluated by setting $z = \exp(j\omega T)$ in (4.78) or, graphically, from Figure 4.15. A sketch of the magnitude response is given in Figure 4.16.

This example was treated in detail to illustrate the technique. In practice, a digital computer program would be written to design the continuous prototype, to make the lowpass-to-bandstop transformation from the equations in Chapter 2, and to apply the bilinear transform to the result. In general, designers may either write their own program for a specific application or use one of the many programs published in such sources as [5]. Interactive computer programs for designing digital filters by means of the bilinear transform are given on the diskette. The designer may use a Butterworth, Chebyshev type I, elliptic, or Bessel approximation. Applying the program ELLIPD to Example 4.3 gave an elliptic filter and resulted in non-monotonic passbands. The magnitude response of the elliptic filter is plotted in Figure 4.16 for comparison. Note that a fourth-order elliptic filter meets the specifications.

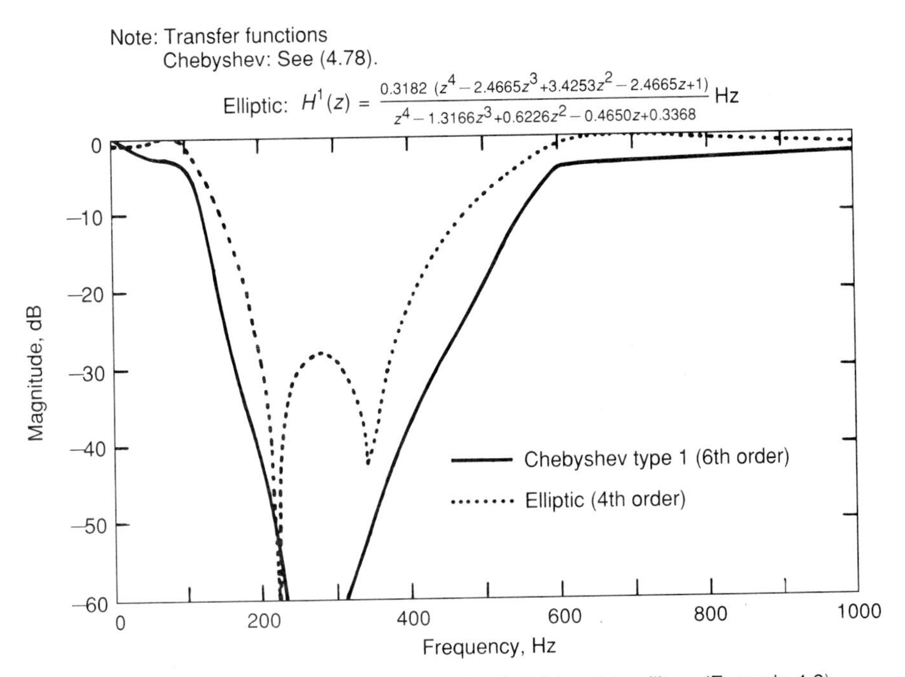

FIGURE 4.16 Magnitude responses for digital bandstop filters (Example 4.3).

The bilinear transform is the technique most widely used to realize recursive digital filters from continuous filter functions in the four basic types (lowpass, highpass, bandpass, and bandstop). The frequency warping problem can be handled by pre-warping the continuous filter as we have shown. The other disadvantage (the bilinear transform's not producing an impulse-invariant design) is not important in many applications. The fact that there is no aliasing makes this an attractive method. In addition, it lends itself readily to computer programming because it requires only a simple substitution for s.

We will consider one more transform method that is easy to use and often results in a satisfactory design: the *matched z-transform.*

4.4 DESIGNING DIGITAL FILTERS BY MEANS OF THE MATCHED z-TRANSFORM

The matched z-transform is based on *heuristic* reasoning derived partly from experience gained with the standard z-transform. We shall list the rules for going from a suitable continuous filter function $H(s)$ to a digital filter function $H(z)$. Then we shall outline the reasoning behind them and discuss some possible shortcomings. The continuous function must be in factored form,

$$H(s) = \frac{K\prod_{i=1}^{m}(s - s_i)}{\prod_{i=1}^{n}(s - p_i)} \tag{4.79}$$

where $n \geq m$. Then $H(s)$ is transformed according to the following rules to get the digital filter function $H(z)$.

1. Replace real poles and zeros p_i and s_i, by $\exp(Tp_i)$, and $\exp(Ts_i)$, respectively.
2. Replace complex poles and zeros, say p_i or $s_i = -a \pm jb$, by $r \exp(\pm j\theta)$, where $r = \exp(-aT)$ and $\theta = bT$.
3. All zeros of $H(s)$ at $s = \infty$ are mapped in $H(z)$ to the point $z = -1$.
4. Select the gain constant of $H(z)$ to match the gain of $H(s)$ at passband center or at some other critical point.

Recall that when we use the z-transform to convert from $H(s)$ as described in Section 4.2, a real pole of $H(s)$ at $s = s_i = -a$, say, results in a pole of $H(z)$ at $z = e^{-aT}$, whereas a pair of complex conjugate poles at $s = -a \pm jb$ results in a pair of complex poles at $z = re^{\pm j\theta}$, where $r = e^{-aT}$ and $\theta = bT$. Thus, from rules 1 and 2, the matched z-transform maps the poles of $H(s)$ in the same manner as the z-transform. Hence a stable continuous filter results in a stable digital filter by the matched z-transform method.

The mapping of the zeros is different, however. With the z-transform, real and complex zeros of $H(s)$ are not converted in the same manner as the poles. With the matched z-transform, by rules 1 and 2, they are. Care must be taken to ensure that ω_s is high enough so that both poles *and* zeros of $H(s)$ lie within the primary strip $[-j\omega_s/2, j\omega_s/2]$ of the left half of the s-plane (see Figure 3.9). Otherwise, there will be aliasing upon making the transformation. Like the z-transform, the matched z-transform is restricted to bandlimited functions (unless a guardband filter is used) because of the aliasing problem. Furthermore, the matched z-transform maps an all-pole continuous filter into an all-pole digital filter that may not adequately represent the frequency characteristics of the former.

If $n > m$ in (4.79), then as s approaches ∞,

$$H(s) \to \lim_{s \to \infty} K/s^L \tag{4.80}$$

where $L = n - m$. Thus, $H(s)$ will have L zeros at infinity. Rule 3 states that these zeros result in a zero of order L in $H(z)$ at $z = -1$. The reasoning behind this is that $s \to \infty$ is the limiting frequency for the continuous filter, whereas $\omega = \omega_s/2$, which corresponds to point $z = -1$ on the circumference of the unit circle, is the highest operating frequency for the digital filter. It seems logical to map these high-frequency zeros accordingly.

Rule 4 deals with the prescribed gain constant of the digital filter. It is often required that the gain of a filter be unity (zero dB) in the passband. For example, if (4.79) represents a lowpass filter, then at zero frequency ($s = 0$), for unity gain, K has the value

$$K = \frac{\prod_{i=1}^{n} p_i}{\prod_{i=1}^{m} s_i} \tag{4.81}$$

Applying these four rules to $H(s)$ gives

$$H(z) = K_D(z+1)^L \frac{\prod_{i=1}^{m} (z - e^{s_i T})}{\prod_{i=1}^{n} (z - e^{p_i T})} \tag{4.82}$$

where $L = n - m$, and K_D has a value that ensures that $H(z)|_{z=1} = 1$.

Recall that $z = 1$ is the zero-frequency point in the z-plane. Therefore,

$$K_D = \frac{\prod_{i=1}^{n}(1 - e^{p_i T})}{2^L \prod_{i=1}^{m}(1 - e^{s_i T})} \tag{4.83}$$

We shall consider a simple example of the transformation.

EXAMPLE 4.4

Use the matched z-transform to design a digital lowpass filter from the continuous lowpass filter

$$H(s) = \frac{s + 4}{(s + 1)(s^2 + 2s + 4)}$$

The digital filter should have unity gain at $\omega = 0$. The computer time step is 0.1 s.

Solution

From rule 1, the real pole and the zero transform to $e^{-T} = e^{-0.1} = 0.9048$ and $e^{-4T} = e^{-0.4} = 0.6703$, respectively. For the complex poles, $a = 1$ and $b = \sqrt{3}$. By rule 2, these are replaced by $z = re^{\pm j\theta}$, where $r = e^{-aT} = e^{-0.1} = 0.9048$ and $\theta = bT = 0.1732$ rad (9.9 degrees), which gives a second-order term $z^2 - 2r(\cos\theta)z + r^2 = z^2 - 1.78252 + 0.8187$ in the denominator of $H(z)$. Because $n = 3$ and $m = 1$, we have $L = 2$, so by rule 3, there are two zeros at $z = -1$. Therefore,

$$H(z) = \frac{K_D(z + 1)^2(z - 0.6703)}{(z - 0.9048)(z^2 - 1.78252 + 0.8187)}$$

where K_D is computed from the requirement that $H(z)|_{z=1} = 1$. Hence

$$K_D = \frac{(1 - 0.9048)(1 - 1.7825 + 0.8187)}{2^2(1 - 0.6703)} = 0.0026$$

4.5 SUMMARY OF TRANSFORMS FROM CONTINUOUS FILTERS TO DIGITAL FILTERS

We have considered five different methods for transforming a continuous filter into a digital filter. These are summarized, along with their advantages and disadvantages, in Table 4.3.

TABLE 4.3 Summary of Transform Methods from $H(s)$ to $H(z)$

Name	Transformation	Advantages	Disadvantages	Remarks
z-transform	$H(z) = \sum$ residues of $H(p)\dfrac{1}{1 - e^{Tp}z^{-1}}$ at poles of $H(p)$	1. Impulse-invariant 2. Same critical frequencies in continuous and digital filters 3. Stable digital filter from stable continuous filter 4. Easy to apply (but not as easy as the other methods that follow)	Possibility of aliasing, hence restricted to band-limited functions	Probably the second-best method
Forward rectangular approximation	Replace s by $(z-1)/T$	Easy to apply	Unstable digital filter can result from stable continuous filter	Not recommended
Backward rectangular approximation	Replace s by $(z-1)/Tz$	Easy to apply	Pole locations and attainable frequencies are restricted in the digital filter	Not recommended
Bilinear transform (trapezoidal approximation)	Replace s by $c\dfrac{z-1}{z+1}$	1. No aliasing, hence not restricted to bandlimited functions 2. Stable digital filter from stable continuous filter 3. Easy to apply	1. Frequency warping 2. Not impulse-invariant	Best all-around method
Matched z-transform	See rules in Section 4.4	1. Stable digital filter from stable continuous filter 2. Same critical frequencies in continuous and digital filters 3. Easy to apply	1. Not suitable for transforming all-pole filters 2. Possibility of aliasing 3. Not inpulse-invariant 4. Heuristic approach	Not as versatile as bilinear or z-transform but can yield useful designs in certain cases

4.6 GENERAL TRANSFORM TECHNIQUES

There are two approaches to the design of recursive digital filters from continuous filters. These are indicated in Figure 4.17 and designated method 1 and 2.

The first approach, wherein the frequency transformation is applied to the continuous lowpass prototype approximation before digitization, has been used in this text. The second method involves carrying out the frequency transformation on a digitized version of the prototype approximation. It was proposed by Constantinides [6] and is further described in [1] and [4]. The resulting frequency transformations are more involved than those on the continuous filter, as is apparent in Table 4.4, which is taken from [6]. In the exercises, the reader is invited to use this table to design a filter for the specifications given in Example 4.2. The digitized lowpass filter $H_n(z_n)$ with cutoff frequency ω_c can be designed by any of the approximation methods. The attenuation between passband and stopband must be compatible with that required for the resulting digital filter $H(z)$ that is obtained from $H_n(z_n)$ by using the appropriate transformation from Table 4.4. In the opinion of the author, using method 2 entails the loss of some intuitive feeling, compared with method 1. However, method 2 is certainly an effective technique. Which approach one chooses is really a matter of preference.

Gold and Radar [3] and later Constantinides [7] suggested that when designing lowpass and highpass digital filters using a bilinear transformation, one should choose the constant c so as to eliminate the need for making a frequency transformation of the continuous lowpass prototype approximation. Furthermore, they pointed out that bandpass and bandstop digital filters can be obtained directly from the continuous lowpass prototype by suitable bi-

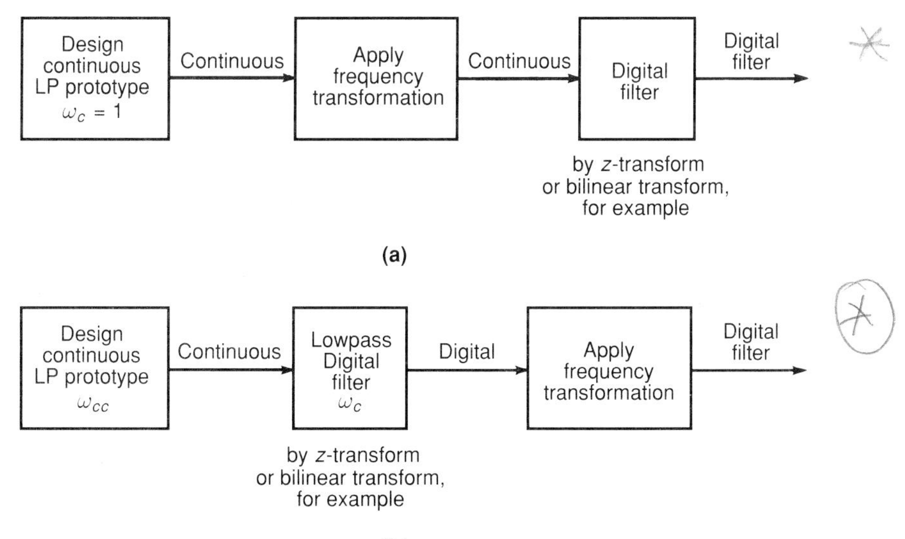

FIGURE 4.17 Two methods for designing digital filters from continuous filter functions. (a) General method 1. (b) General method 2.

TABLE 4.4 Frequency Transformation from Digitized Approximation to Lowpass Prototype with Cutoff Frequency ω_c [6]

Filter Type	Replace z_n^{-1} in Prototype by	Associated Design Formulas	Remarks
Lowpass	$\dfrac{z^{-1}-\alpha}{1-\alpha z^{-1}}$	$\alpha = \dfrac{\sin\left(\dfrac{\omega_c-\omega_u}{2}\right)T}{\sin\left(\dfrac{\omega_c+\omega_u}{2}\right)T}$	ω_u = cutoff frequency of desired lowpass filter
Highpass	$-\dfrac{z^{-1}+\alpha}{1+\alpha z^{-1}}$	$\alpha = -\dfrac{\cos\left(\dfrac{\omega_c+\omega_\ell}{2}\right)T}{\cos\left(\dfrac{\omega_c-\omega_\ell}{2}\right)T}$	ω_ℓ = cutoff frequency of desired highpass filter
Bandpass	$-\left(\dfrac{z^{-2}-\dfrac{2\alpha k}{k+1}z^{-1}+\dfrac{k-1}{k+1}}{\dfrac{k-1}{k+1}z^{-2}-\dfrac{2\alpha k}{k+1}z^{-1}+1}\right)$	$\alpha = \cos\omega_0 T = \dfrac{\cos\left(\dfrac{\omega_u+\omega_\ell}{2}\right)T}{\cos\left(\dfrac{\omega_u-\omega_\ell}{2}\right)T}$ $k = \cot\left[\left(\dfrac{\omega_u\omega_\ell}{2}\right)T\right]\tan\dfrac{\omega_c T}{2}$	ω_u, ω_ℓ = upper and lower frequencies of desired passband
Bandstop	$\left(\dfrac{z^{-2}-\dfrac{2\alpha}{1+k}z^{-1}+\dfrac{1-k}{1+k}}{\dfrac{1-k}{1+k}z^{-2}-\dfrac{2\alpha}{1+k}z^{-1}+1}\right)$	$\alpha = \cos\omega_0 T = \dfrac{\cos\left(\dfrac{\omega_u+\omega_\ell}{2}\right)T}{\cos\left(\dfrac{\omega_u-\omega_\ell}{2}\right)T}$ $k = \tan\left[\left(\dfrac{\omega_u-\omega_\ell}{2}\right)T\right]\tan\dfrac{\omega_c T}{2}$	ω_u, ω_ℓ = upper and lower frequencies of desired stopband

quadratic z-transformations, which are given below. These transformations are easier to use than those in Table 4.4 and are therefore recommended.

Looking at the lowpass case first, we note that if ω_{D_c} is the cutoff frequency of the digital filter, and ω_c is the cutoff frequency of the continuous filter, then they are related by

$$\omega_c = c \tan(\omega_{D_c} T/2) \tag{4.84}$$

Instead of choosing $c = 1$ or $2/T$, we can select it as the solution of (4.84) —that is,

$$c = \omega_c \cot(\omega_{D_c} T/2) \tag{4.85}$$

In particular, if $\omega_c = 1$, then

$$c = \cot(\omega_{D_c} T/2) \tag{4.86}$$

Thus the intermediate step of going from lowpass continuous prototype to lowpass continuous filter with $\omega_c \neq 1$ can be eliminated by choosing c according to (4.86) so that the bilinear transformation becomes

$$s = \cot(\omega_{D_c} T/2) \frac{z-1}{z+1} \tag{4.87}$$

and, in general, the frequencies for the lowpass digital filters are related to those for the continuous filter by

$$\omega = \cot(\omega_{D_c} T/2) \tan(\omega_D T/2) \tag{4.88}$$

The transformation is illustrated in Figure 4.18. We note that the magnitude-squared response for the continuous Butterworth filter, with cutoff frequency ω_c,

$$M^2(\omega) = \frac{1}{1 + (\omega/\omega_c)^{2n}} \tag{4.89}$$

gives, under this transformation, the magnitude-squared response of the digital Butterworth filter as

$$M^2(\omega T) = \frac{1}{1 + [\cot(\omega_{D_c} T/2) \tan(\omega_D T/2)]^{2n}} \tag{4.90}$$

For the highpass design, by our previous method, we replaced s_n in the lowpass prototype with ω_c/s to get a continuous highpass filter with cutoff frequency ω_c. Then we transformed the latter filter into a digital filter by replacing s with $c(z-1)/(z+1)$. The frequency transformation in the continuous domain can be eliminated by inverting the bilinear expression (4.57) and choosing a suitable value for the multiplying constant. Let

$$s = c' \frac{z+1}{z-1} \tag{4.91}$$

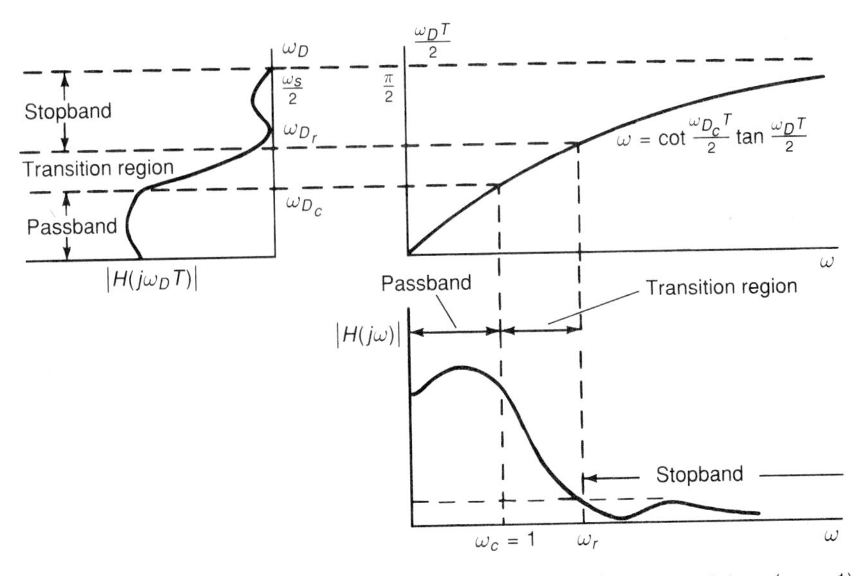

FIGURE 4.18 Direct bilinear transformation from continuous lowpass prototype ($\omega_c = 1$) to digital lowpass filter of arbitrary cutoff frequency ω_{D_c}.

This operation preserves the transformation. That is,

$j\omega$ axis $\rightarrow$ circumference of unit circle
left half of the s-plane $\rightarrow$ interior of unit circle
right half of the s-plane $\rightarrow$ exterior of unit circle

as before. The relationship between frequencies in the s-plane and those in the z-plane is given by

$$\omega = c' \cot(\omega_D T/2) \tag{4.92}$$

Hence

$$c' = \omega_c \tan(\omega_{D_c} T/2) \tag{4.93}$$

where ω_c and ω_{D_c} are the cutoff frequencies of the continuous and digital filters, respectively. In particular, for $\omega_c = 1$,

$$c' = \tan(\omega_{D_c} T/2) \tag{4.94}$$

Then transformation (4.91) becomes

$$s = \tan(\omega_{D_c} T/2)(z+1)/(z-1) \tag{4.95}$$

and the relationship between s-plane and z-plane frequencies is

$$\omega = \tan(\omega_{D_c} T/2)\cot(\omega_D T/2) \tag{4.96}$$

The transformation is illustrated in Figure 4.19.

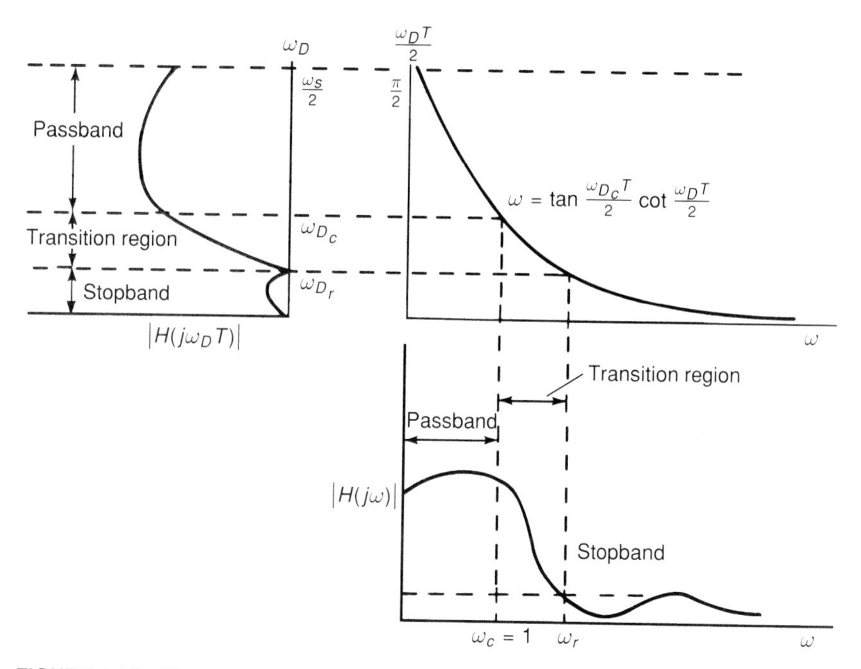

FIGURE 4.19 Direct bilinear transformation of a continuous lowpass prototype approximation ($\omega_c = 1$) to a digital highpass filter of arbitrary cutoff frequency ω_{D_c}.

EXAMPLE 4.5

Use transformation (4.91) to design a highpass filter to meet the specifications given in Example 4.2.

Solution

In this case, $\omega_{D_c} = 2000\pi$ and $T = 2 \times 10^{-4}$ s. Therefore,

$$c' = \tan(\omega_{D_c}T/2) = 0.7265$$

We must pre-warp the critical frequencies ω_{D_i} according to Table 4.5.

As in Example 4.2, a first-order Butterworth lowpass prototype will suffice:

$$H_n(s) = \frac{1}{s_n + 1}$$

Replace s by $c'(z+1)/(z-1)$ to get

$$\begin{aligned} H(z) &= \frac{z-1}{(1+c')[z-(1-c')/(1+c')]} \\ &= \frac{0.5792(z-1)}{z-0.1584}, \end{aligned}$$

TABLE 4.5 Critical Frequencies for Digital and Continuous Highpass Filters

$\omega_{D_i}, i = 1, 2$ (rad/s)	$\omega_i = \tan(\omega_{D_c}T/2)\cot(\omega_{D_i}T/2)$ (rad/s)
700π	3.2504
$200\pi(= \omega_{D_c})$	1.0000

for the transfer function of the required highpass filter, which is the same as we obtained before.

In the case of a bandpass or bandstop design, to eliminate the necessity of a frequency transformation of the continuous prototype, replace s_n in the latter by

$$s_n = c\frac{1 - 2\alpha z^{-1} + z^{-2}}{1 - z^{-2}} \tag{4.97}$$

for bandpass and by

$$s_n = c'\frac{1 - z^{-2}}{1 - 2\alpha z^{-1} + z^{-2}} \tag{4.98}$$

for bandstop (see [6] and [8]). The constants c, c' and α are defined below in terms of filter specifications. Of course, the transformation is now "biquadratic" rather than bilinear. Consider the bandpass transformation (4.97). When z varies on the circumference of the unit circle—that is, $z = \exp(j\omega_D T)$, then

$$s = c\frac{\exp(j2\omega_D T) - 2\alpha\exp(j\omega_D T) + 1}{\exp(j2\omega_D T) - 1} = jc\frac{\alpha - \cos\omega_D T}{\sin\omega_D T} \tag{4.99}$$

Because the right-hand side of (4.99) is purely imaginary, it follows that

$$\omega = c\frac{\alpha - \cos\omega_D T}{\sin\omega_D T} \tag{4.100}$$

When $\omega = 0$,

$$\omega_D = \omega_{D_0} = \frac{1}{T}\cos^{-1}\alpha \tag{4.101}$$

which is the center frequency of the passband. If the specified cutoff frequencies of the passband are ω_{D_1} and ω_{D_2}, with $\omega_c = 1$ being the cutoff frequency of the lowpass prototype, then

$$-\omega_c = c\frac{\alpha - \cos\omega_{D_1}T}{\sin\omega_{D_1}T} \tag{4.102}$$

and

$$\omega_c = c\frac{\alpha - \cos\omega_{D_2}T}{\sin\omega_{D_2}T} \tag{4.103}$$

It follows that

$$-\frac{\alpha - \cos \omega_{D_1} T}{\sin \omega_{D_1} T} = \frac{\alpha - \cos \omega_{D_2} T}{\sin \omega_{D_2} T} \tag{4.104}$$

After doing some algebra, we can evaluate α and c as

$$\alpha = \frac{\cos[(\omega_{D_1} + \omega_{D_2})T/2]}{\cos[(\omega_{D_2} - \omega_{D_1})T/2]} \tag{4.105}$$

and

$$c = \omega_c \cot[(\omega_{D_2} - \omega_{D_1})T/2] \tag{4.106}$$

The transformation (4.100) is illustrated in Figure 4.20. Multiband bandpass filter transformations performed by using the same technique are given in [6]. Similarly, for the bandstop design,

$$\omega = \frac{c' \sin \omega_D T}{\alpha - \cos \omega_D T} \tag{4.107}$$

where

$$\alpha = \frac{\cos[(\omega_{D_2} + \omega_{D_1})T/2]}{\cos[(\omega_{D_2} - \omega_{D_1})T/2]} \tag{4.108}$$

as for the bandpass transformation. Also,

$$c' = \omega_c \tan[(\omega_{D_2} - \omega_{D_1})T/2] \tag{4.109}$$

Multiband bandstop transformations are given in [6].

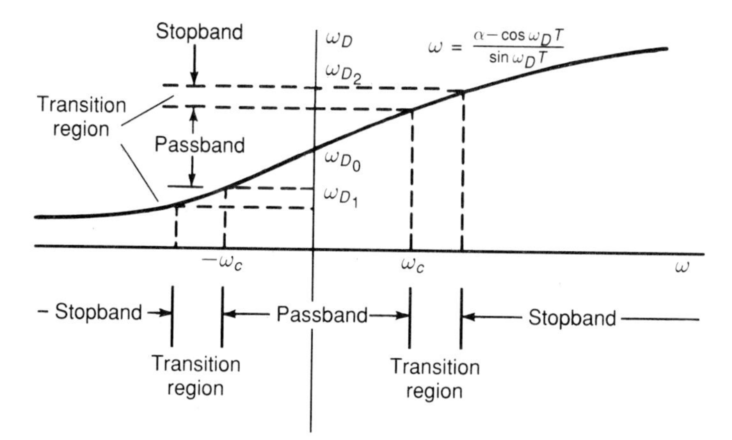

FIGURE 4.20 Frequency transformation for deriving a digital bandpass filter from a continuous lowpass filter.

4.7 DIRECT DESIGN OF RECURSIVE DIGITAL FILTERS

The foregoing transform methods for designing digital filters are best suited to filters having piecewise-constant magnitude response, such as lowpass, highpass, bandpass, and bandstop filters. They are not so useful for designing digital filters with arbitrary magnitude response. For example, as a result of the frequency distortion, the bilinear transform is not suitable for designing a differentiator because the desired magnitude response varies with frequency.

Several investigators, notably Steiglitz [2] and Deczky [9], have proposed computer-aided optimization design procedures for choosing the coefficients of a recursive digital filter the magnitude response of which approximates an arbitrary specified magnitude characteristic. We will discuss the Steiglitz method.

Suppose that samples of an arbitrary magnitude response are specified; that is, we are given

$$A_n \text{ at frequency } \omega_n T, \quad n = 1, 2, \ldots, N$$

where N is the number of *discrete* frequencies, because this is a computer-aided design method. The samples do not need to be uniformly spaced, but they generally are. Let $H(z)$ be the transfer function of the required digital filter, in cascade form.

$$H(z) = C \prod_{k=1}^{K} \frac{1 + a_k z^{-1} + b_k z^{-2}}{1 + c_k z^{-1} + d_k z^{-2}} \tag{4.110}$$

We want to choose the real constants C, a_k, b_k, c_k and d_k, $1 \leq k \leq K$, so that $|H(j\omega_n T)|$ approximates A_n as closely as possible. To make the rather vague phrase "as closely as possible" more definite, we select the $(4K + 1)$ constants so as to minimize the squared error

$$Q = \sum_{n=1}^{N} [|H(j\omega_n T)| - A_n]^2 \tag{4.111}$$

Values of the constants that minimize Q are called optimal values. For brevity, let

$$\theta = \{a_1, b_1, c_1, d_1, \ldots, a_K, b_K, c_K, d_K\}$$

and

$$H_n = H(j\omega_n T)/C$$

Then the error criterion (4.111) becomes

$$Q(\theta, C) = \sum_{n=1}^{N} \{C|H_n| - A_n\}^2 \tag{4.112}$$

By taking the partial derivative of Q with respect to C, and setting it equal to zero, we can solve for the value of C that minimizes Q. Thus

$$\frac{\partial Q}{\partial C} = 2\sum_{n=1}^{N} |H_n|[C|H_n| - A_n] = 0 \tag{4.113}$$

From this, the optimal value for C is

$$C^* = \frac{\displaystyle\sum_{n=1}^{N} |H_n| A_n}{\displaystyle\sum_{n=1}^{N} |H_n|^2} \tag{4.114}$$

It is not so easy to determine the optimal value for θ (denoted by θ^*), because Q is a nonlinear function of θ. Steiglitz used the Fletcher–Powell algorithm [10] to solve for θ^*. This algorithm is an iterative search method that performs a one-dimensional minimization at each cycle along a direction determined by the gradient $\partial Q/\partial\theta$. The latter is needed for application of the method and is given by

$$\frac{\partial Q(\theta, C^*)}{\partial \theta_k} = \frac{\partial Q(\theta, C^*)}{\partial |H_n|} \cdot \frac{\partial |H_n|}{\partial \theta_k} \tag{4.115}$$

where θ_k can denote a_k, b_k, c_k, or d_k. From (4.111), with C^* replacing C, the gradient (4.115) can be written

$$\frac{\partial Q(\theta, C^*)}{\partial \theta_k} = 2C^* \sum_{n=1}^{N} \{|H_n|C^* - A_n\} \frac{\partial |H_n|}{\partial \theta_k} \tag{4.116}$$

The partial derivatives required for (4.116) can be evaluated as follows:

$$\begin{aligned} \frac{\partial |H_n|}{\partial a_k} &= \frac{1}{|H_n|} Re\left\{ \bar{H}_n \frac{\partial H_n}{\partial a_n} \right\} \\ &= \frac{1}{|H_n|} Re\left\{ \bar{H}_n H_n \frac{z^{-1}}{1 + a_k z^{-1} + b_k z^{-2}} \right\}_{z=e^{j\omega_n T}} \\ &= |H_n| Re\left\{ \frac{z^{-1}}{1 + a_k z^{-1} + b_k z^{-2}} \right\}_{z=e^{j\omega_n T}} \end{aligned} \tag{4.117}$$

where $Re\{\quad\}$ denotes the real part of the term in parentheses, and $\bar{H}_n$ is the conjugate of H_n. Similarly,

$$\frac{\partial |H_n|}{\partial b_k} = |H_n| Re\left\{ \frac{z^{-2}}{1 + a_k z^{-1} + b_k z^{-2}} \right\}_{z=e^{j\omega_n T}} \tag{4.118}$$

$$\frac{\partial |H_n|}{\partial c_k} = -|H_n| Re\left\{ \frac{z^{-1}}{1 + c_k z^{-1} + d_k z^{-2}} \right\}_{z=e^{j\omega_n T}} \tag{4.119}$$

$$\frac{\partial |H_n|}{\partial d_k} = -|H_n| Re \left\{ \frac{z^{-2}}{1 + c_k z^{-1} + d_k z^{-2}} \right\}_{z=e^{j\omega_n T}} \tag{4.120}$$

A computer program is written to search for θ^* by means of the Fletcher–Powell algorithm, using the partial derivatives (4.117) through (4.120). An initial value is assumed for θ. This initial guess affects the number of iterations required to find θ^*, and few guidelines are available. It is perhaps best to set equal to zero the initial values of the coefficients in all the K sections except one and to choose the values for the coefficients of the remaining section to suit the type of magnitude function required, e.g. an all-pole function ($a_k = b_k = 0$, $c_k \neq 0$, $d_k \neq 0$) to represent a lowpass function. The search for θ^* is ended when the value of θ does not differ from θ for the previous iteration by more than a prescribed tolerance—say, 10^{-5}.

When the minimization procedure is terminated, the result is usually a digital filter transfer function the magnitude response of which approximates very closely the prescribed values A_n. However, it often turns out that some of the poles and zeros lie *outside* the unit circle, because their locations were not restricted in the minimization process. If the poles are not brought inside the unit circle, the filter will be unstable. If the zeros are not brought inside the unit circle, we will have a non-minimum phase transfer function. (For systems with the same magnitude characteristic, the range in phase angle of the minimum phase transfer function is minimum for all such systems.)

This is where Steiglitz had a clever idea. He simply inverted the poles and zeros outside the unit circle; for example, a pair of complex poles at $z = re^{\pm j\theta}$ becomes $z = (1/r)e^{\pm j\theta}$. This does not change the shape of the magnitude response, as the following simple case illustrates.

Suppose that a real pole is outside the unit circle at p_1, $|p_1| > 1$. Replace this by a pole at $1/p_1$. The original transfer function was

$$H(z) = \frac{1}{z - p_1} \frac{N(z)}{D(z)} \tag{4.121}$$

After the inversion, the new transfer function is

$$H'(z) = \frac{1}{z - 1/p_1} \cdot \frac{N(z)}{D(z)} = \frac{z - p_1}{z - 1/p_1} \cdot H(z) \tag{4.122}$$

The magnitude of the multiplying factor $(z - p_1)/(z - 1/p_1)$ is equal to the ratio of the lengths of the vectors from the poles to a point on the unit circle, $|z| = 1$. From the geometry of Figure 4.21, the multiplying factor squared is

$$\left(\frac{z - p_1}{z - 1/p_1} \right)^2 = \frac{\sin^2 \alpha + (p_1 - \cos \alpha)^2}{\sin^2 \alpha + (\cos \alpha - 1/p_1)^2} = p_1^2 \tag{4.123}$$

Therefore, the multiplying factor is p_1. This is a constant, so the pole inversion does not change the shape of the magnitude response. Showing that this is also true for complex poles and zeros is left as an exercise.

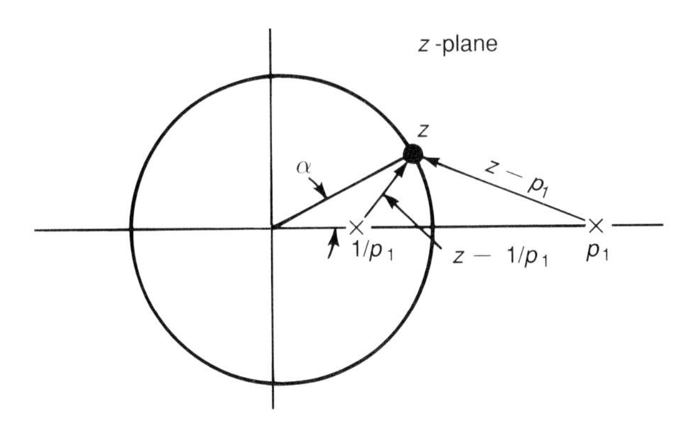

FIGURE 4.21 Geometry for the multiplying factor in pole inversion.

Steiglitz noted that if the optimization program is started again after inverting the poles and zeros as necessary, and if the previously obtained convergence values of the coefficient θ are used as starting values for the new iterations, it is often possible to decrease the squared error Q even further—and so improve the design.

Thus the procedure consists of three steps:

1. Use the optimization program to minimize Q without regard to location of the poles and zeros.
2. After convergence in step 1, invert all poles and zeros that lie outside the unit circle.
3. Using the convergence values for θ and C found in step 1 as starting points, make further iterations of the optimization program. After convergence, use the new optimal values for θ and C to form the digital filter transfer function.

Some examples of the use of Steiglitz's method are given in his original paper [2]. One of these was for a filter with a desired magnitude response that increases linearly with frequency—an ideal differentiator. For this example, 21 samples of the desired magnitude response were chosen with

$$\omega_n = (n-1)(0.05)\omega_s/2, \quad n = 1, 2, \ldots, 21 \tag{4.124}$$

so that the corresponding amplitude samples were

$$A_n = \omega_n T/\pi \tag{4.125}$$

as shown in Figure 4.22. A one-section design ($K = 1$) converged after 96 iterations [2] to give

$$H(z) = \frac{0.36637364(1 - 0.32917379z^{-1} - 0.67082621z^{-2})}{1 + 0.85938970z^{-1} + 0.10210106z^{-2}}$$

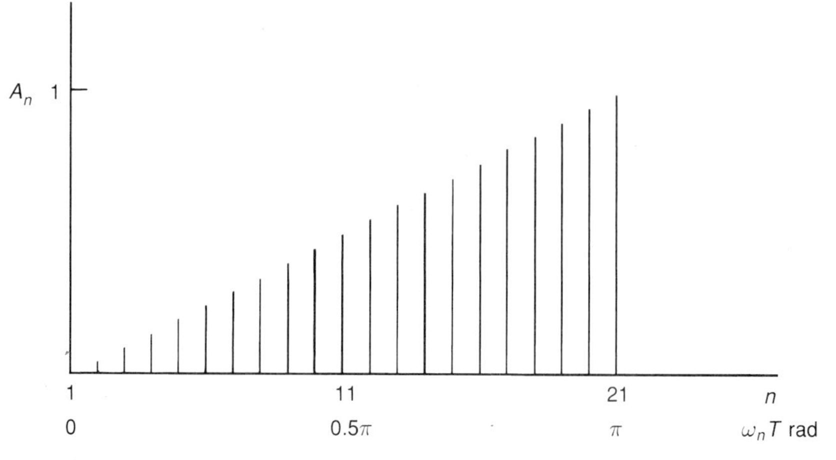

FIGURE 4.22 Samples of desired magnitude response for ideal differentiator.

$$= \frac{0.36637364(1 - z^{-1})(1 + 0.67082621z^{-1})}{(1 + 0.14240300z^{-1})(1 - 0.71698670z^{-1})} \tag{4.126}$$

The error in the magnitude response, compared with the ideal response in Figure 4.22, is shown in Figure 4.23(a). It is evident that the error is less than 1% over the operating range of the filter. The phase response resulting from the design is shown in Figure 4.23(b). This does not deviate excessively from an ideal linear phase lag (also shown), although the method has no provisions for specification for the phase response. The one-section design does not have any poles or zeros outside the unit circle. Nevertheless, Steiglitz used the results of the one-section design as the starting point for a two-section design. The result was an even more accurate approximation to the ideal response. The phase response was not changed appreciably.

The Steiglitz procedure was generalized by Deczky [9]. Instead of minimizing the squared error, (4.111), he used the error raised to the pth power, $p \geq 2$, and suitably weighted—that is,

$$Q = \sum_{n=1}^{N} \mathrm{w}_n [|H(j\omega_n T)| - A_n]^p \tag{4.127}$$

where w_n is the weight at frequency ω_n. When $p = 2$ and $\mathrm{w}_n = 1$ for all n, this reduces to the Steiglitz case. As an alternative error criterion, Deczky proposed minimizing the error in *group delay* raised to the pth power, $p \geq 2$, and weighted—that is,

$$Q = \sum_{n=1}^{N} \mathrm{w}_n [\tau(\omega_n T) - \tau_d(\omega_n T)]^p \tag{4.128}$$

where the group delay of a filter is defined as the negative of the rate of

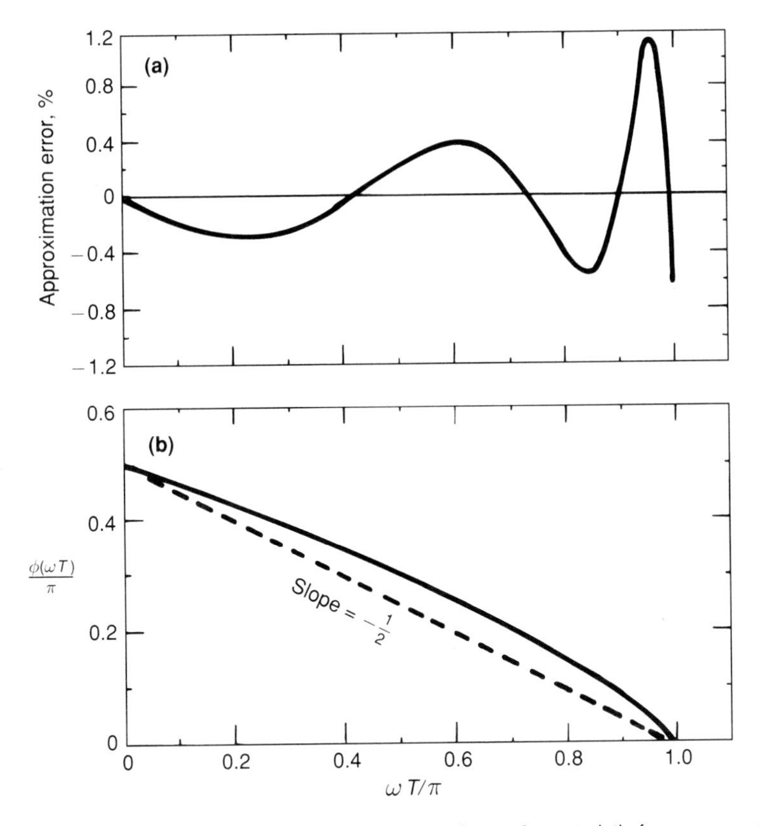

FIGURE 4.23 (a) Magnitude response error and (b) phase characteristic for a one-section differentiator (after Steiglitz).

change of its phase response with respect to frequency:

$$\tau(\omega T) = -\frac{d[\phi(\omega T)]}{d(\omega T)} \tag{4.129}$$

The desired or specified group delay is $\tau_d(\omega T)$. The desired filter is in the cascade form given in (4.110), and we derive the $(4K + 1)$ coefficients as before by the Fletcher–Powell method, minimizing either (4.127) or (4.128).

Other direct design methods have been proposed that do not require minimization of an error criterion. One such class deals with designing a filter the magnitude-squared response of which, $|H(j\omega T)|^2$, approximates a specified magnitude-squared function. Results obtained do not differ much from those given by the bilinear transform of a corresponding continuous filter transfer function. In addition, this direct approach is more complex mathematically. Hence the bilinear transform method is preferred where a continuous prototype is available.

4.8 FILTERING SHORT SEQUENCES

When a step sequence is applied as input to a stable recursive filter, the filter output passes through a transient stage (see, for example, Figure 3.12) before it reaches a steady state. The time required for the output to come within a certain tolerance of the final value is called the settling time, as for continuous filters (Chapter 2). Suppose, now, that an arbitrary sequence is to be filtered, say, by passing a certain band of frequencies. Until steady-state conditions are reached, the filtering will be ineffective. This presents no problem for a sequence that is long compared to the filter settling time. However, for a sequence $x(k)$, $0 \leq k \leq M$, where length of which is of the same order as the settling time, little of the sequence may be left to be filtered by the time steady-state conditions are attained [Figure 4.24(a)]. If the short sequence has been stored prior to filtering, the problem can be alleviated by initialization of the filter. For example, for a first-order filter where one initial condition on the output is required, we can set $y(0)$ equal to $x(0)$ to avoid the initial step on the filter. For a second-order filter, two initial conditions are required. One way to provide these is to generate a new sequence $h(k)$ from the given sequence $x(k)$, as shown in Figure 4.24(b), and to append it in front of $x(k)$:

$$h(k) = 2x(0) - x(-k), \quad -M \leq k \leq 0 \tag{4.130}$$

The sequence to be filtered is now doubled in length, compared with $x(k)$. If filtering starts at $k = -M$, the transient should have died out by the time $k = 0$, where the values and slopes of $h(k)$ and $x(k)$ are matched and, therefore, $x(k)$ can be filtered effectively. For higher-order filters, whether they consist of parallel or cascaded banks of first- and second-order sections or are realized in direct form, this technique usually gives good results. Consider the following example.

EXAMPLE 4.6

A 50-point sequence $x(k)$ is measured with samples spaced at intervals of 0.01 s. It is known to contain a 40-Hz sinusoid superimposed on a low-frequency signal. We wish to remove the unwanted 40-Hz component.

Solution

We design a lowpass filter to the following specifications:

Cutoff frequency = 10 Hz

Magnitude reponse down at least 50 dB at 40 Hz

Sampling frequency = 100 Hz

The DFIL program on the diskette gives a third-order recursive Butterworth filter to meet these specifications. The difference equation of the filter is

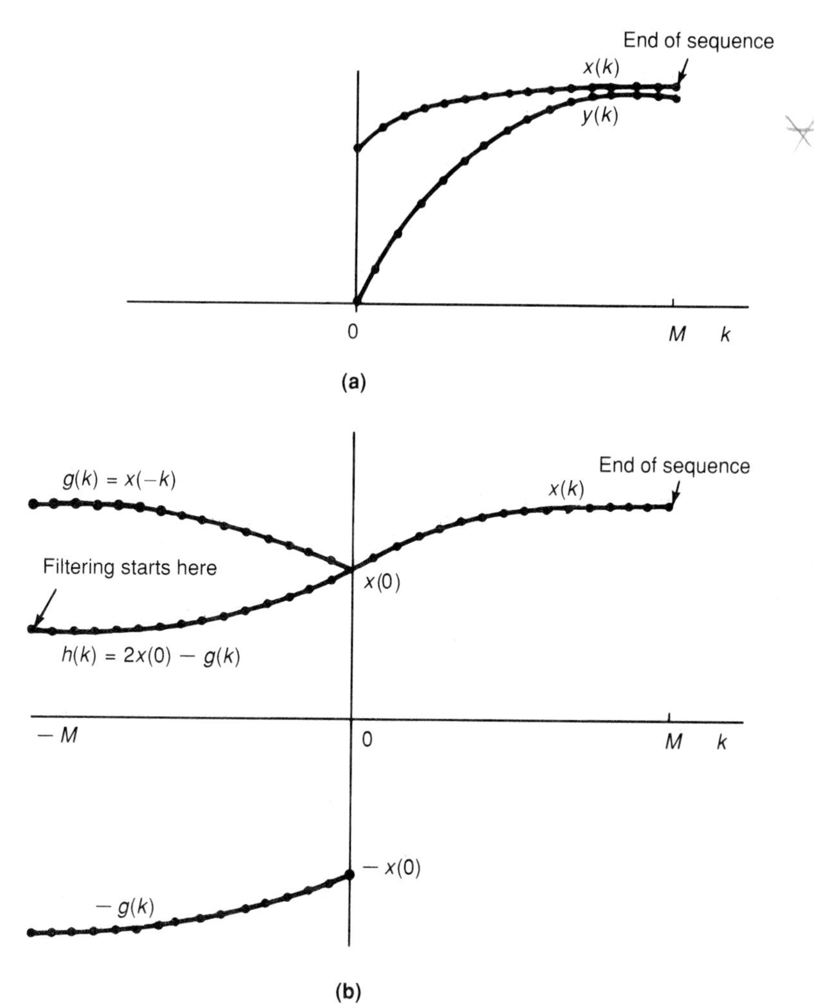

FIGURE 4.24 Recursive filtering of a short sequence. (a) Input and output with no modification. (b) Expanded input sequence to reduce transient.

$$y(k) = \sum_{i=1}^{3} A_i y(k-i) + \sum_{i=0}^{3} B_i x(k-i)$$

where

$$\begin{array}{ll} A_1 = 1.7600 & B_0 = 0.0181 \\ A_2 = 1.1829 & B_1 = 0.0543 \\ A_3 = 0.2781 & B_2 = 0.0543 \\ & B_3 = 0.0181 \end{array}$$

The original sequence $x(k)$ and the filter output $y(k)$ are shown on the right-hand side of Figure 4.25. It is evident that the transient stage occupies

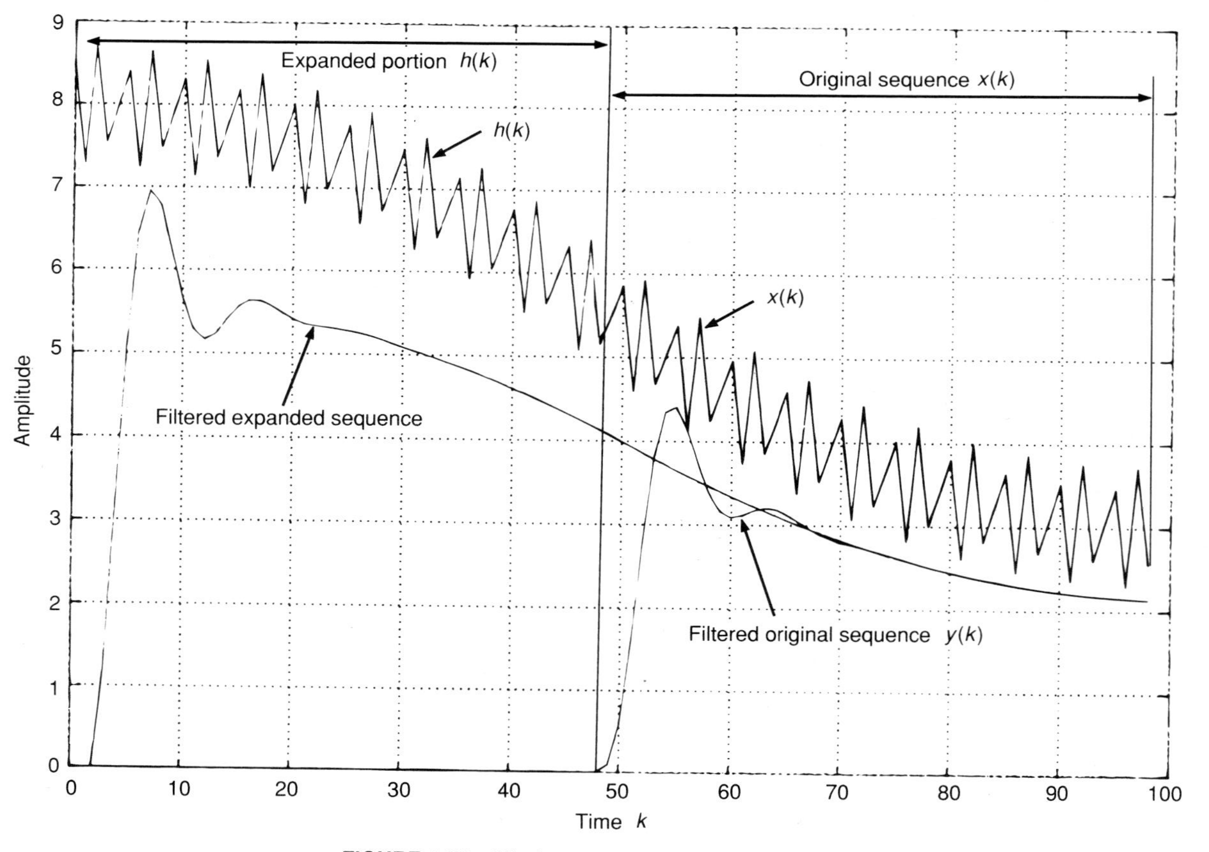

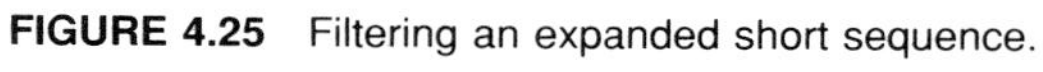

FIGURE 4.25 Filtering an expanded short sequence.

more than one-third of the output. However, if we expand the sequence as described above to obtain $h(k)$ according to (4.130), concatenate $h(k)$ to $x(k)$, and filter the expanded sequence, the transient has died out before we reach the $x(k)$ portion, and the whole of $x(k)$ is filtered properly, as shown in Figure 4.25.

4.9 FILTERING AN ARBITRARY SEQUENCE

The programs DFIL and ELLIPD on the diskette can be used to design a recursive digital filter to given specifications (see Appendix 4A). The program outputs include the filter poles and zeros, the coefficients of the difference equation for the filter, and the filter frequency response. The difference equation algorithm can be programmed to filter an arbitrary sequence either in real time or from stored data. We will illustrate the procedure with an example.

EXAMPLE 4.7

Discrete measurements are made of a 0.5-Hz square wave at 0.05-s intervals. It is desired to pass the 1.5-Hz sinusoidal component of the wave and to discard the remainder. Use one of the filter programs on the diskette.

Solution

In Example 2.1 we saw that, with a suitable choice of origin, a square wave can be expanded in a Fourier sine series with odd harmonics only. Thus the series will have the form

$$w(t) = C\left(\sin x + \frac{1}{3}\sin 3x + \frac{1}{5}\sin 5x + \cdots\right)$$

where C is a constant, $x = 2\pi f_1 t$, and f_1 is the fundamental frequency (0.5 Hz in this case). We want to obtain the third harmonic, the 1.5-Hz component.

We note that although, in principle, the frequencies present can approach infinity, in practice the higher-frequency components are multiplied by the factor $1/n$, where n is the harmonic number and rapidly becomes negligible. Nevertheless, by sampling at only a 20-Hz rate, we risk aliasing of the components above 10 Hz.

In view of the remarks in the last paragraph, we will consider two different filtering arrangements.

1. Use a bandpass filter to pass the 1.5-Hz component, and assume that aliasing is negligible.

2. Use an anti-aliasing lowpass filter to remove the components above 10 Hz, and sample and bandpass filter the remainder.

Under method 1, we design a bandpass filter to the following specifications:

Sampling frequency = 20 Hz

Passband = 1.0 to 2.0 Hz

At least 30-dB attenuation at 0.5 Hz and 2.5 Hz

Passband ripple = 1.0 dB

Using the program ELLIPD, we find that a sixth-order elliptic filter meets the specifications. Figure 4.26 shows the magnitude response for the filter. The square wave was filtered with the corresponding sixth-order difference equation,

$$y(k) = \sum_{i=1}^{6} A_i y(k-i) + \sum_{i=0}^{6} B_i w(k-1)$$

where the coefficients have the following values:

$A_1 = 0.510708981 \times 10$ $\qquad B_0 = 0.167092404 \times 10^{-1}$

$A_2 = -0.114221154 \times 10^2$ $\qquad B_1 = -0.549752641 \times 10^{-1}$

$A_3 = 0.142551649 \times 10^2$ $\qquad B_2 = 0.604067306 \times 10^{-1}$

$A_4 = -0.104637099 \times 10^2$ $\qquad B_3 = -0.371019669 \times 10^{-1}$

$A_5 = 0.42866400 \times 10$ $\qquad B_4 = -0.604067306 \times 10^{-1}$

$A_6 = -0.769760671$ $\qquad B_5 = 0.549752641 \times 10^{-1}$

$B_6 = -0.1670924004 \times 10^{-1}$

The result of the filtering is shown in Figure 4.27. After an initial transient, the output settles into a 1.5-Hz sinusoidal oscillation that is the third harmonic of the square wave.

Our success in using method 1 to obtain the desired output implies that aliasing is not a problem here. Nevertheless, to illustrate the technique for more general sequences, we will apply the second method also.

With method 2 we first design an anti-aliasing continuous lowpass filter to the following specifications:

Cutoff frequency = 4 Hz

At least 40 dB down at 8 Hz

Passband ripple = 1 dB

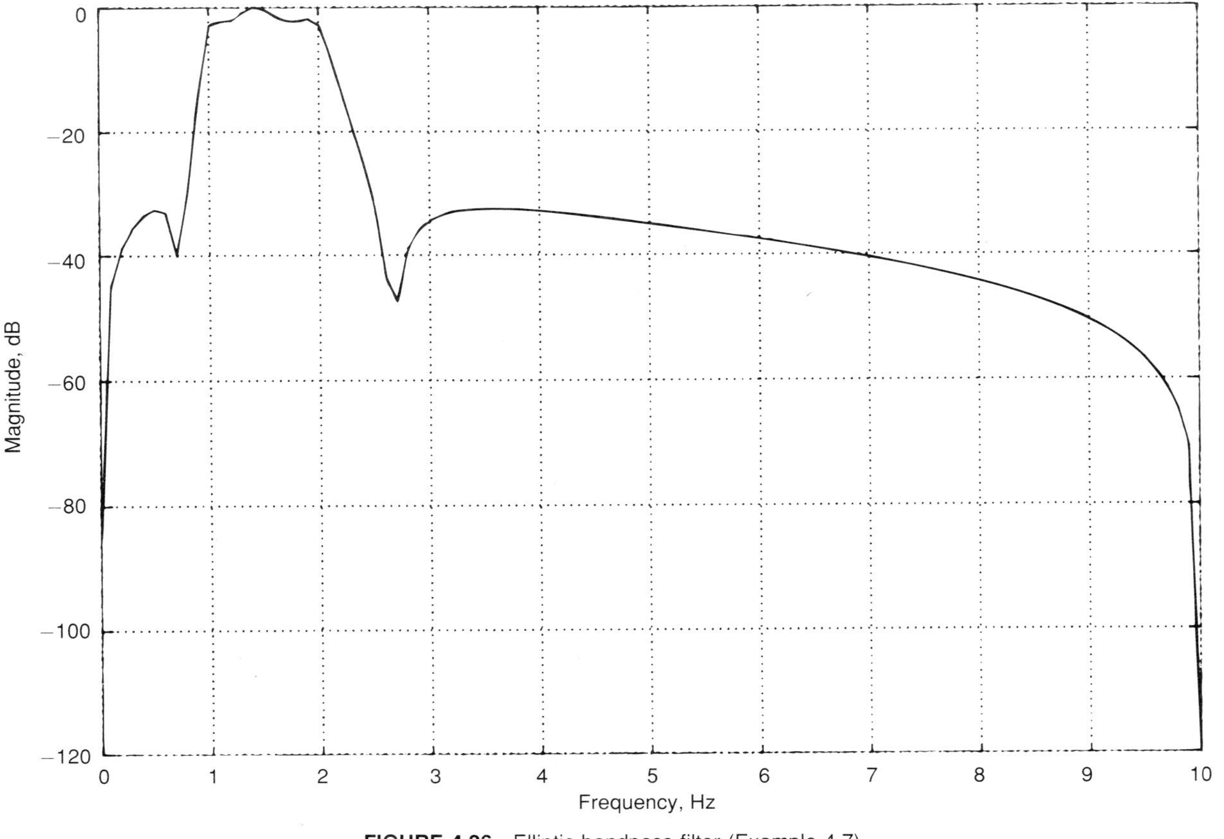

FIGURE 4.26 Elliptic bandpass filter (Example 4.7).

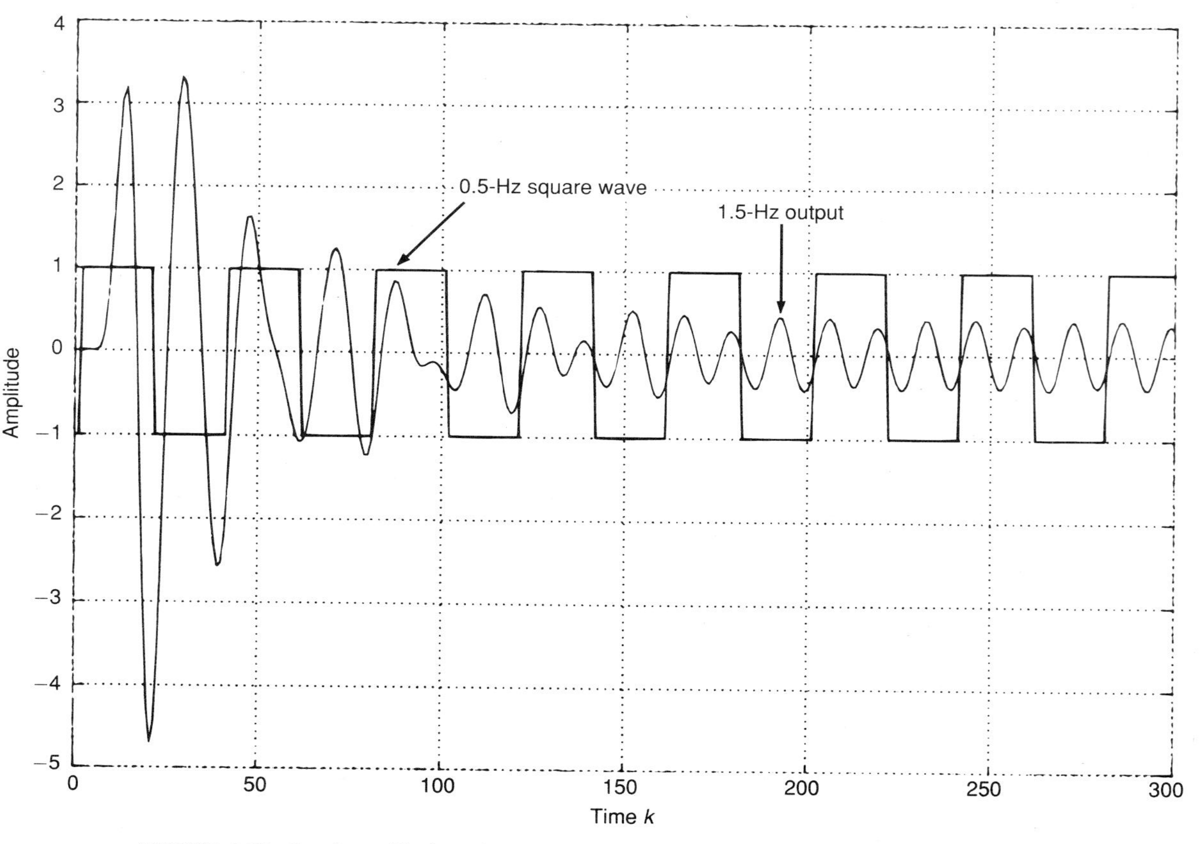

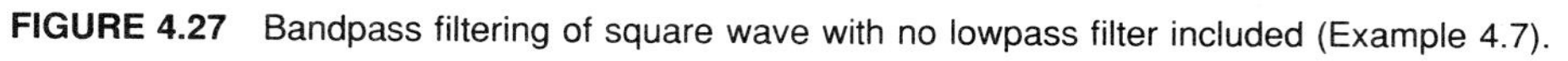

FIGURE 4.27 Bandpass filtering of square wave with no lowpass filter included (Example 4.7).

According to the ELLIP program, a fourth-order continuous elliptic filter satisfies these requirements.

The output of the lowpass filter is then sampled at 20 Hz and filtered by the bandpass filter designed for method 1. The results are shown in Figure 4.28. It is evident from the output of the lowpass filter that most of the square wave energy is contained in the lower-frequency components (below 10 Hz), which confirms that not much aliasing will occur. After an initial transient stage, the output of the bandpass filter gives the desired third harmonic.

In this example, we did not try to match initial conditions for the filter with the input sequence, as we did in the last section. Instead, all initial conditions on the filters were set equal to zero. This situation could arise when filtering in real time—that is, not on stored data. As a result, there is an unavoidable transient stage with a recursive filter, but this does not pose a problem if the sequence is long enough.

4.10 SUMMARY

Several methods are described for converting a suitable continuous filter frequency function $H(z)$ into a recursive digital filter design $H(z)$ that satisfies given specifications. These methods include

1. An impulse-invariant design obtained by applying the z-transform to $H(s)$. Use of this technique is restricted to bandlimited functions because of aliasing.
2. Numerical approximations to integration that give three different transforms according to the forward difference, the backward difference, and the trapezoidal rules, respectively. Only the bilinear transform given by the trapezoidal rule is recommended. Although there is no aliasing with the bilinear transform, warping, or distortion, occurs between the s-plane and z-plane frequencies. As a result, it is necessary to pre-warp the critical frequencies in $H(s)$ to meet the specification requirements on $H(z)$. The frequency warping is not very detrimental for filters that have piecewise-constant magnitude responses (such as lowpass, highpass, bandpass, and bandstop filters). The bilinear transform is probably the method used most often for obtaining such filters (which account for most of the practical applications) from continuous functions.
3. The matched z-transform, which is a heuristic technique, is very simple to use when a suitable continuous function $H(s)$ is available in factored form. It often results in a useful design but, like the z-transform, is restricted to bandlimited functions.

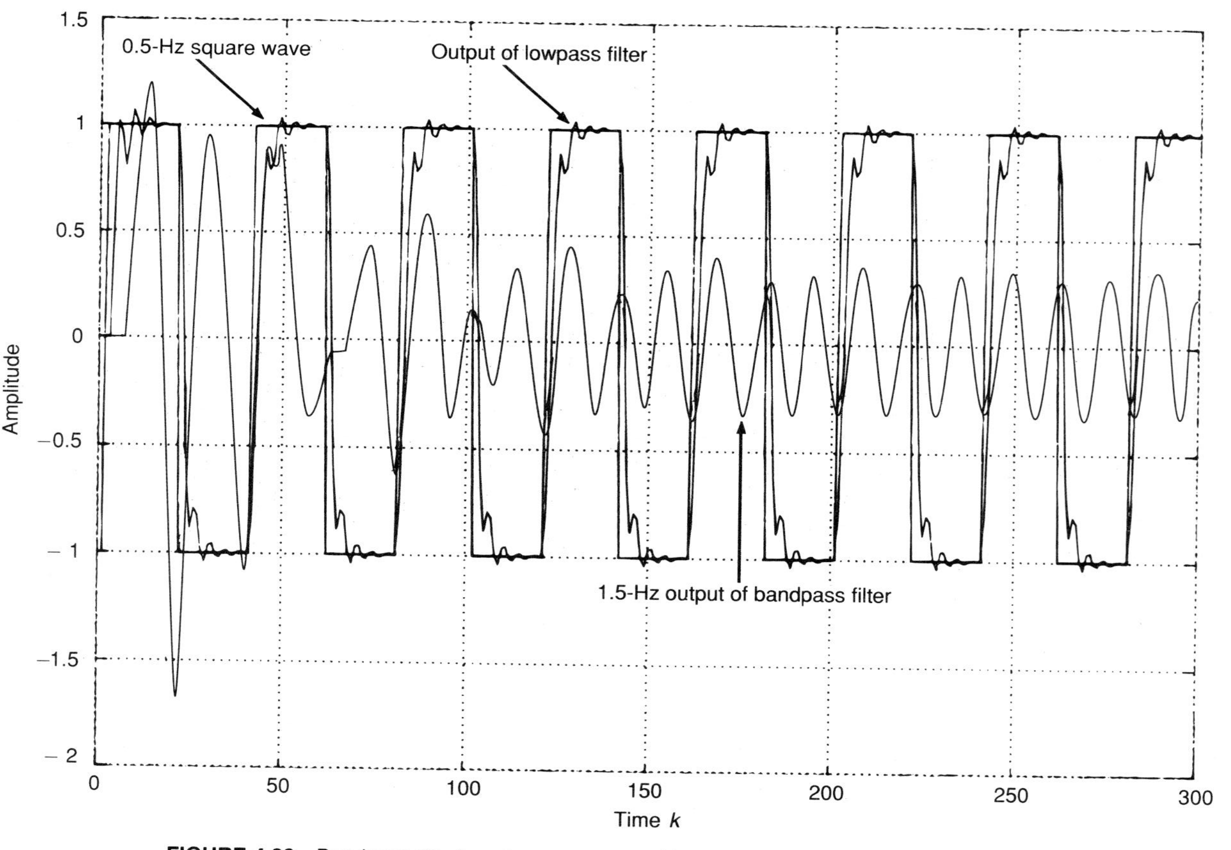

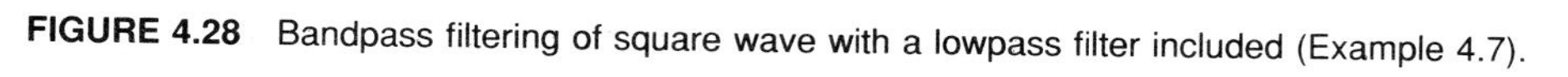

FIGURE 4.28 Bandpass filtering of square wave with a lowpass filter included (Example 4.7).

In a different class of methods from the foregoing, the continuous lowpass prototype is digitized, and the frequency transformations are carried out in the digital domain. This yields suitable filters at the expense, perhaps, of some loss of intuition during the design stage, compared with carrying out the frequency transformations in the continuous domain.

All these techniques for designing digital filters from continuous functions are best suited for filters with piecewise constant magnitude responses. They are not so useful for designing a filter the magnitude response of which varies with frequency, such as a differentiator. For cases where the magnitude function is arbitrary, direct methods are available that minimize some function of the error between the actual magnitude response and the desired response. This nonlinear problem requires a computer-aided design.

REFERENCES FOR CHAPTER 4

1. L. R. Rabiner and B. Gold, *Theory and Application of Digital Signal Processing* (Englewood Cliffs, NJ: Prentice Hall, 1975), Chap. 4.
2. K. Steiglitz, "Computer-Aided Design of Recursive Digital Filters," *IEEE Trans. on Audio Electroacoustics* 18, (June 1970): 123–129.
3. B. Gold and C. M. Rader, *Digital Processing of Signals* (New York: McGraw-Hill, 1969), Chap. 3.
4. A. V. Oppenheim and R. W. Schafer, *Digital Signal Processing* (Englewood Cliffs, NJ: Prentice-Hall, 1975), Chap. 5.
5. IEEE Digital Signal Processing Committee, *Selected Computer Programs for Digital Signal Processing* (New York: IEEE Press, 1979).
6. A. G. Constantinides, "Spectral Transformations for Digital Filters," *Proc. IEEE*, 117, no. 8 (1970): 1585–1590.
7. R. E. Bogner and A. G. Constantinides, eds., *Introduction to Digital Filtering* (New York: Wiley, 1975), Chap. 4.
8. A. Peled and B. Liu, *Digital Signal Processing* (New York: Wiley, 1976), Chap. 2.
9. A. G. Deczky, "Synthesis of Recursive Digital Filters Using the Minimum P-Error Criterion," *IEEE Trans. on Audio Electroacoustics*, AU-20, no. 4, 257–263, 1972.
10. R. Fletcher and M. J. D. Powell, "A Rapidly Convergent Descent Method for Minimization," *Computer Jour.* 6, no. 2 (1963): 163–168.
11. N. K. Bose, *Digital Filters: Theory and Applications* (New York: Elsevier Science Publishing, 1985).
12. A. Antoniou, *Digital Filters: Analysis and Design* (New York: McGraw-Hill, 1979).

EXERCISES FOR CHAPTER 4

Note: In each of the following filter design problems, we need *at least* the transfer function representing the design filter. In cases where frequency response plots (magnitude and phase) are requested, plots over the operating range (zero to half the sampling frequency) will suffice. These may be aided by the accompanying diskette programs.

1. Design an impulse-invariant lowpass digital filter to meet the following specifications:
   a) Cutoff frequency = 100 Hz
   b) Attenuation of at least 10 dB at 150 Hz
   c) Ripple in passband not to exceed 0.5 dB
   d) Maximum gain of unity in the passband

   The sampling frequency is 1000 Hz. Plot the magnitude response and the phase response.

2. From the second-order Butterworth lowpass prototype continuous filter, design an impulse-invariant lowpass digital filter with cutoff frequency of 50 Hz. Compare the magnitude response, in decibels, for sampling frequencies of 200 Hz and 500 Hz, respectively, with that of the corresponding continuous filter. Plot the phase responses.

3. By taking the z-transform of a suitable continuous function, design a highpass digital filter to meet the following specifications:
   a) Cutoff frequency = 400 Hz
   b) Attenuation at least 20 dB at 200 Hz
   c) Passband ripple not to exceed 1 dB

4. Design an impulse-invariant lowpass digital filter to meet the following specifications:
   a) Cutoff frequency = 500 Hz
   b) Attenuation of at least 18 dB at 1000 Hz
   c) Phase response approximately linear

   The sampling frequency is 4000 Hz. Plot the magnitude response and the phase response.

5. Prove that, with the transformation implied by the backward rectangular rule, the left half of the s-plane maps into the interior of the circle $|z - 1/2| = 1/2$ in the z-plane and that the right half of the s-plane maps into the exterior of that circle.

6. Design a bandpass digital filter to meet the following specifications:
   a) Passband from 100 to 200 Hz
   b) Down 20 dB at 50 Hz and 400 Hz
   c) Monotonic in passband and stopband
   d) Maximum gain of unity in the passband

   The sampling frequency is 2000 Hz. Plot the magnitude response and the phase response.

7. Repeat Exercise 1, using the bilinear transform instead of the z-transform to obtain the digital filter transfer functions.

8. Use the bilinear transform to design a highpass digital filter to meet the specifications in Exercise 3.

9. Design a bandstop digital filter via the bilinear transform to meet the following specifications:

a) Passband = 40 to 100 Hz
b) Attenuation at least 20 dB at 60 Hz and 70 Hz
c) Ripple in passband not to exceed 1 dB
d) Maximum gain of unity in the passband
e) Sampling frequency = 500 Hz

Plot the frequency responses.

10. Use the program ELLIPD on the diskette to design an elliptic digital filter to meet the requirements in Exercise 3 if a passband ripple of not more than 1 dB is permissible.

11. Use the program ELLIPD on the diskette to design an elliptic digital filter to satisfy the requirements of Exercise 9.

12. Use the continuous filter function obtained in Exercise 9 to design a digital filter by means of the matched z-transform. Verify whether the matched z-transform design meet the specifications. Compare frequency responses with the corresponding results obtained via the bilinear transform.

13. Transform the continuous filter function

$$H(s) = \frac{0.125s + 1}{s + 1}$$

into a digital filter function using the following transformations:

a) z-transform
b) Forward rectangular approximation
c) Backward rectangular approximation
d) Bilinear transform
e) Matched z-transform

Compute and plot the frequency responses for the various digital filters if $f_s =$ 50 Hz. Compare your results with the frequency response of the continuous filter $H(s)$ in the same range (zero to $f_s/2$).

14. Using the appropriate transformation from Table 4.4, design a highpass filter to meet the specifications of Example 4.2.

15. Show that if a pair of complex poles in a transfer function $H(z)$ lie outside the unit circle and are inverted, the shape of the magnitude response is not changed.

APPENDIX 4A

Diskette Program Examples

Examples of the use of two diskette programs are given. The programs are:

1. DFIL
2. ELLIPD

They are arranged for interactive operation. In these programs, a digital filter is designed to given specifications by taking the bilinear transform of a suitable continuous filter function.

The program DFIL lets the user choose a Butterworth, Chebyshev type 1, or Bessel digital filter design. A lowpass, highpass, bandpass, or bandstop filter can be selected. Prompts are given for the passband and stopband specifications. The poles and zeros of the filter design and the corresponding difference equation coefficients are displayed. The magnitude response and the phase response are stored in a file DFIL.RSP created by the program. In the example, a Chebyshev bandpass filter is requested. A copy of the computer printout is shown below. A sixth-order filter is required to meet specifications. Magnitude and phase plots are given in Figures 4A.1 and 4A.2, respectively.

An elliptic digital filter is given by the program ELLIPD. Interactive operation is similar to that of DFIL. The frequency responses are in a file ELLIPD.RSP. The same specifications as given for the DFIL example resulted in a fourth-order elliptic bandpass design. A copy of the computer printout and magnitude and phase plots (Figures 4A.3 and 4A.4) are shown on the following pages.

```
DFIL
ENTER CLASS OF FILTER (1: BUTTERWORTH, 2: CHEBYSHEV, 3: BESSEL):
2

CHEBYSHEV FILTER DESIGN

ENTER TYPE OF FILTER (LP,HP,BP,BS):
```

```
BP
ENTER SAMPLING FREQUENCY (HZ):
20000.
ENTER LOWER CUTOFF FREQUENCY (HZ):
4000.
ENTER UPPER CUTOFF FREQUENCY (HZ):
5000.
ENTER MINIMUM ATTENUATION (DB) IN STOP BANDS:
40.
ENTER HIGHEST FREQUENCY (HZ) IN LOWER STOP BAND
WHERE MINIMUM ATTENUATION CAN OCCUR:
2000.
ENTER LOWEST FREQUENCY (HZ) IN UPPER STOP BAND
WHERE MINIMUM ATTENUATION CAN OCCUR:
7000.
ENTER PASSBAND RIPPLE (DB):
2.
FILTER OF ORDER  6 REQUIRED TO MEET SPECIFICATIONS.
THERE ARE  3 ZEROS AT THE ORIGIN.
CONTINUOUS FILTER POLE LOCATIONS:
P( 1) =   -.115664E+04 + J   -.395007E+05
P( 2) =   -.860988E+03 + J    .294038E+05
P( 3) =   -.201763E+04 + J   -.340352E+05
P( 4) =   -.201763E+04 + J    .340352E+05
P( 5) =   -.115664E+04 + J    .395007E+05
P( 6) =   -.860988E+03 + J   -.294038E+05
Pause - Please enter a blank line (to continue) or a DOS command.
DIGITAL FILTER OF ORDER  6 REQUIRED TO MEET SPECIFICATIONS.
ZERO LOCATIONS:
Z( 1) =   1.000000 + J   .000000
Z( 2) =   1.000000 + J   .000000
Z( 3) =   1.000000 + J   .000000
Z( 4) =  -1.000000 + J   .000000
Z( 5) =  -1.000000 + J   .000000
Z( 6) =  -1.000000 + J   .000000
POLE LOCATIONS:
P( 1) =   .011787 + J  -.971078
P( 2) =   .149641 + J  -.931235
P( 3) =   .011787 + J   .971078
P( 4) =   .289903 + J   .928222
P( 5) =   .149641 + J   .931235
P( 6) =   .289903 + J  -.928222
Pause - Please enter a blank line (to continue) or a DOS command.
MAGNITUDE AND PHASE RESPONSES IN FILE DFIL.RSP
FILTER GAIN CONSTANT K FOR MAX.MAG OF UNITY =   .113597E-02
Pause - Please enter a blank line (to continue) or a DOS command.
DIGITAL FILTER DIFFERENCE EQUATION:
Y(K) = A1 Y(K-1) + A2 Y(K-2) +...+ B0 X(K) + B1 X(K-1) +...
```

```
                          B( 0) =      .001136
A( 1) =      .902661      B( 1) =      .000000
A( 2) =    -2.972611      B( 2) =     -.003408
A( 3) =     1.675251      B( 3) =      .000000
A( 4) =    -2.754585      B( 4) =      .003408
A( 5) =      .773207      B( 5) =      .000000
A( 6) =     -.793394      B( 6) =     -.001136
Stop - Program terminated.


ELLIPD

ELLIPTIC DIGITAL FILTER DESIGN

ENTER TYPE OF FILTER (LP,HP,BP,BS):
BP
ENTER SAMPLING FREQUENCY (HZ):
20000.
ENTER LOWER CUTOFF FREQUENCY (HZ):
4000.
ENTER UPPER CUTOFF FREQUENCY (HZ):
5000.
ENTER MINIMUM ATTENUATION (DB) IN STOP BANDS:
40.
ENTER PASSBAND RIPPLE (DB):
2.
ENTER HIGHEST FREQUENCY (HZ) IN LOWER STOP BAND
WHERE MINIMUM ATTENUATION CAN OCCUR:
2000.
ENTER LOWEST FREQUENCY (HZ) IN UPPER STOP BAND
WHERE MINIMUM ATTENUATION CAN OCCUR:
7000.
ELLIPTIC FILTER OF ORDER  4 REQUIRED TO MEET SPECIFICATIONS.
DIGITAL FILTER ZERO LOCATIONS:
Z( 1) =    .847712 + J    -.530456
Z( 2) =   -.729897 + J    -.683557
Z( 3) =   -.729897 + J     .683557
Z( 4) =    .847712 + J     .530456
DIGITAL FILTER POLE LOCATIONS:
P( 1) =   .026859 + J    .938013
P( 2) =   .026859 + J   -.938013
P( 3) =   .266701 + J   -.902228
P( 4) =   .266701 + J    .902228
Pause - Please enter a blank line (to continue) or a DOS command.
MAGNITUDE AND PHASE RESPONSES IN FILE ELLIPD.RSP
FILTER GAIN CONSTANT K FOR MAX.MAG OF UNITY =    .233468E-01
Pause - Please enter a blank line (to continue) or a DOS command.
DIGITAL ELLIPTIC FILTER DIFFERENCE EQUATION:
Y(K) = A1 Y(K-1) + A2 Y(K-2) +...+ B0 X(K) + B1 X(K-1) +...
```

```
                              B( 0) =      .023347
A( 1) =      .587120          B( 1) =     -.005501
A( 2) =    -1.794387          B( 2) =     -.011089
A( 3) =      .517256          B( 3) =     -.005501
A( 4) =     -.779449          B( 4) =      .023347
Stop - Program terminated.
```

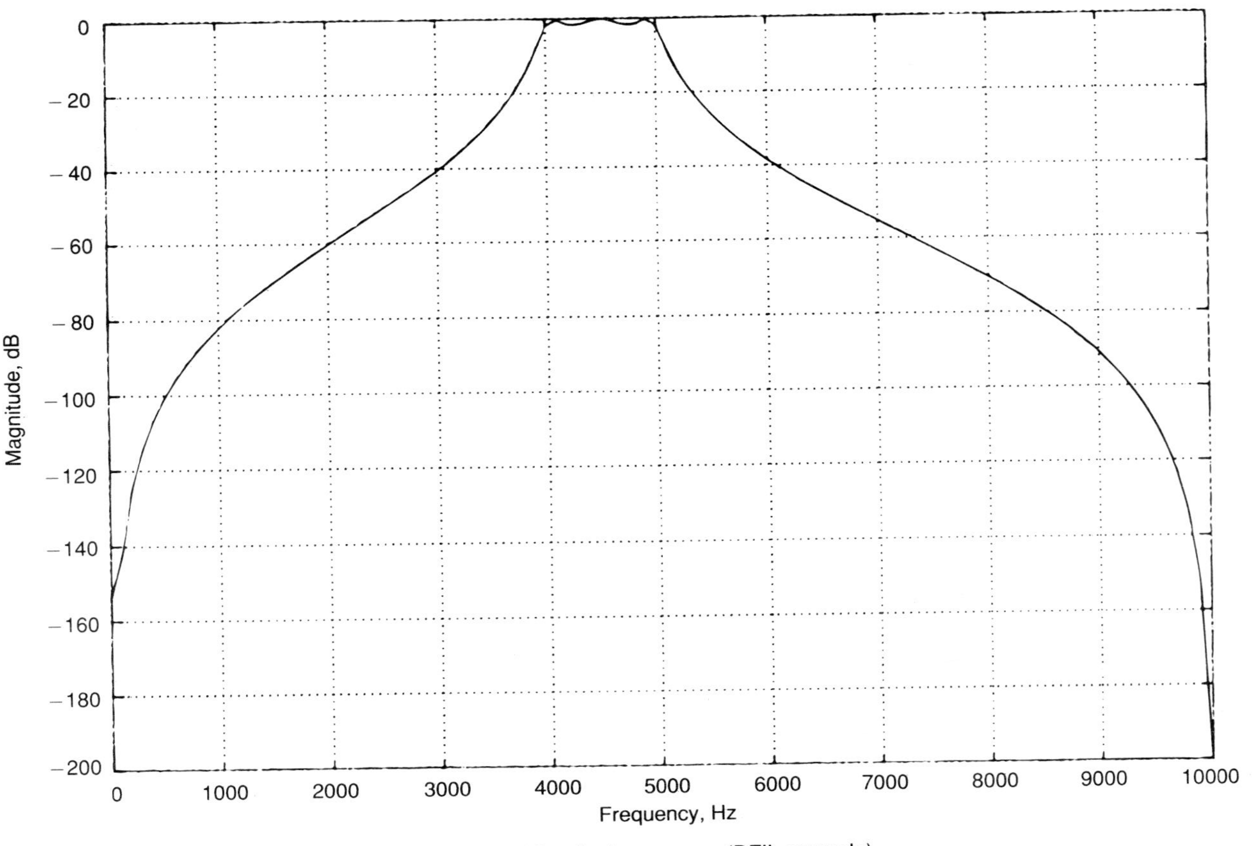

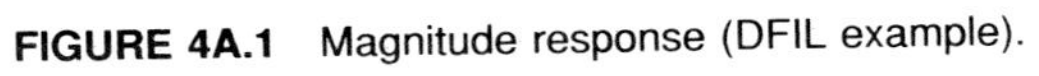

FIGURE 4A.1 Magnitude response (DFIL example).

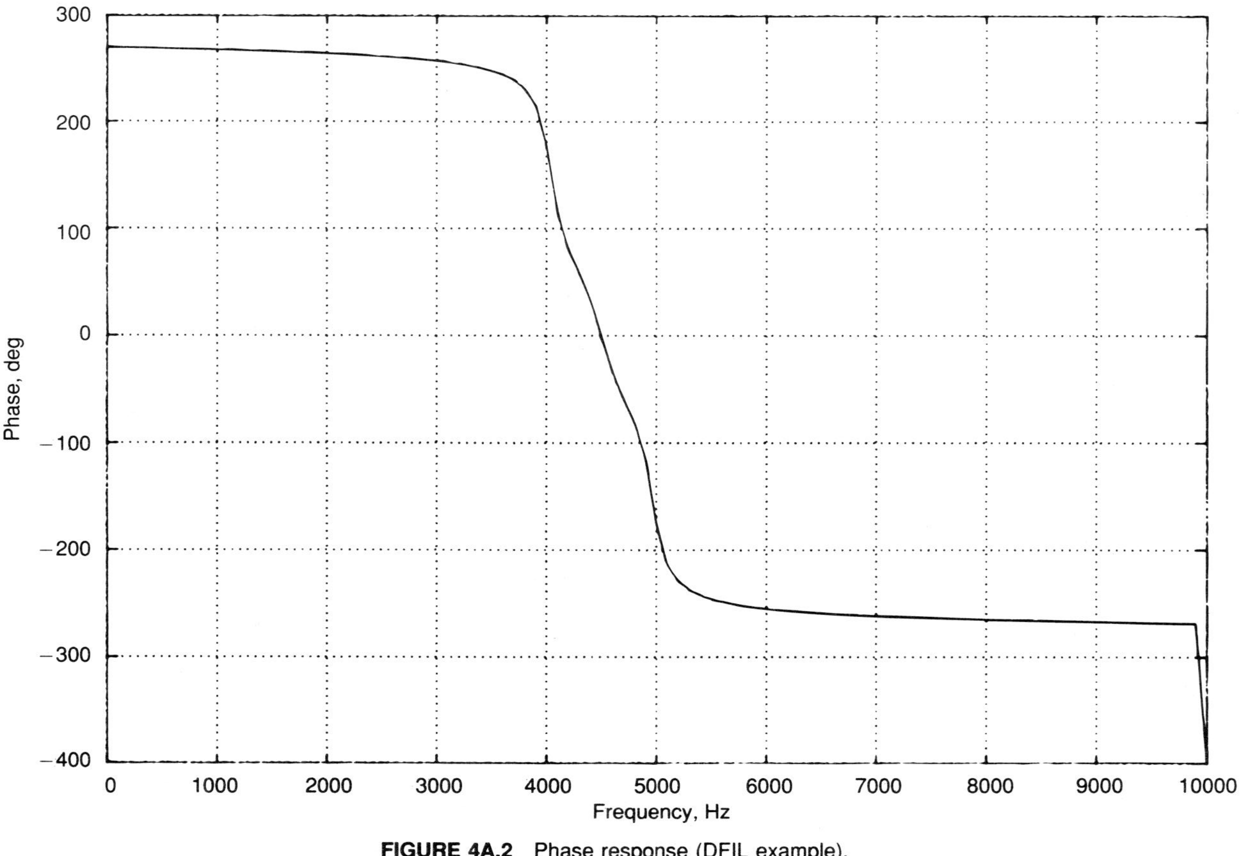

FIGURE 4A.2 Phase response (DFIL example).

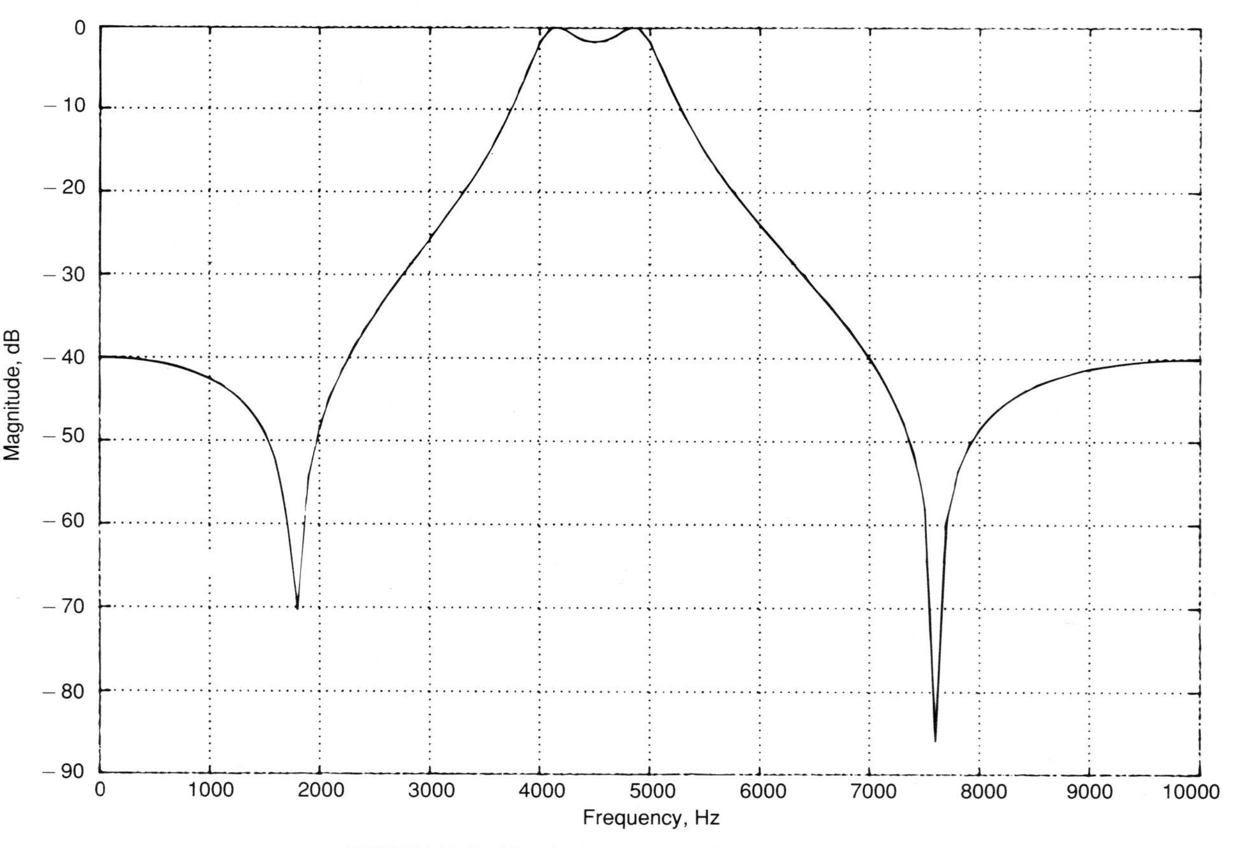

FIGURE 4A.3 Magnitude response (ELLIPD example).

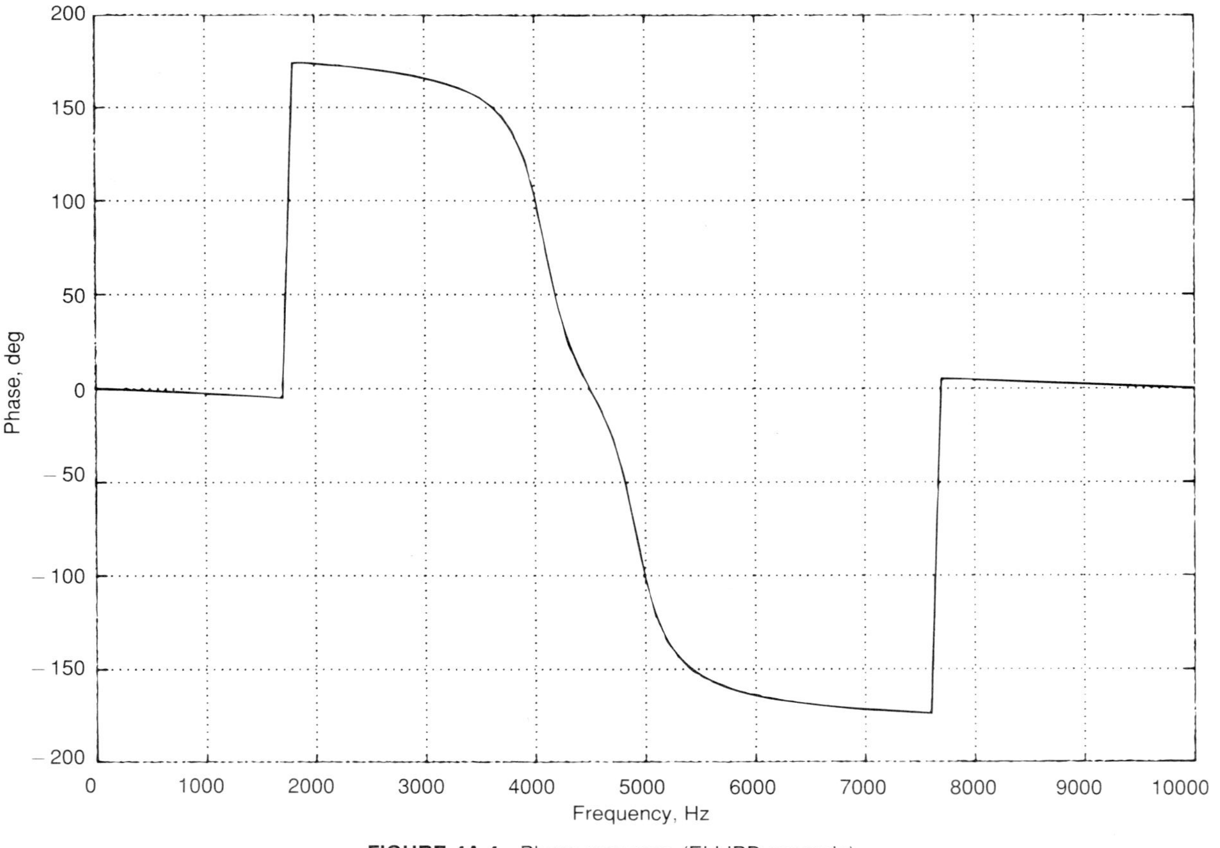

FIGURE 4A.4 Phase response (ELLIPD example).

CHAPTER 5

FINITE IMPULSE RESPONSE (FIR) AND NONRECURSIVE FILTERS

5.1 INTRODUCTION

As we noted in Chapter 4, the design of recursive filters can be carried out in a systematic manner by simply transforming suitable continuous filter functions. The nonrecursive filter can also be regarded as having a continuous analog—namely, the transversal filter or tapped delay line. In this device, a voltage signal $x(t)$ is fed in at one end of the delay line, and the line is tapped at N evenly spaced points along its length. The delay between taps is T seconds. The signals from the taps are fed through N weighting resistors $R_0, R_1, \ldots, R_{N-1}$ and summed to form the output $y(t)$.

$$y(t) = \sum_{i=0}^{N-1} R_i x(t - iT) \tag{5.1}$$

In the realizable form of the nonrecursive digital filter, its output is formed from a weighted linear combination of current input and previous values of the input. As a difference equation, it may be written

$$y_k = \sum_{i=0}^{m} B_i x_{k-i} \tag{5.2}$$

where the coefficients B_i are constant and $y_k = y(kT)$. The similarity to (5.1) is apparent. However, it has not been found useful to take a transversal filter design as a basis for designing nonrecursive filters, because the latter have much wider application and can be designed more easily from frequency specifications. Hence we will not consider transversal filters further.

Because (5.2) is also in the form of a convolution summation, the pulse response of a nonrecursive filter is

$$h_k = B_k, \qquad k = 0, 1, \ldots, m \tag{5.3}$$

That is, the pulse response is finite. As we discussed in Chapter 1, the terms *nonrecursive filter* and *finite impulse response* (FIR) *filter* are almost synonymous. It is evident from (5.2) that we can *always* implement an FIR filter in nonrecursive form. In some cases, we can also implement an FIR filter in recursive form, but only by cheating a little: We ignore pole-zero cancellations that would have made the filter nonrecursive. Some examples are given below.

Taking the z-transform of (5.2) gives the transfer function of the filter as

$$H(z) = \frac{Y(z)}{X(z)} = \sum_{i=0}^{m} B_i z^{-i} \tag{5.4}$$

Some basic properties of the nonrecursive filter can be inferred from the foregoing.

1. Because there is no feedback [see Eq. 5.2], the filter cannot be unstable; that is, a bounded input results in a bounded output. This inherent stability property is also manifested in the absence of poles in the transfer function (5.4), except possibly at the origin.
2. The filter has finite memory [see (5.2)], because it "forgets" all inputs before the mth previous one. This is sometimes a desirable property. For example, if a digital algorithm were used as a filter in a fire-control system, it would be an advantage for the filter to have finite memory for acquisition transients.

The filter defined in (5.2) is realizable or causal, because the output depends on the current and previous input values, and so an output cannot precede an input. Thus this filter can be used to process a sequence of data as it is received—that is, in real time. There are many applications, however, where data are collected and stored for later processing. In other words, the processing is not done in real time. The "current" time instant can be located arbitrarily as the data are processed, so the current output of the filter may depend on past, current, and future input values. Such a filter is said to be "nonrealizable" (that is, in real time), but it is perfectly feasible to build for the conditions noted. The nonrealizable filter has the form

$$y_k = \sum_{i=-m_1}^{m_2} B_i x_{k-i} \tag{5.5}$$

or

$$H(z) = \sum_{i=-m_1}^{m_2} B_i z^{-i} \tag{5.6}$$

Often, the number of future input values equals the number of past input values, or $m_1 = m_2$ in (5.5) and (5.6). The frequency response of a nonrecursive filter is found by setting z equal to $e^{j\omega T}$ in (5.4) or (5.6). Thus, for the latter,

$$H(j\omega T) = \sum_{i=-m_1}^{m_2} B_i e^{-j\omega T} = M(\omega T)\exp(j\phi(\omega T)) \tag{5.7}$$

Many common procedures carried out in data processing represent nonrecursive filtering operations. We will discuss several examples in the following sections before treating the actual design of nonrecursive filters to given specifications.

5.2 SOME BASIC FIR FILTERS

5.2.1 Symmetric Filter, Sum Filter, and Difference Filter

The basic symmetric filter is defined by

$$y_k = \sum_{i=-m}^{m} B_i x_{k-i} \tag{5.8}$$

or

$$H(z) = \sum_{i=-m}^{m} B_i z^{-i} \tag{5.9}$$

where $B_i = B_{-i}$. Therefore, because $z = \exp(j\omega T)$ and $\cos\omega T = [\exp(j\omega T) + \exp(-j\omega T)]/2$, we have

$$H(j\omega T) = B_0 + 2\sum_{i=1}^{m} B_i \cos j\omega T, \quad |\omega| \leq \omega_s/2 \tag{5.10}$$

Because $H(j\omega T)$ is real, the phase shift between input and output is either zero or π (or a multiple thereof). The values of the coefficients B_i depend on the particular function $H(j\omega T)$ being represented.

EXAMPLE 5.1

As a simple example of a symmetric filtering operation, suppose data are smoothed three samples at a time so that

$$y_k = B_{-1}x_{k+1} + B_0 x_k + B_1 x_{k-1} \tag{5.11}$$

with $B_{-1} = B_0 = B_1 = B$. Then the frequency response is

$$H(j\omega T) = B(1 + 2\cos\omega T), \quad |\omega| \leq \omega_s/2 \tag{5.12}$$

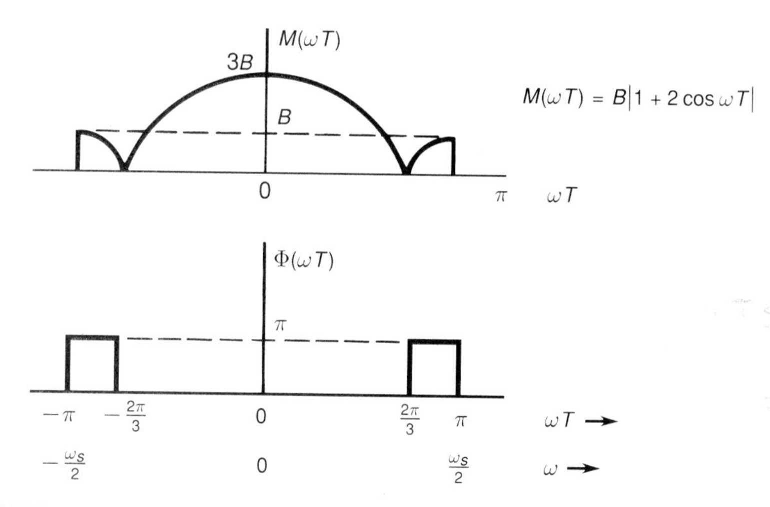

FIGURE 5.1 Magnitude response and phase response for a smoothing-by-threes filter.

The magnitude response and the phase response are shown in Figure 5.1. If $B = 1/3$, it is evident that this operation is the same as taking the average, or mean, value of three successive samples in the sequence $x(k)$.

The symmetric filter is nonrealizable in real time. Simple examples of realizable filtering operations are given by the sum filter and the difference filter. The sum filter is defined by

$$y_k = B_0 x_k + B_1 x_{k-1} \tag{5.13}$$

with pulse response $h_0 = B_0, h_1 = B_1$, and $h_k = 0$ otherwise. From (5.13), the transfer function is

$$H(z) = B_0 + B_1 z^{-1} \tag{5.14}$$

Setting z equal to $\exp(j\omega T)$, we obtain the frequency response for $B_0 = B_1 = B$ as

$$H(j\omega T) = 2B \exp\left(-j\frac{\omega T}{2}\right) \cos\frac{\omega T}{2}, \quad |\omega| \leq \frac{\omega_s}{2} \tag{5.15}$$

The magnitude response and the phase response are shown in Figure 5.2(a) and (b). The magnitude response is down 3 dB at one-quarter of the sampling frequency. The phase response is linear.

The difference filter is

$$y_k = B_0 x_k - B_1 x_{k-1} \tag{5.16}$$

with pulse response $h_0 = B_0, h_1 = B_1$, and $h_k = 0$ otherwise. The transfer function is

$$H(z) = B_0 - B_1 z^{-1} \tag{5.17}$$

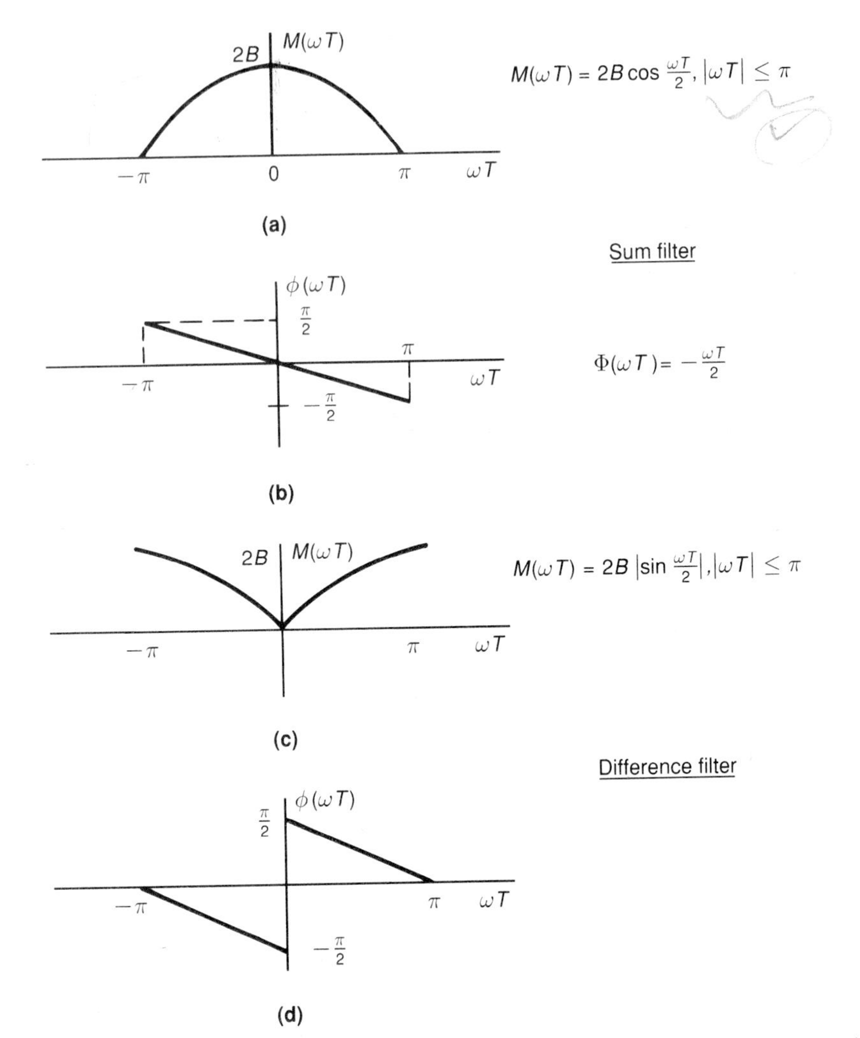

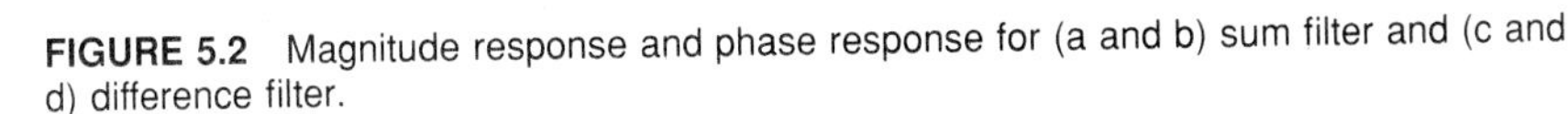

FIGURE 5.2 Magnitude response and phase response for (a and b) sum filter and (c and d) difference filter.

For $B_0 = B_1 = B$, the frequency response is

$$H(j\omega T) = 2B \exp\left(-j\frac{\omega T - \pi}{2}\right) \sin\frac{\omega T}{2}, \qquad |\omega| \leq \frac{\omega_s}{2} \tag{5.18}$$

The magnitude response and the phase response are shown in Figure 5.2(c) and (d). It is evident from the shape of the magnitude responses in Figure 5.2 that the sum operation and the difference operation represent crude lowpass and highpass filters, respectively. Cascade combinations of sum and difference filters can be made to achieve certain desired frequency responses. An example is included as an exercise at the end of this chapter.

5.2.2 Moving-Average Filter

Performing some common processing operations on a sequence of data may be considered as filtering the raw data. For example, taking the mean of a finite sequence $x_1, x_2, \ldots, x_N$,

$$\overline{x} = \frac{1}{N} \sum_{i=1}^{N} x_i \tag{5.19}$$

is in effect a very simple lowpass filtering operation that passes a single frequency component—the zero-frequency term, or the mean. Of course, all the data must be processed before the single number $\overline{x}$ is obtained.

For long data sequences, the mean, as defined by (5.19) may have little practical significance. Instead, what is required is a running estimate of the mean, or a so-called moving average. Taking the moving average involves averaging the sequence of length N in successive sub-sequences of length $m < N$. Applying the technique with $m = 2r + 1 = 5$ is illustrated in Figure 5.3. In effect, a "window" of length m is moved along the data.

The moving-average operation is defined as follows:

$$y_k = \overline{x}_k = \frac{1}{m} \sum_{i=-r}^{r} x_{k-i}, \quad r + 1 \leq k \leq N - r \tag{5.20}$$

It is evident that the mean is now a function of time, but $2r$ data points are lost (r points at the beginning and r points at the end of the raw or unfiltered data). The frequency characteristics of this operation can be determined by taking the z-transform of (5.20). The transfer function is

$$H(z) = \frac{1}{m} \sum_{i=-r}^{r} z^{-i} \tag{5.21}$$

so if we neglect pole-zero cancellation, this function can be written for *recursive* implementation as

$$mH(z) = z^r \frac{1 - z^{-(2r+1)}}{1 - z^{-1}} = \frac{z^m - 1}{z^r(z - 1)} \tag{5.22}$$

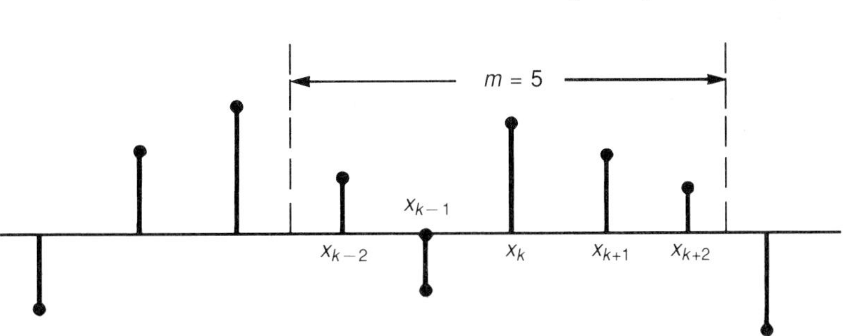

FIGURE 5.3 The moving-average operation.

A pole-zero diagram for this function is given in Figure 5.4. The presence of uncanceled zeros on the unit circle at angles $\theta = \omega T = 2k\pi/m, k = 1, 2, \ldots, m-1$, indicates that the frequency response goes to zero at the frequencies $\omega = 2k\pi/(mT) = k\omega_s/m$, which are in the stopband. Setting z equal to $e^{j\omega T}$ in (5.21) gives the frequency response as

$$H(j\omega T) = \frac{1}{m}\frac{\sin\frac{m\omega T}{2}}{\sin\frac{\omega T}{2}} \tag{5.23}$$

A plot of this function is shown in Figure 5.5. The main-lobe width measured from $-\omega_s/m$ to ω_s/m, is $2\omega_s/m$. Examination of Figure 5.5 reveals that the moving-average operation can be regarded as a lowpass filtering with cutoff frequency $\omega_c = \omega_s/(2m)$, at which point the magnitude response is down about 3 to 4 dB, depending on the value of m.

The moving-average operation is useful for removing low-frequency trends from data. A suitable arrangement is shown in Figure 5.6. The situation is also depicted in Figure 5.7 where the data are shown in continuous form for ease of illustration.

Suppose, for example, that the solid curve in Figure 5.7(a) represents terrain elevation data in the foothills of a mountain range. Because of the inaccuracy of the data compared with the true terrain, we can consider this curve a realization of a random process (see Chapter 7). A missile or aircraft trying to follow the contours of the ground (an operation called terrain following) might have little problem following the general low-frequency

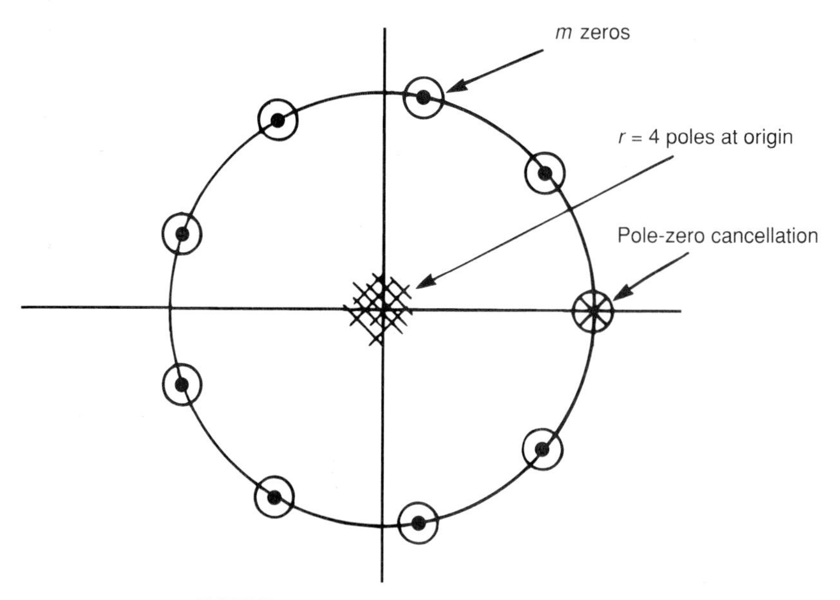

FIGURE 5.4 Pole-zero diagram for $mH(z)$, $m = 9$.

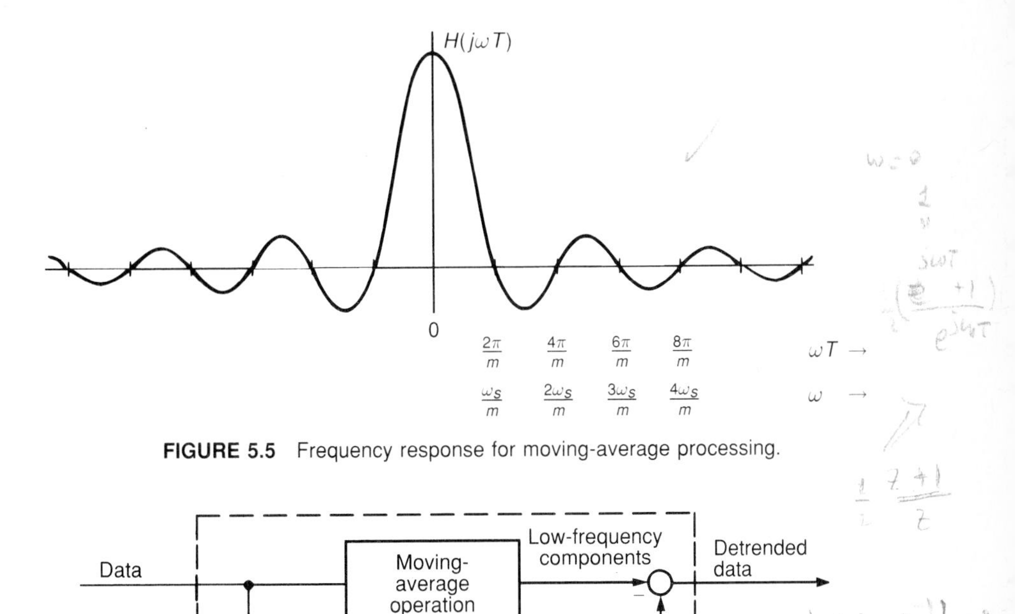

FIGURE 5.5 Frequency response for moving-average processing.

FIGURE 5.6 The moving-average operation as a highpass filter.

trend of the terrain [shown as a dashed curve in Figure 5.7(a)], but it might crash because of the higher-frequency fluctuations about the trend. When we remove the trend from the data, the remaining detrended data might be analyzed as a stationary random process with zero mean and constant variance (Chapter 7). Alternatively, moving estimates of both mean and variance can be obtained by nonrecursive or recursive filtering, as shown in Section 5.7. Armed with this information, and also taking into account the statistics of the data measurement errors, the data analyst is in a position to make recommendations about how high the missile or aircraft should fly above the ground.

A realizable moving-average operation that will filter data in real time is defined as

$$y_k = \frac{1}{m}\sum_{i=0}^{m-1} x_{k-i}, \quad m-1 \le k \le N \tag{5.24}$$

where $(m-1)$ data points are lost; that is, m data points must be available before the first filter output is obtained. The transfer function is

$$H(z) = \frac{1}{m}\frac{1-z^{-m}}{1-z^{-1}} \tag{5.25}$$

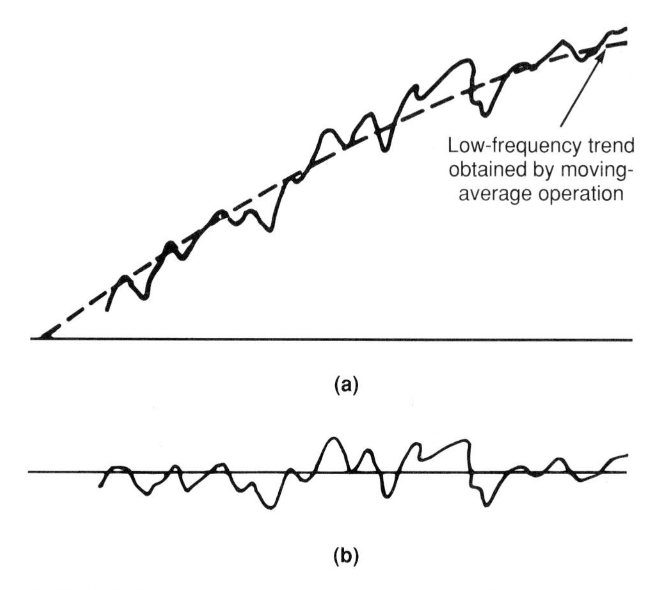

FIGURE 5.7 Detrending data. (a) Raw data with trend removal. (b) Detrended data.

If we do not cancel the pole and zero at $z = 1$, the operation can be written in recursive form,

$$y_k = y_{k-1} + \frac{1}{m}(x_k - x_{k-m}) \tag{5.26}$$

which is computationally more efficient than the nonrecursive form (5.24). Derivation of the frequency response for this filter is left as an exercise.

5.2.3 Comb Filters and Modified Comb Filters

A comb filter takes its name from the shape of the magnitude response, which, with evenly spaced zeros, resembles a comb. If we take the magnitude response for the moving-average filter from Figure 5.5, we see that with the exception of the passband centered at the origin, it resembles a comb filter. We will call such a filter, with pole-zero cancellation to form a passband, a modified comb filter. Comb filters are useful for passing or eliminating specific frequencies and their harmonics. They have the disadvantage that there is not, in general, a large attenuation in the stopband. In addition, considerable storage may be required for the delayed input values.

A *pure* comb filter is given by the difference equation.

$$y_k = x_k - x_{k-N} \tag{5.27}$$

where the delay $N > 0$ is an integer number of samples. The transfer function is

$$H(z) = 1 - z^{-N} = (z^N - 1)/z^N \tag{5.28}$$

This filter has N poles at the origin and N zeros evenly spaced about the unit circle at

$$z = \exp(j2\ell\pi/N), \quad \ell = 0, 1, \ldots, N-1 \tag{5.29}$$

or at frequencies

$$f_\ell = \ell f_s/N \tag{5.30}$$

Substituting $z = e^{j\omega T}$ in (5.28) gives the frequency response as

$$H(j\omega T) = 2\exp[-j(N\omega T - \pi)/2]\sin(N\omega T/2) \tag{5.31}$$

The form of the magnitude response is given in Figure 5.8, where the characteristic comb shape is evident. The centers of the passbands lie halfway between the zeros of the response—that is, at frequencies $(2\ell+1) f_s/(2N), \ell = 0, 1, \ldots, N-1$. From (5.31), it is evident that the phase is piecewise-linear. The comb filter can serve as a crude bandstop filter to remove a certain frequency and its harmonics. For example, the readings from an instrument mounted on a vibrating support might contain the fundamental frequency of the vibrations and its harmonics as well as the desired readings. However, the difference comb filter, as (5.27) is called, may not be suitable because it also eliminates the zero frequency term. Instead, we could use the sum comb filter defined by

$$y_k = x_k + x_{k-N} \tag{5.32}$$

This filter has N poles at the origin and N zeros at

$$z = \exp[j(2\ell+1)\pi/N], \quad \ell = 0, 1, \ldots, N-1 \tag{5.33}$$

or at frequencies

$$f_\ell = (2\ell+1) f_s/(2N) \tag{5.34}$$

The frequency response is

$$H(j\omega T) = 2\exp(-j(N\omega T/2]\cos(N\omega T/2) \tag{5.35}$$

Derivation of these expressions is left as an exercise. An example of the use of the sum comb filter follows.

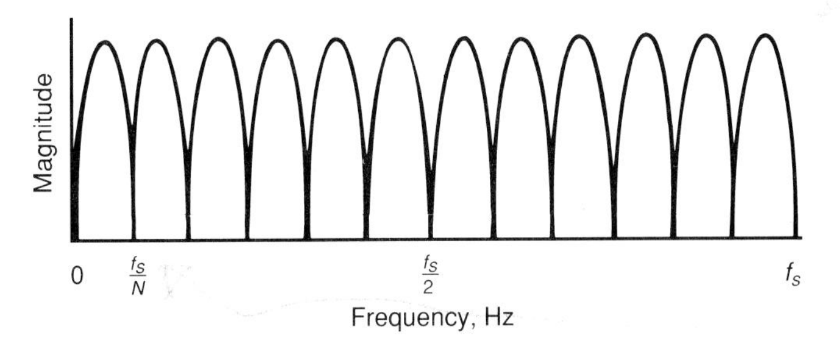

FIGURE 5.8 Magnitude response for difference comb filter ($N = 12$).

EXAMPLE 5.2

As a result of aircraft body-bending modes, readings from an aircraft accelerometer are corrupted by a frequency component at 500 Hz and its odd harmonics. The computer time step is 10^{-4} s. Design a comb filter to remove the unwanted frequencies.

Solution

From (5.34) with $\ell = 0$, we have the number of delays as

$$N = f_s/(2\, f_0) = 10$$

for $f_0 = 500$ and $f_s = 10000$ Hz. Therefore, the required filter algorithm is

$$y_k = x_k + x_{k-10}$$

Within its operating range up to frequency $f_s/2$, this filter will eliminate the frequencies 500, 1500, 2500, 3500, and 4500 Hz.

We can use more complex arrangements of modified comb filters for specific harmonic filtering. Assume that, in general, there are $(r-1)$ N-unit delays, with $r \geq 2$. Then, referring to Figure 5.9, we have

$$y_k = x_k + x_{k-N} + x_{k-2N} + \cdots + x_{k-(r-1)N} \tag{5.36}$$

Take z-transforms of both sides to get

$$Y(z) = (1 + z^{-N} + z^{-2N} + \cdots + z^{-(r-1)N})X(z) \tag{5.37}$$

Therefore, the transfer function is

$$H(z) = \sum_{i=0}^{r-1} z^{-Ni} = \frac{1 - z^{-Nr}}{1 - z^{-N}} \tag{5.38}$$

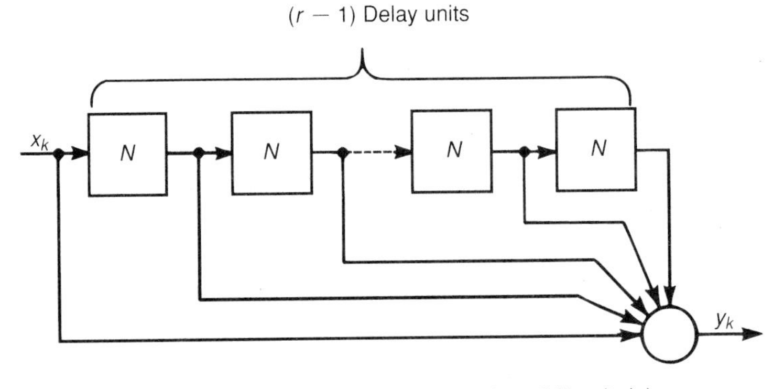

FIGURE 5.9 Modified comb filter with $(r-1)N$-unit delays.

This reduces to $1 + z^{-N}$, the transfer function of the sum comb filter, for $r = 2$. For $r > 2$, we will write (5.38) as

$$H(z) = \frac{z^{Nr} - 1}{z^{N(r-1)}(z^N - 1)} \tag{5.38A}$$

Equation (5.38A) implies that there are Nr zeros on the unit circle in the z-plane, and Nr poles. Of the latter, $N(r-1)$ are at the origin and N on the unit circle, where they cancel N of the zeros. The zeros are located at

$$z = \exp\left(j\frac{2m\pi}{Nr}\right), \quad m = 0, 1, 2, \ldots, (Nr - 1) \tag{5.39}$$

The canceling poles are at

$$z = \exp\left(j\frac{2k\pi}{N}\right), \quad k = 0, 1, 2, \ldots, (N - 1) \tag{5.40}$$

There is pole-zero cancellation when $m = kr$. A pole-zero plot drawn for $N = 6$ and $r = 4$ is shown in Figure 5.10(a). The frequencies at which the poles and zeros cancel are the centers of the passbands of the filter. From (5.39) and (5.40), noting that $z = \exp(j\omega t)$ on the unit circle, we get the frequencies at which the zeros occur as

$$f_m = \frac{m}{Nr} f_s \tag{5.41}$$

whereas the canceling poles occur at

$$f_k = \frac{k}{N} f_s \tag{5.42}$$

where m and k have the values indicated in (5.39) and (5.40), respectively.

The frequency response of the filter is obtained by setting z equal to $\exp(j\omega T)$ in (5.38), which results in

$$H(j\omega T) = \exp(-j(Nr - N)\omega T/2)\frac{\sin(Nr\omega T/2)}{\sin(N\omega T/2)} \tag{5.43}$$

Thus the filter has piecewise-linear phase, as indicated by the exponential on the right-hand side of (5.43). An illustrative plot of the magnitude response for $N = 6$ and $r = 4$ is given in Figure 5.10(b). The passbands centered about the pole-zero cancellation frequencies are evident. The filter passes a narrow band of frequencies centered about zero, f_s/N, and their harmonics. Hence it is useful if we want to amplify these and reduce background noise (we define *noise* as any *unwanted* signal or portion of a signal).

From (5.43), the magnitude response has the form $|\sin(rU)/\sin U|$, where $U \triangleq N\omega T/2$. The denominator is a factor of the numerator. The resulting polynomials for some lower values of r are given in Table 5.1.

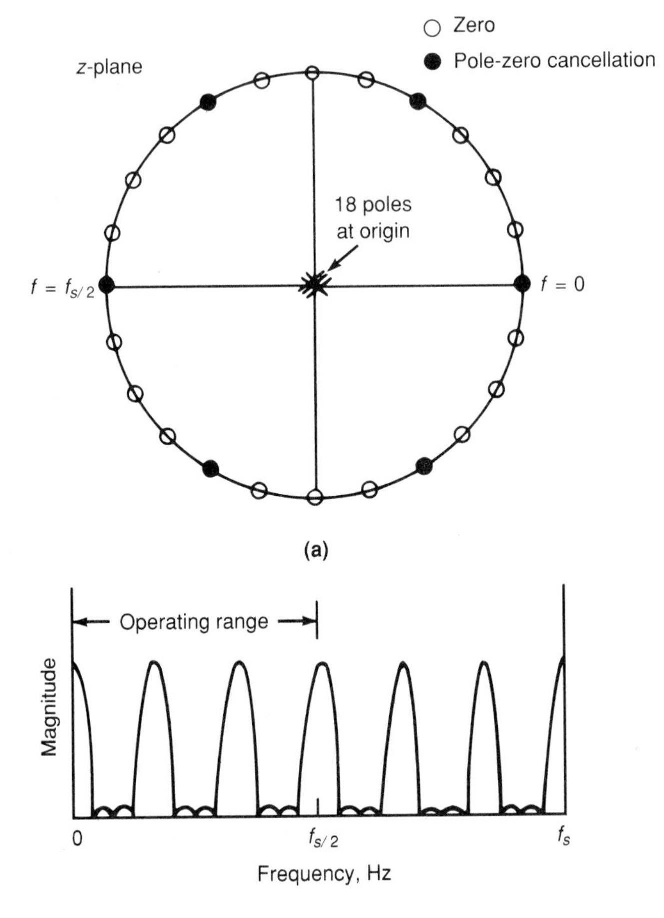

FIGURE 5.10 Modified sum comb filter pole-zero plot and magnitude response ($N = 6$, $r = 4$). (a) Pole-zero plot. (b) Magnitude response.

TABLE 5.1 Argument of Modified Comb Filter Magnitude Response, $r = 2$ to 6

r	$\sin rU/\sin U, \quad U = N\omega T/2$	First Side Lobe Down by (dB)
2	$2\cos U$	0
3	$3 - 4\sin^2 U$	9.5
4	$8\cos^3 U - 4\cos U$	11.4
5	$5 - 20\sin^2 U + 16\sin^4 U$	12.1
6	$32\cos^5 U - 32\cos^3 U + 6\cos U$	12.6

Engin Yaz

The passbands are called main lobes. The reduced oscillations, or ripples, between the uncanceled zeros (see Figure 5.10) are called side lobes. The center of the first side lobe lies approximately halfway between the first two uncanceled zeros. Therefore, from (5.41), the corresponding frequency is

$$f = \frac{3}{2Nr} f_s \tag{5.44}$$

At that frequency,

$$U = \frac{N\omega T}{2} = \frac{3\pi}{2r} \tag{5.45}$$

From this and (5.43), the magnitude of the first side lobe is

$$M(\omega T) = |-\text{cosec}\ (3\pi/2r)| \tag{5.46}$$

It is evident from Table 5.1 that the maximum gain in the passband centered at $\omega T = 0$ is numerically equal to r. Hence, using (5.46), we can compute the attenuation in the stopband, as given in the last column of Table 5.1. Even for a 9-stage filter ($r = 10$), the attenuation is only 13.1 dB, so there is not much advantage in using more than, say, 3 or 4 stages. In summary, the designed passband center frequency determines N [see (5.42)], and the specified attenuation fixes r. The computer time step must be selected to suit the particular frequency being passed and to ensure that N is integer. Consider the following example.

EXAMPLE 5.3

A signal is believed to contain frequency components at 10, 20, and 30 Hz, buried in background noise. Select the values of N, r, and computer time step for a modified comb filter such as we have described to amplify these frequencies and reduce the noise. The minimum attenuation in the stopband should be at least 10 dB.

Solution

From the attenuation requirement, a 3-stage filter ($r = 4$) will suffice, according to Table 5.1. The highest frequency of interest is 30 Hz, although this is not necessarily the highest frequency in the signal. For lack of further information on the highest frequency present, we will sample at 4 times the highest known frequency (see Chapter 3)—that is, $f_s = 4 \times 30 = 120$ Hz—so the computer time step is 1/120 = 0.0083 s. From (5.42), with $k = 1$ the first nonzero frequency passband is centered at

$$f_\ell = f_s/N$$

so if $f_\ell = 10$ Hz and $f_s = 120$ Hz, we have $N = 12$. Accordingly, the difference equation for the filter is

$$y_k = x_k + x_{k-12} + x_{k-24} + x_{k-36}$$

in nonrecursive form or, from (5.38),

$$y_k = y_{k-12} + x_k - x_{k-48}$$

in recursive form.

If we do not want the passband centered at $\omega = 0$, we can arbitrarily introduce an additional zero at $z = 1$ in (5.38). Doing so negates the pole-zero cancellation there and gives

$$H(z) = \frac{(1 - z^{-Nr})(1 - z^{-1})}{1 - z^{-N}}$$

from which the recursive form of the filter becomes

$$\begin{aligned} y_k &= y_{k-N} + x_k - x_{k-1} - x_{k-Nr} + x_{k-Nr-1} \\ &= y_{k-12} + x_k - x_{k-1} - x_{k-48} + x_{k-49} \end{aligned}$$

for this example.

If only one passband is required in the operating range, we must modify the procedure. In general, for a filter with difference equation

$$y_k = \sum_{i=0}^{N-1} x_{k-i} \tag{5.47}$$

the transfer function has the form

$$H(z) = \frac{1 - z^{-N}}{1 - z^{-1}} \tag{5.48}$$

This is a lowpass filter with passband centered at $z = 1$ (that is, zero frequency), where the pole-zero cancellation occurs. In the last example, we noted that multiple pole-zero cancellations resulted in multiple passbands. The transfer function (5.48) suggests that if we want a *highpass* filter, we need a pole-zero cancellation at $z = -1$, so the transfer function of such a filter should have the form

$$H(z) = \frac{1 + z^{-N}}{1 + z^{-1}} \tag{5.49}$$

where N is odd, to ensure that there is a zero to be canceled at $z = -1$. This transfer function would result from the difference equation

$$\begin{aligned} y_k &= x_k - x_{k-1} + x_{k-2} - x_{k-3} + \cdots + x_{k-N-1} \\ &= \sum_{i=0}^{N-1} (-1)^i x_{k-i} \end{aligned} \tag{5.50}$$

Thus the impulse response is

$$h_i = (-1)^i, \quad i = 0, N - 1 \tag{5.51}$$

and the zeros are on the unit circle at

$$z = \exp(j(2\ell + 1)\pi/N), \quad \ell = 0, 1, \ldots, N-1 \tag{5.52}$$

Upon substituting $z = \exp(j\omega T)$ in (5.49), we get the frequency response

$$H(j\omega T) = \exp[-j(N-1)\omega T/2]\frac{\cos(N\omega T/2)}{\cos(\omega T/2)} \tag{5.53}$$

For odd values of N, the magnitude response $|\cos NU/\cos U|$, where $U \triangleq \omega T/2$, has a common factor $\cos U$ in numerator and denominator. The resulting expressions for $N = 3$ and $N = 5$ are given in Table 5.2.

The maximum gain in the passband (at $\omega T = \pi$) is numerically equal to N. Figure 5.11 shows the pulse response, pole-zero plot, and frequency responses for $N = 5$. The attenuation leaves something to be desired. Even for $N = 9$, the first side lobe is down by only 13 dB.

Lynn [1] suggested that we can increase the attenuation by using higher-order zeros and canceling poles. By this method, instead of (5.49), we have

$$H(z) = \left[\frac{1 + z^{-N}}{1 + z^{-1}}\right]^n \tag{5.54}$$

where n is integer. The frequency response becomes

$$H(j\omega T) = \exp[-jn(N-1)\omega T/2]\frac{\cos^n(N\omega T/2)}{\cos^n(\omega T/2)} \tag{5.55}$$

For example, for $N = 5$ and $n = 2$, the maximum gain in the passband is N^2, and the first side lobe is down 24.3 dB—n times its value in Table 5.2. The difference equation in nonrecursive form is

$$\begin{aligned} y_k = x_k - 2x_{k-1} + 3x_{k-2} - 4x_{k-3} + 5x_{k-4} - 4x_{k-5} \\ + 3x_{k-6} - 2x_{k-7} + x_{k-8} \end{aligned} \tag{5.56}$$

It is obtained by dividing the denominator of (5.54) into its numerator. The difference equation in recursive form is

$$y_k = -2y_{k-1} - y_{k-2} + x_k + 2x_{k-5} + x_{k-10} \tag{5.57}$$

The pulse response and pole-zero plot are shown in Figure 5.12. The pulse response exhibits a certain type of symmetry, which is one of the require-

TABLE 5.2 Argument of Highpass Filter Magnitude Response

N	$\cos NU/\cos U, \quad U = \omega T/2$	First Side Lobe Down by (dB)
3	$4\cos^2 U - 3$	9.5
5	$16\cos^4 U - 20\cos^2 U + 5$	12.1

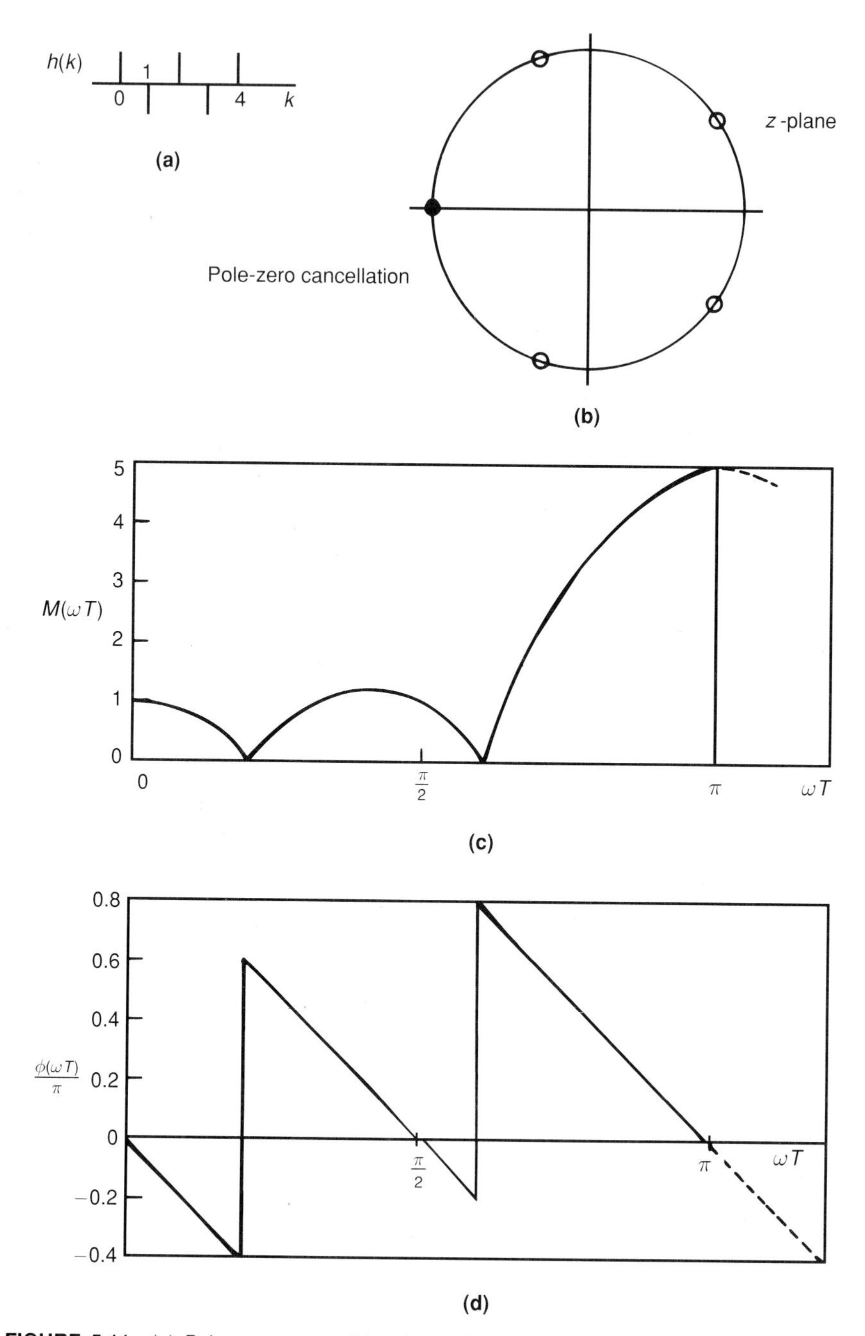

FIGURE 5.11 (a) Pulse response, (b) pole-zero plot, (c) magnitude response, and (d) phase response for pole-zero cancellation highpass filter ($N = 5$).

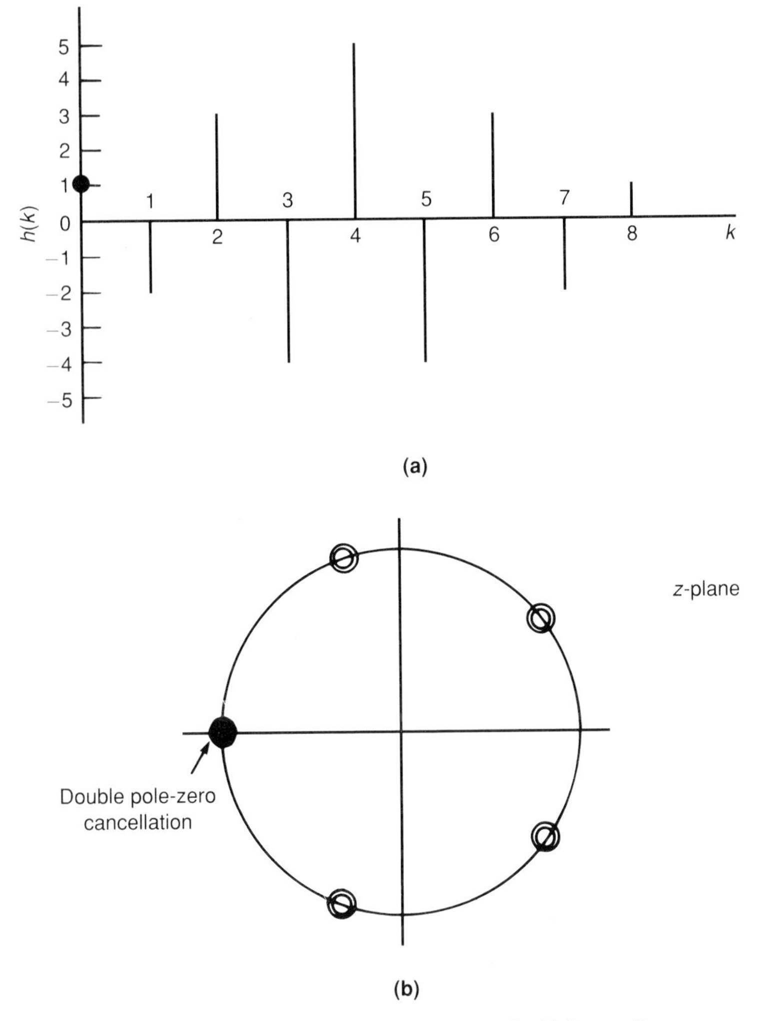

FIGURE 5.12 (a) Pulse response and (b) pole-zero plot for highpass filter, $H(z) = [(1 + z^{-5})/(1 + z^{-1})]^2$

ments for a linear-phase filter, as we shall see in the next section. The attenuation (the ratio of first side lobe to main lobe) is directly proportional to the order n of the zeros. However, the number of computations also goes up with n, so a trade-off is necessary.

This method is best suited to highpass filters with narrow passbands relative to the operating range. If we define the cutoff frequency as the frequency at which the magnitude response in the passband is down 3 dB relative to its maximum value, Table 5.2 yields $\omega_c T = 0.69\pi$ for $n = 3$ and 0.82π for $N = 5$. Because $N = 3$ is the smallest number of delays possible, we cannot design a highpass filter by this method if the width of its passband exceeds $\pi - 0.69\pi = 0.31\pi$. For $N = 3$ and $n = 2$, the width of the passband decreases to 0.22π. We next consider an example that reveals the limitations of this method of highpass filter design.

EXAMPLE 5.4

Design, by the pole-zero cancellation method, a highpass filter that has a cutoff frequency of 50 Hz and has the first side lobe down at least 15 dB.

Solution

From Table 5.2, the $N = 3$, $n = 1$, filter has 9.5-dB attenuation. Therefore, the $N = 3$, $n = 2$, filter has 19-dB attenuation and meets the specification. The transfer function is given by (5.54). If the cutoff frequency is 50 Hz and occurs at $0.78\, f_s/2$, then $f_s = 128$ Hz and the required computer time step is $T = 1/f_s = 0.0078$ seconds. Thus we can filter frequencies only up to 64 Hz, giving a 14-Hz width of passband. Even more restrictive is the necessity of a particular time step, which may not be feasible in many applications.

In [1], Lynn extended the procedure to the design of bandpass filters with passbands created by pole-zero cancellation, much as we did when we examined the multistage multiple-passband modified comb filter. Changing notation slightly, we write (5.38) as

$$H(z) = \frac{1 - z^{-N}}{1 - z^{-M}} \tag{5.58}$$

where $N = rM$ and $r > 1$ is an integer. This gives N zeros on the unit circle and M poles there to cancel M of the zeros and create $M/2$ passbands in the operating range (see Figure 5.10, for example). If we want only one passband in the operating range, we must negate all but two of the pole-zero cancellations by introducing additional zeros at the unwanted cancellation locations. For example, if $N = 12$ and $M = 4$, (5.58) becomes

$$H(z) = \frac{1 - z^{-12}}{1 - z^{-4}} \tag{5.59}$$

This filter will have pole-zero cancellations—and hence passbands—at $\omega T = 0$, $\pi/2$, π, and $3\pi/2$. If we want to eliminate the passbands at 0 and π, we introduce additional zeros at $z = 1$ and $z = -1$, respectively. The new transfer function becomes

$$\begin{aligned} H'(z) &= \frac{(1 - z^{-12})(1 + z^{-1})(1 - z^{-1})}{1 - z^{-4}} \\ &= \frac{1 - z^{-12}}{1 + z^{-2}} \end{aligned} \tag{5.60}$$

so that now, the only passband in the operating range is at $z = e^{j\pi/2}$, or $\omega T = \pi/2$.

Similarly, we can obtain single passbands at other locations. Instead of starting with (5.58), we will use the form of the result (5.60) to write the desired transfer function as

$$H(z) = \frac{1 - z^{-N}}{1 + z^{-P}} \tag{5.61}$$

where $N = qP$ and $q > 1$ is an integer. Suppose, for example, that $N = 24$ and we desire a single passband centered at $\omega T = \pi/6$—that is, $z = \exp(j\pi/6)$. If we select $P = 6$, we will have six pole-zero cancellations: two at the desired locations $\omega T = \pm\pi/6$, two at $\pm\pi/2$, and two at $\pm 5\pi/6$. To eliminate the last two pairs, we must introduce zeros at those points. Therefore, the transfer function for the specified filter becomes

$$\begin{aligned} H(z) &= \frac{(1 - z^{-24})(1 + z^{-2})(1 - \sqrt{3}z^{-1} + z^{-2})}{1 + z^{-6}} \\ &= \frac{(1 - z^{-24})(1 - \sqrt{3}z^{-1} + z^{-2})}{1 - z^{-2} + z^{-4}} \end{aligned} \tag{5.62}$$

from which the difference equation may be written, in either recursive or nonrecursive form.

We can also take higher-order zeros and canceling poles for the bandpass design. By considering the lengths of vectors drawn from the various poles and zeros to a frequency point of interest on the unit circle, Lynn developed, for various types of filters, the ratios of main lobe to first side lobe given in Table 5.3.

TABLE 5.3 Ratio of Main Lobe to Side Lobe for Pole-Zero Cancellation Filters (after Lynn [1])

Type of filter	Main Lobe / First Side Lobe	Gain of Filter at Center of Passband	Remarks
Lowpass and highpass	$\left(N \sin \frac{3\pi}{2N}\right)^n$	N^n	N = number of zeros on unit circle; n = order of each zero
Bandpass with center frequency $\omega T = \pi/2$	$\left(\frac{N}{2} \sin \frac{3\pi}{N}\right)^n$	$\left(\frac{N}{2}\right)^n$	same as above
Bandpass with center frequency $\omega T = \pi/3$ or $2\pi/3$	$\dfrac{\left[N \sin \frac{3\pi}{2N} \sin\left(\frac{\pi}{3} - \frac{3\pi}{2N}\right)\right]^n}{\sin^n \frac{\pi}{3}}$ $= \left[1.15N \sin \frac{3\pi}{2N} \sin\left(\frac{\pi}{3} - \frac{3\pi}{2N}\right)\right]^n$		same as above

Although comb filters and modified comb filters are easy to design and involve multiplication by integer coefficients, they are special-purpose filters subject to several restrictions.

1. In general, the passbands are narrow compared with the operating range of the filter.
2. Considerable storage may be required for the delayed samples.
3. For specified critical frequencies, the time step required to ensure an integer number of delays may not suit the sampling period of the signal being processed.

A more general and more systematic method for designing nonrecursive filters to approximate an arbitrary specified frequency response is discussed in Section 5.4. First, however, we will consider the characteristics of filters with linear phase.

5.3 FIR FILTERS WITH LINEAR PHASE

In many applications, it is desirable that a digital filter have a linear-phase characteristic. In particular, is it important for speech and music processing and for data transmission wherein nonlinear phase would give unacceptable frequency distortion. FIR filters can be designed to give exact linear phase, whether they are implemented nonrecursively or recursively, as we saw, for example, in the last section.

The requirement of linear phase imposes certain conditions on the form of the filter pulse response. These may be determined as follows. Consider the causal finite-duration sequence $h_k, k = 0, \ldots, N-1$. Let this sequence be the pulse response of the filter. Its frequency response is

$$H(j\omega T) = \sum_{k=0}^{N-1} h_k e^{-jk\omega T} \tag{5.63}$$

The frequency response may also be written

$$H(j\omega T) = |H(j\omega T)|e^{j\phi(\omega T)} \tag{5.64}$$

If the filter is to have linear phase, then

$$\phi(\omega T) = -\alpha\omega T \tag{5.65}$$

where α is a constant, not necessarily integer. Then (5.64) becomes

$$H(j\omega T) = |H(j\omega T)|e^{-j\alpha\omega T} \tag{5.66}$$

Comparing (5.63) and (5.66) indicates that

$$\sum_{k=0}^{N-1} h_k e^{jk\omega T} = |H(j\omega T)|e^{-j\alpha\omega T} \tag{5.67}$$

Equating the real and the imaginary parts in (5.67) gives two equations:

$$\sum_{k=0}^{N-1} h_k \cos k\omega T = |H(j\omega T)| \cos \alpha\omega T \tag{5.68}$$

and

$$\sum_{k=0}^{N-1} h_k \sin k\omega T = |H(j\omega T)| \sin \alpha\omega T \tag{5.69}$$

Hence, combining (5.68) and (5.69), we get

$$\sum_{k=0}^{N-1} h_k \cos k\omega T \sin \alpha\omega T - \sum_{k=0}^{N-1} h_k \sin k\omega T \cos \alpha\omega T = 0 \tag{5.70}$$

or

$$\sum_{k=0}^{N-1} h_k \sin[(\alpha - k)\omega T] = 0 \tag{5.71}$$

The only nontrivial solutions to this equation for α and h_k are

$$\alpha = \frac{N-1}{2}$$
$$h_k = h_{N-k-1} \tag{5.72}$$

Uniqueness of the solutions is guaranteed by the Fourier series form of (5.71). Thus (5.72) implies that for each value of N, there is only one value of α for which linear phase can be obtained exactly. It also indicates that the pulse response must have a certain type of symmetry. Examples of pulse responses satisfying these conditions for N odd and for N even are given in Figure 5.13. We recall that the filter depicted in Figure 5.12 had that kind of symmetry. The condition (5.65) for linear phase implies not only that the phase delay $\phi(\omega T)/(\omega T)$ is constant but also that the group delay $d\phi(\omega T)/d\omega T$ is constant, the latter being equal to α.

We can take advantage of the symmetry in the pulse response by shifting the origin to the center of symmetry (Figure 5.13). For N odd, let $r = k - \alpha$ so that

$$H(j\omega T) = \sum_{r=-\alpha}^{\alpha} h'_r e^{-j(r+\alpha)\omega T} = e^{-j\alpha\omega T} \sum_{r=0}^{\alpha} a_r \cos(r\omega T) \tag{5.73}$$

where $a_0 = h_\alpha$ and $a_r = 2h_{\alpha-r}$, $r > 0$. Thus the magnitude response is

$$M(\omega T) = \left| \sum_{r=0}^{\alpha} a_r \cos(r\omega T) \right| \tag{5.74}$$

and the phase response is

$$\phi(\omega T) = -\alpha\omega T + c \tag{5.75}$$

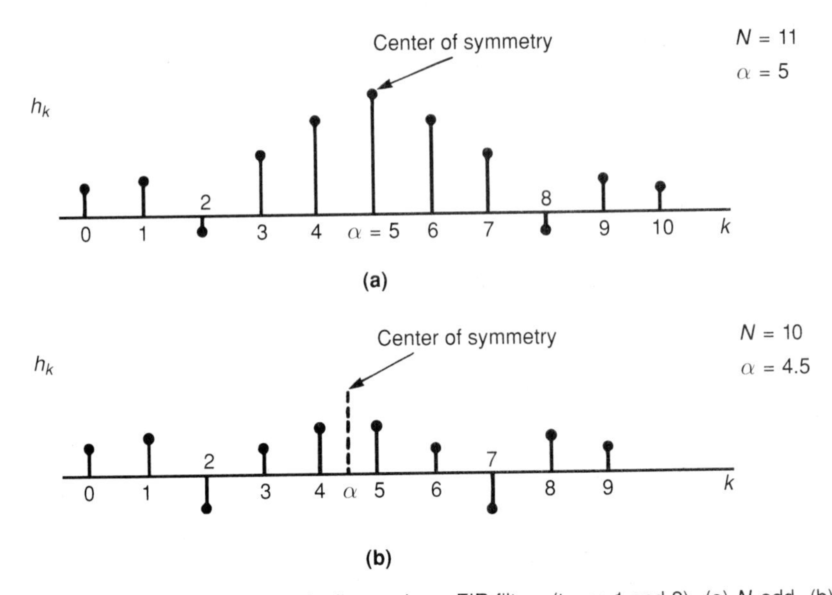

FIGURE 5.13 Pulse responses for linear-phase FIR filters (types 1 and 2). (a) N odd. (b) N even.

where c is either 0 or π, depending on the sign of the summation. This is called a type-1 filter. If $c = 0$, the phase is linear; if $c = \pi$, the phase is piecewise-linear with step changes of $\pm\pi$. Note that c is a function of frequency ω, but we omit the functional dependence for brevity.

For N even, let $r = k - \alpha - \frac{1}{2} = k - N/2$. Then

$$H(j\omega T) = e^{-j\alpha\omega T} \sum_{r=1}^{N/2} a_r \cos\left[\left(r - \frac{1}{2}\right)\omega T\right] \tag{5.76}$$

where $a_r = 2h_{N/2-r}$ in this case. The magnitude response is

$$M(\omega T) = \left|\sum_{r=1}^{N/2} a_r \cos\left[\left(r - \frac{1}{2}\right)\omega T\right]\right| \tag{5.77}$$

and the phase response is

$$\phi(\omega T) = -\alpha\omega T + c \tag{5.78}$$

where c is either 0 or π, depending on the sign of the summation. This is called a type-2 filter. At $\omega T = \pi$, the end of the operating range, we see from (5.77) that $M(\omega T) = 0$ for all values of r, so we cannot use a type-2 filter to represent a highpass or bandstop filter or any other filter the magnitude of which is not zero at $\omega T = \pi$.

Let us now consider the zero locations for type-1 and type-2 filters. For N odd, if we write (5.73) as

$$H(j\omega T) = e^{-j\alpha\omega T} \sum_{r=0}^{\alpha} a_r[(\exp(jr\omega T) + \exp(-jr\omega T))/2] \tag{5.79}$$

and replace $e^{j\omega T}$ by z, we get the transfer function

$$H(z) = z^{-\alpha} \sum_{r=0}^{\alpha} a_r(z^r + z^{-r})/2 \tag{5.80}$$

[We have often gone from $H(z)$ to $H(j\omega T)$ by setting z equal to $\exp(j\omega T)$. It is just as valid, of course, to go in the reverse direction.] It follows that

$$H(z^{-1}) = z^{\alpha} \sum_{r=0}^{\alpha} a_r(z^{-r} + z^r)/2 \tag{5.81}$$

Because (5.80) and (5.81) differ only by the delay terms $z^{-\alpha}$ and z^{α}, it follows that $H(z)$ and $H(z^{-1})$ have the same zeros. Therefore, if $H(z)$ has a pair of complex conjugate zeros at $z = re^{\pm j\theta}$, it also has a reciprocal pair at $z = 1/re^{\pm j\theta}$—reciprocal, that is, with respect to the unit circle. If the zeros of $H(z)$ are *on* the unit circle, as in the case of a comb filter, each is its own reciprocal. Real zeros have real reciprocals. A typical plot of the zeros for a type-1 filter is shown in Figure 5.14(a).

For N even, using (5.76), we see that, again, $H(z) = H(z^{-1})$ and that its zeros off the unit circle are in reciprocal pairs. In contrast to the N-odd case, there is a first- or higher-order zero at $z = -1 \quad (\omega T = \pi)$, as noted already. A typical plot of possible zero locations for the N-even linear-phase filter is given in Figure 5.14(b).

We noted above that condition (5.65) implied both linear phase and linear group delay. It also required a certain symmetry—see (5.72)—for the filter pulse response. We saw earlier that in the case of the modified comb filters, such as (5.56), this same kind of symmetry could give *piecewise-linear* phase over the operating range. We can also get piecewise-linear phase and constant group delay by changing condition (5.65). Suppose that instead of imposing condition (5.65), we require the filter phase to be

$$\phi(\omega T) = K - \alpha\omega T \tag{5.82}$$

where K (in radians) is a constant. This will give constant group delay α and piecewise-linear phase delay. Proceeding as before from (5.63) to (5.72), but using (5.82) instead of (5.65), we get the required conditions on α, K

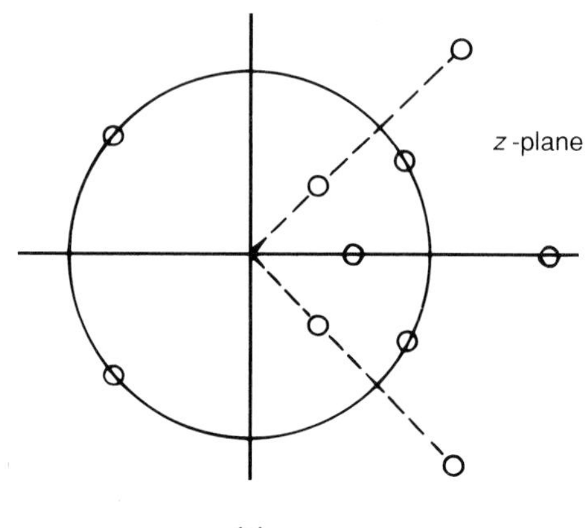

(a)

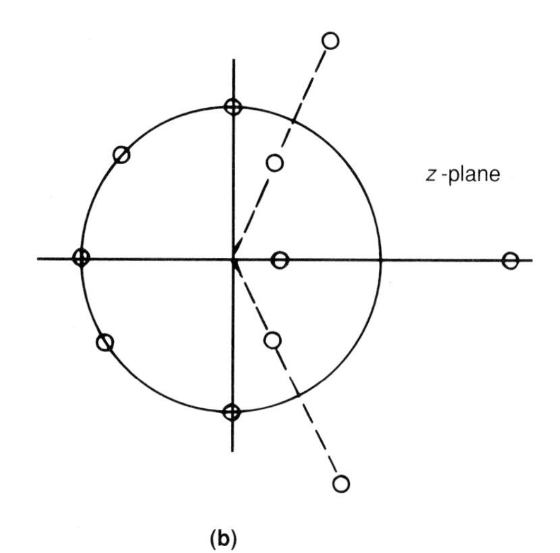

(b)

FIGURE 5.14 Typical zero locations for linear-phase digital filters with symmetric pulse response (types 1 and 2). (a) N odd. (b) N even.

and h_k as

$$\alpha = \frac{N-1}{2}$$
$$K = \pm\pi/2 \tag{5.83}$$
$$h_k = -h_{N-k-1}$$

We see that the delay α has not changed compared with (5.72), but now the pulse response h_k is anti-symmetric with respect to its center sample. Typical examples of such pulse responses for N odd and N even are shown in Figure 5.15.

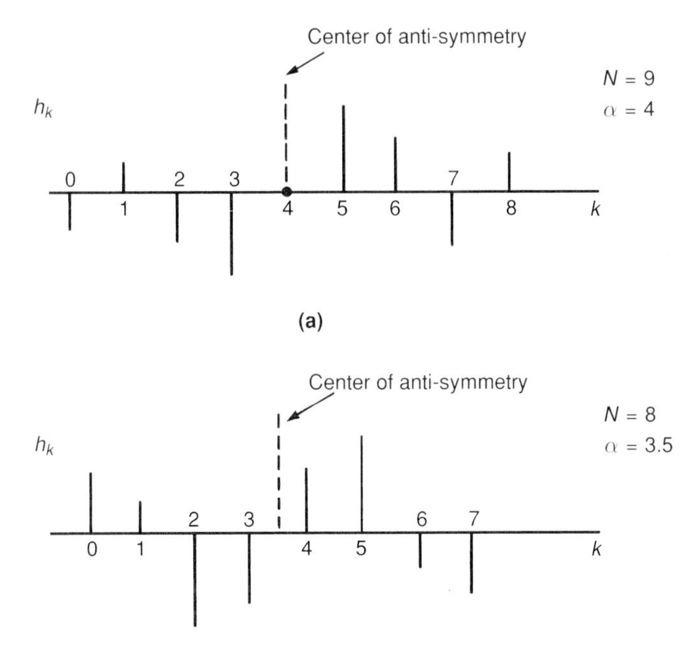

FIGURE 5.15 Anti-symmetric pulse responses for piecewise-linear-phase filters (types 3 and 4). (a) N odd. (b) N even.

As in the symmetric case, we can take advantage of this anti-symmetry. For N odd, let $r = k - \alpha$ so that

$$H(j\omega T) = \sum_{r=-\alpha}^{\alpha} h'_r e^{-j(r+\alpha)\omega T}$$

$$= je^{-j\alpha\omega T} \sum_{r=1}^{\alpha} a_r \sin(r\omega T) \tag{5.84}$$

$$= e^{-j(\alpha\omega T + \pi/2)} \sum_{r=1}^{\alpha} a_r \sin(r\omega T) \tag{5.85}$$

where $a_r = 2h_{\alpha-r}$ and h_k satisfies (5.83). The magnitude response is

$$M(\omega T) = \left| \sum_{r=1}^{\alpha} a_r \sin(r\omega T) \right| \tag{5.86}$$

and the phase response is

$$\phi(\omega T) = -\alpha\omega T + \pi/2 + c \tag{5.87}$$

where c is either 0 or π, depending on the sign of the summation. This is called a type-3 filter. The magnitude response (5.85) is zero at $\omega T = 0$ and $\omega T = \pi$—that is, at both ends of the operating range. Because of this and because the frequency response (5.84) is purely imaginary, this type of filter is useful for approximating an ideal differentiator or Hilbert transformer (see Section 5.6.4).

For N even, let $r = k - \alpha - 1/2 = k - N/2$. Then

$$H(j\omega T) = je^{-j\alpha\omega T} \sum_{r=1}^{N/2} a_r \sin\left[\left(r - \frac{1}{2}\right)\omega T\right] \tag{5.88}$$

where $a_r = 2h_{N/2-r}$. The magnitude response is

$$M(\omega T) = \left| \sum_{r=1}^{N/2} a_r \sin\left[\left(r - \frac{1}{2}\right)\omega T\right] \right| \tag{5.89}$$

and the phase response is

$$\phi(\omega T) = -\alpha\omega T + \pi/2 + c \tag{5.90}$$

where c is either 0 or π, depending on the sign of the summation. This is called a type-4 filter. At $\omega T = 0$, $M(\omega T) = 0$ according to (5.89), and because of the purely imaginary nature of the frequency response in (5.88), this type of filter is suitable for approximating functions with similar behavior, such as an ideal differentiator or Hilbert transformer.

As for the symmetric pulse response cases, $H(z)$ and $H(z^{-1})$ are the same except for the delay terms $z^{-\alpha}$ and z^{α}. There is an additional negative-sign difference for the anti-symmetric case. Therefore $H(z)$ has the same zeros as $H(z^{-1})$. Typical zero locations for N odd and N even for the anti-symmetric case are shown in Figure 5.16.

The characteristics of the four types of linear-phase FIR filters are summarized in Table 5.4. The next question is how to determine the filter unit pulse response h_k for a desired frequency response. One method for doing so is described in Section 5.4.

5.4 FOURIER SERIES METHOD FOR THE DESIGN OF NONRECURSIVE FILTERS

5.4.1 Expansion of the Desired Function

Let us now consider a systematic approach to the design of nonrecursive filters the frequency response of which is specified. This approach is called the Fourier series, or windowed, method. The latter term will be explained shortly.

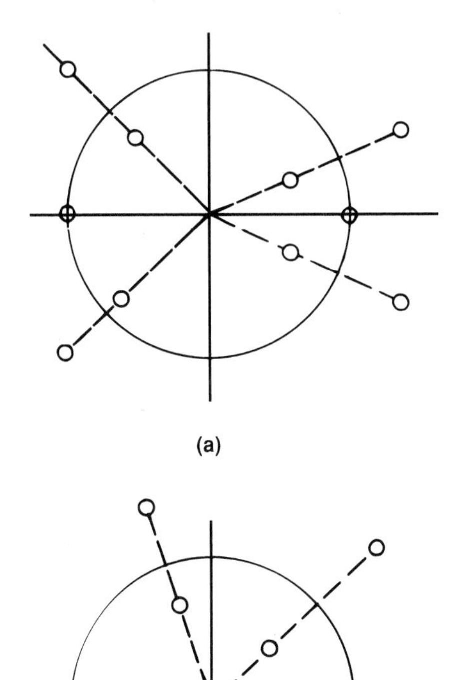

(a)

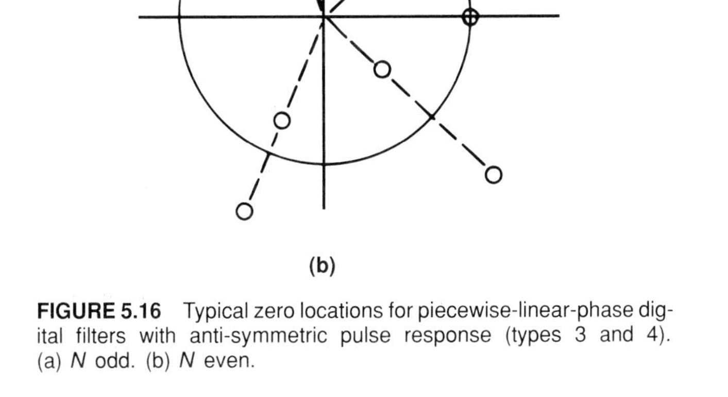

(b)

FIGURE 5.16 Typical zero locations for piecewise-linear-phase digital filters with anti-symmetric pulse response (types 3 and 4). (a) *N* odd. (b) *N* even.

Because

$$H[\,j(\omega + \omega_s)T] = H(\,j\omega T) \tag{5.91}$$

the frequency response function is a periodic function of frequency with period ω_s. Hence $H(\,j\omega T)$ can be expanded in a Fourier series

$$H(\,j\omega T) = \sum_{n=-\infty}^{\infty} c_n e^{-jn\omega T} \tag{5.92}$$

where

$$c_n = \frac{1}{\omega_s} \int_{-\omega_s/2}^{\omega_s/2} H(\,j\omega T) e^{jn\omega T}\, d\omega \tag{5.93}$$

TABLE 5.4 Summary of Characteristics of Linear-Phase FIR Filters

Type	Even or Odd Number of Samples N	Symmetry of Pulse Response h_n	Frequency Response $H(j\omega T)$	Magnitude Response $M(\omega T)$	Phase Response $\phi(\omega T)$	Relation Between Coefficients	Suitable for
1	odd	symmetric	$e^{-j\alpha\omega T}\sum_{n=0}^{\alpha} a_n \cos(n\omega T)$	$\left\lvert\sum_{n=0}^{\alpha} a_n \cos(n\omega T)\right\rvert$ $M(0) = \left\lvert\sum_{n=0}^{\alpha} a_n\right\rvert$ $M(\pi) = \left\lvert\sum_{n=0}^{\alpha} (-1)^n a_n\right\rvert$	$-\alpha\omega T + c$	$a_0 = h_\alpha$ $a_n = 2h_{\alpha-n},$ $n > 0$	lowpass, highpass, bandpass, bandstop
2	even	symmetric	$e^{-j\alpha\omega T}\sum_{n=1}^{N/2} a_n \cos\left[\left(n - \frac{1}{2}\right)\omega T\right]$	$\left\lvert\sum_{n=1}^{N/2} a_n \cos\left[\left(n - \frac{1}{2}\right)\omega T\right]\right\rvert$ $M(0) = \left\lvert\sum_{n=0}^{N/2} a_n\right\rvert$ $M(\pi) = 0$	$-\alpha\omega T + c$	$a_n = 2h_{N/2-n}$	lowpass, bandpass
3	odd	anti-symmetric	$e^{-j(\alpha\omega T+\pi/2)}\sum_{n=1}^{\alpha} a_n \sin(n\omega T)$	$\left\lvert\sum_{n=1}^{\alpha} a_n \sin(n\omega T)\right\rvert$ $M(0) = 0$ $M(\pi) = 0$	$-\alpha\omega T + \frac{\pi}{2} + c$	$a_n = 2h_{\alpha-n}$	differ-entiator, Hilbert trans-former
4	even	anti-symmetric	$e^{-j(\alpha\omega T+\pi/2)}\sum_{n=1}^{N/2} a_n \sin\left[\left(n - \frac{1}{2}\right)\omega T\right]$	$\left\lvert\sum_{n=1}^{N/2} a_n \sin\left[\left(n - \frac{1}{2}\right)\omega T\right]\right\rvert$ $M(0) = 0$ $M(\pi) = \left\lvert\sum_{n=1}^{N/2} (-1)^n a_n\right\rvert$	$-\alpha\omega T + \frac{\pi}{2} + c$	$a_n = 2h_{N/2-n}$	differ-entiator, Hilbert trans-former

Notes:
1. $\alpha = \frac{N-1}{2}$
2. $c = c(\omega) = 0$ or π, depending upon the sign of the summation

A well-known theorem in Fourier series analysis [2] states that the Fourier coefficients c_n given by (5.93) minimize the integral-squared error between an expansion of the form (5.92) and the function H represented by the expansion.

On the unit circle, we have $z = e^{j\omega T}$. Therefore, (5.92) can be written in the form

$$H(z) = \sum_{n=-\infty}^{\infty} h_n z^{-n} \tag{5.94}$$

where $h_n = c_n$, as given in (5.93). Thus if a particular form for $H(j\omega T)$ is specified, the corresponding coefficients c_n can be calculated for a nonrecursive filter, provided that the integral (5.93) can be evaluated. Two drawbacks to this approach, in its present form, are apparent. First, working with an infinite number of terms is not practical. Second, with negative values of n, the resulting filter is nonrealizable for real-time processing. Before treating the modifications necessary to achieve a satisfactory design, we will examine in more detail how the required function $H(j\omega T)$ is specified.

We can write

$$H(j\omega T) = M(\omega T)\exp[-j\phi(\omega T)] \tag{5.95}$$

in terms of magnitude and phase, or we can write

$$H(j\omega T) = R(\omega T) + jI(\omega T) \tag{5.96}$$

as the real and imaginary parts. If both magnitude and phase are specified as functions of ωT, then R and I are also specified. For a real pulse response $h(n)$, the real part R is an even periodic function of ωT and, hence, can be expanded in a Fourier cosine series (see Chapter 2).

$$R(\omega T) = \frac{a_0}{2} + \sum_{i=1}^{\infty} a_n \cos(n\omega T) \tag{5.97}$$

where

$$a_n = \frac{4}{\omega_s}\int_0^{\omega_s/2} R(\omega T)\cos(n\omega T)\,d\omega, \quad n = 0, 1, \ldots \tag{5.98}$$

The imaginary part I is an odd periodic function of ωT and so may be expanded in a Fourier sine series.

$$I(\omega T) = \sum_{i=1}^{\infty} b_n \sin(n\omega T) \tag{5.99}$$

where

$$b_n = \frac{4}{\omega_s}\int_0^{\omega_s/2} I(\omega T)\sin(n\omega T)\,d\omega \tag{5.100}$$

Therefore, from (5.96), the frequency response can be written

$$H(j\omega T) = \frac{a_0}{2} + \sum_{n=1}^{\infty} a_n \cos n\omega T + j \sum_{n=1}^{\infty} b_n \sin n\omega T \tag{5.101}$$

From the exponential representation of sines and cosines, it follows that

$$H(z)\Big|_{z=e^{j\omega T}} = \frac{a_0}{2} + \frac{1}{2}\sum_{n=1}^{\infty}[(a_n + b_n)z^n + (a_n - b_n)z^{-n}]\Big|_{z=e^{j\omega T}} \tag{5.102}$$

To obtain a practical filter, truncate the summations after N terms, where N is finite. Then

$$H_N(z) = \frac{a_0}{2} + \frac{1}{2}\sum_{n=1}^{N}[(a_n + b_n)z^n + (a_n - b_n)z^{-n}] \tag{5.103}$$

The positive powers of z indicate that this filter is nonrealizable. Divide (5.103) across by z^N, obtaining

$$z^{-N}H_N(z) = \frac{a_0}{2}z^{-N} + \frac{1}{2}\sum_{n=1}^{N}[(a_n + b_n)z^{n-N} + (a_n - b_n)z^{-n-N}] \tag{5.104}$$

The right-hand side of (5.104) is realizable because it contains no positive powers of z. The z^{-N} term multiplying $H_N(z)$ corresponds to a delay of N samples, or NT seconds. Define

$$H_R(z) = z^{-N}H_N(z) \tag{5.105}$$

Then the output of the realizable filter $H_R(z)$ is delayed by N samples with respect to the output of the nonrealizable filter $H_N(z)$. In the frequency domain, this corresponds to a linear phase shift of $-N\omega T$ radians. In many applications, this phase shift can be tolerated if the magnitude response is of primary importance. Then the realizable filter given by the right-hand side of (5.104) can be written in the form

$$H_R(z) = \sum_{k=0}^{2N} h_k z^{-k} \tag{5.106}$$

where

$$\begin{aligned} h_k &= (a_{N-k} + b_{N-k})/2, \quad 0 \le k \le N-1 \\ h_N &= a_0/2 \\ h_k &= (a_{k-N} - b_{k-N})/2, \quad N+1 \le k \le 2N \end{aligned} \tag{5.107}$$

and the coefficients a_k and b_k are computed from the desired frequency response function according to (5.98) and (5.100).

In cases where the magnitude response $M(\omega T)$ alone of the desired filter is specified and the phase is not of concern, it is possible to simplify the design by letting the phase be zero so that $M(\omega T)$ equals $R(\omega T)$ as given in (5.97), and $I(\omega T)$ is zero. Then the realizable filter (5.106) has symmetrical coefficients:

$$\begin{aligned} h_k &= a_{N-k}/2, \quad 0 \le k \le N-1 \\ h_N &= a_0/2 \\ h_k &= a_{k-N}/2, \quad N+1 \le k \le 2N \end{aligned} \tag{5.108}$$

The coefficients a_k are computed from (5.98). Because of the symmetry, the number of multiplier coefficients can be reduced by a factor of 2. The filter arrangement of $N = 3$ is shown in Figure 5.17.

To illustrate the procedure for filter design, let us consider three examples.

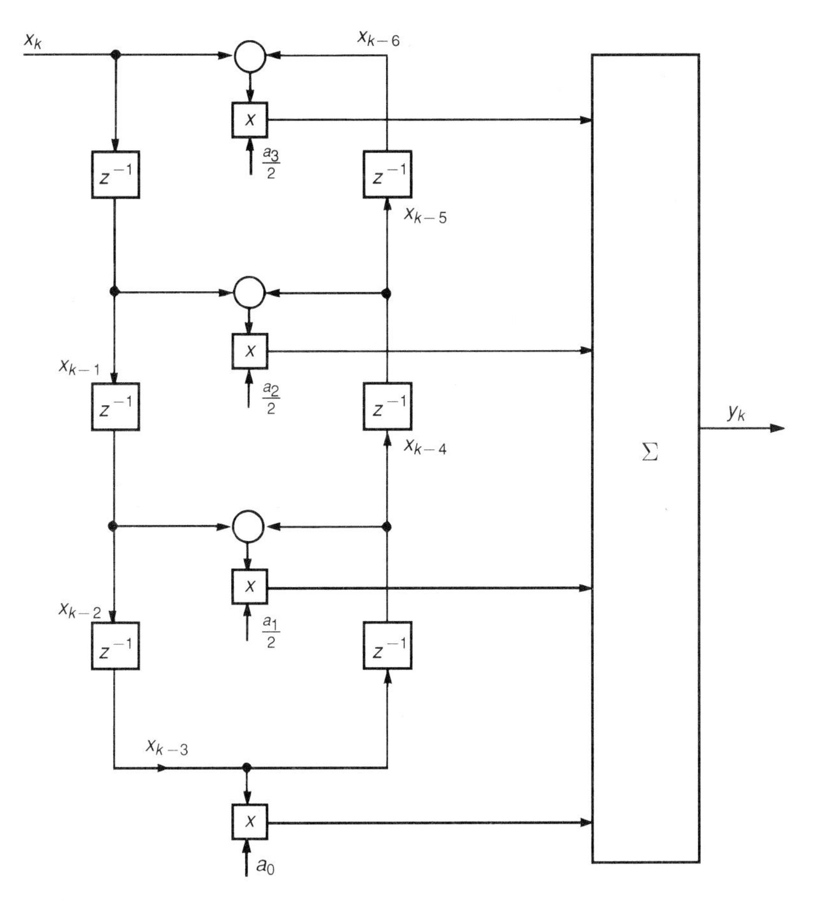

FIGURE 5.17 Nonrecursive filter layout for a specified magnitude response using the Fourier series method ($N = 3$).

EXAMPLE 5.5

Suppose the specified magnitude response is parabolic—that is,

$$M(\omega T) = 1 - \left(\frac{\omega T}{\pi}\right)^2, \quad |\omega T| \leq \pi \tag{5.109}$$

The object is to design a nonrecursive filter the magnitude response of which approximates that specified.

Solution

Expanding the magnitude response function in a Fourier series and evaluating the coefficients a_n from (5.98) gives

$$M(\omega T) = \frac{2}{3} + \frac{4}{\pi^2}\sum_{n=1}^{\infty}(-1)^{n+1}\cos(n\omega T)/n^2 \tag{5.110}$$

so the transfer function of the truncated nonrealizable filter is

$$H_N(z) = \frac{2}{3} + \frac{2}{\pi^2}\sum_{n=1}^{N}(-1)^{n+1}(z^n + z^{-n})/n^2 \tag{5.111}$$

Divide across by z^N to get the transfer function of the realizable filter,

$$H_R(z) = \frac{2}{3}z^{-N} + \frac{2}{\pi^2}\sum_{n=1}^{N}(-1)^{n+1}(z^{n-N} + z^{-n-N})/n^2 \tag{5.112}$$

From this we see that the pulse response has the symmetry of a linear-phase filter given by (5.72).

Suppose the series is truncated at $N = 3$. Then, from (5.112) with $z = e^{j\omega T}$, the frequency response is

$$H_R(j\omega T) = e^{-j3\omega T}\left[\frac{2}{3} + \frac{4}{\pi^2}\cos\omega T - \frac{1}{\pi^2}\cos 2\omega T + \frac{4}{9\pi^2}\cos 3\omega T\right] \tag{5.113}$$

The magnitude response and the phase response are plotted in Figure 5.18. The phase is linear, as implied by (5.113). The comparison between the desired magnitude response of (5.109) and that given by (5.113) indicates reasonable correspondence except at the ends. If more terms were taken—say, N equal to 20—the correspondence at the ends would be much better.

It should be noted that because the given magnitude response is a continuous function of frequency in Example 5.5, we do not have to contend with the Gibbs phenomenon that arises in a truncated Fourier series approximation at a point of discontinuity (see Section 5.4.2). Example 5.7 illustrates this point, but first we will consider an example where the desired frequency response is purely imaginary.

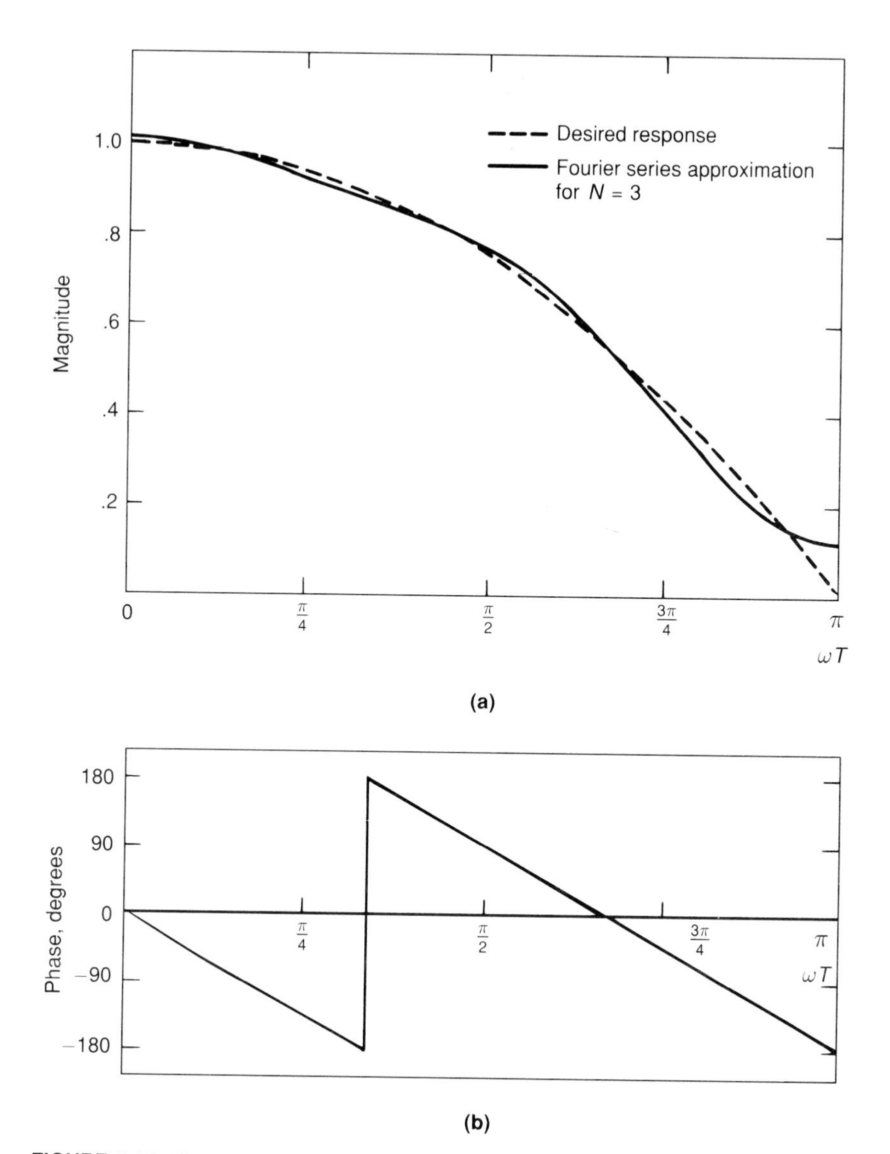

FIGURE 5.18 Frequency responses for filter from Fourier series approximation of Example 5.5. (a) Magnitude response. (b) Phase response.

EXAMPLE 5.6

It is desired to approximate an ideal differentiator,

$$H(j\omega T) = j\omega T, \quad -\pi \leq \omega T \leq \pi \tag{5.114}$$

as illustrated in Figure 5.19.

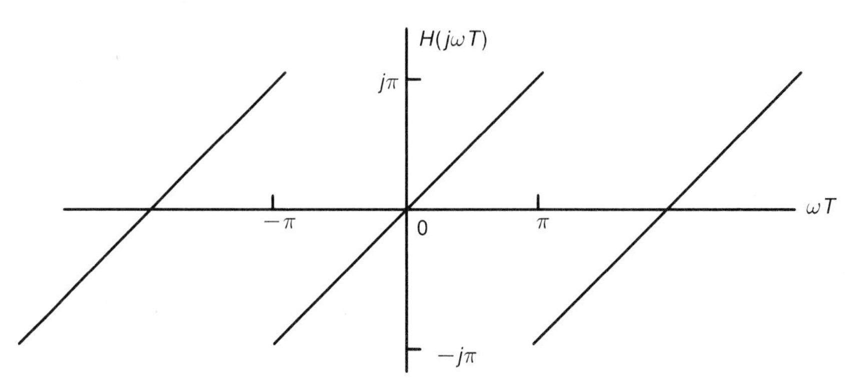

FIGURE 5.19 Frequency response of an ideal differentiator.

Solution

In this case,

$$H(j\omega T) = jI(\omega T) \tag{5.115}$$

where

$$I(\omega T) = \omega T, \quad -\pi \leq \omega T \leq \pi \tag{5.116}$$

Insert this value of $I(\omega T)$ into (5.100), and change the variable of integration from ω to ωT to get the coefficients of the Fourier series expansion

$$b_n = \frac{2}{\pi} \int_0^{\pi} x \sin nx \, dx \tag{5.117}$$

where $x = \omega T$. Integrating by parts yields

$$\begin{aligned} b_n &= \frac{2}{\pi} \left\{ -\frac{x}{n} \cos nx \Big|_0^{\pi} + \frac{1}{n} \int_0^{\pi} \cos nx \, dx \right\} \\ &= -2 \frac{\cos n\pi}{n} \\ &= -2(-1)^n/n, \quad n = 1, 2, \ldots \end{aligned} \tag{5.118}$$

Also $b_0 = 0$, because $I(\omega T)$ is odd. Therefore, from (5.104), we get the transfer function of the realizable filter as

$$H_R(z) = \frac{1}{2} \sum_{n=1}^{N} b_n (z^{n-N} - z^{-n-N}) \tag{5.119}$$

because $a_n = 0, n = 0, 1, \ldots$. Insert the values of b_n to get

$$H_R(z) = -\sum_{n=1}^{N} (-1)^n (z^{n-N} - z^{-n-N})/n \tag{5.120}$$

Now we must choose the truncation value N. If we take $N = 3$ and include the $b_0 = 0$ term, we get

$$H_R(z) = \frac{1}{2}\{b_0(z^{-3} - z^{-3}) + b_1(z^{-2} - z^{-4}) + b_2(z^{-1} - z^{-5}) + b_3(z^0 - z^{-6})\}$$

$$= (z^{-2} - z^{-4}) - \frac{1}{2}(z^{-1} - z^{-5}) + \frac{1}{3}(1 - z^{-6}) \tag{5.121}$$

The pulse response is shown in Figure 5.20. It has seven taps, or coefficients, and exhibits anti-symmetry about its center point, in accordance with (5.83). From (5.121), by inspection, we can write the difference equation,

$$y_k = \frac{1}{3}x_k - \frac{1}{2}x_{k-1} + x_{k-2} - x_{k-4} + \frac{1}{2}x_{k-5} - \frac{1}{3}x_{k-6} \tag{5.122}$$

The frequency response of the approximation is

$$H(j\omega T) = jI(\omega T)$$
$$= j\sum_{n=1}^{3} b_n \sin n\omega T \tag{5.123}$$

from (5.100). Inserting the values for b_n from (5.118), we get

$$H(j\omega T) = j\left[2\sin\omega T - \sin 2\omega T + \frac{2}{3}\sin 3\omega T\right] \tag{5.123A}$$

This is plotted in Figure 5.21. Evidently, it is not a very good approximation to the ideal. We truncated the Fourier series too early. If we took many more terms—say $N = 20$, implying 41 coefficients—we would get a closer approximation. However, no matter how many terms we take, the approximation with an odd number of coefficients will always go to zero at $\omega T = \pi$ [see also (5.85)]. This is the average of the values of the ideal function at the discontinuity at $\omega T = \pi$ (see Figure 5.19). Extension of the solution to include more coefficients is self-evident and is left as an exercise.

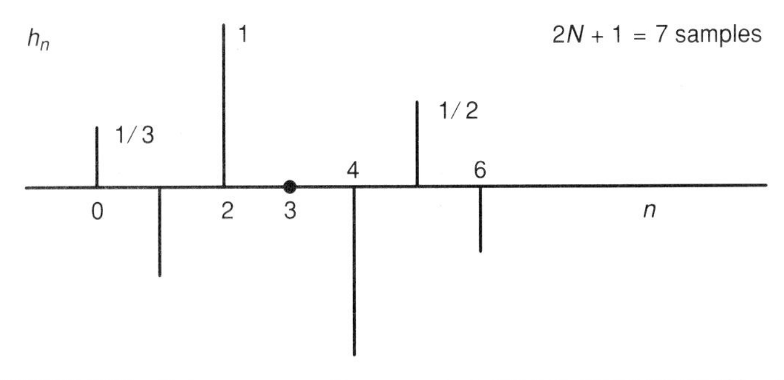

FIGURE 5.20 Pulse response for a seven-coefficient differentiator ($N = 3$).

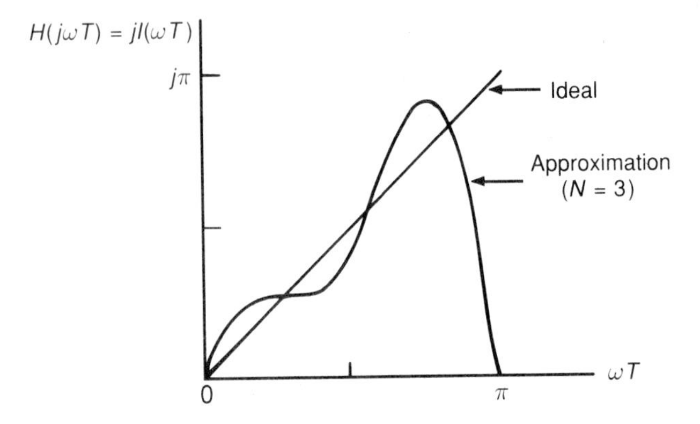

FIGURE 5.21 Frequency response for a seven-coefficient differentiator ($N = 3$).

We note that the Fourier series approximation is a weighted sum of sinusoids, so we would expect the approximation to have ripples if we take only a finite number of terms, even if the function being approximated is smooth, as in this example. For functions with discontinuities (see the next example), these ripples become so large that the method would be unsuitable if designed windows were not used to smooth the oscillations.

EXAMPLE 5.7

In this example, it is desired to obtain a Fourier series approximation to an ideal bandpass filter

$$M(\omega T) = \begin{cases} 1, & \frac{\pi}{3} \le |\omega T| \le \frac{2\pi}{3} \\ 0, & \text{otherwise} \end{cases} \tag{5.124}$$

Solution

Proceeding as before, we expand $M(\omega T)$ in a Fourier cosine series.

$$M(\omega T) = \frac{1}{3} + \frac{4}{\pi} \sum_{n=1}^{\infty} \cos\frac{n\pi}{2} \sin\frac{n\pi}{6} \cos(n\omega T)/n \tag{5.125}$$

Truncating after N terms gives

$$H_N(z) = \frac{1}{3} + \frac{2}{\pi} \sum_{n=1}^{N} \cos\frac{n\pi}{2} \sin\frac{n\pi}{6} (z^n + z^{-n})/n \tag{5.126}$$

and the realizable filter is

$$H_R(z) = \frac{1}{3} z^{-N} + \frac{2}{\pi} \sum_{n=1}^{N} \cos\frac{n\pi}{2} \sin\frac{n\pi}{6} (z^{n-N} + z^{-n-N})/n \tag{5.127}$$

The frequency response is obtained by setting z equal to $e^{j\omega T}$.

$$H_R(j\omega T) = e^{-jN\omega T}\left[\frac{1}{3} + \frac{4}{\pi}\sum_{n=1}^{N}\cos\frac{n\pi}{2}\sin\frac{n\pi}{6}\cos(n\omega T)/n\right] \quad (5.128)$$

The absolute value of the term in the square brackets in (5.128) is the magnitude response $M(\omega T)$. It is plotted for $N = 14$ in Figure 5.22.

5.4.2 The Gibbs Phenomenon

The oscillations evident in the magnitude response of Figure 5.22 are not satisfactory, because in effect the stopband is down only about 20 dB. These oscillations are due to the fact that, by means of a finite number of terms of a Fourier series, a given function can never be approximated in the vicinity of a discontinuity with a tolerance less than a characteristic overswing of 9%, no matter how many terms one is prepared to use. This is known as the Gibbs phenomenon. A good treatment of the subject is given in Guillemin's text [3]. Let us now discuss the reason for the Gibbs phenomenon.

Truncating a Fourier series of the form (5.92) to

$$H_N(j\omega T) = \sum_{n=-N}^{N} c_n e^{-jn\omega T} \quad (5.129)$$

is equivalent to multiplying the coefficients of the original series by a function

$$w_R(n) = \begin{cases} 1, & -N \le n \le N \\ 0, & \text{otherwise} \end{cases} \quad (5.130)$$

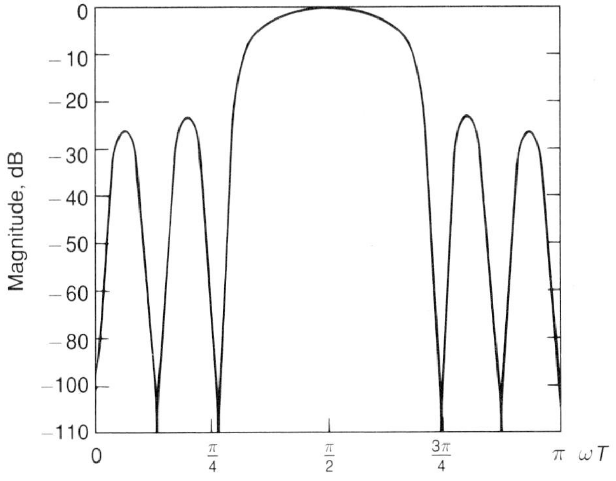

Magnitude, dB

0, −10, −20, −30, −40, −50, −60, −80, −100, −110

0, $\frac{\pi}{4}$, $\frac{\pi}{2}$, $\frac{3\pi}{4}$, π ωT

FIGURE 5.22 Magnitude response of Fourier series approximation to ideal bandpass filter (Example 5.7 without window function).

This is called a (two-sided) rectangular window function, and it occurs naturally as a result of the truncation operation. The subscript R on w denotes "rectangular."

From (5.130), the rectangular window has the z-transform

$$W_R(z) = z^N \frac{1 - z^{-M}}{1 - z^{-1}} \tag{5.131}$$

where $M = 2N + 1$. Set z equal to $e^{j\omega T}$ to get the frequency response

$$W_R(j\omega T) = \sin(M\omega T/2) / \sin(\omega T/2) \tag{5.132}$$

(The frequency response of a window function is sometimes called a *spectral window*.) It has the same form as the frequency response for the (nonrealizable) moving-average operation (5.23). Thus Figure 5.5 is valid for (5.132) with m replaced by M.

In practice, a one-sided rectangular window

$$w_R(n) = \begin{cases} 1, & 0 \le n \le N \\ 0, & \text{otherwise} \end{cases} \tag{5.130A}$$

represents the truncation of a series, such as (5.102), to $(N + 1)$ terms. This window has the same-shaped frequency response as (5.132) but has a phase-lag factor of $\exp[-j\omega(N - 1)T/2]$.

Multiplying the filter coefficients by $w_R(n)$ in the time domain—that is, $h_N(n) = h(n)w_R(n)$—corresponds to convolution in the frequency domain. Therefore, by the complex convolution theorem (3.56),

$$H_N(j\omega T) = \frac{1}{2\pi} \int_{-\pi}^{\pi} H(j\nu T) W_R[j(\omega - \nu)T] \, d\nu T \tag{5.133}$$

where $H(j\omega T)$ is the frequency response of the untruncated or desired filter. Because both H and W_R are periodic functions, the operation defined in (5.133) is often called *periodic* convolution. Periodic convolution gives a continuous function of frequency, in contrast to the discrete function given by circular convolution that is associated with the discrete Fourier transform (treated in Chapter 6). We note that as N approaches infinity, the frequency response (5.132) of the rectangular window approaches a delta function that, when used in the integral (5.133), implies that $H_N(j\omega T)$ approaches $H(j\omega T)$.

The periodic-convolution operation (5.133) is illustrated in Figure 5.23 for the function sought in Example 5.7 and the rectangular window due to truncation. For clarity, real functions of frequency are shown. The convolution operation is similar to the linear convolution depicted in Figure 1.12, but the resulting function $H_N(j\omega T)$ is periodic. The convolution of H with the oscillatory spectral window W_R results in ripples in the response H_N, as shown in Figure 5.23(d). As more terms are kept in the Fourier expansion (that is, as N becomes larger), the frequency of the ripples increases, as shown in Figure 5.24. However, the amplitude of the overswing relative

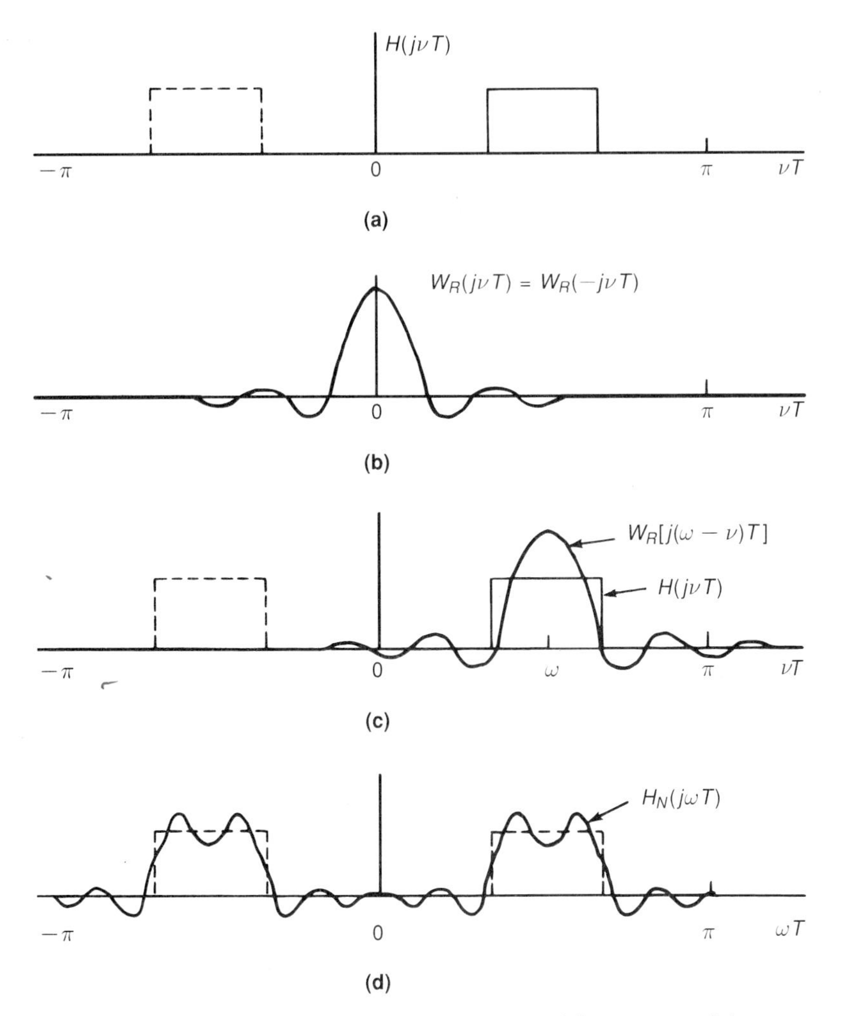

FIGURE 5.23 Periodic convolution of the ideal spectrum and the spectrum of the rectangular window.

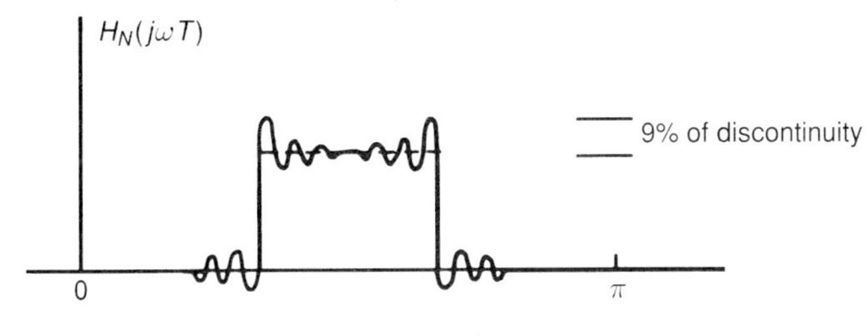

FIGURE 5.24 The Gibbs phenomenon.

to the magnitude of the discontinuity remains the same (about 9% of the discontinuity). This is the Gibbs phenomenon. In the limit as N approaches infinity, the ripples are compressed into a single vertical line at the discontinuity (still showing the 9% error). However, this line occupies no space in the frequency range, so we can say that, for practical purposes, the Fourier series completely represents the given function as N approaches infinity.

5.4.3 Window Functions

Because we must truncate the Fourier series, it would appear that we must live with large ripples in the stopband when we want to approximate frequency functions with discontinuities, such as ideal lowpass, highpass, bandpass, and bandstop functions (Figure 2.6). As we have seen, the problem lies with the oscillating nature of the rectangular spectral window, which is due to truncation. In particular, the area under the side lobes of this window are not negligible compared with the area under the main lobe (see Figure 5.5). This observation motivated many researchers to design window functions to weight the Fourier coefficients and provide a more gradual truncation of the series, hence reducing the oscillations at the discontinuity. Desirable window characteristics are

1. Small width of the main lobe of the window frequency response containing as much of the total area, or energy, as possible.
2. Side lobes of the frequency response that decrease in energy rapidly as ωT approaches π.

Having a narrow main lobe and having small side lobes are conflicting goals. Hence a trade-off between these objectives is necessary.

As a first attempt at a less abrupt truncation of the series, suppose that we weight every coefficient out to the $(N-1)$st with unity, as for the rectangular window, but weight the Nth coefficient with one-half. Hamming [4] called this a *modified* rectangular window, and we will denote it by $w_{RM}(n)$. Thus

$$w_{RM}(n) = \begin{cases} 1, & -(N-1) \le n \le N-1 \\ \frac{1}{2}, & n = -N, N \\ 0, & \text{otherwise} \end{cases} \tag{5.134}$$

Then

$$W_{RM}(z) = \sum_{n=-(N-1)}^{N-1} z^{-n} + \frac{1}{2}(z^N + z^{-N}) \tag{5.135}$$

When we set z equal to $e^{j\omega T}$, the separate term in (5.135) becomes $\cos N\omega T$. We can find the frequency response of the modified rectangular window readily by subtracting $\cos N\omega T$ from $W_R(j\omega T)$ given in (5.132).

The result is

$$W_{RM}(j\omega T) = \sin(M\omega T/2)/\sin(\omega T/2) - \cos N\omega T$$
$$= \sin(N\omega T)\cos(\omega T/2)/\sin(\omega T/2) \tag{5.136}$$

because $M = 2N + 1$. The extra factor $\cos \omega T/2$ goes to zero at $\omega T = \pi$ and produces some improvement in the oscillation problem as noted in [4], and shown in Figure 5.25. However, a more drastic modification—or more gradual smoothing—of the rectangular window is needed.

The triangular window proposed by Bartlett [4] for spectral estimation has the form

$$w_B(n) = \begin{cases} 1 - \frac{|n|}{N}, & |n| \leq N \\ 0, & \text{otherwise} \end{cases} \tag{5.137}$$

This is a two-sided window for a two-sided expansion such as (5.92), truncated at $\pm N$, with $M = 2N + 1$ terms. For a filter such as (5.104) with the Fourier series truncated at $n = N$, a one-sided window is used that has the same form as (5.137) but with $0 \leq n \leq N$. In deriving the frequency response of the window, it is mathematically less cumbersome to consider the two-sided version. From (5.137), the z-transform of the window is

$$W_B(z) = \sum_{n=-N}^{N} \left(1 - \frac{|n|}{N}\right) z^{-n} \tag{5.138}$$

Set z equal to $e^{j\omega T}$ to get the frequency response

$$W_B(j\omega T) = \left(\frac{\sin N\omega T/2}{N\omega T/2}\right)^2 \tag{5.139}$$

The frequency response for the Bartlett window is shown in Figure 5.26. Note that the main-lobe width is twice that of the rectangular window, and the side lobes are never negative.

A more tailored window proposed by von Hann,

$$w_{vH}(n) = \begin{cases} \frac{1}{2}\left(1 + \cos \frac{\pi n}{N}\right), & |n| \leq N \\ 0, & \text{otherwise} \end{cases} \tag{5.140}$$

is sometimes called the raised-cosine window. It has the frequency response

$$W_{vH}(j\omega T) = \frac{\sin(N\omega T)\cos(\omega T/2)}{2\sin(\omega T/2)} \left\{ \frac{1}{1 - [\sin \omega T/2/\sin \pi/(2N)]^2} \right\} \tag{5.141}$$

We leave the derivation of this expression as an exercise for the mathematically inclined. The main thing to note is that the main-lobe width for the von Hann spectral window is twice that of the rectangular spectral window, but its side lobes are about 20 dB lower.

Hamming [5] noted that the von Hann window and the modified rectangular window have opposite signs in the side lobes. He reasoned that a

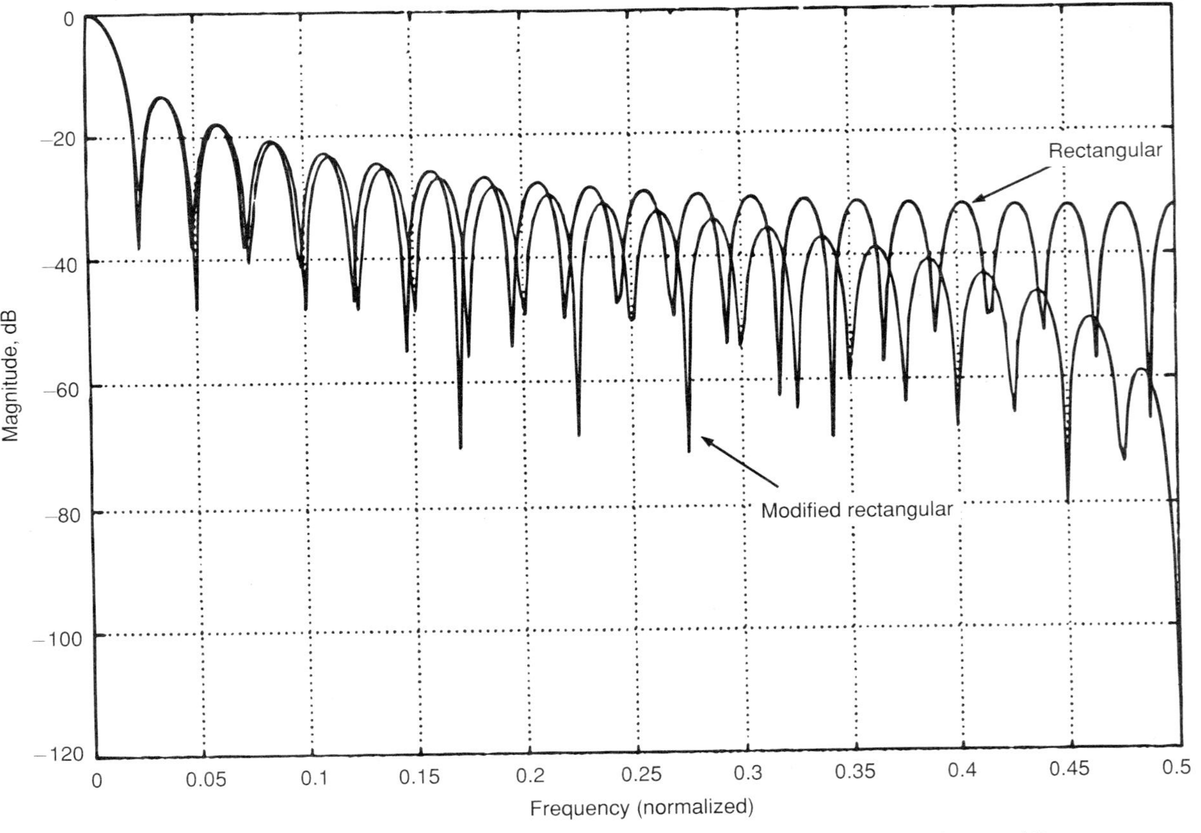

FIGURE 5.25 Spectra for rectangular and modified rectangular windows ($M = 2N + 1 = 41$).

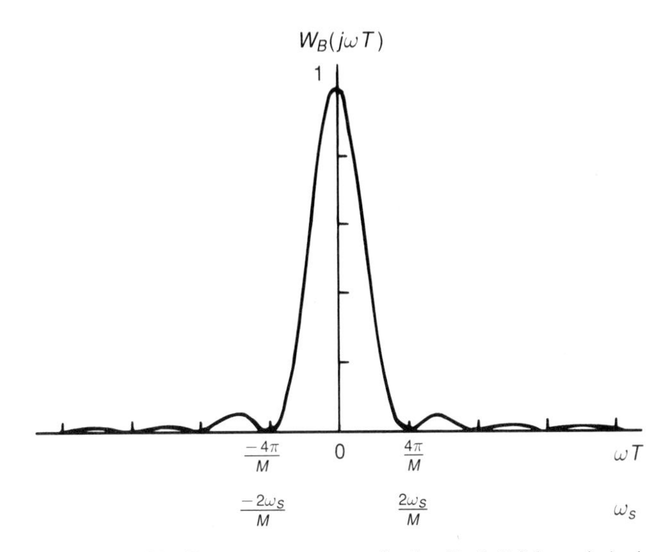

FIGURE 5.26 Frequency response for the Bartlett (triangular) window.

small amount of the modified window added to the von Hann window could be used to reduce the maximum that occurs in the side lobes. This resulted in the Hamming window, which has the generalized form

$$w_H(n) = \begin{cases} \alpha + (1-\alpha)\cos\frac{\pi n}{N}, & |n| \le N \\ 0, & \text{otherwise} \end{cases} \tag{5.142}$$

where $\alpha = 0.54$ (if $\alpha = 0.50$, we have the von Hann window). In deriving his window, Hamming determined what value of α minimizes the value of the maximum side lobe. He found that it is a function of N that varies from 0.56 at $N = 5$ and 0.54 at $N = 10$ to slightly less than 0.54 for larger values of N. A good representative value is $\alpha = 0.54$, as we have said. The rectangular, Bartlett, von Hann, and Hamming windows are compared in Figure 5.27.

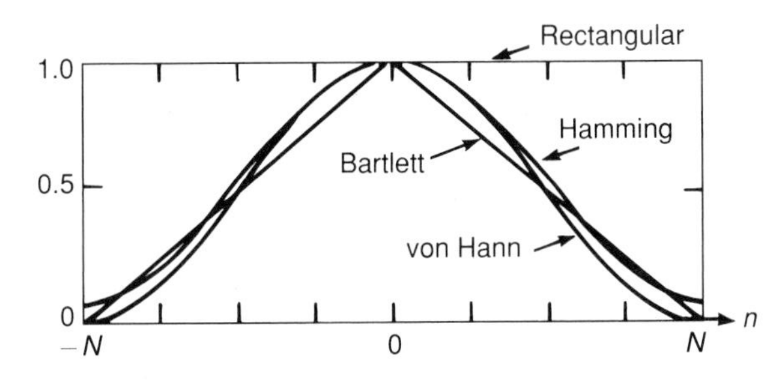

FIGURE 5.27 Rectangular, Bartlett, von Hann, and Hamming window functions.

The frequency response of the Hamming window can be obtained by noting that the window may be represented as the product of a rectangular window and an infinite-duration window of the form given in (5.142) but defined for all values of n. Thus (5.142) may be replaced by

$$w_H(n) = w_R(n) \cdot \left[\alpha + (1-\alpha)\cos\frac{\pi n}{N}\right] \tag{5.143}$$

where $w_R(n)$ is a two-sided M-point rectangular window with $M = 2N + 1$. The frequency response of the Hamming window is obtained as the periodic convolution of the frequency response of the rectangular window with three delta functions corresponding to the transform of the infinite-duration window. Hence

$$\begin{aligned} W_H(j\omega T) &= W_R(j\omega T) * \left[\alpha\delta(\omega T) + \frac{1-\alpha}{2}\delta\left(\omega T - \frac{\pi}{N}\right)\right. \\ &\qquad \left. + \frac{1-\alpha}{2}\delta\left(\omega T + \frac{\pi}{N}\right)\right] \\ &= \alpha W_R(j\omega T) + \frac{1-\alpha}{2} W_R[j(\omega T - \pi/N)] \\ &\quad + \frac{1-\alpha}{2} W_R[j(\omega T + \pi/N)] \end{aligned} \tag{5.144}$$

This function is plotted in Figure 5.28. (The frequency response for the von Hann window is very similar). Note that the weighted translated rectangular windows in (5.144) tend to cancel the side lobes of the untranslated rectangular window, which results in a function with most (99.96%) of its power concentrated in the main lobe. However, the main lobe of the Hamming window is twice as wide as that of the rectangular window. The peak side-lobe ripple is down about 40 dB from the main-lobe peak. Because the Hamming window provides less oscillation in the side lobes than the von Hann window and has the same main-lobe width, it is generally used instead. With the possible exception of the Kaiser window function (5.146), it is the most widely used window function.

Many other window functions have been proposed [7]. We will mention but two more: the Blackman window and the Kaiser window.

The Blackman window [7] is given by

$$w_{BL}(n) = \begin{cases} 0.42 - 0.5\cos\frac{\pi n}{N} + 0.08\cos\frac{2\pi n}{N}, & |n| \le N \\ 0, & \text{otherwise} \end{cases} \tag{5.145}$$

The additional cosine term (compared with the Hamming and von Hann windows) reduces the Gibbs phenomenon ripple even further, but the main-lobe width increases from $8\pi/M$, to $12\pi/M$, as shown in Table 5.5, where window function performances are compared. The attenuation of the first side lobe compared to the main lobe is a function of M, the width of the window ($M = 2N+1$), but the approximate values in the last column of the

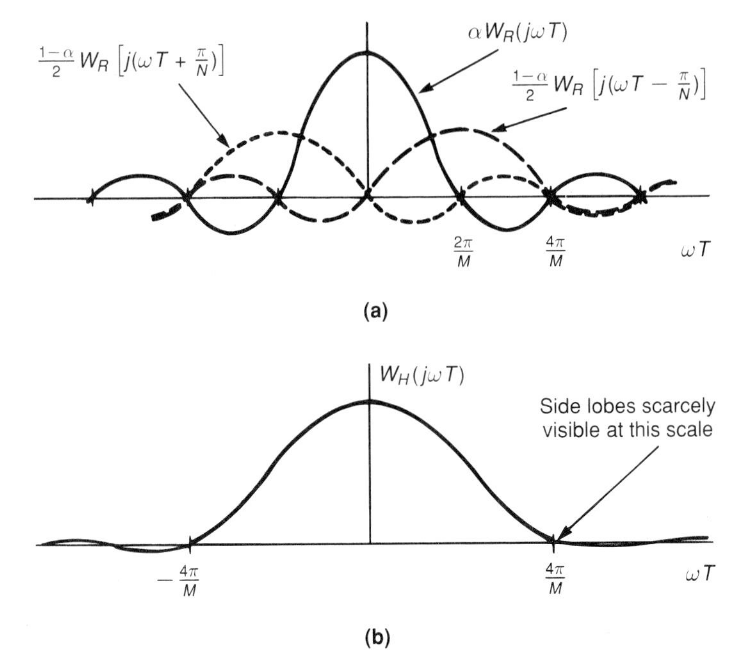

FIGURE 5.28 Frequency response for the Hamming window.

table are typical. We conclude from the table that the additional smoothing of filter magnitude response that it is possible to achieve by introducing a designed window function is accompanied by a wider transition region at a discontinuity because of the increase in the main-lobe width of the window. Spectra for several window functions are compared in Figure 5.29.

TABLE 5.5 Comparison of Window Function Performances

Window Type	Width of Main Lobe		Approximate Amount by Which the First Side Lobe Is Down Compared with the Main Lobe (dB)
	ω	ωT	
Rectangular	$\frac{2\omega_s}{M}$	$\frac{4\pi}{M}$	13
Bartlett	$\frac{4\omega_s}{M}$	$\frac{8\pi}{M}$	26
von Hann	$\frac{4\omega_s}{M}$	$\frac{8\pi}{M}$	32
Hamming	$\frac{4\omega_s}{M}$	$\frac{8\pi}{M}$	40
Blackman	$\frac{6\omega_s}{M}$	$\frac{12\pi}{M}$	60

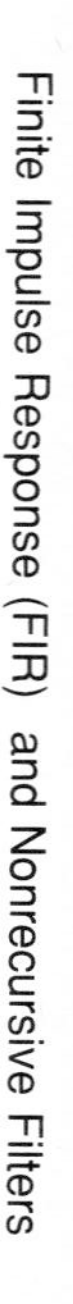

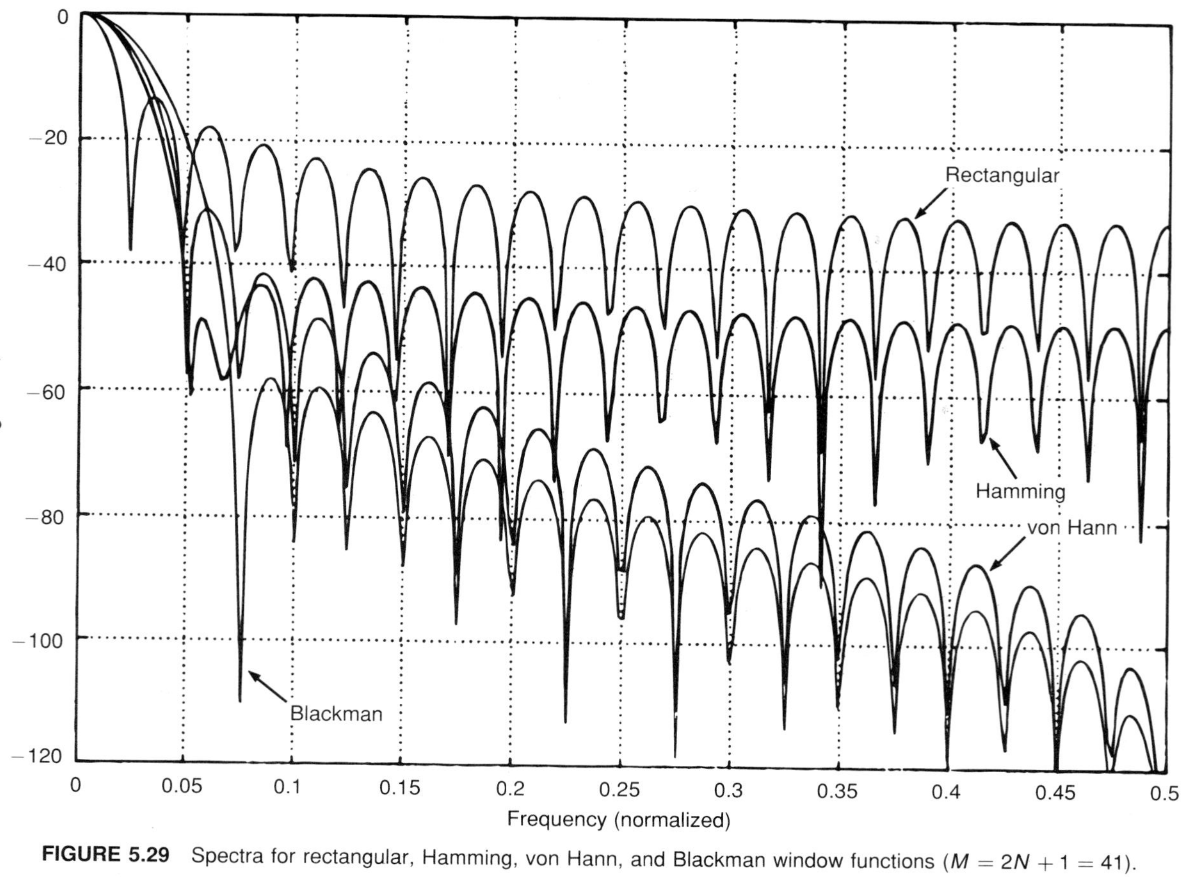

FIGURE 5.29 Spectra for rectangular, Hamming, von Hann, and Blackman window functions ($M = 2N + 1 = 41$).

Noting the trade-off between main-lobe width and side-lobe attenuation shown in Table 5.5, Kaiser proposed an adjustable window that will give the smallest main-lobe width for a given peak side-lobe amplitude [8], [9]. The Kaiser window has the form

$$w_k(n) = \frac{I_0(\beta)}{I_0(\alpha)}, \quad |n| \leq N \tag{5.146}$$

where I_0 is the zeroth-order Bessel function of the first kind. It can be evaluated from the rapidly converging series

$$I_0(x) = 1 + \sum_{k=1}^{\infty} \left[\frac{1}{k!} \left(\frac{x}{2} \right)^k \right]^2 \tag{5.147}$$

Also,

$$\beta = \alpha \sqrt{1 - \left(\frac{n}{N} \right)^2} \tag{5.148}$$

and α is an independent parameter that can be adjusted to vary the ripple ratio (the ratio of maximum side-lobe amplitude to main-lobe amplitude, in percent). We note from (5.146) that $\alpha = 0$ implies rectangular windowing. Typical values of α range from 1 to 10. The ripple ratio decreases with increasing α. Time and frequency characteristics of special window functions are summarized in Table 5.6. A computer program listing by Rabiner et al. [10] gives a windowed design of FIR filters with a choice of seven different window functions. A program on the diskette can be used to design windowed approximations to the ideal lowpass, highpass, bandpass, and bandstop filters.

Returning to Example 5.7, let us introduce a Hamming window to reduce the oscillations evident in Figure 5.22. The window has the form

$$w_H(n) = 0.54 + 0.46 \cos \frac{\pi n}{N} \tag{5.149}$$

The windowed frequency response is

$$H_R(\omega T) = e^{-jN\omega T} M_R(\omega T) \tag{5.150}$$

where

$$M_R(\omega T) = \frac{1}{3} + \frac{4}{\pi} \sum_{n=1}^{N} \left[0.54 + 0.46 \cos \frac{\pi n}{N} \right] \cos \frac{\pi n}{2} \sin \frac{n\pi}{6} \cos(n\omega T)/n \tag{5.151}$$

The magnitude response is plotted for $N = 14$ in Figure 5.30. The stopband is now down about 50 dB. However, it is evident that the window has the effect of widening the transition region of the filter, resulting in a less sharp cutoff (compare the unwindowed function of Figure 5.22).

A second systematic technique for the design of nonrecursive filters is called the *frequency-sampling method*. The idea is to approximate a desired continuous frequency response by uniformly sampling the latter and then

TABLE 5.6 Time and Frequency Description of Window Functions

Type of Window	Sequence $w(n)$ (two-sided)	Spectrum $W(j\omega T)$ ($X \triangleq \frac{\omega T}{2}$, $M = 2N + 1$)
Rectangular	$1, \quad \|n\| \le N$ $0, \quad$ otherwise	$\frac{1}{M}\frac{\sin MX}{\sin X}$
Modified rectangular	$1, \quad \|n\| \le N - 1$ $\frac{1}{2}, \quad \|n\| = N$ $0, \quad$ otherwise	$(\sin 2NX \cos X / \sin X)/(2N)$
Bartlett (triangular)	$1 - \frac{\|n\|}{N}, \quad \|n\| \le N$ $0, \quad$ otherwise	$\left(\frac{\sin NX}{NX}\right)^2$
Generalized Hamming	$\alpha + (1 - \alpha)\cos \pi n/N, \quad \|n\| \le N$ $0, \quad$ otherwise (Hamming: $\alpha = 0.54$, von Hann: $\alpha = 0.50$)	$\frac{1}{2N\alpha}\left[\alpha \frac{\sin MX}{\sin X} + 0.5(1-\alpha)\frac{\sin(MX - M\pi/2N)}{\sin(X - \pi/2N)} + 0.5(1-\alpha)\frac{\sin(MX + M\pi/2N)}{\sin(X + \pi/2N)}\right]$
Blackman	$0.42 + 0.5\cos \pi n/N + 0.08\cos 2\pi n/N, \quad \|n\| \le N$ $0, \quad$ otherwise	$\frac{1}{0.84N}\left[0.42\frac{\sin MX}{\sin X} + 0.25\frac{\sin(MX - M\pi/2N)}{\sin(X - \pi/2N)} + 0.25\frac{\sin(MX + M\pi/2N)}{\sin(X + \pi/2N)} + 0.04\frac{\sin(MX - M\pi/N)}{\sin(X - \pi/N)} + 0.04\frac{\sin(MX + M\pi/N)}{\sin(X + \pi/N)}\right]$
Kaiser	$I_0(\beta)/I_0(\alpha), \quad \|n\| \le N$ $0, \quad$ otherwise See (5.147) and (5.148).	No closed form is available.

Note: The spectra are normalized to have unity gain at zero frequency.

interpolating to get a continuous frequency function that equals the desired function at the samples and is, generally, in error between the samples. The design problem is to reduce this intersample error as much as possible. Before we can discuss this frequency sampling method further, however, we must wait until the discrete Fourier transform is introduced in the next chapter.

5.5 CHEBYSHEV APPROXIMATION FOR DESIGNING LINEAR-PHASE FIR FILTERS

An efficient procedure for designing linear-phase FIR filters was developed by Parks and McClellan [11], [12], [14], [15]. This technique involves approximating a desired magnitude response so that the weighted error between

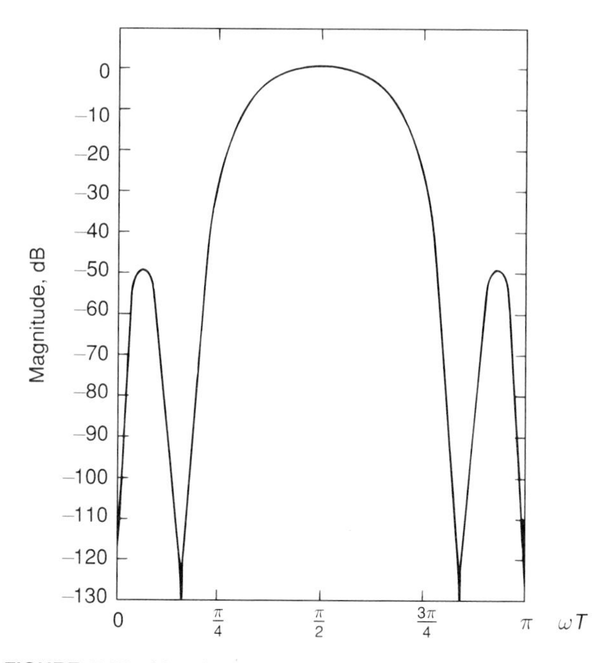

FIGURE 5.30 Magnitude response of Fourier series approximations to an ideal bandpass filter (Example 5.7 with Hamming window).

the approximation and the desired response is optimized in the Chebyshev approximation sense—that is, the maximum error is minimized. The Parks–McClellan algorithm is widely used for FIR filter design.

The frequency responses for the four different types of linear-phase FIR filters are given in (5.73), (5.76), (5.88), and (5.89). For convenience in using the Parks–McClellan program, let us switch to their notation in this section. Define a real-valued *amplitude* function $A(\omega T)$ that can be either positive or negative so that

$$M(\omega T) = |A(\omega T)| \tag{5.152}$$

and the frequency response becomes

$$H(j\omega T) = M(\omega T)e^{j\phi(\omega T)} = A(\omega T)e^{j\theta(\omega T)}$$

Note that the $\pm\pi$ step discontinuities that occur in $\phi(\omega T)$ each time $A(\omega T)$ changes sign are not present in $\theta(\omega T)$. Of course, in general, $A(\omega T)$ cannot be expressed in decibels.

Define the normalized frequency variable $F = \omega T/2\pi$ so that over the operating range $0 \le \omega T \le \pi$, we have $0 \le F \le \frac{1}{2}$. The amplitude functions $A(F)$ for the four types of linear-phase filters are

Type 1
$$A(F) = \sum_{k=0}^{\alpha} a_k \cos(2\pi kF) \tag{5.153}$$

Type 2
$$A(F) = \sum_{k=1}^{N/2} a_k \cos\left[\left(k - \frac{1}{2}\right) 2\pi F\right] \tag{5.154}$$

Type 3
$$A(F) = \sum_{k=1}^{\alpha} a_k \sin(2\pi k F) \tag{5.155}$$

Type 4
$$A(F) = \sum_{k=1}^{N/2} a_k \sin\left[\left(k - \frac{1}{2}\right) 2\pi F\right] \tag{5.156}$$

where $\alpha = (N-1)/2$. By using basic trigonometric identities, all four expressions can be put in the form

$$A(F) = Q(F) \sum_{k=0}^{n-1} c_k \cos(2\pi k F) \tag{5.157}$$

where $n - 1 = \alpha$ for types 1 and 3, and $n = N/2$ for types 2 and 4. Also,

$$Q(F) = \begin{cases} 1 \\ \cos \pi F \\ \sin 2\pi F \\ \sin \pi F \end{cases} \tag{5.158}$$

for types 1, 2, 3, and 4, respectively. With $h_k = h_{N-k-1}$ for the symmetric filters and $h_k = -h_{N-k-1}$ for the anti-symmetric filter, the relationship between the filter coefficients h_k and the coefficients c_k in (5.157) can be written as shown in Table 5.7.

TABLE 5.7 Filter Coefficients in Terms of Design Coefficients

Type	Relationships Between Coefficients
1	$h_{\alpha-k} = c_k/2,\quad k = 1, 2, \ldots, \alpha = (N-1)/2$ $h_\alpha = c_0$
2	$h_0 = c_{n-1}/2$ $h_{n-k} = (c_{k-1} + c_k)/4,\quad k = 2, 3, \ldots, n-1 = N/2 - 1$ $h_{n-1} = (c_0 + c_1/2)/2$
3	$h_0 = c_{\alpha-1}/4$ $h_1 = c_{\alpha-2}/4$ $h_{\alpha-k} = (c_{k-1} - c_{k+1})/4,\quad k = 2, 3, \ldots, \alpha - 2 = (N-1)/2 - 2$ $h_{\alpha-1} = (c_0 - c_2/2)/2$
4	$h_0 = c_{n-1}/4$ $h_{n-k} = (c_{k-1} - c_k)/4,\quad k = 2, 3, \ldots, n-1 = N/2 - 1$ $h_{n-1} = (c_0 - c_1/2)/2$

To illustrate the Parks–McClellan technique, we will assume that the desired amplitude function $D(F)$ has the form shown in Figure 5.31; that is, it is an approximation to an ideal lowpass filter with a transition region separating the passband and stopband. It is evident that if we want to approximate $D(F)$ with an amplitude function $A(f)$ that is the weighted linear sum of cosines as given in (5.157), the error $E = D - A$ will oscillate, or have ripples. Tolerances $\pm\delta_1$ and $\pm\delta_2$ are placed on the respective errors in passband and stopband, as indicated in Figure 5.31. Any ripples must lie within these tolerances. We want to select the n coefficients c_k in (5.157) to ensure this.

Many sets of coefficients will give an approximating amplitude function within the error bands. But one such set of coefficients minimizes the maximum error in the amplitude response, giving the Chebyshev approximation. Following Parks and McClellan, we will pose the problem formally.

1. Let $\mathcal{F}$ be a subset of the normalized frequency axis $[0, \frac{1}{2}]$. In the lowpass example, $\mathcal{F}$ could be the union of the passband and stopband frequencies but excludes the transition region. In general, $\mathcal{F}$ could contain all the frequencies in multiple passbands and stopbands.
2. A desired amplitude function $D(F)$ and a positive weight function $W(F)$ are both defined and continuous on $\mathcal{F}$. The weight function permits different error bounds to be selected in different frequency bands, for example.
3. An amplitude function $A(F)$ is in the form

$$A(F) = Q(F) \sum_{k=0}^{n-1} c_k \cos(2\pi k F) \tag{5.159}$$

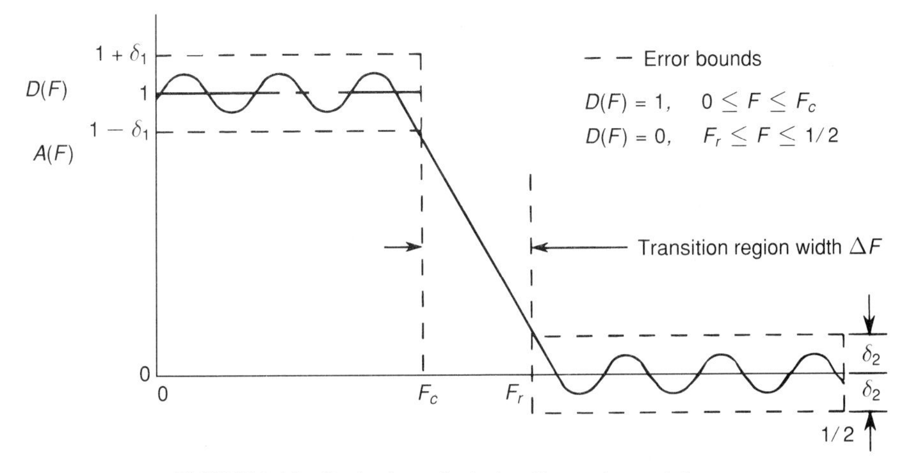

FIGURE 5.31 Desired amplitude function and error tolerances.

Choose coefficients c_k so as to minimize the weighted error

$$\|E(F)\| = \max_{F \in \mathcal{F}} W(F)|D(F) - A(F)| \tag{5.160}$$

The amplitude function $A(F)$ so obtained is said to be the best weighted Chebyshev approximation to the desired function $D(F)$ *on* $\mathcal{F}$.

To identify the optimal solution, Parks and McClellan used the following theorem:

Alternation Theorem

If $A(F)$ is a linear combination of n cosines as in (5.159), then a necessary and sufficient condition for $A(F)$ to be the *unique* optimal weighted Chebyshev approximation to a specified continuous function $D(F)$ on $\mathcal{F}$ is that the weighted-error function $E(F) = W(F)[D(F) - A(F)]$ have *at least* $(n+1)$ extremal frequencies on $\mathcal{F}$ and that these points be such that, with $F_1 < F_2 < \cdots < F_{n+1}$,

$$E(F_i) = -E(F_{i+1}), \quad i = 1, 2, \ldots, n$$

Thus the theorem implies that the best approximation must have an equiripple error function. Furthermore, the uniqueness of the optimal approximation guarantees that if, for example, a solution is found with $(n+2)$ extremal frequencies, there cannot be one with $(n+1)$ extremal frequencies for the same specifications.

Although the alternation theorem does not indicate directly how to find the coefficients c_k of the optimal solution, it does identify the solution when it is found. If the extremal frequencies of the optimal solution can be found, it is possible to interpolate an amplitude function through points (at those frequencies) that lie alternately on the upper and lower error bounds in the passbands and stopbands. Therefore, the problem of designing the filter becomes that of finding the extremal frequencies for the optimal solution.

The Remez exchange algorithm [13] can be used to find the extremal frequencies. Suppose that the error function $E(F) = D(F) - A(F)$ takes on the values $\pm\delta$ on an arbitrarily chosen trial set T_0 of frequency points F_m, $m = 1, \ldots, n+1$. We can solve the $(n+1)$ linear equations

$$E(F_m) = (-1)^m \delta, \quad m = 1, \ldots, n+1 \tag{5.161}$$

for the $(n+1)$ unknowns δ and the coefficients c_k, $k = 0, 1, \ldots, n-1$, required for $A(F)$. If we substitute these values for c_k into $A(F)$ and evaluate $A(F)$ over $\mathcal{F}$, we find that, in general, the resulting error function $E(F)$ has one or more frequencies, not included in the first trial set, at which $E(F)$ has a larger value than $\delta = \delta_0$. Then we choose a new set T_1 of frequency points F_m, $m = 1, 2, \ldots, n+1$, which are the points at which $A(F)$ based on T_0 took on $(n+1)$ extreme values; in other words, set T_0 is exchanged for T_1. The $(n+1)$ equations (5.161) are solved for new values of $\delta \, (= \delta_1 > \delta_0)$, and c_k, $k = 0, 1, \ldots, n-1$. The procedure is repeated until the amplitude of the oscillation δ on the trial set of frequencies T_r equals the maximum error on $\mathcal{F}$. The trial set at which this occurs is the set of extremal frequencies.

The following simple example illustrates how the Remez exchange algorithm works.

EXAMPLE 5.8

We want to approximate the function

$$D(F) = F$$

by a function of the form

$$A(F) = \sum_{k=0}^{n-1} c_k \cos 2\pi F$$

where $n = 2$ and $F \in [0, \frac{1}{2}]$. The approximation should minimize the Chebyshev error

$$\max_{F \in (0, \frac{1}{2})} |E(F)| = \max_{F \in (0, \frac{1}{2})} |D(F) - A(F)|$$

Solution

This problem could be considered that of approximating the amplitude response of an ideal differentiator with the sum of a constant c_0 and a cosine of amplitude c_1. Choose a trial set of frequencies T_0 on which $E(F)$ can oscillate. Because there are three unknowns c_0, c_1, and δ, we need three linear equations of the form (5.161) to solve for them. For this problem, these equations are

$$F_m = c_0 + c_1 \cos 2\pi F_m + (-1)^m \delta, \quad m = 0, 1, 2$$

As the first trial set, take

$$T_0 = [F_0, F_1, F_2] = [0, 0.25, 0.5]$$

so that the three equations are

$$\begin{bmatrix} 1 & 1 & 1 \\ 1 & 0 & -1 \\ 1 & -1 & 1 \end{bmatrix} \begin{bmatrix} c_0 \\ c_1 \\ \delta \end{bmatrix} = \begin{bmatrix} 0 \\ 0.25 \\ 0.5 \end{bmatrix}$$

The solutions are $c_0 = 0.25$, $c_1 = -0.25$, and $\delta = 0$. It is evident from Figure 5.32 that the error extremes occur at 0.125 and 0.375 and that, therefore, the corresponding frequencies are not in T_0. Accordingly, we exchange frequencies so that the next trial set is $T_1 = [0, 0.125, 0.375]$. The new set of equations is

$$\begin{bmatrix} 1 & 1 & 1 \\ 1 & 1/\sqrt{2} & -1 \\ 1 & -1/\sqrt{2} & 1 \end{bmatrix} \begin{bmatrix} c_0 \\ c_1 \\ \delta \end{bmatrix} = \begin{bmatrix} 0 \\ 0.125 \\ 0.375 \end{bmatrix}$$

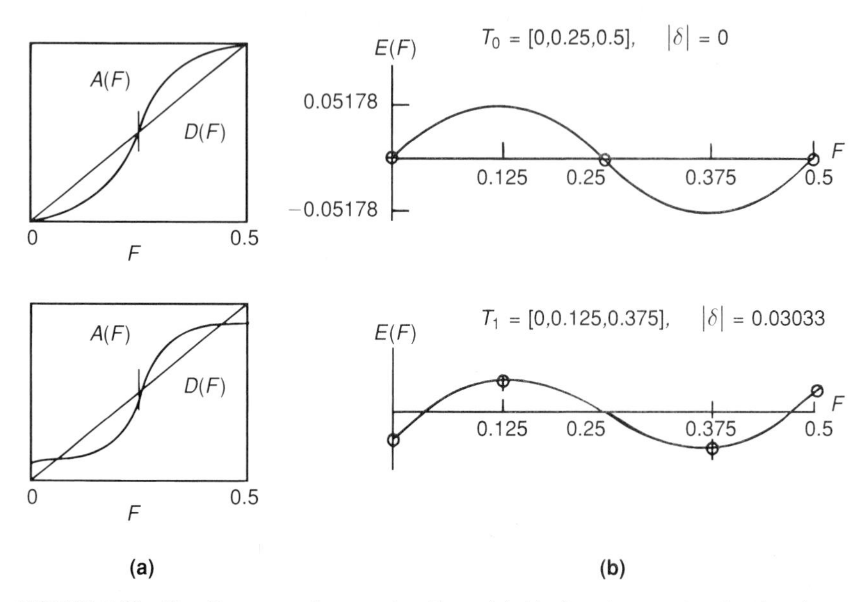

FIGURE 5.32 The Remez exchange algorithm. (a) Ideal and approximating functions. (b) Error function.

The solutions are $c_0 = 0.25$, $c_1 = -0.21967$, and $\delta = 0.03033$. Now we note that the extreme values of error $E(F)$ occur on T_1. There is an additional extreme error equal to the other three on the boundary point $F = \frac{1}{2}$, so we have $n + 2 = 4$ extremal frequencies in this case (see Figure 5.32).

A FORTRAN program for the Parks–McClellan algorithm is found in the IEEE publication *Programs for Digital Signal Processing* [16] and [17]. A bandpass filter of length 23 was designed by means of this program. Formulas for estimating the filter length N are given in [15]. This reference also includes a more detailed description of the method. The filter specifications are given in Figure 5.33, together with the resulting impulse response and extremal frequencies. A plot of the magnitude response is shown in Figure 5.34. There are 13 (that is, $n + 1$) extremal points marked on the figure, which is the minimum number according to the alternation theorem.

The next three sections (Sections 5.6, 5.7, and 5.8) deal with some digital filters for special applications. The first introduces the discrete Hilbert transform, an understanding of which one needs to appreciate a particular digital filter called the Hilbert transformer. The next section offers an introductory treatment of decimation and interpolation—lowering and raising sampling rates—and the associated filtering required. Finally, Section 5.8 presents some *ad hoc* methods of getting a moving estimate of the variance of a sequence.

```
            FINITE IMPULSE RESPONSE (FIR)
          LINEAR-PHASE DIGITAL FILTER DESIGN
              REMEZ EXCHANGE ALGORITHM

                   BANDPASS FILTER

                 FILTER LENGTH =  23

           ***** IMPULSE RESPONSE *****
          H( 1) = -0.24062216E-01 = H( 23)
          H( 2) =  0.16347874E-01 = H( 22)
          H( 3) = -0.57457656E-01 = H( 21)
          H( 4) =  0.37831277E-01 = H( 20)
          H( 5) =  0.11173493E+00 = H( 19)
          H( 6) =  0.25013860E-01 = H( 18)
          H( 7) =  0.27231481E-02 = H( 17)
          H( 8) =  0.63863754E-01 = H( 16)
          H( 9) = -0.11668825E+00 = H( 15)
          H(10) = -0.26963013E+00 = H( 14)
          H(11) =  0.83750188E-01 = H( 13)
          H(12) =  0.40265179E+00 = H( 12)

                          BAND  1        BAND  2        BAND  3
LOWER BAND EDGE       0.0            0.1300000      0.3300000
UPPER BAND EDGE       0.1000000      0.3000000      0.5000000
DESIRED VALUE         0.0            1.0000000      0.0
WEIGHTING             1.0000000      1.0000000      1.0000000
DEVIATION             0.1495056      0.1495056      0.1495056
DEVIATION IN DB     -16.5608359      1.2102146    -16.5068359

EXTREMAL FREQUENCIES--MAXIMA OF THE ERROR CURVE
   0.0          0.0703123    0.1000000    0.1300000    0.1586454
   0.2107279    0.2706228    0.3000000    0.3300000    0.3560413
   0.4029155    0.4523939    0.5000000
```

FIGURE 5.33 Bandpass filter design by the Parks–McClellan algorithm.

5.6 DISCRETE HILBERT TRANSFORMS

5.6.1 Introduction

For causal, or positive-time, sequences (sequences that are zero for $n < 0$), certain mathematical relationships exist between the real and imaginary parts of their Fourier transforms (or magnitude and phase). As a result, when one part is specified, the other is uniquely determined. In a second case, if a sequence is such that its Fourier transform, considered as a periodic function, is zero for half a period (that is, $-\pi \leq \omega T \leq 0$), then it must be a *complex* sequence, and relationships exist between the real sequences that form the

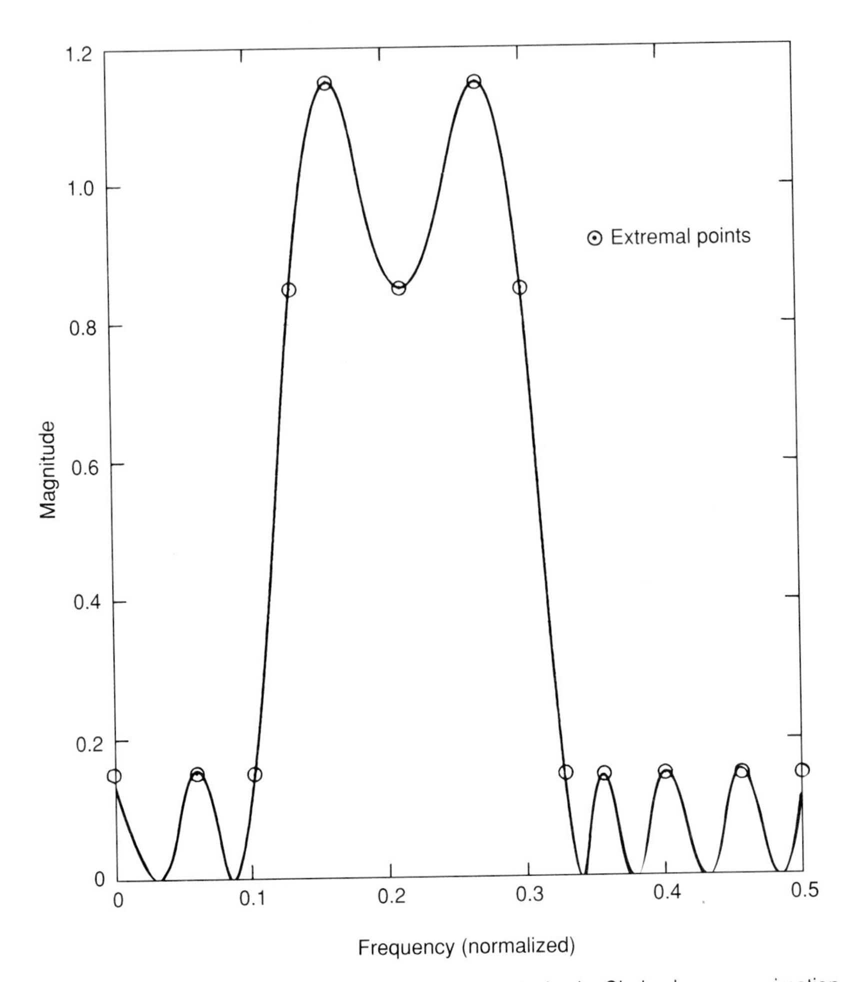

FIGURE 5.34 Magnitude response for bandpass filter design by Chebyshev approximation.

real and imaginary parts of the complex sequence such that one part is uniquely determined by the other. The real and imaginary parts of the DFT of a finite sequence that is zero for the last half of its length (see Chapter 6) can also be related. A fourth relationship links the real and imaginary parts of the logarithm of the Fourier transform if the inverse transform of the logarithm is zero for negative time.

Such mathematical relationships are collectively called Hilbert transforms. We will consider situations 1 and 2. The former is obviously of interest where, for example, the magnitude response of a desired filter is specified. It enables us to determine in advance whether such a specification puts unreasonable restrictions on the phase response. The second case has application in the processing of a signal that contains a band of frequencies only (a so-called bandpass signal) [18].

5.6.2 Mathematical Preliminaries

Before discussing the Hilbert transforms in detail, it is necessary to review some mathematical concepts pertaining to sequences. Similar relationships hold for continuous signals.

Even and Odd Sequences An arbitrary sequence $f(n)$ can be represented by the sum of an *even* sequence $f_e(n)$—that is, $f_e(n) = f_e(-n)$—and an *odd* sequence $f_0(n) = -f_0(-n)$. Thus,

$$f(n) = f_e(n) + f_0(n) \tag{5.162}$$

The sequences f_e and f_0 can be determined as follows: From (5.162),

$$\begin{aligned} f(-n) &= f_e(-n) + f_0(-n) \\ &= f_e(n) - f_0(n) \end{aligned} \tag{5.163}$$

Add (5.163) to (5.162) and subtract (5.163) from (5.162) to get, respectively,

$$f_e(n) = \frac{1}{2}[f(n) + f(-n)] \tag{5.164}$$

$$f_0(n) = \frac{1}{2}[f(n) - f(-n)] \tag{5.165}$$

An example of the breakdown of a given sequence into even and odd components appears in Figure 5.35.

Suppose we have a causal sequence—that is,

$$f(n) = 0, \quad n < 0 \tag{5.166}$$

The breakdown of a simple causal sequence is illustrated in Figure 5.36.

It follows that, for a causal sequence, $f_e(n)$ and $f_0(n)$ are related by

$$f_0(n) = \begin{cases} f_e(n), & n > 0 \\ 0, & n = 0 \\ -f_e(n), & n < 0 \end{cases} \tag{5.167}$$

$$f_e(n) = \begin{cases} f_0(n), & n > 0 \\ f_e(0), & n = 0 \\ -f_0(n), & n < 0 \end{cases} \tag{5.168}$$

Equations (5.167) and (5.168) may be written in a shorter form:

$$f_0(n) = g(n)f_e(n) \tag{5.169}$$

$$f_e(n) = g(n)f_0(n) + f_e(0)\delta(n) \tag{5.170}$$

where

$$g(n) = \begin{cases} 1, & n > 0 \\ 0, & n = 0 \\ -1, & n < 0 \end{cases} \tag{5.171}$$

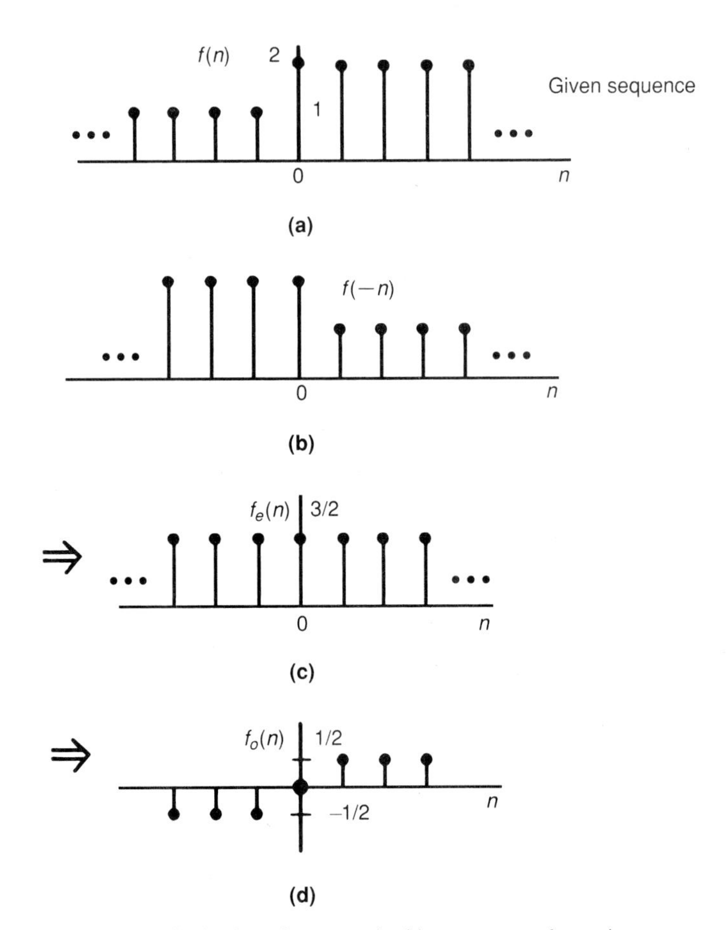

FIGURE 5.35 Derivation of even and odd components for a given sequence.

and

$$\delta(n) = \begin{cases} 1, & n = 0 \\ 0, & n \neq 0 \end{cases} \tag{5.172}$$

The Fourier transform of $g(n)$ is

$$\begin{aligned} G(j\omega T) &= \sum_{n=-\infty}^{\infty} g(n)e^{-jn\omega T} \\ &= -\sum_{n=-\infty}^{-1} e^{-jn\omega T} + \sum_{n=1}^{\infty} e^{-jn\omega T} \\ &= -\sum_{n=1}^{\infty} e^{jn\omega T} + \sum_{n=1}^{\infty} e^{-jn\omega T} \end{aligned}$$

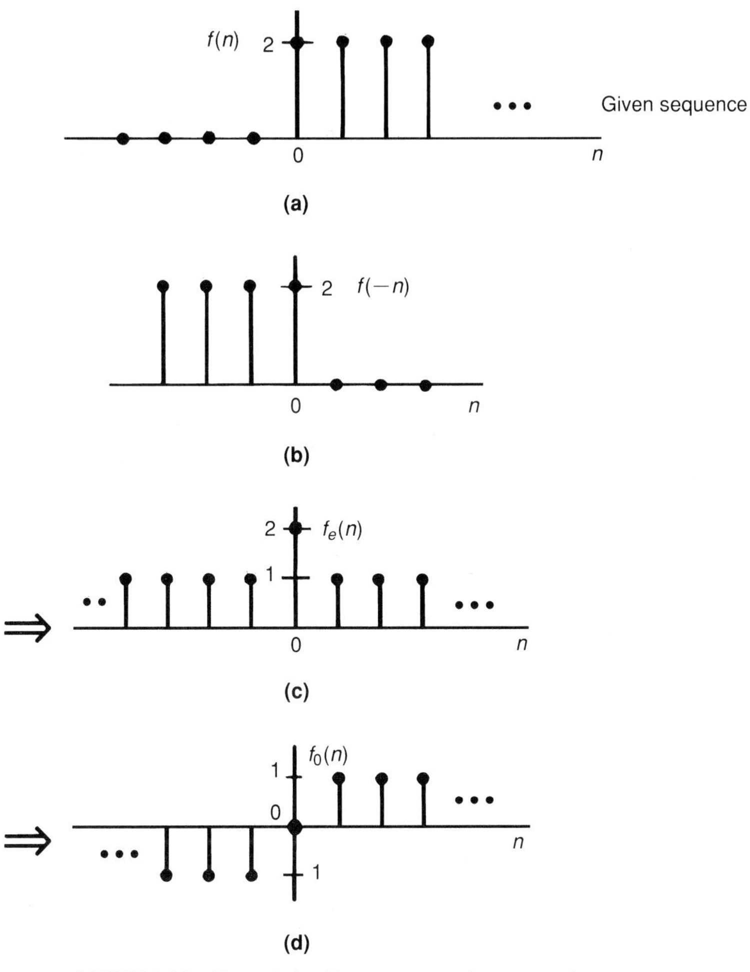

FIGURE 5.36 Even and odd components for a causal sequence.

$$= \frac{-1}{1 - e^{j\omega T}} + \frac{1}{1 - e^{-j\omega T}}$$

$$= -j \cot \frac{\omega T}{2} \tag{5.173}$$

This result is used in Section 5.6.3.

Fourier Transforms of Even and Odd Components If $F_R(j\omega T)$ and $F_I(j\omega T)$ are the real and imaginary parts, respectively, of $F(j\omega T)$, the Fourier transform of a sequence $f(n)$—that is,

$$F(j\omega T) = F_R(j\omega T) + jF_I(j\omega T) \tag{5.174}$$

then $F_R(j\omega T)$ and $F_I(j\omega T)$ are the Fourier transforms of the even and odd sequences $f_e(n)$ and $f_0(n)$, respectively.

The proof of the last statement is straightforward.

$$F(j\omega T) = \sum_{n=-\infty}^{\infty} f(n)e^{-jn\omega T}$$

$$= \sum_{n=-\infty}^{\infty} [f_e(n) + f_0(n)][\cos n\omega T - j \sin n\omega T]$$

$$= \sum_{n=-\infty}^{\infty} f_e(n) \cos n\omega T - j \sum_{n=-\infty}^{\infty} f_0(n) \sin n\omega T$$

because odd functions summed over symmetric limits are zero. But

$$\mathcal{F}\{f_e(n)\} = \sum_{n=-\infty}^{\infty} f_e(n) \cos n\omega T$$

and

$$\mathcal{F}\{f_0(n)\} = \sum_{n=-\infty}^{\infty} f_0(n)[\cos(n\omega T) - j \sin n\omega T]$$

$$= -j \sum_{n=-\infty}^{\infty} f_0(n) \sin n\omega T$$

Therefore, the statement is correct. This completes the mathematical preliminaries necessary for a discussion of cases 1 and 2 of Hilbert transform relations.

5.6.3 Hilbert Transform Relationships for Real and Imaginary Parts of Fourier Transforms (Case 1)

We can use the complex-convolution theorem to relate the real and imaginary parts of $F(j\omega T)$ for a causal sequence. Take the Fourier transforms of (5.169) and (5.170),

$$F_I(j\phi) = \frac{1}{2\pi} \int_{-\pi}^{\pi} F_R(e^{j\theta}) \cot \frac{\phi - \theta}{2} \, d\theta \qquad (5.175)$$

and

$$F_R(j\phi) = \frac{1}{2\pi} \int_{-\pi}^{\pi} F_I(e^{j\theta}) \cot \frac{\phi - \theta}{2} \, d\theta + f_e(0) \qquad (5.176)$$

where $\phi = \omega T$. Thus, for causal sequences, the imaginary part of the frequency response (that is, of the z-transform on the unit circle) is uniquely given by the real part, and the real part is uniquely related to the imaginary part (to within a constant). Equations (5.175) and (5.176) are called discrete Hilbert transforms.

5.6.4 Hilbert Transform Relationships for Real and Imaginary Parts of Complex Sequences (Case 2)

For the case of a causal sequence, the Hilbert transform expressions (5.175) and (5.176) define unique relationships between the real and imaginary parts of the Fourier transform of the sequence. If, by analogy, we define a "causal" frequency response $F(j\omega T)$ as one for which

$$F(j\omega T) = 0, \quad -\pi \leq \omega T < 0 \tag{5.177}$$

then we can derive Hilbert transform relationships between the real and imaginary parts of the corresponding *complex* sequence $f(n)$ the frequency response of which satisfies (5.177). Such complex sequences arise, for example, in the processing of signals from single-sideband modulation systems.

That $f(n)$ must be complex if (5.177) is true can be verified as follows. If $f(n)$ were real, then the complex conjugate $F^*(j\omega T)$ with negative argument corresponding to the lower half of the unit circle $-\pi \leq \omega T < 0$ would be

$$F^*(-j\omega T) = \sum_{n=-\infty}^{\infty} f(n)e^{-jn\omega T} = F(j\omega T) \tag{5.178}$$

Hence if $f(n)$ were real and (5.177) were satisfied, (5.178) would imply that $F(j\omega T) = 0$ for all ω. Therefore, $f(n)$ must be complex; that is,

$$f(n) = f_R(n) + jf_I(n) \tag{5.179}$$

We want to derive relationships between the real sequences $f_R(n)$ and $f_I(n)$. We note that

$$F(j\omega T) = F_R(j\omega T) + jF_I(j\omega T) \tag{5.180}$$

where $F_R(j\omega T)$ and $F_I(j\omega T)$ are the Fourier transforms of $f_R(n)$ and $f_I(n)$, respectively, because the latter are real sequences. Equation (5.177) implies that

$$F_R(j\omega T) = -jF_I(j\omega T), \quad -\pi \leq \omega T < 0$$

or

$$F_I(j\omega T) = jF_R(j\omega T), \quad -\pi \leq \omega T < 0 \tag{5.181}$$

Also, because

$$F^*(-j\omega T) = F_R(j\omega T) - jF_I(j\omega T) \tag{5.182}$$

and the nonzero portions of $F^*(-j\omega T)$ and $F(j\omega T)$ do not overlap, it follows that

$$F_R(j\omega T) - jF_I(j\omega T) = 0, \quad 0 \leq \omega T < \pi$$

or

$$F_I(j\omega T) = -jF_R(j\omega T), \quad 0 \leq \omega T < \pi \tag{5.183}$$

From (5.181) and (5.183), it is apparent that the real sequence $f_I(n)$ can be obtained from the real sequence $f_R(n)$ by filtering the latter with a filter the frequency response of which is

$$H(j\omega T) = \begin{cases} -j, & 0 \leq \omega T < \pi \\ j, & -\pi \leq \omega T < 0 \end{cases} \tag{5.184}$$

The pulse response of the filter is

$$\begin{aligned} h(n) &= \frac{1}{2\pi}\left(\int_0^{\pi} -je^{jn\omega T}\, d\omega T + \int_{-\pi}^{0} je^{jn\omega T}\, d\omega T\right) \\ &= \begin{cases} \dfrac{2}{\pi n}\sin^2(n\pi/2), & n \neq 0 \\ 0, & n = 0 \end{cases} \end{aligned} \tag{5.185}$$

Equations (5.184) and (5.185) define the *ideal* Hilbert transformer. It is nonrealizable because $h(n) \neq 0$ for $n < 0$ (see Figure 5.37). Realizable approximations to the ideal Hilbert filter can be designed, however, just as realizable approximations to the ideal lowpass filter are feasible. For example, the Fourier series design method described in Section 5.4 is suitable. The frequency response is purely imaginary, so an expansion similar to that in Example 5.6 for the ideal differentiator can be used. Because of the sharp cutoff in the frequency response (5.184), a window function is necessary.

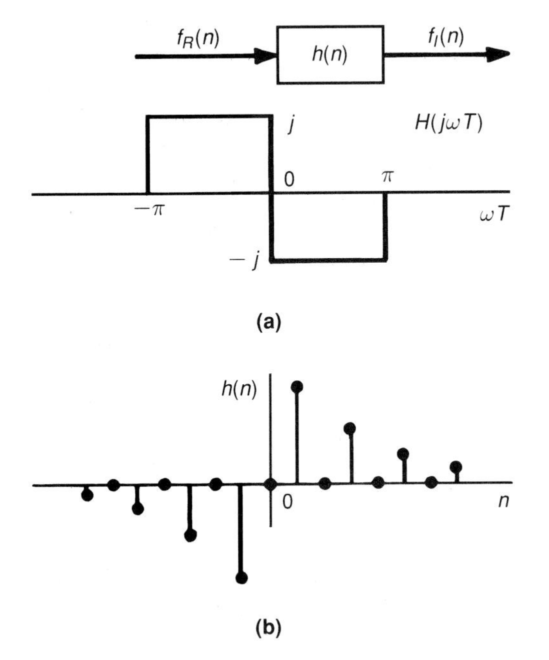

FIGURE 5.37 Ideal Hilbert transformer. (a) Frequency response. (b) Pulse response.

From (5.185), the convolutional relationship between $f_I(n)$ and $f_R(n)$ can be written:

$$f_I(n) = \sum_{\substack{m=-\infty \\ m\neq n}}^{\infty} f_R(n-m)\frac{\sin^2 \frac{m\pi}{2}}{m\pi/2} \tag{5.186}$$

or, alternatively, $f_R(n)$ can be obtained from $f_I(n)$ by means of

$$f_R(n) = -\sum_{\substack{m=-\infty \\ m\neq n}}^{\infty} f_I(n-m)\frac{\sin^2 \frac{m\pi}{2}}{m\pi/2} \tag{5.187}$$

Equations (5.186) and (5.187) are the Hilbert transform relationships between signals $f_I(n)$ and $f_R(n)$.

The foregoing can be used to reduce the sampling rate required for a real bandpass signal $f_R(n)$. Suppose that the original sampling rate is ω_s and the spectrum $F_R(j\omega T)$ has the form shown in Figure 5.38(a), where B is the bandwidth of $f_R(n)$ [18]. For simplicity of illustration, let $\omega_s = 8B$. Use a Hilbert transformer to synthesize the complex signal

$$f(n) = f_R(n) + jf_I(n) \tag{5.188}$$

with the spectrum as shown in Figure 5.38(b). Note that the spectrum of the complex signal $f(n)$ does not have the negative lobe in the range $-\omega_s/2 \leq \omega < 0$ [see (5.177)]. It is evident that the sampling rate for the complex signal $f(n)$ can be reduced to $B/2\pi$ complex samples per second without aliasing or overlap of the spectral lobes (see also the next section). The spectrum of the reduced-rate signal $g(n)$ is shown in Figure 5.38(c). To recover the original signal $f_R(n)$ at sampling rate ω_s, use a bandpass filter [Figure 5.38(d)]. The real part of the bandpass filter output is $f_R(n)$.

In the next section, we will discuss in more detail the concept of representing a bandpass signal in terms of a lowpass signal (a signal with low-frequency components only) via the Hilbert transform.

5.6.5 Representing a Bandpass Signal in Terms of a Lowpass Signal, and Bandpass Sampling

Consider a complex *lowpass* sequence

$$g(n) = g_R(n) + jg_I(n) \tag{5.189}$$

where $g_I(n)$ is the Hilbert transform of $g_R(n)$. For illustration purposes, let the real parts of the spectra of the three signals involved be as shown in Figure 5.39.

Now consider a complex sequence

$$f(n) = g(n)e^{jn\omega_c T} = f_R(n) + jf_I(n) \tag{5.190}$$

where $f_R(n)$ and $f_I(n)$ are real sequences and ω_c is a constant. The corre-

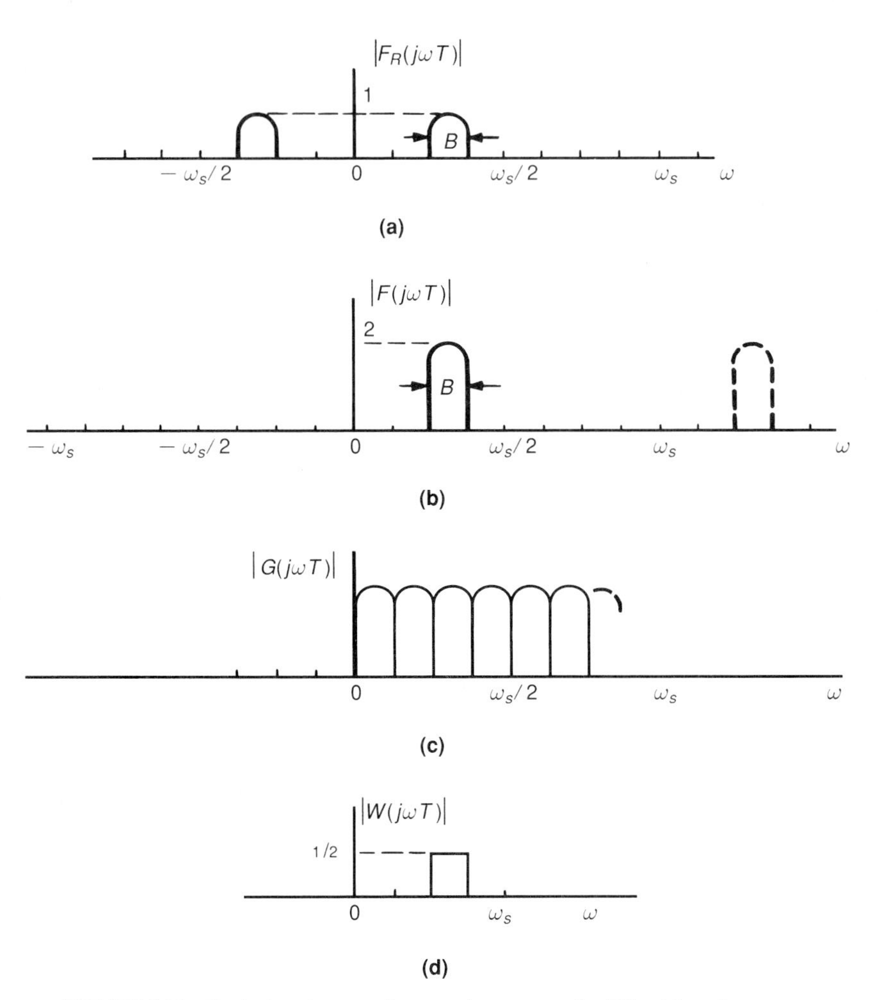

FIGURE 5.38 Reducing the sampling rate by means of a Hilbert transformer.

sponding Fourier transform is

$$F(j\omega T) = G[j(\omega - \omega_c)T] \tag{5.191}$$

so the spectra of $F(j\omega T)$, $F_R(j\omega T)$, and $F_I(j\omega T)$ are translated by frequency ω_c as shown in Figure 5.40.

The sequence $f_I(n)$ is the Hilbert transform of $f_R(n)$. If the complex lowpass signal is expressed as

$$g(n) = A(n)e^{j\phi(n)} \tag{5.192}$$

then

$$f_R(n) = A(n)\cos[\omega_c n + \phi(n)] \tag{5.193}$$

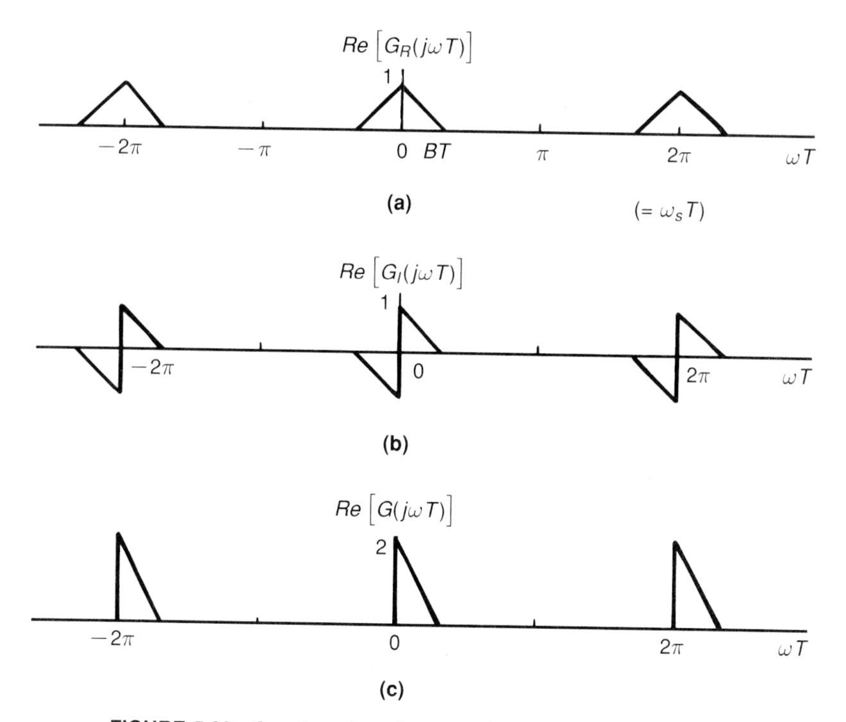

FIGURE 5.39 Spectrum for a lowpass signal and its decomposition.

and

$$f_I(n) = A(n) \sin [\omega_c n + \phi(n)] \tag{5.194}$$

which express the bandpass signals in terms of the lowpass signal.

With reference to Figure 5.40(b), if $F_R(j\omega T)$ is a bandpass signal, the sampling rate could be reduced to $2\omega_h = 2(\omega_c + B)$. (If $\omega_h T = \pi/2$, the sampling rate could be half the original rate.) The sampling rate can be reduced much more if we take the Hilbert transform of $f_R(n)$, obtaining $f_I(n)$, and then form $f(n)$, the spectrum of which is given in Figure 5.40(a). Because in the interval $(-\pi, \pi)$, $F(j\omega T)$ is zero everywhere except for $\omega_c T \leq \omega T \leq (\omega_c + B)T$, and the complex lowpass signal is

$$g(n) = f(n) e^{-jn\omega_c T} \tag{5.195}$$

we can sample at a rate $\omega_{s_I} = B$.

To summarize, if we want to process a real signal $f_R(n)$ the spectrum of which is nonzero only in the frequency range $\omega_c \leq \omega \leq \omega_c + B$, then instead of sampling at a rate $\omega_s > 2(\omega_c + B)$, we can form the complex signal $f(n) = f_R(n) + jf_I(n)$, where $f_I(n)$ is the Hilbert transform of $f_R(n)$, and sample $f(n)$ at a much lower rate, $\omega_s = B$. The original bandpass signal $f_R(n)$ can then be recovered as described in the last section.

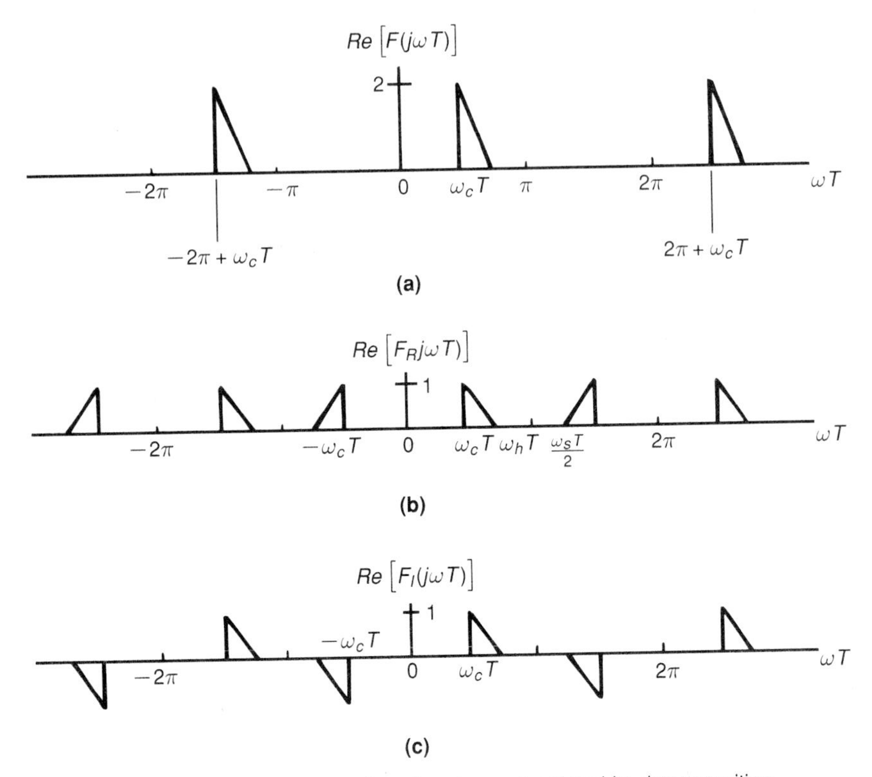

FIGURE 5.40 Spectrum for a bandpass signal and its decomposition.

5.7 DECIMATION AND INTERPOLATION

5.7.1 Introduction

In many digital signal-processing systems, it is necessary to convert from a given sampling rate to a different sampling rate. If the new rate is higher than the original, the change is called interpolation; if lower, it is called decimation. The latter term is somewhat of a misnomer, because in this context, it means a rate reduction by any factor, not just by a factor of 10.

Applications include conversion between different signal code formats (such as conversion from delta modulation to pulse code modulation, which operate at different sampling rates); efficient transmission of speech at low bit rates, with speech regeneration at a higher sampling rate; and narrowband filtering wherein a digital filter can be implemented in an efficient manner by using a reduction in sampling rate followed by a rate increase (see [19]).

A brief introduction to the subject is given here. The treatment is confined to fundamental concepts and follows that given in Crochiere and Rabiner's paper [20]. Much more detailed coverage is provided in the text by the same authors [21] and in [22].

5.7.2 Decimation by an Integer Factor

Given a sequence $x(n)$ sampled at a rate f_s, we want to reduce the sampling rate to f_s/M, where M is integer. To avoid aliasing at the lower sampling rate f_s/M, it is necessary to filter the original sequence $x(n)$ with a lowpass filter, obtaining a new sequence $u(n)$ that contains most of the low-frequency characteristics of $x(n)$. Form a new sequence $y(n)$ with sampling rate f_s/M by extracting every Mth sample of $u(n)$. The remaining samples are discarded. Diagramatically, we can represent decimation as shown in Figure 5.41. In terms of frequency responses (simplified for ease of illustration), the situation is as shown in Figure 5.42.

Let us elaborate. With reference to Figure 5.41,

$$u(n) = \sum_{i=0}^{\infty} h(i)x(n-i) \tag{5.196}$$

where $h(i)$ is the unit pulse response of the lowpass filter. Then

$$\begin{aligned} y(m) &= u(mM), \quad m = 0, 1, 2, \ldots \\ &= \sum_{i=0}^{\infty} h(i)x(mM-i) \\ &= \sum_{i=0}^{\infty} h(mM-i)x(i) \end{aligned} \tag{5.197}$$

We note that it is necessary to compute only every Mth output of the filter.

The spectrum of the signal $x(n)$ shown in Figure 5.42(a) indicates that there are frequency components present up to half the original sampling rate f_s (see Chapter 3). If we wish to decimate by a factor of M, it must be assumed that we are not interested in any frequencies above $f_s/2M$ and that therefore, in principle, an ideal lowpass filter with that cutoff frequency could be used to prevent aliasing at the lower sampling rate. Figure 5.42(b) shows the spectrum for the filtered signal $u(n)$ when an approximation to the ideal lowpass filter is used. The lobes of the spectrum are repeated with a period of f_s. When the sampling frequency is reduced to f_s/M, the same spectral lobes repeat with period f_s/M, as shown in Figure 5.42(c).

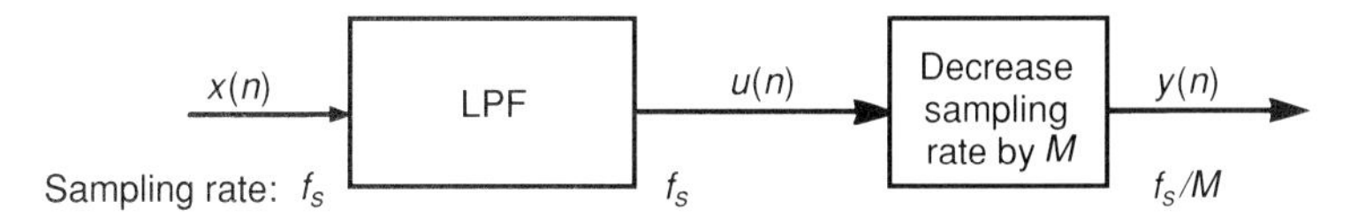

FIGURE 5.41 Decimation of a signal $x(n)$.

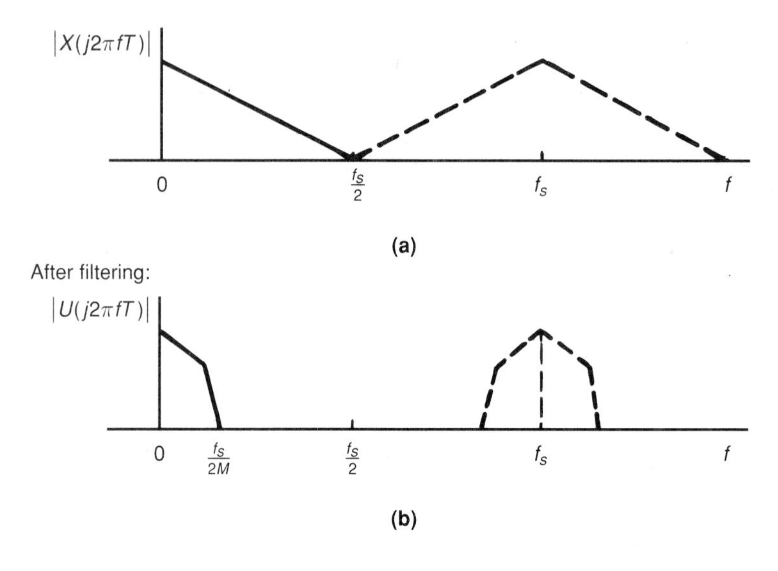

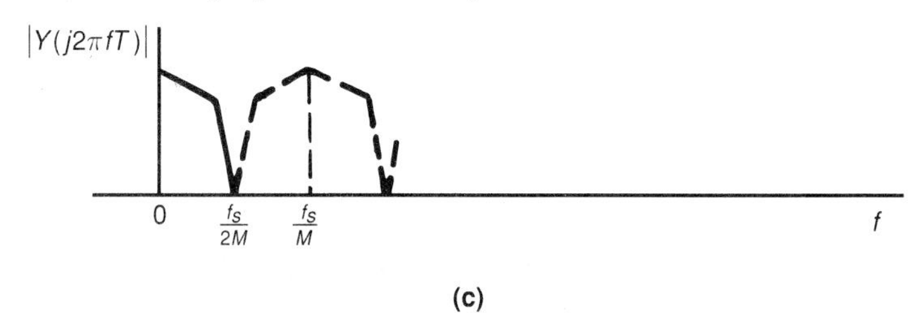

FIGURE 5.42 Changes in magnitude response caused by decimation.

EXAMPLE 5.9

Suppose that a signal containing speech has been sampled at 100 kHz and we are interested only in the frequency content up to 3 kHz. Reduce the sampling rate appropriately.

Solution

It will suffice to decimate the signal to a sampling rate of 10 kHz. To prevent aliasing, we design a lowpass filter to the following specifications:

Sampling frequency $f_s = 100$ kHz

Cutoff frequency $f_c = 3$ kHz

Maximum ripple in stopband $= 0.5$ dB

Minimum attenuation in stopband ≈ 40 dB at $f_r = 5$ kHz

Using the diskette program ELLIPD, we find that a fourth-order elliptic recursive filter will satisfy these requirements. This requires about 10 multiplications and 8 additions per output sample with 20 storage registers if 2 second-order sections are used in parallel.

A linear-phase nonrecursive filter designed to the same specifications via McClellan's program [17] required 120 taps. Because the filter has symmetric coefficients (see Section 5.5), we can reduce the number of multiplications by a factor of 2. In addition, with a nonrecursive filter, we can take advantage of the fact that we have to compute only every Mth output point. We are decimating by 10 in this case, so only 6 multiplications and 12 additions are required per output sample. Because speech is being processed, the linear-phase characteristic of the nonrecursive filter is more desirable in this example.

5.7.3 Interpolation by an Integer Factor

In this case we want to increase the sampling rate f_s of the given sequence $x(n)$ by an integer factor L. We do this by inserting $(L-1)$ zero-valued samples after each sample of $x(n)$. This creates a sequence $u(n)$ with sampling rate Lf_s, the frequency components of which are periodic with period equal to the original sampling frequency f_s, as shown in Figure 5.43. To eliminate these periodic components and retain only the baseband frequencies, it is necessary to filter $u(n)$ with an appropriate lowpass filter (cutoff frequency $f_s/2$). The resulting signal $y(n)$ with sampling rate Lf_s is the desired interpolated signal.

To elaborate further, in Figure 5.43,

$$u(\ell) = \begin{cases} x(\ell/L), & \ell = 0, L, 2L, \ldots \\ 0, & \text{otherwise} \end{cases} \tag{5.198}$$

Figure 5.43(a) shows the spectrum of the original signal $x(n)$. When the sampling frequency is increased by padding $x(n)$ with zeros, the spectrum appears to be unchanged [Figure 5.43(b)]. However, the new baseband of frequencies (0 to $Lf_s/2$) contains not only the original baseband frequencies (0 to $f_s/2$) but also periodic repetitions or *images* of these frequencies, as shown in Figure 5.43. This is not desirable, and the unwanted images are eliminated by means of a lowpass anti-imaging filter that approximates the ideal lowpass characteristic designed for a sampling rate of Lf_s but with cutoff frequency $f_s/2$ [Figure 5.43(c)]. It can be shown that the gain of the filter must be L to ensure that the amplitude of $y(l)$ is correct. We note that if an Nth-order nonrecursive filter is used for this purpose, only N/L multiplications are required because of the number of zero samples in $u(\ell)$.

5.7.4 Decimation or Interpolation by a Factor L/M

If we desire a sampling-rate conversion by L/M (where M and L are integer), we can achieve this by first interpolating by L and then decimating by M, as shown in Figure 5.44. Because the two lowpass filters are in cas-

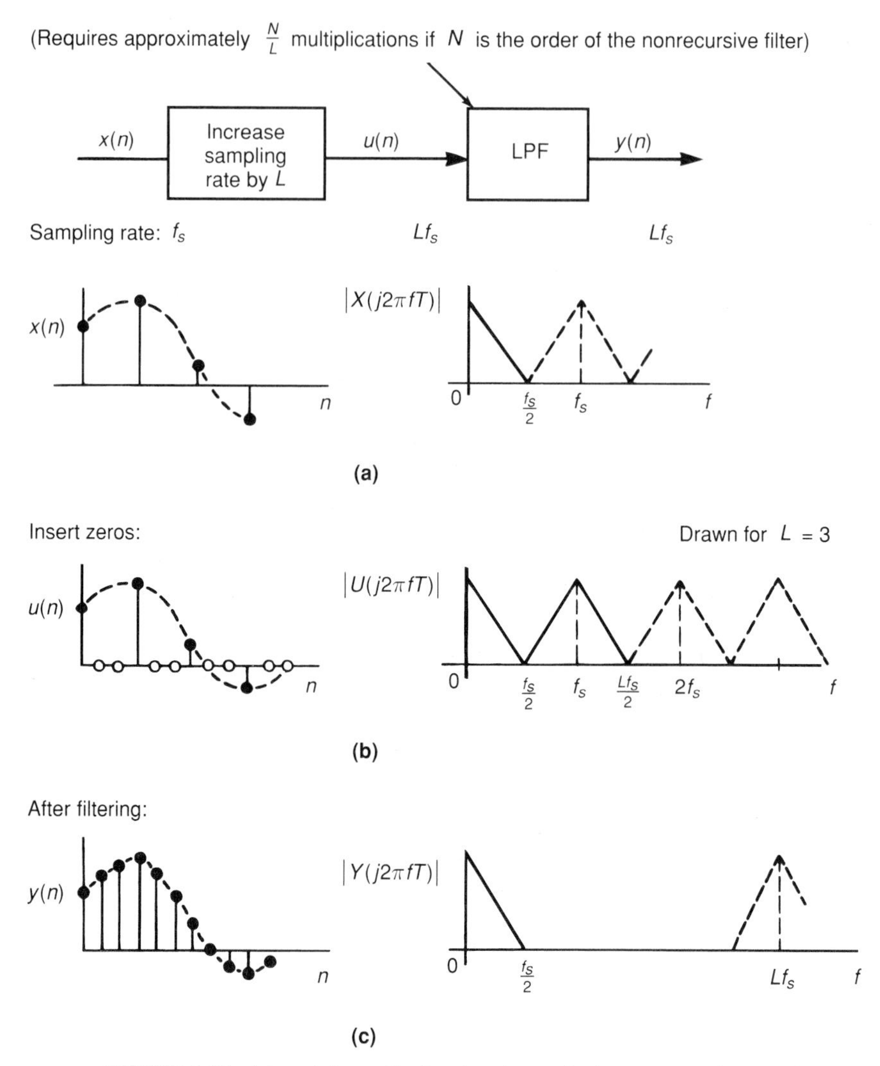

FIGURE 5.43 Interpolation with signal and magnitude response changes.

cade and are operating at the same sampling rate Lf_s, it is more efficient to combine them in one composite lowpass filter, as indicated in Figure 5.45. If $L > M$, the unit is an interpolator; if $L < M$, it is a decimator. It may be advantageous to use a nonrecursive filter.

$$v(n) = \sum_{m=0}^{\infty} h(m)u(n-m) \tag{5.199}$$

It is apparent that the filter computations need be made only for every Mth output point. Also, it is known that $u(n)$ is nonzero only for every Lth input point, so approximately N/L multiplications are necessary for each of

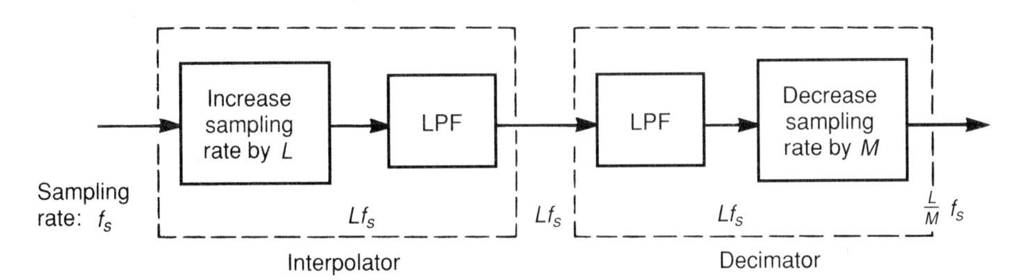

FIGURE 5.44 Sampling-rate change by the factor L/M.

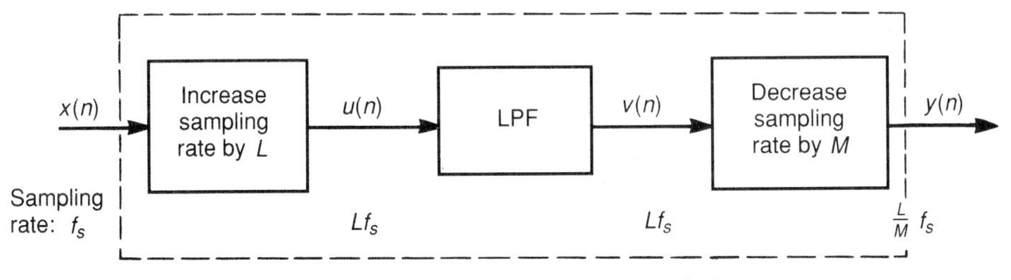

FIGURE 5.45 General interpolator/decimator.

the retained output samples. Thus savings in computation are often to be had by using a nonrecursive instead of a recursive filter. However, a recursive filter generally requires less storage for a given sharpness of cutoff.

If there is a large reduction in sampling rate, it is usually more effective to reduce f_s in a series of decimation stages. This reduces the filtering requirement at each stage. The optimum design (in the sense of minimum computation requirements) of multistage decimators and interpolators is treated in a paper by Crochiere and Rabiner [20].

5.8 MOVING ESTIMATES OF VARIANCE BY TWO DIFFERENT FILTERING METHODS

For this application, we need to understand what is meant by the mean and the variance of a random variable or a random process. Chapter 7 contains a review of these and other basic concepts in probability theory that are fundamental to estimation.

In processing a finite data sequence $x_i, i = 1, \ldots, M$, it is often desirable to estimate its variance about the mean, especially when the data result from some random phenomenon. This is one means of compressing the data while extracting meaningful information. A static estimate of variance is provided by

$$s^2 = \frac{1}{M}\sum_{i=1}^{M}(x_i - \overline{x})^2 \tag{5.200}$$

where the mean is estimated as follows:

$$\overline{x} = \frac{1}{M} \sum_{i=1}^{M} x_i \tag{5.201}$$

We could use $1/(M-1)$ as the multiplying factor in (5.200) to provide an unbiased estimate, but this is of no practical significance for moderate or large M.

Frequently, the data cannot be considered to result from a stationary random process—that is, a process with time-invariant statistics. In such cases, the notion of finding a constant mean or variance to describe the sequence is optimistic at best. Furthermore, with long data sequences, the storage necessary for computing (5.200) might be excessive. Instead, it is more meaningful to make a moving estimate of the variance, as we did with the average or mean in Section 5.2.2. Realizable moving estimates of mean and variance over a window of width N are given by

$$\overline{x}_k = \frac{1}{N} \sum_{i=r}^{k} x_i \tag{5.202}$$

and

$$s_k^2 = \frac{1}{N} \sum_{i=r}^{k} (x_i - \overline{x}_k)^2 \tag{5.203}$$

where $r = k - N + 1$. These are nonrecursive filtering operations, albeit nonlinear in the case of s_k^2. It is necessary to store the first N values of x_i before obtaining an estimate of s_k^2.

If storage limitations are a factor, it might be desirable to make recursive estimates instead of performing the nonrecursive moving-average operations described in (5.202) and (5.203). A recursive estimate of mean and variance is obtained via two coupled first-order recursive digital filters. For the mean, we have

$$\overline{x}_k = A\overline{x}_{k-1} + (1-A)x_k, \quad k \geq 2 \tag{5.204}$$

with initial condition $\overline{x}_1 = x_1$ and $|A| < 1$. Multiplying x_k by $(1-A)$ ensures that, for constant x_k, $\overline{x}_k$ would converge to that value. A recursive estimate of variance is provided by

$$s_k^2 = As_{k-1}^2 + (1-A)(x_k - \overline{x}_k)^2 \tag{5.205}$$

with initial conditions $s_1^2 = 0$ and $\overline{x}_k$ obtained from (5.204).

The value of the coefficient A relates the decay of the recursive filter to the window width N of the non-recursive filter. That is,

$$A = \exp(-T/NT) = \exp(-1/N) \tag{5.206}$$

where NT is approximately equal to the time constant of the corresponding continuous filter. For example, for $N = 100$, $A = .990050$, and for $N = 400$, $A = 0.997503$.

Figure 5.46 compares the standard deviation estimate s_k computed nonrecursively according to (5.203) and that same parameter computed recursively from (5.205) for the same data sequence, which in this case was digitized elevation data (see Section 5.2.2). It is evident from the figure that there is reasonable correspondence between the two estimates, but in general it may be necessary to adjust A for a better fit. There is an initial transient of about N samples in the recursive estimate, whereas the nonrecursive estimate starts at N cells or samples. The recursive estimate tends to be smoother and has the advantage, which we have noted, of requiring less storage space.

5.9 SUMMARY

Basic nonrecursive filtering processes are described. Moving-average operations, both realizable and nonrealizable, provide useful means for obtaining time-varying estimates of signal statistics. Although the moving-average filters have finite pulse responses, they can be realized either nonrecursively or recursively. Modified comb filters can be designed as lowpass, highpass, or bandpass filters by means of pole-zero cancellation. They can have single or multiple passbands and can be realized either nonrecursively or recursively. To offset their design simplicity, they are marred by low attenuation from passband to stopband and, in general, require considerable storage for the delayed samples.

The Fourier series method is a systematic technique for designing a nonrecursive filter with a desired frequency response. Window functions are needed in the resulting design to reduce oscillations, due to the Gibbs phenomenon, at discontinuities of the desired frequency response. However, the window function widens the transition region of the filter. The Chebyshev approximation method developed by Parks and McClellan is described. It is widely used for FIR filter design.

Computer programs to design digital filters by means of this and other techniques are listed in [16]. These programs eliminate much of the tedium associated with the design of higher-order filters. One such program for windowed design, FSAPPROX, is given on the diskette.

An introduction to Hilbert transforms covers the relationships between the real and imaginary parts of the Fourier transform of causal sequences. It also indicates how the imaginary part of a complex sequence is obtained from the real part by means of a Hilbert transformer, with application to the sampling of bandpass signals. A brief treatment of the basics of decimation and interpolation follows. The choice of lowpass filter type required, that is, recursive or nonrecursive, is discussed next. Means of obtaining nonrecursive and recursive moving estimates of variance are described.

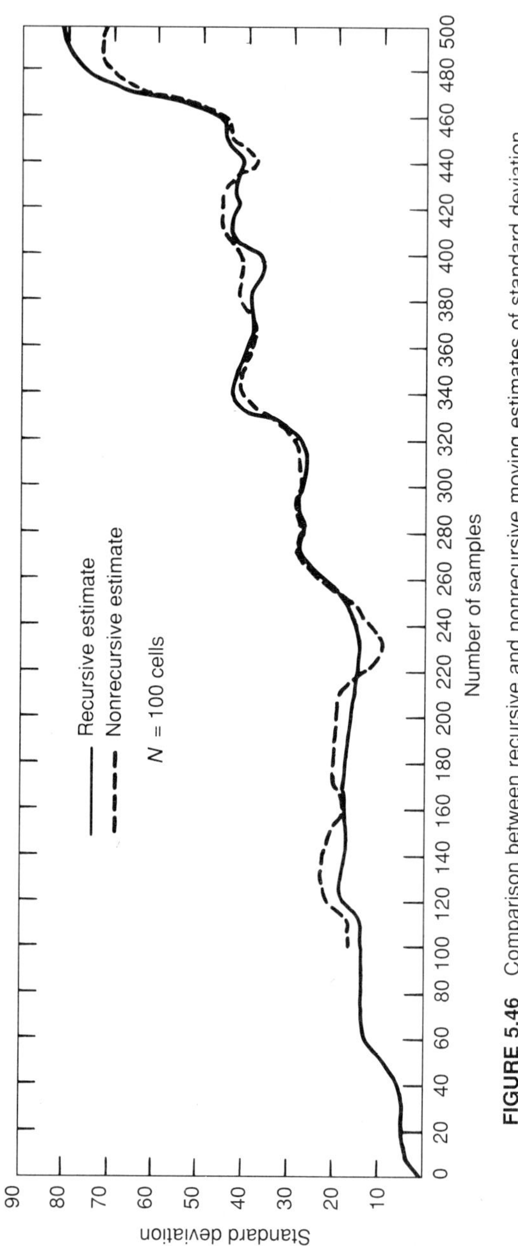

FIGURE 5.46 Comparison between recursive and nonrecursive moving estimates of standard deviation.

REFERENCES FOR CHAPTER 5

1. R. E. Bogner and A. G. Constantinides, ed., *Introduction to Digital Filtering* (New York: Wiley, 1975), Chap. 9.
2. R. V. Churchill, *Fourier Series and Boundary Value Problems* (New York: McGraw-Hill, 1941).
3. E. A. Guillemin, *The Mathematics of Circuit Analysis* (New York: John Wiley, 1958), Chap. 7.
4. M. S. Bartlett, *An Introduction to Stochastic Processes with Special Reference to Methods and Applications* (Cambridge, England: Cambridge University Press, 1953).
5. R. W. Hamming, *Digital Filters* (Englewood Cliffs, NJ: Prentice-Hall, 1977).
6. R. B. Blackman and J. W. Tukey, *The Measurement of Power Spectra* (New York: Dover, 1958).
7. A. Antoniou, *Digital Filters: Analysis and Design* (New York: McGraw-Hill, 1979), Chap. 9.
8. J. F. Kaiser, "Design Methods for Sampled-Data Filters," *Proc. First Allerton Conf. on Circuit and System Theory*, November 1963: 221–236.
9. J. F. Kaiser, "Digital Filters," Chap. 7 in *System Analysis by Digital Computer*, ed. F. F. Kuo and J. F. Kaiser (New York: Wiley, 1966).
10. L. R. Rabiner, C. A. McGonegal, and D. Paul, "FIR Windowed Design Program—WINDOW," in *Programs for Digital Signal Processing* (New York: IEEE Press, 1979).
11. T. W. Parks and J. H. McClellan, "Chebyshev Approximation for Nonrecursive Digital Filters with Linear Phase," *IEEE Trans. on Circuit Theory* **19**, no. 4 (March 1972): 189–194.
12. J. H. McClellan and T. W. Parks, "A Unified Approach to the Design of Optimum FIR Linear-Phase Digital Filters," *IEEE Trans. on Circuit Theory*, **20**, no. 6, (November 1973): 697–701.
13. E. Ya Remez, "General Computational Methods of Tchebycheff Approximation," *Atomic Energy Transl.* 4491 (1957): 1–85.
14. T. W. Parks and C. S. Burrus, *Digital Filter Design* (New York: Wiley, 1987).
15. L. R. Rabiner, J. H. McClellan, and T. W. Parks, "FIR Digital Filter Design Techniques Using Weighted Chebyshev Approximation," *Proc. IEEE* 63 (April 1975): 595–610.
16. Digital Signal Processing Committee, IEEE Acoustics, Speech, and Signal Processing Society, *Programs for Digital Signal Processing* (New York: IEEE Press, 1979).
17. J. H. McClellan, T. W. Parks, and L. R. Rabiner, "A Computer Program for Designing Optimum FIR Linear Phase Digital Filters," *IEEE Trans. Audio Electroacoustics* AU–21 (December 1973): 506–525.
18. L. R. Rabiner and B. Gold, *Theory and Application of Digital Signal Processing* (Englewood Cliffs, NJ: Prentice-Hall, 1975), Chap. 3.

19. L. R. Rabiner and R. E. Crochiere, "A Novel Implementation for Narrow-Band FIR Digital Filters," *IEEE Trans. on Acoustics, Speech, and Signal Processing* ASSP–23, no. 5 (October 1975): 457–464.

20. R. E. Crochiere and L. R. Rabiner, "Optimum FIR Digital Filter Implementations for Decimation, Interpolation, and Narrow-Band Filtering," *IEEE Trans. on Acoustics, Speech and Signal Processing* ASSP–23, no. 5 (October 1975): 444–456.

21. R. E. Crochiere and L. R. Rabiner, *Multirate Digital Signal Processing* (Englewood Cliffs, NJ: Prentice-Hall, 1983).

22. A. Peled and B. Liu, *Digital Signal Processing* (New York: Wiley, 1976), Chap. 2.

EXERCISES FOR CHAPTER 5

1. A nonrecursive filter consists of a difference filter and a sum filter in series—that is, $m_k = x_k - x_{k-1}$ and $y_k = m_k + m_{k-1}$, respectively. For the overall filter,

a) Derive the transfer function $Y(z)/X(z)$.

b) Compute and sketch the magnitude response and the phase response for $|\omega| \leq \omega_s/2$.

c) Make a pole-zero plot.

d) Write the pulse response.

2. Find and sketch the magnitude response and the phase response ($|\omega| \leq \omega_s/2$) for the nonrecursive filter

$$y_k = x_k + x_{k-2}$$

3. If the clock period or computer time step is 10 ms and the input is a sampled sinusoid of amplitude unity and frequency 25 kHz, find the output sequence for

a) The filter of Exercise 1

b) The filter of Exercise 2

4. Determine the frequency response for the realizable moving-average operation defined by (5.24). Sketch the responses for $m = 9$ and make a pole-zero plot.

5. Derive the expression (5.31) for the frequency response of the difference comb filter.

6. Derive the expressions (5.33) through (5.35) for the sum comb filter. Make a pole-zero plot and sketch the magnitude response.

7. Verify the first entry in Table 5.3, for ratio of main lobe to first side lobe and gain of filter at center of passband.

8. A comb filter with 12 delays is modified by pole-zero cancellation to yield a bandpass filter with passband centered at 1000 Hz. The sampling frequency is 6000 Hz. For the resulting filter, write the transfer function, make a pole-zero plot, sketch the magnitude response, and write the difference equation in both nonrecursive and recursive form.

9. Design a comb filter to the specifications of Exercise 8, but use second-order zeros and canceling poles. Write the transfer function, and the difference equation in recursive form.

10. Establish conditions (5.83) for an FIR filter for constant group delay and piecewise-linear phase.

11. Suppose that we modify the rectangular window as follows:

$$w_{RM}(n) = \begin{cases} 1, & -(N-2) \leq n \leq N-2 \\ \frac{1}{2}, & n = -(N-1), N+1 \\ \frac{1}{4}, & n = -N, N \\ 0, & \text{otherwise} \end{cases}$$

Determine the frequency response $W_{RM}(j\omega T)$. Plot this and compare it with that for the ideal rectangular window. What improvement, if any, is there in the ratio of maximum side lobe to main lobe?

12. Verify the expression (5.139) for the frequency response of the Bartlett window.

13. Confirm the Blackman window spectrum given in Table 5.6.

14. The desired magnitude response of a digital filter in the range $|\omega T| \leq \pi$ is

$$M(\omega T) = 1 - |\omega T|/\pi$$

Use the Fourier series method to design a suitable approximation. Plot the magnitude response. Is a window function necessary?

15. Use the Fourier series method to design an approximation to the ideal differentiator as in Example 5.6, but truncate at $N = 10$. Write the difference equation for the resulting design.

16. The ideal Hilbert transformer (5.184) has a purely imaginary frequency response. Design an approximation using the Fourier series method. Plot the magnitude response for $N = 10$ and for $N = 50$.

17. Use the Fourier series method in conjunction with a Hamming window to design an approximation to an ideal lowpass filter with magnitude response

$$M(\omega T) = \begin{cases} 1, & |\omega T| \leq \frac{\pi}{3} \\ 0, & \text{otherwise} \end{cases}$$

Compare the response with that obtained from an unwindowed design, for $N = 12$.

18. Design an approximation to an ideal highpass filter with magnitude response.

$$M(\omega T) = \begin{cases} 0, & 0 \leq |\omega T| \leq \frac{\pi}{3} \\ 1, & \text{otherwise} \end{cases}$$

by the Fourier series method. Take $N = 11$ and use a suitable window function to smooth the Gibbs phenomenon effect. Plot the magnitude response for the windowed and unwindowed functions.

19. Plot the magnitude response of the ideal bandpass filter of Example 5.7, but use a Blackman window instead of a Hamming window.

20. Use the Kaiser window for the ideal bandpass filter of Example 5.7 with $\alpha = 4$ and with $\alpha = 8$. How does the filter magnitude response compare with that found by using the Hamming window (Figure 5.30)?

APPENDIX 5A

Diskette Program Example

The FORTRAN program FSAPPROX designs an FIR filter to given specifications via the windowed method. It is arranged for interactive operation.

The user is prompted for filter type, length, cutoff frequencies, and window type. The resulting filter coefficients are displayed. The magnitude response and the phase response are stored in a file FSAPPROX.RSP created by the program. An example of a bandpass filter design is given. The magnitude response is plotted in Figure 5A.1.

```
           LINEAR PHASE NONRECURSIVE FILTER DESIGN
ENTER TYPE OF FILTER (LP,HP,BP,BS):
BP
ENTER VALUE OF N (2 - 512)
(N IS TRUNCATION INDEX FOR FOURIER SERIES):
14
  H -- HAMMING
  V -- VON HANN
  B -- BLACKMAN
  K -- KAISER
ENTER WINDOW TYPE:
B
ENTER NORMALIZED LOWER CUTOFF FREQUENCY (0.0 - 0.5):
0.16667
ENTER NORMALIZED UPPER CUTOFF FREQUENCY (0.0 - 0.5):
.33333
CENTER COEFFICIENT C0 .166666 E + 00
FIRST HALF OF (SYMMETRIC) FILTER COEFFICIENTS:
 B(  1) =   0.000000395
 B(  2) =  -0.253711641
 B(  3) =  -0.000000669
 B(  4) =   0.098404594
 B(  5) =   0.000000237
```

```
B(  6) =  -0.000000124
B(  7) =   0.000000137
B(  8) =  -0.016309848
B(  9) =  -0.000000124
B( 10) =   0.004986995
B( 11) =   0.000000019
B( 12) =  -0.000000005
B( 13) =   0.000000002
B( 14) =   0.000000001
Pause - Please enter a blank line (to continue) or a DOS command.
MAGNITUDE AND PHASE RESPONSES IN FILE FSAPPROX.RSP
Stop - Program terminated.
```

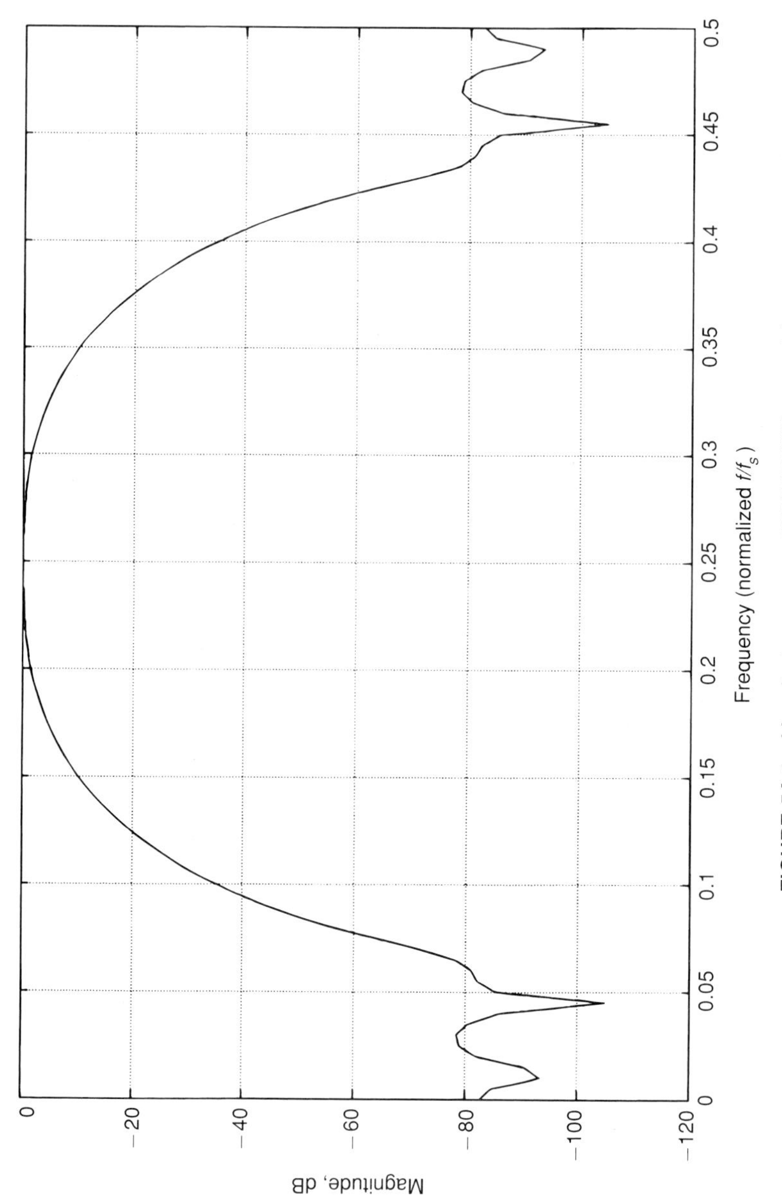

FIGURE 5A.1 Magnitude response (FSAPPROX example).

CHAPTER 6

THE DISCRETE FOURIER TRANSFORM AND THE FAST FOURIER TRANSFORM

6.1 INTRODUCTION

We have considered frequency representations of sequences and discrete linear time-invariant systems in terms of Fourier and z-transforms and have found them to be very useful for filter design. These transforms are *continuous* functions of frequency, although the sequences and systems they represent are, of course, discrete in time. Often, in practice, the sequence to be represented is of finite duration—that is, it has only a finite number of nonzero values. For use in this case, there is an alternative representation called the discrete Fourier transform (DFT) that is *discrete* in frequency. It is, itself, a sequence of finite length. The discrete Fourier transform lends itself readily to numerical computation of a Fourier transform on a digital computer, where one is forced into a discrete frequency representation, because the DFT corresponds to samples of the Fourier transform of the sequence or to samples of the z-transform about the unit circle. These samples are equally spaced in frequency.

Many of the properties of the DFT are analogous to those of the Fourier transform and the z-transform. Associated with the DFT is a circular convolution of two sequences instead of a linear convolution. By expanding the sequences in an appropriate manner, we can get a linear convolution from a circular convolution. The DFT of a circular convolution of two sequences is the product of their DFTs—a fact that is just as useful for analysis as was the corresponding property for the continuous frequency transforms.

The DFT has found particular application in the software implementation of digital filters and other signal-processing algorithms because of the development of the fast Fourier transform (FFT). This is a collective name for a number of algorithms used to evaluate the DFT numerically. They are computationally efficient because they eliminate the redundant operations involved in direct evaluation of the DFT.

6.2 THE DISCRETE FOURIER TRANSFORM

In previous chapters we have used the z-transform to obtain the frequency characteristics of discrete signals and systems. Recall that for a sequence $x(n) = x(nT)$ defined for all integer n, the z-transform is

$$X(z) = \sum_{n=-\infty}^{\infty} x(n)z^{-n} \tag{6.1}$$

On the unit circle, $z = e^{j\omega T}$, and the z-transform becomes the Fourier transform of the sequence,

$$X(j\omega T) = \sum_{n=-\infty}^{\infty} x(n)e^{-jn\omega T} \tag{6.2}$$

For a finite-length sequence $x(n) = 0, 1, \ldots, N-1$, (6.2) can be written

$$X(j\omega T) = \sum_{n=0}^{N-1} x(n)e^{-jn\omega T} \tag{6.3}$$

Both (6.2) and (6.3) are *continuous* functions of frequency ω. To obtain a *discrete* function of frequency suitable for computation on a digital computer, we replace ω in (6.3) by $k\Omega$, where k is integer and Ω is the frequency increment in radians per second. In other words, we define the discrete Fourier transform (DFT) of a sequence of N samples $X(n), n = 0, 1, \ldots, N-1$, as another sequence, discrete in frequency:

$$X(k) = \sum_{n=0}^{N-1} x(n)e^{-jnk\Omega T} \tag{6.4}$$

For brevity, we often express this as

$$X(k) = DFT\{x(n)\} \tag{6.5}$$

How do we choose the frequency increment Ω? Consider $x(n), n = 0, 1, \ldots, N-1$, as one period of a periodic sequence $\tilde{x}(n) = \tilde{x}(nT)$. Then, as shown in Figure 6.1, the period is NT s, which means that the fundamental frequency, or frequency resolution in radians per second, is

$$\Omega = \frac{2\pi}{NT} \tag{6.6}$$

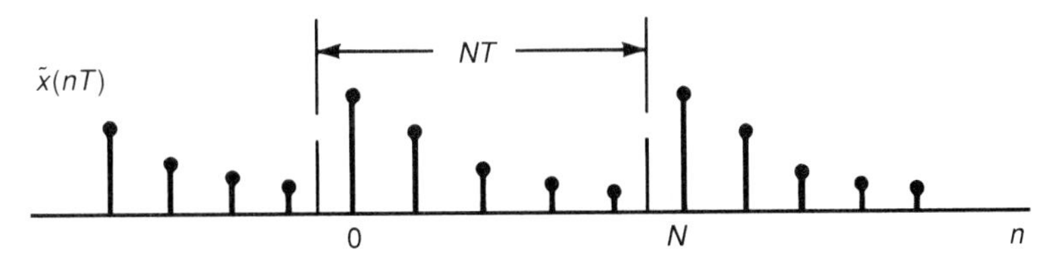

FIGURE 6.1 Periodic sequence $\tilde{x}(n)$.

Hence the DFT (6.4) will provide frequency samples at $0, \Omega, 2\Omega$, and so on. For brevity, and without loss of generality, we can let $T = 1$ so that (6.4) and (6.6) become

$$X(k) = \sum_{n=0}^{N-1} x(n)e^{-jnk\Omega} \tag{6.7}$$

and

$$\Omega = \frac{2\pi}{N} \tag{6.8}$$

respectively. We shall prove that only N distinct values of $X(k)$ can be computed. We will be dealing with periodic sequences, so it is convenient to make use of the modulo function [1]. The function "y modulo N" is defined as

$$((y))_N = y - rN \tag{6.9}$$

where r is the largest integer the magnitude of which does not exceed $|y/N|$. The sign of r is the sign of y/N. The subscript N on $((y))$ will be omitted unless some ambiguity arises from our considering sequences of different lengths. Note that

$$\begin{aligned} ((y)) &= y - rN \\ &= y, \quad \text{if } y < N \\ &= 0, \quad \text{if } y = rN \\ &< N, \quad \text{in general} \end{aligned} \tag{6.10}$$

The variable y is not necessarily integer, but in many of the cases that follow, we will be concerned with the modulo function of an integer.

From the properties of the modulo function, the exponent on the right-hand side of (6.7) can be written as

$$\begin{aligned} e^{-j\Omega nk} &= e^{-j\Omega n[((k))+rN]} \\ &= e^{-j\Omega n((k))}e^{-j\Omega nrN} \\ &= e^{-j\Omega n((k))} \end{aligned} \tag{6.11}$$

because $\Omega N = 2\pi$. It follows that

$$X(k) = X((k)) \tag{6.12}$$

and, because $((k)) < N$ in general, there are only N distinct values of $X(k)$. Therefore, (6.7) may be written

$$X(k) = \sum_{n=0}^{N-1} x(n)e^{-jnk\Omega}, \quad k = 0, 1, \ldots, N-1 \tag{6.13}$$

This is a sequence of N numbers. From (6.12),

$$X(k) = X(k + rN) \tag{6.14}$$

where r is an integer ranging from minus infinity to plus infinity, so the DFT can also be considered a periodic sequence of numbers with period N.

The z-transform of a finite sequence $x(n), n = 0, 1, \ldots, N-1$, at N points evenly spaced at $\omega = k\Omega$ on the unit circle in the z-plane is

$$X(z)\big|_{z=e^{jk\Omega}} = \sum_{n=0}^{N-1} x(n)e^{-jnk\Omega} \tag{6.15}$$

which, by definition, is the DFT. This is a third interpretation of the DFT.

EXAMPLE 6.1

Find the DFT of the finite sequence

$$x(n) = a^n, \quad n = 0, 1, \ldots, N-1, 0 < a < 1$$

Solution

From the definition (6.7),

$$\begin{aligned} X(k) &= \sum_{n=0}^{N-1} a^n e^{-jnk\Omega} \\ &= \sum_{n=0}^{N-1} (ae^{-jk\Omega})^n \\ &= \frac{1 - a^N}{1 - ae^{-jk\Omega}} \end{aligned}$$

From this, we get the magnitude,

$$M(k) = (1 - a^N)/(1 + a^2 - 2a\cos k\Omega)^{1/2}$$

and the phase,

$$\phi(k) = \tan^{-1}[-a\sin k\Omega/(1 - a\cos k\Omega)]$$

In Figure 6.2, these are plotted over one period for $N = 9$ and $a = 0.5$.

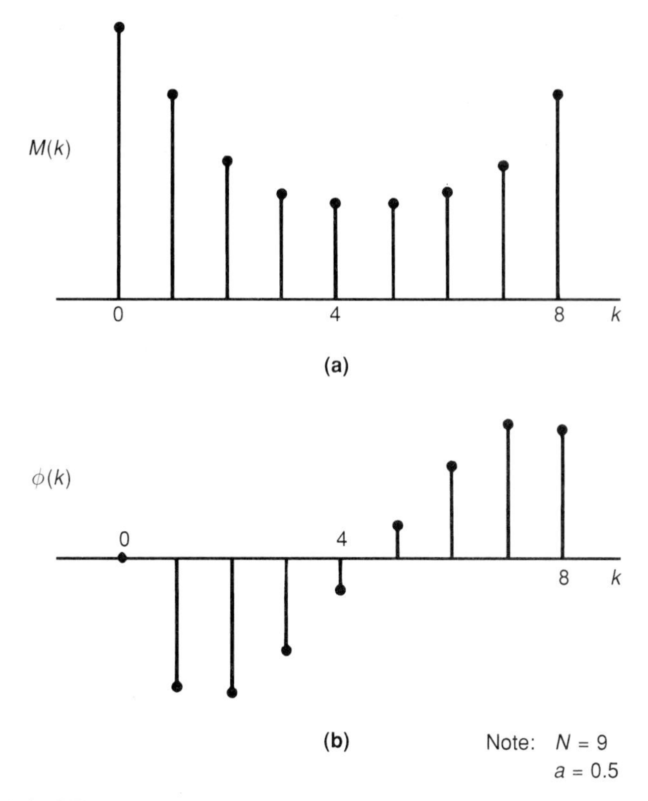

FIGURE 6.2 (a) Magnitude and (b) phase sequences from the DFT of $x(n) = a^n; n = 0; 1; \ldots; N-1$.

6.3 THE INVERSE DISCRETE FOURIER TRANSFORM

The inverse discrete Fourier transform (IDFT) is used to map the DFT back into the sequence from which it was derived. As an aid to the derivation of a suitable expression for the IDFT, consider first the DFT of a complex sinusoidal sequence of frequency $q\Omega$:

$$\tilde{x}(n) = e^{jq\Omega n} \tag{6.16}$$

The DFT is

$$\tilde{X}(k) = \sum_{n=0}^{N-1} e^{jn\Omega(q-k)} \tag{6.17}$$

$$= \frac{1 - e^{jN\Omega(q-k)}}{1 - e^{j\Omega(q-k)}} \tag{6.18}$$

Equation (6.18) is valid whether or not q is integer. However, (6.18) can be simplified if it is. Suppose q is integer and, in addition, $k = ((q))$.

$$e^{j\Omega(q-k)} = e^{j\Omega(q-q+rN)} = 1 \tag{6.19}$$

It follows from (6.17) that

$$\tilde{X}(k) = \sum_{n=0}^{N-1} 1 = N \tag{6.20}$$

Now if q is integer and $k \neq ((q))$, then

$$e^{jN\Omega(q-k)} = e^{j(q-k)2\pi} = 1 \tag{6.21}$$

which implies from (6.18) that $\tilde{X}(k) = 0$.

To summarize, for integer q, if

$$\tilde{x}(n) = e^{jq\Omega n} \tag{6.22}$$

then

$$\tilde{X}(k) = \begin{cases} N & \text{if } k = ((q)) \\ 0 & \text{if } k \neq ((q)) \end{cases} = N\delta_{((q)),k} \tag{6.23}$$

We can use this result to prove the following theorem.

Theorem 6.1 (Inverse DFT)

If

$$DFT\{x(n)\} = X(k) = \sum_{n=0}^{N-1} x(n)e^{-jnk\Omega}, \quad k = 0, 1, 2, \ldots, N-1 \tag{6.24}$$

then there exists an inverse DFT such that

$$DFT^{-1}\{X(k)\} = x(m) = \frac{1}{N}\sum_{k=0}^{N-1} X(k)e^{jmk\Omega}, \quad m = 0, 1, \ldots, N-1 \tag{6.25}$$

Proof

Substitute $X(k)$ from (6.24) into the right side of (6.25).

$$\begin{aligned} \frac{1}{N}\sum_{k=0}^{N-1} X(k)e^{jmk\Omega} &= \frac{1}{N}\sum_{k=0}^{N-1}\left[\sum_{n=0}^{N-1} x(n)e^{-jnk\Omega}\right]e^{jmk\Omega} \\ &= \frac{1}{N}\sum_{n=0}^{N-1} x(n) \underbrace{\sum_{k=0}^{N-1} e^{j(m-n)k\Omega}}_{N\delta_{((m)),n} \text{ (from (6.23))}} \\ &= x((m)) \\ &= x(m) \end{aligned} \tag{6.26}$$

because $m < N$. This completes the proof.

In general, as with the DFT, the IDFT $x(m)$ can be considered to be one period of a periodic sequence $\tilde{x}(m)$. For convenience, the DFT pair (6.13) and (6.25) are often written

$$X(k) = \sum_{n=0}^{N-1} x(n)W^{nk}, \quad k = 0, 1, \ldots, N-1 \tag{6.27}$$

$$x(n) = \frac{1}{N}\sum_{k=0}^{N-1} X(k)W^{-nk}, \quad n = 0, 1, \ldots, N-1 \tag{6.28}$$

where $W = e^{-j\Omega}$ depends on N. It is written W_N to avoid confusion when sequences of different lengths are being considered. Generally, in this chapter, we reserve n as the primary index for the time samples and k as the primary index for the frequency samples, but we will use additional indices for both, where necessary.

6.4 PROPERTIES OF THE DFT

Like the Fourier, Laplace, and z-transforms, the DFT has several important properties that enhance its utility for analyzing finite time sequences and designing digital filters with finite pulse responses. Some of these properties will now be considered.

6.4.1 Linearity

If

$$x_3(n) = a_1x_1(n) + a_2x_2(n) \tag{6.29}$$

where x_1, x_2, and x_3 are N-point sequences and a_1 and a_2 are constants, then

$$X_3(k) = a_1X_1(k) + a_2X_2(k) \tag{6.30}$$

where X_1, X_2, and X_3 are the DFTs of x_1, x_2, and x_3, respectively.

Proof

By definition,

$$\begin{aligned} X_3(k) &= \sum_{n=0}^{N-1} x_3(n)e^{-jnk\Omega} \\ &= \sum_{n=0}^{N-1} [a_1x_1(n) + a_2x_2(n)]e^{-jnk\Omega} \end{aligned}$$

$$= a_1 \sum_{n=0}^{N-1} x_1(n)e^{-jnk\Omega} + a_2 \sum_{n=0}^{N-1} x_2(n)e^{-jnk\Omega}$$

$$= a_1 X_1(k) + a_2 X_2(k) \qquad \text{Q.E.D.}$$

6.4.2 Symmetry

If $x(n), n = 0, 1, \ldots, N-1$, is a *real* sequence with DFT $X(k)$, then

$$X(N-k) = X^*(k) \tag{6.31}$$

where X^* is the complex conjugate of X.

Proof

Because

$$X(k) = \sum_{n=0}^{N-1} x(n)e^{-jnk\Omega}$$

it follows that

$$X(N-k) = \sum_{n=0}^{N-1} x(n)e^{-jn(N-k)\Omega} = \sum_{n=0}^{N-1} x(n)e^{jnk\Omega}$$

because $N\Omega = 2\pi$ and $e^{-j2\pi n} = 1$. Therefore,

$$X(N-k) = \left[\sum_{n=0}^{N-1} x(n)e^{-jnk\Omega}\right]^*$$

because $x(n)$ is real. Hence

$$X(N-k) = X^*(k) \qquad \text{Q.E.D.}$$

We conclude from this property that

1. *Real part:* $Re\{X(k)\} = Re\{X(N-k)\}$ (6.32)
2. *Imaginary part:* $Im\{X(k)\} = -Im\{X(N-k)\}$ (6.33)
3. *Magnitude:* $M(k) = M(N-k)$ (6.34)
4. *Phase:* $\phi(k) = -\phi(N-k)$ (6.35)

It is evident from the foregoing that to obtain the DFT of a real N-point sequence, it suffices to compute $(N+1)/2$ or $(N/2+1)$ frequency samples for N odd or even, respectively.

In Section 6.2, we saw that the finite sequence $x(n)$ can be considered to be one period of a periodic function with period N. Also, the DFT, $X(k) = X((k))$, is periodic with period N. Both $x(n)$ and $X(k)$ can be considered as N evenly spaced samples about two circles: one circle for time samples, the other for frequency samples. The following definitions apply to either. We can define an *even* function in this context as one that is symmetric about the zero point on the circle (see Figure 6.3)—that is,

$$x_{N-n} = x_n \tag{6.36}$$

An *odd* function is one that is anti-symmetric about the zero point—that is,

$$x_{N-n} = -x_n \tag{6.37}$$

PSyadj=-3pt

By these definitions, the magnitude and phase components of the DFT of a real sequence $x(n)$ are even and odd, respectively. That is,

$$M_{N-k} = M_k \tag{6.38}$$

and

$$\phi_{N-k} = -\phi_k = 2\,\ell\pi - \phi_k \tag{6.39}$$

in general, if we allow for the periodicity of the DFT. Here ℓ is integer, and for brevity we write M_k for $M(k)$ and ϕ_k for $\phi(k)$. We shall often use this shorter notation when there is no chance of confusion.

6.4.3 Circular Shifting

If $x(n), n = 0, \ldots, N-1$, is a finite sequence, then the circular shift of this sequence by n_0 samples, say, has meaning only if we consider $x(n)$ as one period of a periodic sequence $\tilde{x}(n) = x((n))$, as shown in Figure 6.4. We shift $\tilde{x}(n)$ through n_0 samples and recover the shifted finite sequence $x_s(n)$

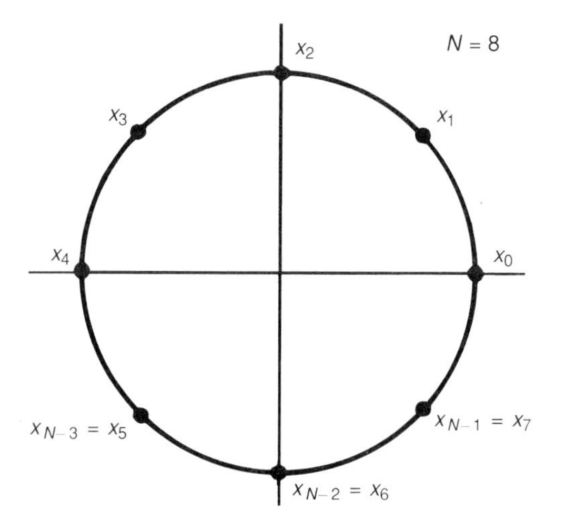

FIGURE 6.3 Evenly spaced samples for the definition of even and odd sequences. Even sequence: $x_n = x_{N-n}$. Odd sequence: $x_n = -x_{N-n}$.

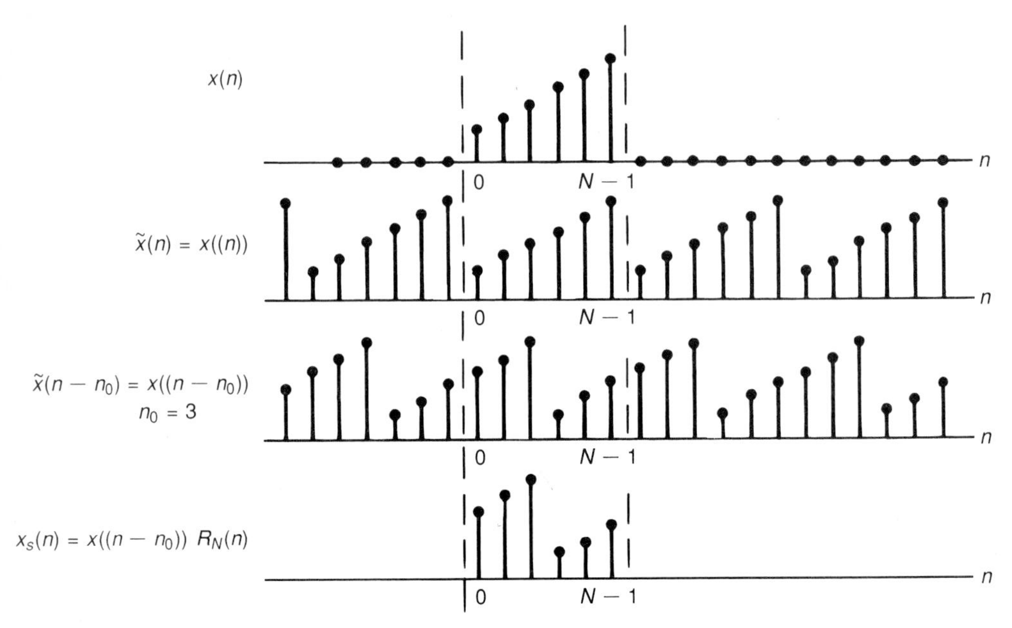

FIGURE 6.4 Circular shift of a finite sequence $x(n)$.

by multiplying $x((n - n_0))$ by the rectangular window sequence

$$R_N(n) = \begin{cases} 1, & 0 \leq n \leq N-1 \\ 0, & \text{otherwise} \end{cases} \tag{6.40}$$

as indicated in the figure. If

$$x_s(n) = x((n - n_0))R_N(n) \tag{6.41}$$

then the DFT of the shifted sequence is

$$DFT\{x_s(n)\} = e^{-jn_0 k\Omega}\tilde{X}(k)R_N(k) \tag{6.42}$$

where

$$\tilde{X}(k) = DFT\{\tilde{x}(n)\} \tag{6.43}$$

and

$$R_N(k) = \begin{cases} 1, & 0 \leq k \leq N-1 \\ 0, & \text{otherwise} \end{cases} \tag{6.44}$$

Proof

Let us first consider the DFT of the periodic sequence $\tilde{x}(n - n_0)$.

$$\text{DFT }\{\tilde{x}(n-n_0)\} = \sum_{n=0}^{N-1} \tilde{x}(n-n_0)e^{-jnk\Omega}$$

$$= e^{-jn_0 k\Omega} \sum_{n=0}^{N-1} \tilde{x}(n-n_0)e^{-j(n-n_0)k\Omega} \tag{6.45}$$

In the summation, let $m = n - n_0$. Then

$$\text{DFT } \{\tilde{x}(n - n_0)\} = e^{-jn_0 k\Omega} \sum_{m=-n_0}^{N-1-n_0} \tilde{x}(m) e^{-jmk\Omega} \tag{6.46}$$

Divide the summation into two parts.

$$\sum_{m=-n_0}^{N-1-n_0} x(m) e^{-jmk\Omega} = \sum_{m=-n_0}^{-1} \tilde{x}(m) e^{-jmk\Omega} + \sum_{m=0}^{N-1-n_0} \tilde{x}(m) e^{-jmk\Omega} \tag{6.47}$$

In the first summation on the right-hand side, let $m = r - N$. Then

$$\sum_{m=-n_0}^{-1} \tilde{x}(m) e^{-jmk\Omega} = \sum_{r=N-n_0}^{N-1} \tilde{x}(r - N) e^{-j(r-N)k\Omega} \tag{6.48}$$

Because $\tilde{x}$ is periodic, $\tilde{x}(r - N) = \tilde{x}(r)$. Also, $e^{-j(r-N)k\Omega} = e^{-jrk\Omega}$ because $e^{jNk\Omega} = 1$. Therefore, (6.48) becomes

$$\sum_{m=-n_0}^{-1} \tilde{x}(m) e^{-jmk\Omega} = \sum_{r=N-n_0}^{N-1} \tilde{x}(r) e^{-jrk\Omega} \tag{6.49}$$

Replace the index r by m in the right-hand side of this expression, and insert it in (6.47) to get

$$\begin{aligned} \sum_{m=-n_0}^{N-1-n_0} x(m) e^{-jmk\Omega} &= \sum_{m=N-n_0}^{N-1} \tilde{x}(m) e^{-jmk\Omega} + \sum_{m=0}^{N-1-n_0} \tilde{x}(m) e^{-jmk\Omega} \\ &= \sum_{m=0}^{N-1} \tilde{x}(m) e^{-jmk\Omega} \\ &= \tilde{X}(k) \end{aligned} \tag{6.50}$$

Therefore, from (6.46),

$$\text{DFT } \{\tilde{x}(n - n_0)\} = e^{-jn_0 k\Omega} \tilde{X}(k) \tag{6.51}$$

or

$$\tilde{X}_s(k) = e^{-jn_0 k\Omega} \tilde{X}(k) \tag{6.52}$$

where $\tilde{X}_s$ is the periodic extension of X_s. Now

$$X_s(k) = \tilde{X}_s(k) R_N(k) \tag{6.53}$$

Hence, combining (6.52) and (6.53), we get

$$X_s(k) = e^{-jn_0k\Omega}\tilde{X}(k)R_N(k) \tag{6.54}$$

Q.E.D.

6.4.4 Circular Convolution

In earlier chapters, we noted that the Fourier transform or the z-transform of the linear convolution of two time functions was simply the product of the transforms of the individual functions. A similar result holds for the DFT, but instead of a linear convolution of two sequences, we have a circular convolution. A circular convolution of two N-point sequences $x(n)$ and $h(n)$ is defined as

$$y(n) = \left[\sum_{m=0}^{N-1} x(m)h((n-m))\right] R_N(n) \tag{6.55}$$

For brevity, (6.55) is sometimes written

$$y(n) = x(n)\textcircled{N}h(n) \tag{6.56}$$

Let us consider what (6.55) means. The sequence $h((n-m))$ is a periodic extension of $h(n-m)$. The sequence $y(n)$ would also be periodic if we did not multiply by the function $R_N(n)$ defined in (6.40).

EXAMPLE 6.2

From the expression (6.55), find, graphically, the circular convolutions of the two finite sequences

$$x(n) = \begin{cases} 1, & n = 1, 2 \\ 0, & \text{otherwise} \end{cases}$$

and

$$h(n) = \begin{cases} a^n, & n = 0, 1, \ldots, N-1, a = 0.9, N = 5 \\ 0, & \text{otherwise} \end{cases}$$

Solution

In order for us to take the circular convolution, both sequences must be of the same length, $N = 5$ in this case. Therefore, we consider $x(n)$ to be of length 5 by "padding" it with three zeros, as shown in Figure 6.5(a). The convolution expression (6.55) requires the periodic extension of $h(m)$—that is, $\tilde{h}(m)$ or $h((m))$—as shown in Figure 6.5(c). Reflect this about the origin ($m = 0$) as shown in part (d) of the figure, to get $h((-m))$. The overlap of $h((-m))$ with $x(m)$ gives the $n = 0$ point on $y(n)$ [Figure 6.5(f)]. Shift $h((m))$ one sample to the right to get $h((1-m))$ as in Figure 6.5(e). The overlap of $h((1-m))$ with $x(m)$ gives the point $y(1)$ on part (f) of the

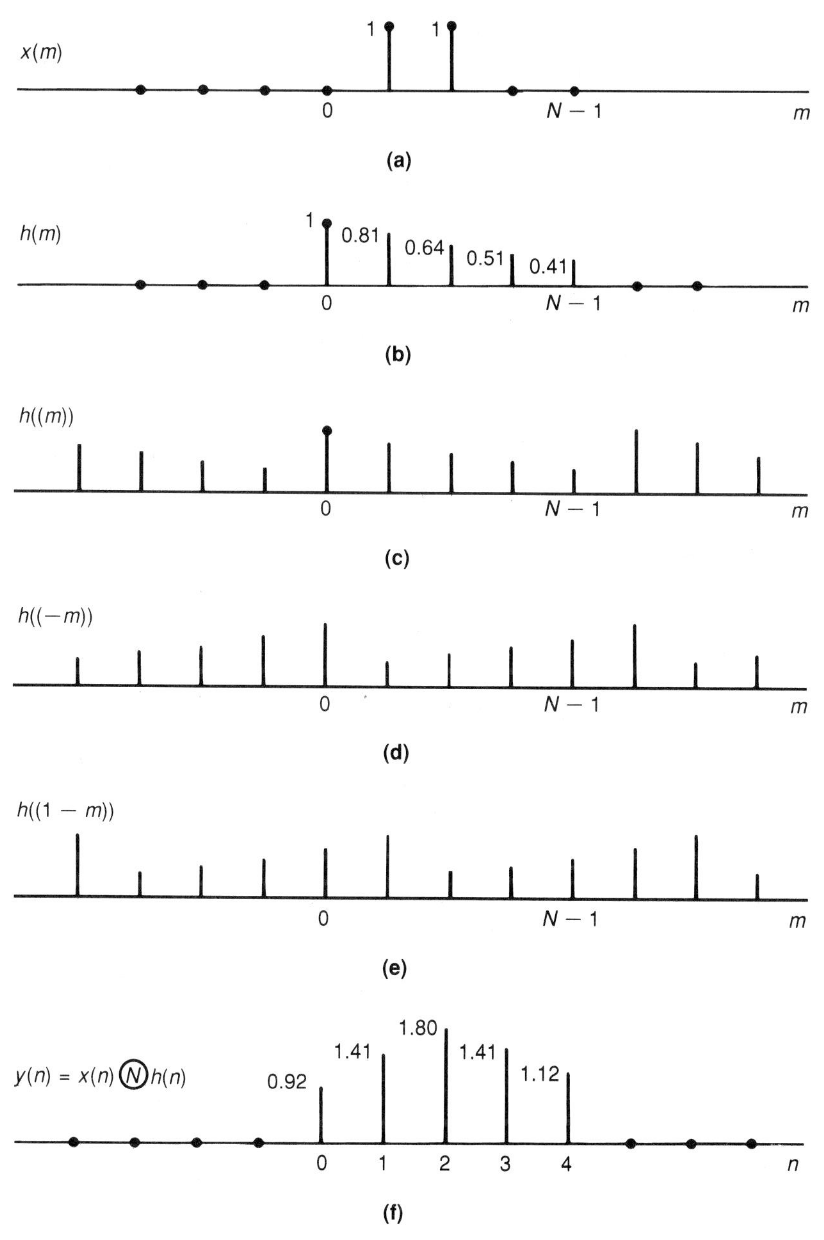

FIGURE 6.5 Circular convolution of two sequences (Example 6.2).

figure. By moving $h((1-m))$ one sample to the right, we get $h((2-m))$ and can compute $y(2)$. The process is continued until the sequence $y(n), n = 0, 1, \ldots, N-1$, is generated as shown in Figure 6.5(f).

The convolution theorem defines the relationship between circular-convolved sequences and their DFT.

Theorem 6.2 (Convolution Theorem for DFTs)

If

$$Y(k) = H(k)X(k) \tag{6.57}$$

then

$$y(n) = \left[\sum_{m=0}^{N-1} x(m)h((n-m))\right] R_N(n) \tag{6.58}$$

$$= \left[\sum_{m=0}^{N-1} h(m)x((n-m))\right] R_N(n) \tag{6.59}$$

where $x(n), h(n)$, and $y(n)$ are N-point sequences and $X(k), H(k)$, and $Y(k)$ are their DFTs.

Proof

We have

$$X(k) = \sum_{m=0}^{N-1} x(m)e^{-jmk\Omega}$$

and

$$H(k) = \sum_{\ell=0}^{N-1} h(\ell)e^{-jlk\Omega}$$

Then the periodic sequence

$$\begin{aligned} y((n)) &= \text{DFT}^{-1}\{Y(k)\} \\ &= \frac{1}{N}\sum_{k=0}^{N-1} Y(k)e^{jnk\Omega} \\ &= \frac{1}{N}\sum_{k=0}^{N-1} X(k)H(k)e^{jnk\Omega} \\ &= \frac{1}{N}\sum_{k=0}^{N-1}\left[\sum_{m=0}^{N-1} x(m)e^{-jmk\Omega}\right]\left[\sum_{\ell=0}^{N-1} h(\ell)e^{-j\ell k\Omega}\right] e^{jnk\Omega} \end{aligned}$$

$$= \frac{1}{N}\sum_{m=0}^{N-1} x(m)\sum_{\ell=0}^{N-1} h(\ell) \underbrace{\sum_{k=0}^{N-1} e^{jk\Omega(n-m-\ell)}}_{N\delta_{((n-m)),\ell} \text{ from (6.23)}}$$

$$= \sum_{m=0}^{N-1} x(m)h((n-m)) \tag{6.60}$$

It follows that the N-point sequence

$$y(n) = y((n))R_N(n)$$
$$= \left[\sum_{m=0}^{N-1} x(m)h((n-m))\right] R_N(n) \tag{6.61}$$

which was to be proved. The proof of (6.59) follows from a simple change of variables.

A summary of the properties of the DFT is given in Table 6.1.

TABLE 6.1 Summary of DFT Properties

Property	If	Then
1. Linearity	$x_3(n) = a_1x_1(n) + a_2x_2(n)$	$X_3(k) = a_1X_1(k) + a_2X_2(k)$
2. Symmetry	$x(n)$ is a real sequence with the DFT $X(k) = M(k)e^{j\phi(k)}$	$X(N-k) = X^*(k)$ $M(N-k) = M(k)$ $\phi(N-k) = -\phi(k)$
3. Circular shifting	$x_S(n) = x((n-n_0))R_N(n)$	$X_S(k) = e^{jn_0k}\tilde{X}(k)R_N(k)$
4. Circular convolution	$y(n) = \sum_{m=0}^{N-1} x(m)h((n-m))R_N(n)$	$Y(k) = H(k)X(k)$ (and conversely)
5. Inverse	$X(k) = \sum_{n=0}^{N-1} x(n)e^{-jnk\Omega}$, $k = 0, 1, \ldots, N-1$ $\Omega = 2\pi/N$	$x(n) = \frac{1}{N}\sum_{k=0}^{N-1} X(k)e^{jnk\Omega}$, $n = 0, 1, \ldots, N-1$

Notes:

1. $x(n)$, $n = 0, 1, \ldots, N-1$, is a finite sequence, real or complex.
2. $X(k) = \text{DFT}\{x(n)\} = \sum_{n=0}^{N-1} x(n)e^{-jnk\Omega}$, $k = 0, 1, \ldots, N-1$
3. $X^*(k)$ is the complex conjugate of $X(k)$.
4. $\tilde{x}(n)$ and $\tilde{X}(k)$ are periodic extensions of $x(n)$ and $X(k)$, respectively.
5. $x((n)) = \tilde{x}(n)$, where $((n)) = n$ modulo N

6.5 LINEAR CONVOLUTION FROM CIRCULAR CONVOLUTION

Usually, in practice, we are interested in the linear convolution of two finite sequences for signal-processing purposes. For example, one of the sequences could be a signal to be filtered, and the other sequence could represent the pulse response of a nonrecursive filter used to process the signal. From Theorem 6.2, we see that multiplying the DFT of both sequences and taking an inverse yields the circular convolution of the sequences. In order to get a linear convolution from a circular convolution, we must make certain modifications.

First, we note that the linear convolution of two N-point sequences is a sequence containing $(2N - 1)$ points, whereas there are only N distinct samples in the circular convolution of two N-point sequences. In general, the two sequences being convolved are of unequal length—say, N-point and M-point, respectively, with $M < N$. In that case, the linear convolution will contain $(N + M - 1)$ points. In order to take a circular convolution, it is necessary to make the sequences of equal length by adding $(N - M)$ zeros to the M-point sequence. The circular convolution then results in an N-point sequence that is $(M - 1)$ points shorter than that given by the linear convolution. We will use a simple example to illustrate the problem and to indicate how it can be solved.

EXAMPLE 6.3

Consider two finite-duration samples $x(n)$ and $h(n)$ defined as follows:

$$x(n) = R_N(n)$$
$$h(n) = R_M(n) \tag{6.62}$$

where $M = 8$ and $N = 20$. For simplicity in illustration, we have chosen unit sequences, but this does not change the principles involved. The 20-point DFTs of the sequences are computed and multiplied together. The inverse DFT, $s(n)$, of the product is evaluated. Specify which points in $s(n)$ correspond to points that would be obtained in a linear convolution of $x(n)$ and $h(n)$.

Solution

For a linear convolution of $x(n)$ and $h(n)$—that is, $y(n) = x(n) * h(n)$—we need at least $M + N - 1 = 8 + 20 - 1 = 27$ points. The linear convolution is shown graphically in Figure 6.6(a). If we take 20-point DFTs, we are short $M - 1 = 7$ points, so the circular convolution will contain aliasing, as shown in Figure 6.6(b). The aliasing involves the first $(M - 1)$ points as shown on the figure. If we make both sequences $x(n)$ and $h(n)$ the same length as the linear convolution (that is, $L = N + M - 1$) by appending $(M - 1)$ zeros

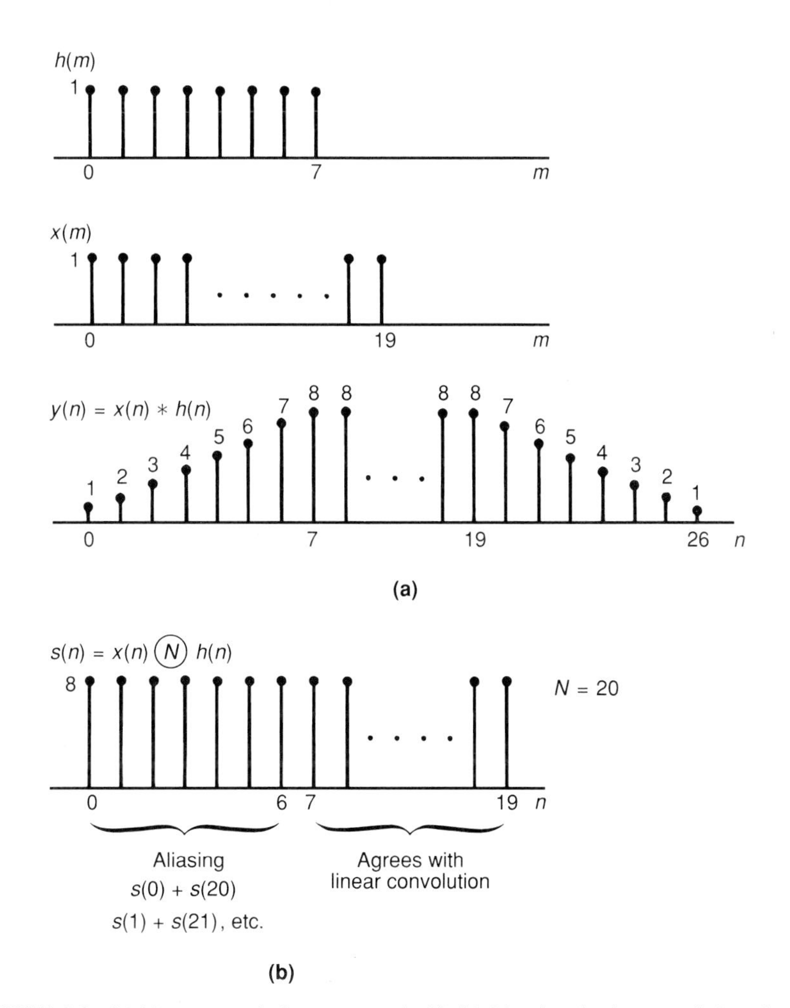

FIGURE 6.6 (a) Linear convolution compared with (b) 20-point circular convolution of two finite sequences.

to $x(n)$ and $(N - 1)$ zeros to $h(n)$, the L-point circular convolution of $x(n)$ and $h(n)$ is identical to their linear convolution. Demonstrating this is left as an exercise.

Thus we have a procedure for obtaining a linear convolution from the circular convolution of two finite sequences: Add zeros to the sequences so that each sequence has a length L, that is at least as long as the sequence corresponding to their linear convolution, and then take the L-point circular convolution of $x(n)$ and $h(n)$. If $h(n)$ is the pulse response of a nonrecursive filter used to filter $x(n)$, it is sometimes easier to carry out the filtering in the frequency domain. Take the L-point DFTs of the augmented sequences $x(n)$ and $h(n)$, multiply them together, and take the L-point IDFT of the product.

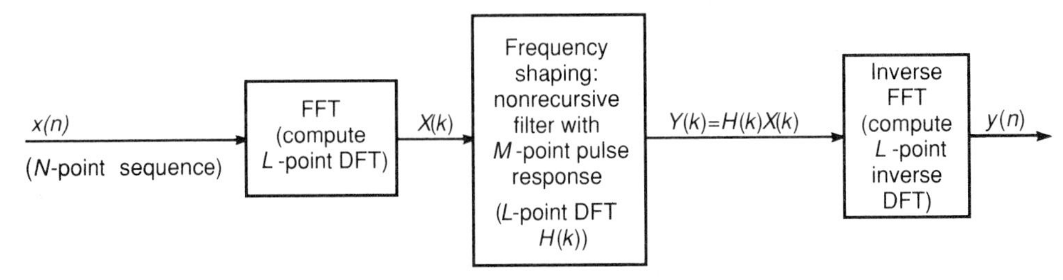

Note: L is an integer that is a power of 2 and is greater than or equal to $M + N - 1$.

FIGURE 6.7 Digital filtering via the DFT.

In practice, because the FFT is generally used for efficient evaluation of the DFT (see Section 6.7), it is desirable that L be a power of 2. Thus L can be taken as the smallest integer that is a power of 2 and not less than the number of points in the linear convolution of $x(n)$ and $h(n)$. The processing arrangement is shown schematically in Figure 6.7. Filtering will be discussed in more detail in Section 6.8.

6.6 PROCESSING LONG SEQUENCES

Suppose that a data sequence $x(n)$ of indefinite length is to be processed in a nonrecursive filter with pulse response $h(n)$ of length M by convolving the two sequences. Because of the length of the data sequence, it would not be practical to store it all before performing the linear convolution. Instead, the data sequence is divided up into shorter sections (see Figure 6.8). These

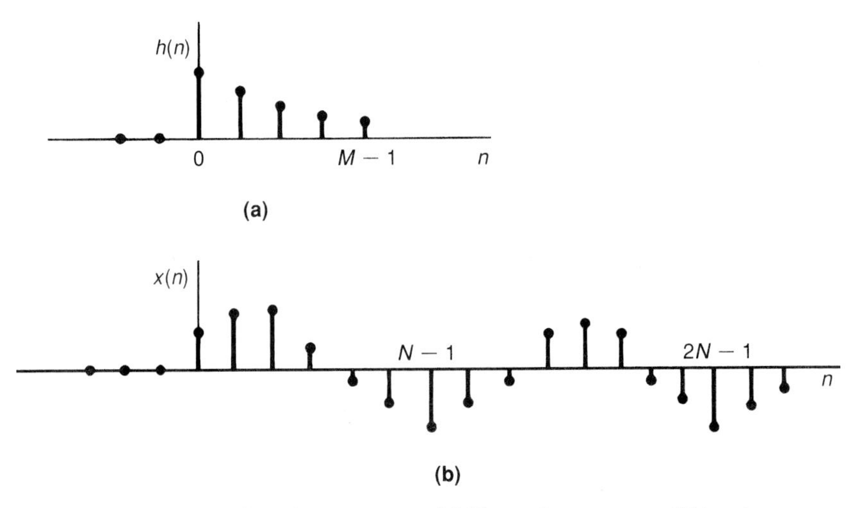

FIGURE 6.8 Filtering a long data sequence. (a) Filter pulse response. (b) Input sequence.

are processed separately and the results combined later to yield the desired filtered data.

Two alternative methods are commonly used for filtering the sectioned data and combining the results [2]. These are the overlap–add method and the overlap–save method. We will consider each in turn.

6.6.1 Overlap–Add

Divide the data into nonoverlapping sections $x_i(n), i = 0, 1, 2, \ldots,$ each containing N points. The linear convolution of $h(n)$ with any section $x_i(n)$ is the same as the L-point circular convolution—that is,

$$x_i(n) \ * \ h(n) = x_i(n) \textcircled{L} h(n) \tag{6.63}$$

where $L = M + N - 1$. The circular convolution requires the augmentation of $h(n)$ and $x_i(n)$ with zeros, as discussed in Section 6.5. Then obtain the filtered output for the data sequence by adding together the partial results (6.63):

$$x(n) \ * \ h(n) = \sum_{i=0}^{P-1} x_i(n) \ * \ h(n) \tag{6.64}$$

where P is the number of sections. The summation indicated by (6.64) can be done on a running basis as the data are received. The term *overlap–add* reflects the fact that each filtered section (6.63) overlaps the next filtered section by $(M - 1)$ points and the fact that they are added together in (6.64). The overlap–add operation is depicted in Figure 6.9.

6.6.2 Overlap-Save

In this method, the data sequence is again divided into N-point sections $x_i(n)$, but this time the sections overlap by $(M - 1)$ points, where M is the number of points in the filter pulse response. If we take an N-point circular convolution of $x_i(n)$ with $h(n)$, the first $(M - 1)$ points will not agree with the linear convolution of $x_i(n)$ and $h(n)$ because of aliasing (see Example 6.3); the remaining $(N - M + 1)$ points, however, will agree with the linear convolution. Hence we discard the first $(M - 1)$ points of the filtered section $x_i(n) \textcircled{N} h(n)$. Because of the overlap in the sections $x_i(n)$, the previous filtered section provides the missing $(M - 1)$ points ("saved" from the prior section) when the filtered sections are abutted together. The only filtered section that does not have a prior section is the first, so $(M - 1)$ points are missing from the filtered data when this method is used. Figure 6.10 illustrates the overlap–save technique. In practice, the method can be modified at the beginning of the data sequence so that no filtered data are lost.

Which method is used is a matter of choice. Both require about the same amount of computation.

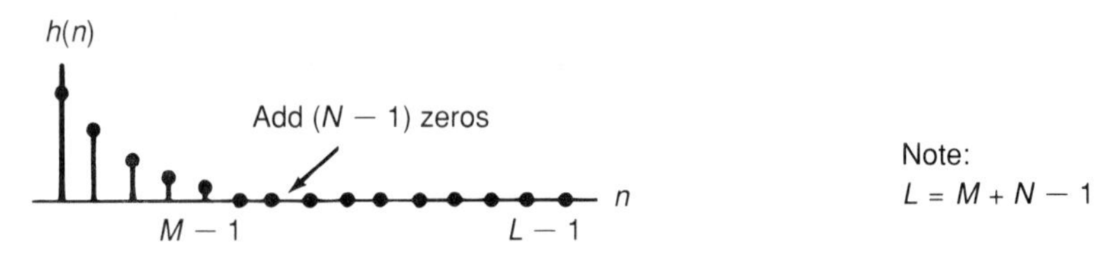

(a) Filter pulse response

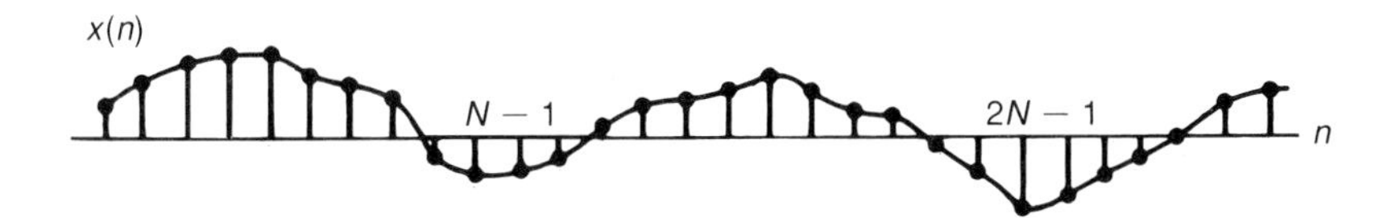

(b) Sequence to be filtered

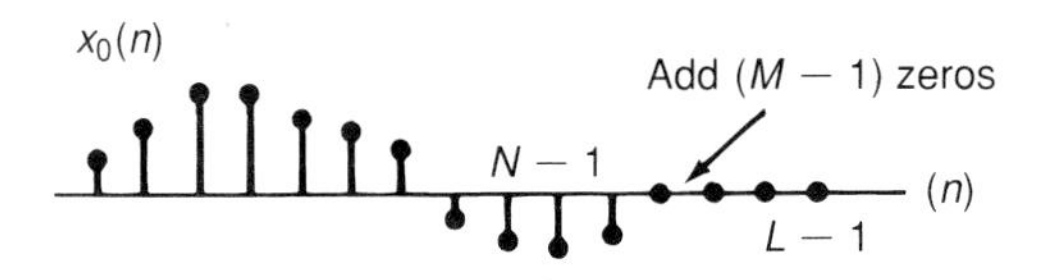

(c) First subsequence

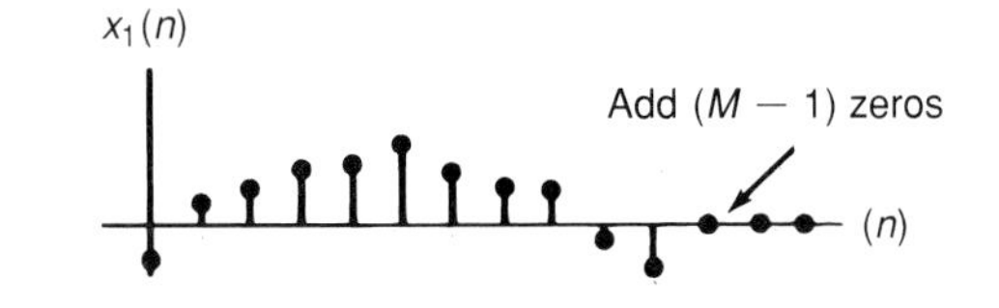

(d) Second subsequence

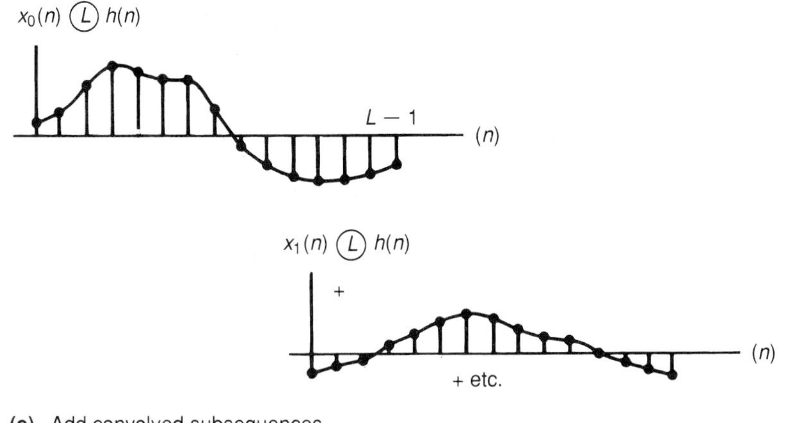

(e) Add convolved subsequences

FIGURE 6.9 The overlap–add method for filtering long sequences.

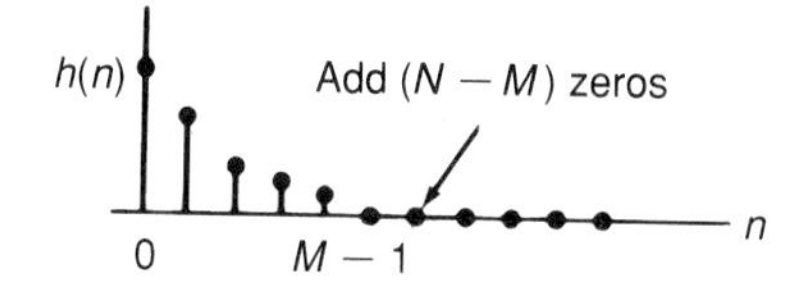

(a) Filter pulse response

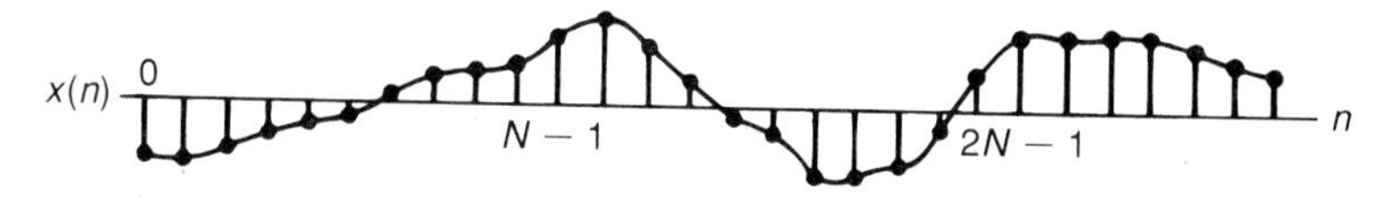

(b) Sequence to be filtered

$x_0(n)$ 0 $N-1$ n

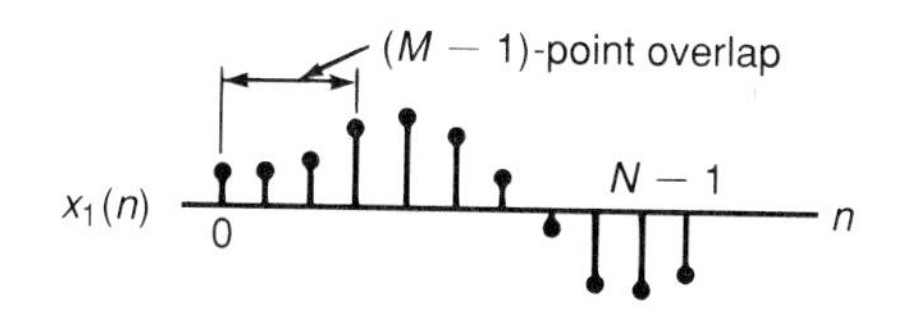

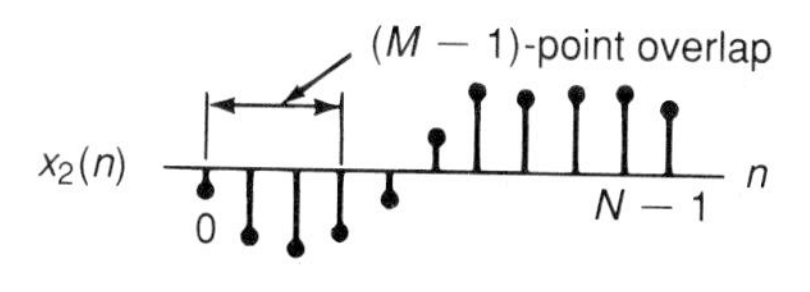

(c) Overlapping subsequences

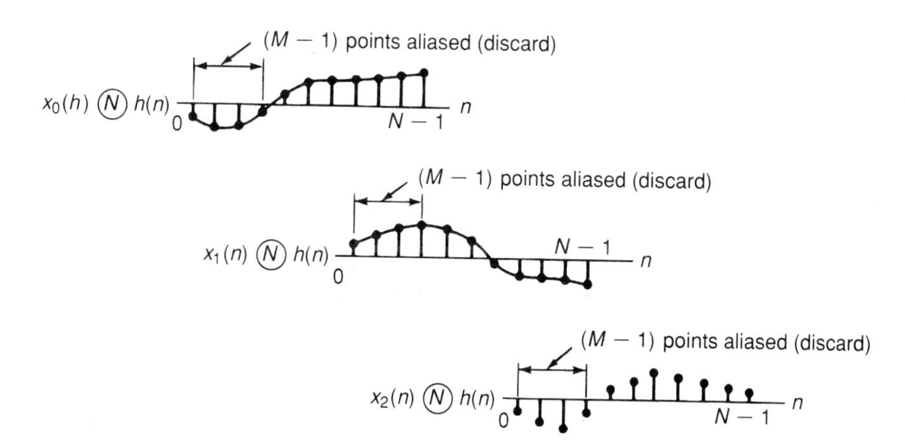

(d) Convolved subsequences with discarded points

FIGURE 6.10 The overlap–save method for filtering long sequences.

6.7 THE FAST FOURIER TRANSFORM

Often, the amount of computation required in practice, determines whether a particular signal-processing algorithm is used. When a complex multiplication is carried out on a digital machine, it requires four real multiplications and two real additions. A complex addition requires two real additions. A complex multiplication plus a complex addition (that is, four real multiplications and four real additions) will be called an operation. Consider the DFT of a finite complex sequence $f(nT)$—for example, two signals in quadrature:

$$F(k) = \sum_{n=0}^{N-1} f(nT)e^{-jnk\Omega T}, \quad k = 0, 1, \ldots, N - 1 \tag{6.65}$$

It is evident that direct evaluation of (6.65) would require on the order of N^2 operations. Thus it would take more than a million operations to compute the DFT of a sequence with more than 1000 samples! This would severely limit the application of the DFT to short sequences were it not for the development of a number of algorithms collectively called fast Fourier transforms (FFT) [4], [5], [6]. These transforms provide an efficient means of evaluating the DFT by eliminating redundant operations. For example, if N is a power of 2, then the FFT makes it possible to calculate the DFT with $N \log_2 N$ operations instead of N^2 operations. For $N = 1024$, this implies about 10^4 instead of 10^6 operations—a saving of 99%. The FFT can also be applied to advantage in cases where N is a power of some integer other than 2 or has arbitrary factors. However, because it is efficient, is easy to understand, and was the first widely publicized FFT proposed, most attention has been focused on the case where N is a power of 2. (As we noted earlier, the given sequence may be augmented with zeros, if necessary, to ensure this.) We will limit our treatment to the case where N is a power of 2.

There are basically two classes of FFT algorithms, each with many modifications. These classes are decimation in time, and decimation in frequency. In decimation in time, the sequence for which we need the DFT is successively divided up into smaller sequences, and the DFTs of these subsequences are combined in a certain pattern to yield the required DFT of the entire sequence with much fewer operations. In the decimation-in-frequency approach, the frequency samples of the DFT are decomposed into smaller and smaller subsequences in a similar manner. We will illustrate both methods with a power-of-2-length sequence. We require the DFT of the N-point sequence $f(nT)$,

$$F_k = \sum_{n=0}^{N-1} f_n W^{nk}, \quad k - 0, 1, \ldots, N - 1 \tag{6.66}$$

where for brevity we have written F_k for $F(k)$, f_n for $f(nT)$, and W for

$e^{-j\Omega}$. Note that $W = e^{-j2\pi/N}$ depends on $N = 2^b$, where $b > 0$ is an integer. If, for example, we were to consider the DFT of a sequence of length $N/2$, then W would be replaced by W^2 in (6.66), and the summation limit would be $N/2 - 1$.

6.7.1 Decimation in Time

By this method of evaluating (6.65), the sequence $f_n, n = 0, 1, \ldots, N-1$ is first divided into two shorter interwoven sequences, g_n and h_n, which contain the even-numbered and odd-numbered samples of f_n, respectively. That is,

$$g_n = f_{2n}, \quad n = 0, 1, \ldots, \frac{N}{2} - 1 \tag{6.67}$$

$$h_n = f_{2n+1}, \quad n = 0, 1, \ldots, \frac{N}{2} - 1 \tag{6.68}$$

For example, for an 8-point sequence f_n, g_n contains samples f_0, f_2, f_4, f_6, and h_n has samples f_1, f_3, f_5, f_7, as shown in Figure 6.11.

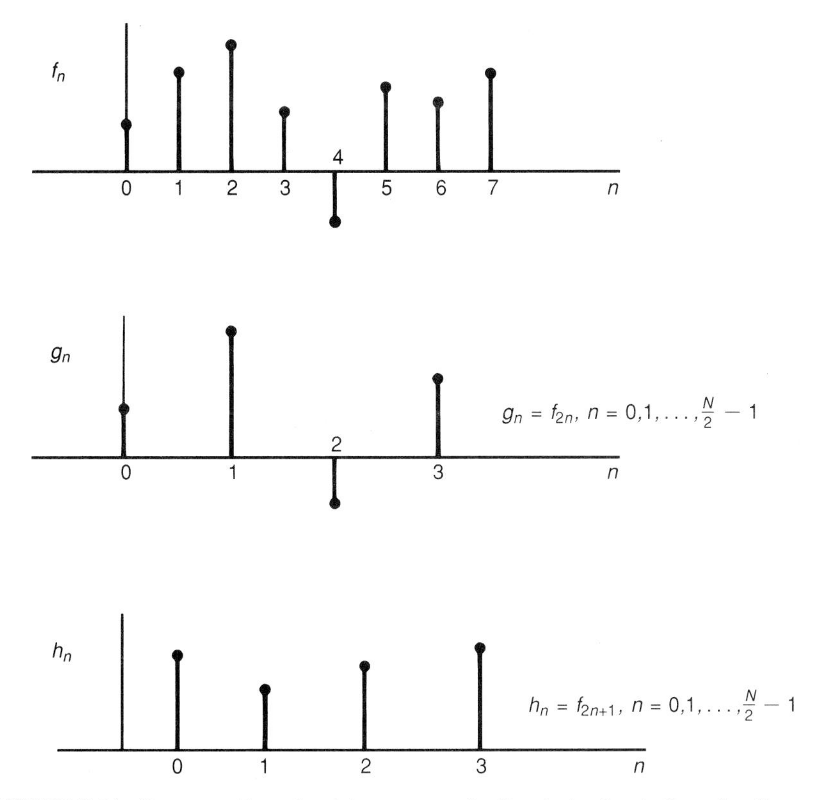

FIGURE 6.11 Decomposition of a data sequence for the decimation-in-time algorithm.

The DFTs of these subsequences may be written

$$G_k = \sum_{n=0}^{N/2-1} g_n (W^2)^{nk}, \quad k = 0, 1, \ldots, \frac{N}{2} - 1 \tag{6.69}$$

$$H_k = \sum_{n=0}^{N/2-1} h_n (W^2)^{nk}, \quad k = 0, 1, \ldots, \frac{N}{2} - 1 \tag{6.70}$$

We can express the DFT of the entire sequence in terms of G_k and H_k as follows:

$$\begin{aligned} F_k &= \sum_{n=0}^{N-1} f_n W^{nk} \\ &= \sum_{n=0}^{N/2-1} [g_n W^{2nk} + h_n W^{(2n+1)k}] \\ &= \sum_{n=0}^{N/2-1} g_n W^{2nk} + W^k \sum_{n=0}^{N/2-1} h_n W^{2nk} \\ &= G_k + W^k H_k, k = 0, 1, \ldots, N-1 \end{aligned} \tag{6.71}$$

In (6.71) index k runs from zero to $N-1$, whereas G_k and H_k have period $N/2$. Hence we may write (6.71) as

$$F_k = \begin{cases} G_k + W^k H_k, & 0 \leq k \leq \frac{N}{2} - 1 \\ G_{k-N/2} + W^k H_{k-N/2}, & \frac{N}{2} \leq k \leq N-1 \end{cases} \tag{6.72}$$

The result of this first step in applying the DFT for $N = 8$ is shown in Figure 6.12. As the foregoing reduction is carried out, G_k and H_k require on the order of $(N/2)^2$ operations each, and N operations are required for multiplying H_k by W^k and adding $W^k H_k$ to G_k. This gives a total of approximately $N + N^2/2$ operations, compared with N^2 operations for direct evaluation of the DFT. Thus there is a saving in computation after only one stage.

To proceed with the method, the $N/2$-point sequences g_n and h_n are both decomposed into two $N/4$-point sequences by taking the even- and odd-numbered samples, as was done with f_n. Thus

$$\begin{aligned} p_n &= g_{2n}, \quad n = 0, 1, \ldots, \frac{N}{4} - 1 \\ q_n &= g_{2n+1}, \quad n = 0, 1, \ldots, \frac{N}{4} - 1 \end{aligned} \tag{6.73}$$

and

$$\begin{aligned} r_n &= h_{2n}, \quad n = 0, 1, \ldots, \frac{N}{4} - 1 \\ s_n &= h_{2n+1}, \quad n = 0, 1, \ldots, \frac{N}{4} - 1 \end{aligned} \tag{6.74}$$

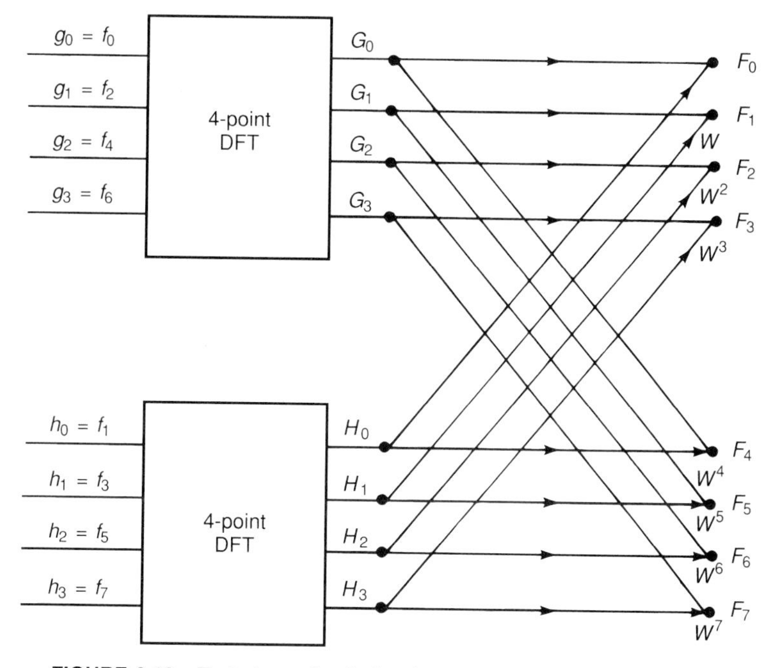

FIGURE 6.12 First stage of a decimation-in-time FFT reduction for $N = 8$.

By the same reasoning as for the first stage, it follows that

$$G_k = \begin{cases} P_k + W^{2k} Q_k, & 0 \le k \le \frac{N}{4} - 1 \\ P_{k-N/4} + W^{2k} Q_{k-N/4}, & \frac{N}{4} \le k \le \frac{N}{2} - 1 \end{cases} \tag{6.75}$$

and

$$H_k = \begin{cases} R_k + W^{2k} S_k, & 0 \le k \le \frac{N}{4} - 1 \\ R_{k-N/4} + W^{2k} S_{k-N/4}, & \frac{N}{4} \le k \le \frac{N}{2} - 1 \end{cases} \tag{6.76}$$

where $P_k, Q_k, R_k,$ and S_k are the $N/4$-point DFTs of $p_n, g_n, r_n,$ and s_n, respectively.

The total number of operations is now reduced to approximately

$$4 \times \left(\frac{N}{4}\right)^2 + N + N = \frac{N^2}{4} + 2N$$

The operations of the first and second steps are depicted in Figure 6.13 for $N = 8$. Note that G_k and H_k are combined as shown in Figure 6.12.

The pattern now becomes apparent. Each of the sequences $p_n, g_n, r_n,$ and s_n is decomposed into two sequences with $N/8$ samples, and these in turn are divided into $N/16$-sample sequences, and so on. The process

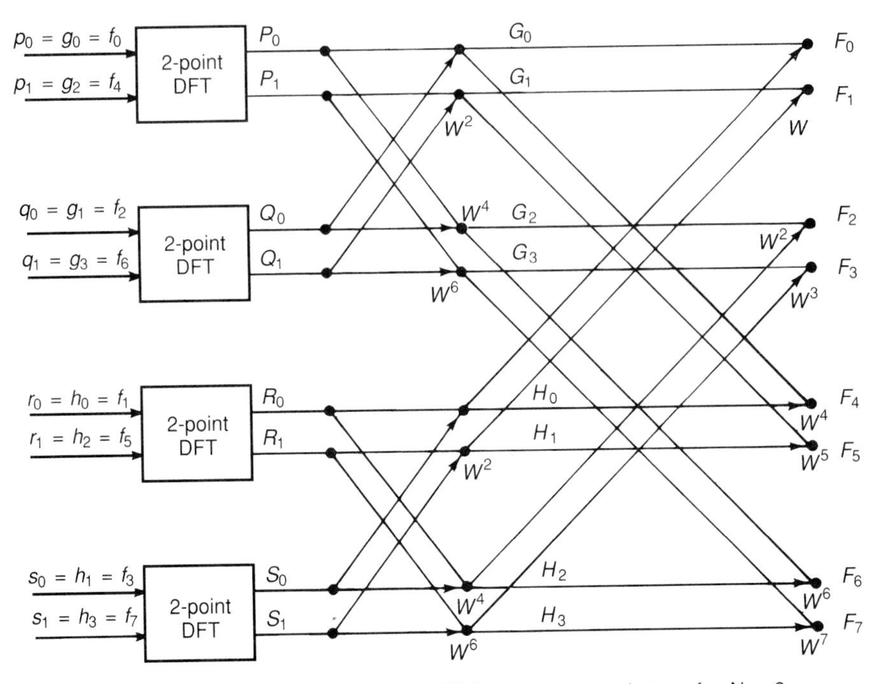

FIGURE 6.13 Decimation-in-time FFT after the second stage for $N = 8$.

ends when the original N-point sequence f_n is divided up into N 1-point "sequences" with a 1-point DFT computed for each. These are combined in an appropriate manner to yield F_k. Combining the DFTs of subsequences with fewer points reduces the redundant operations inherent in the direct computation of an N-point DFT. This is the basis for the FFT. When the DFT computation of the power-of-2-length sequence has been completely reduced to a combination of complex multiplications and additions, the ultimate saving in computation is achieved (approximately $N \log_2 N$ operations are required). For the 8-point example, 3 steps are required to reach this stage, and the operations are combined as shown in Figure 6.14. The resulting number of operations is about $N \log_2 N = 24$, as compared with the $N^2 = 64$ operations required for direct computation of the DFT. Even for this short sequence, then, considerable savings are effected by the FFT.

From Figure 6.14 it is evident that the output sequence is in natural order—$F_0, F_1, F_2, F_3, F_4, F_5, F_6, F_7$—whereas the input sequence is not. In fact, the input sequence $f_0, f_4, f_2, f_6, f_1, f_5, f_3, f_7$ is in bit-reversed order, as can be seen in Table 6.2. If the input is in natural order the output will be in bit-reversed order. We can either process the input sequence with a bit-reversal algorithm to get the output sequence in natural order, or let the input sequence be in natural order and combine the operations of the FFT so as to obtain the output in natural order.

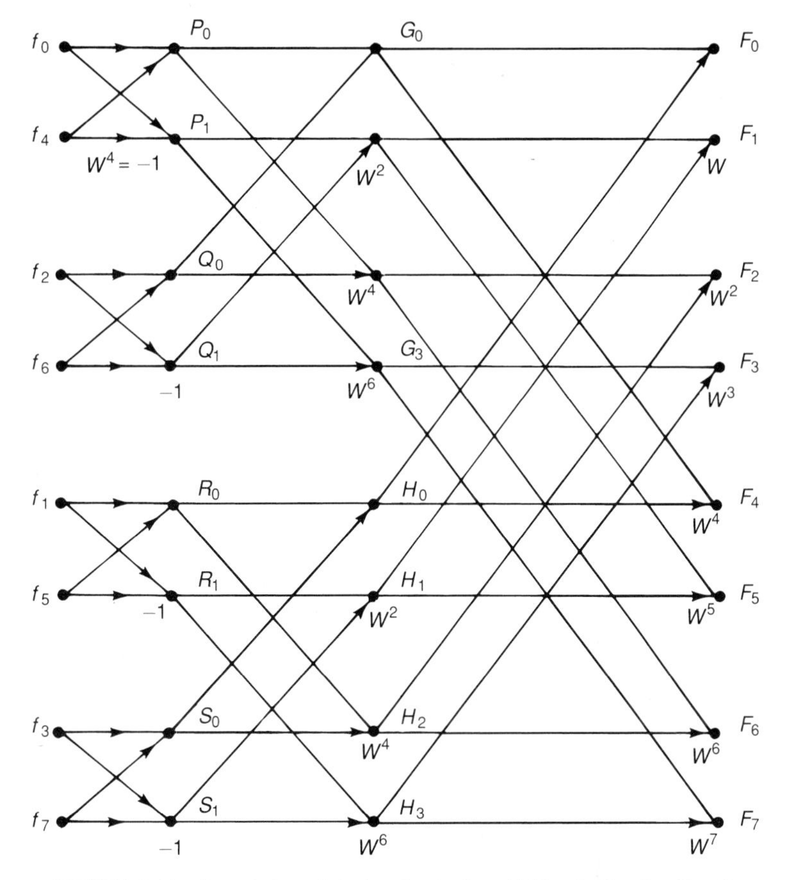

FIGURE 6.14 Completion of decimation-in-time FFT reduction for $N = 8$.

TABLE 6.2 Bit Reversal

Index	Binary Representation	Bit-Reversed Order	Bit-Reversed Index
0	000	000	0
1	001	100	4
2	010	010	2
3	011	110	6
4	100	001	1
5	101	101	5
6	110	011	3
7	111	111	7

6.7.2 Decimation in Frequency

To illustrate this version of the FFT, we again assume that the sequence for which we need the DFT has, when augmented with zeros if necessary, N samples, where N is a power of 2. The sequence is divided into two sequences g_n and h_n, where g_n contains the first $N/2$ samples of f_n, and h_n contains the remaining samples. Thus

$$g_n = f_n, \quad n = 0, 1, \ldots, \frac{N}{2} - 1 \tag{6.77}$$

$$h_n = f_n + \frac{N}{2}, \quad n = 0, 1, \ldots, \frac{N}{2} - 1 \tag{6.78}$$

Then the N-point DFT of f_n is

$$\begin{aligned} F_k &= \sum_{n=0}^{N-1} f_n W^{nk} \\ &= \sum_{n=0}^{N/2-1} f_n W^{nk} + \sum_{n=N/2}^{N-1} f_n W^{nk} \\ &= \sum_{n=0}^{N/2-1} g_n W^{nk} + \sum_{n=0}^{N/2-1} h_n W^{(n+N/2)k} \\ &= \sum_{n=0}^{N/2-1} [g_n + (-1)^k h_n] W^{nk} \quad k = 0, 1, \ldots, N-1 \end{aligned} \tag{6.79}$$

because $W^{N/2} = e^{-j\pi} = -1$.

Now divide the sequence F_k into the even-numbered and the odd-numbered samples (this accounts for the term decimation-in-frequency). For the even-numbered frequencies,

$$F_{2k} = \sum_{n=0}^{N/2-1} (g_n + h_n)(W^2)^{nk}, \quad k = 0, 1, \ldots, \frac{N}{2} - 1 \tag{6.80}$$

and for the odd-numbered frequencies,

$$\begin{aligned} F_{2k+1} &= \sum_{n=0}^{N/2-1} (g_n - h_n) W^{(2k+1)n} \\ &= \sum_{n=0}^{N/2-1} [(g_n - h_n) W^n](W^2)^{nk}, \quad k = 0, 1, \ldots, \frac{N}{2} - 1 \end{aligned} \tag{6.81}$$

Equations (6.80) and (6.81) will be recognized as the $N/2$-point DFTs of $(g_n + h_n)$ and $(g_n - h_n)W^n$, respectively.

Similarly, we can subdivide each $N/2$-point DFT into two $N/4$-point DFTs and continue until finally we have N 1-point DFTs. The complete process is illustrated for $N = 8$ in Figure 6.15. The output sequence is in bit-reversed order for input in natural order, but, as for the decimation-in-

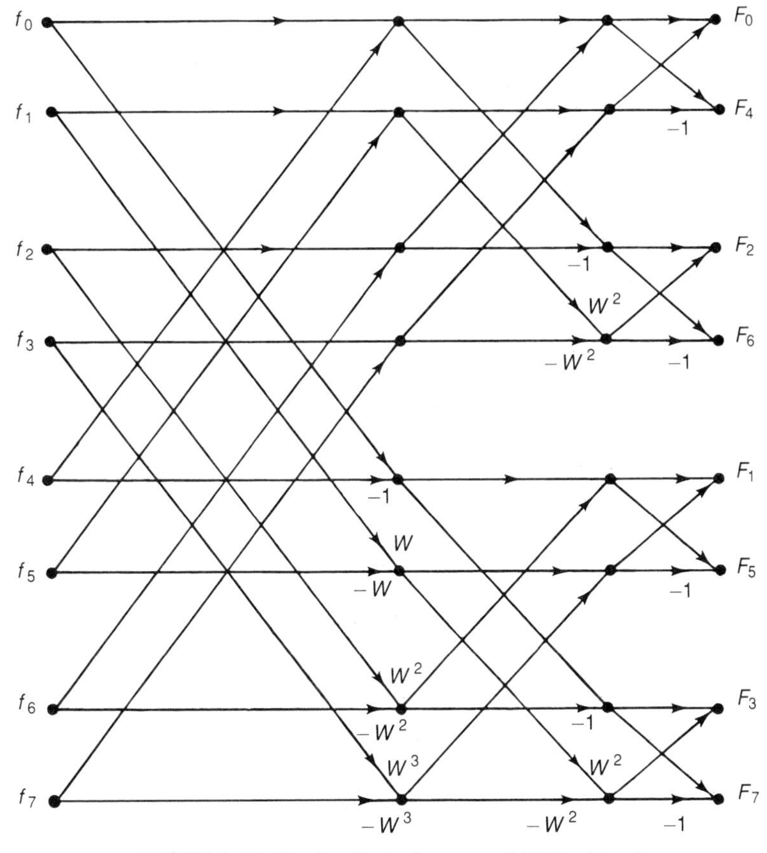

FIGURE 6.15 Decimation-in-frequency FFT for $N = 8$.

time algorithm, obtaining both input and output in natural order does not pose any particular problem.

The number of operations the decimation-in-frequency algorithm entails is also $N \log_2 N$ instead of the N^2 operations required for direct evaluation of the DFT. A computer program for obtaining the DFT of a finite sequence via the FFT is given on the diskette.

6.8 USE OF THE DISCRETE FOURIER TRANSFORM FOR DIGITAL-FILTER DESIGN

6.8.1 Introduction

As the foregoing development implies, the DFT lends itself readily to the design of digital filters with finite pulse responses. Suppose that an Nth-order filter is specified, with a desired frequency response $H(j\omega T)$. Take

N equally spaced samples $H_k, k = 0, 1, \ldots, N-1$, of this frequency characteristic about the unit circle, or, equivalently, sample the characteristic at intervals of $\Omega = 2\pi/NT$ along the frequency axis. Three alternative approaches are now feasible.

1. Take the IDFT according to (6.25) with $H_k = H(k)$ to obtain the finite pulse response $h(n), n = 0, 1, \ldots, N-1$. Then obtain the filter output for a given input sequence $x(n)$ by linearly convolving $x(n)$ with $h(n)$, as described in Section 6.5 for example.
2. A second method is to carry out the filtering entirely in the frequency domain by using the frequency samples H_k, as described in Section 6.8.2.
3. Although methods 1 and 2 both use the frequency samples H_k, the term *frequency sampling* is generally applied to the third method, by which we seek a *continuous* frequency characteristic to approximate the desired response $H(j\omega T)$ by using the samples H_k. This method is treated in Section 6.8.3.

6.8.2 Filtering Entirely in the Frequency Domain

Use an FFT algorithm to take an N-point DFT of the discrete signal $x(n)$ to be filtered. Multiply the resulting frequency samples $X_k = X(k)$ by N evenly spaced samples H_k of the *desired* filter response $H(j\omega T)$. The output Y_k, which gives the filtered frequency samples, is transformed to the time domain by taking the IDFT via an FFT algorithm. (This process is summarized in Figure 6.16.) The value of N is taken large enough to ensure that the product $X_k H_k$ corresponds to a linear convolution of the underlying sequences, as discussed in Section 6.5.

Because the filtering is carried out entirely in the frequency domain, it is not necessary, in this case, to evaluate the filter coefficients. It is evident that the method is better suited to the processing of stored data than to real-time operation.

6.8.3 Frequency Sampling

This method uses the DFT to produce a finite pulse response filter that has a continuous frequency response that is equal to the desired frequency response $H(j\omega T)$ at the equally spaced frequency sampling instances $k\Omega, k =$

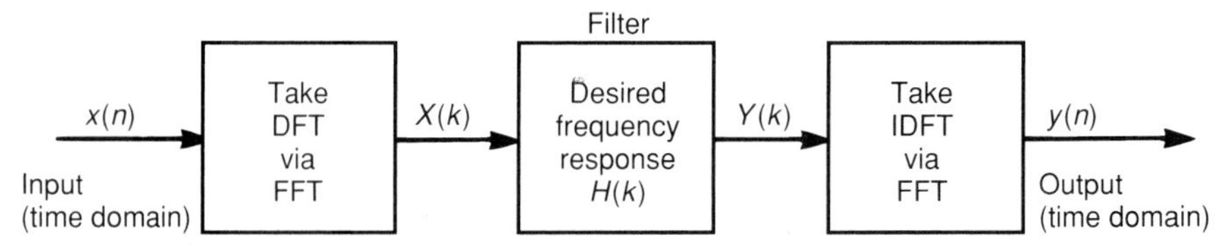

FIGURE 6.16 Filtering in the frequency domain.

$0, 1, 2, \ldots, N-1$, and approximates the desired response between sampling instances. Let

$$H_k = M_k e^{j\phi_k} \tag{6.82}$$

be samples of the desired frequency response at sampling instances $k\Omega$, where M_k is the magnitude and ϕ_k is the phase.

Take the IDFT of H_k to get the unit pulse response

$$h_n = \frac{1}{N}\sum_{k=0}^{N-1} H_k e^{jnk\Omega}, \quad n = 0, 1, \ldots, N-1 \tag{6.83}$$

Setting $a_k = e^{jk\Omega}$ for convenience, the transfer function of the filter is

$$\begin{aligned} H(z) &= \sum_{n=0}^{N-1} h_n z^{-n} \\ &= \sum_{n=0}^{N-1}\left[\frac{1}{N}\sum_{k=0}^{N-1} H_k a_k^n\right] z^{-n} \\ &= \frac{1}{N}\sum_{k=0}^{N-1} H_k \sum_{n=0}^{N-1}(a_k z^{-1})^n \\ &= \frac{1}{N}\sum_{k=0}^{N-1} H_k \frac{1-z^{-N}}{1-a_k z^{-1}} \end{aligned} \tag{6.84}$$

Thus the transfer function can be found from the frequency samples H_k. Because the right side of (6.84) has nonzero poles, $H(z)$ has a recursive form even though the pulse response (6.83) is finite. However, if we write (6.84) as

$$H(z) = \frac{z^N - 1}{N z^{N-1}} \sum_{k=0}^{N-1} \frac{H_k}{z - a_k} \tag{6.85}$$

it is evident that the N zeros from $(z^N - 1)$ cancel the N poles $z = a_k$, so we are left with

$$H(z) = \sum_{k=0}^{N-1} H_k \sum_{\substack{i=0 \\ i\neq k}}^{N-1} (z - a_i)/(N z^{N-1}) \tag{6.86}$$

where $a_i = e^{ji\Omega}$—that is, a nonrecursive form—if the pole-zero cancellation is carried out. This is similar to the moving-average filter and modified comb filters we discussed in Chapter 5, which could be realized either recursively or nonrecursively, depending on whether pole-zero cancellations were made. In any of these cases there is no stability problem, because the effective (noncanceled) poles are at the origin of the z-plane.

To get the frequency response of the filter, let $z = e^{j\omega T}$ in (6.84). For brevity, we continue to let $T = 1$. This does not change the principles involved; a non-unity value for T can be inserted when desired. Then

$$\begin{aligned}
H(j\omega T) &= \frac{1}{N}\sum_{k=0}^{N-1} H_k \frac{1 - e^{-jN\omega}}{1 - e^{jk\Omega}e^{-j\omega}} \\
&= \frac{1}{N}\sum_{k=0}^{N-1} H_k \frac{1 - e^{-jN(\omega - k\Omega)}}{1 - e^{-j(\omega - k\Omega)}} \quad (\text{because } e^{jkN\Omega} = 1) \\
&= \frac{1}{N}\sum_{k=0}^{N-1} H_k \frac{e^{-jN(\omega - k\Omega)/2}\sin[N(\omega - k\Omega)/2]}{e^{-j(\omega - k\Omega)/2}\sin[(\omega - k\Omega/2)]} \\
&= \frac{e^{-jN\omega/2}}{e^{-j\omega/2}N}\sum_{k=0}^{N-1} H_k \frac{e^{jNk\Omega)/2}}{e^{jk\Omega/2}}\frac{\sin[N(\omega - k\Omega)/2]}{\sin[(\omega - k\Omega)/2]} \\
&= \frac{e^{-j(N-1)\omega/2}}{N}\sum_{k=0}^{N-1} H_k e^{j(N-1)k\Omega/2}\frac{\sin[N(\omega - k\Omega)/2]}{\sin[(\omega - k\Omega)/2]}
\end{aligned} \tag{6.87}$$

This indicates that the frequency response of the filter is a linear combination of the frequency samples H_k with frequency interpolation functions of the form

$$S(\omega, k) = \frac{\sin[N(\omega - k\Omega)/2]}{\sin[(\omega - k\Omega)/2]} \tag{6.88}$$

The oscillatory nature of the interpolation $S(\omega, k)$ indicates that between the frequency samples H_k, the designed filter will have ripples that cause errors relative to the desired frequency response $H(j\omega T)$. Some ripple cancellation in adjacent frequency bands results from the superposition of the interpolation functions implied in (6.87). For some purposes, this cancellation will produce a satisfactory design without further modifications.

If it is desired to reduce the interpolation error further, some of the frequency samples can be made unconstrained variables. The values of the unconstrained variables are optimized to minimize some simple function of the approximation error—the peak error, for instance. The unconstrained variables might be chosen as frequency samples lying in a translation region between, say, a passband and stopband in which the response is specified (see [3]).

6.8.4 Filter Design by the Frequency Sampling Method

We will now investigate the conditions that must be met by a practical filter designed via frequency sampling. Suppose that the pulse response $h_n, n = 0, 1, 2, \ldots, N-1$, of the filter is to be a real-valued signal. This implies that the frequency response samples $H_k = \text{DFT}[h_n]$ must satisfy the symmetry requirement

$$H_{N-k} = H_k^* \tag{6.89}$$

where H_k^* is the complex conjugate of H_k. As noted under property 2 in Section 6.4, (6.89) implies that for a real signal h_n with frequency response sequence H_k, the magnitude response is an even function,

$$M_{N-k} = M_k \tag{6.90}$$

and the phase is an odd function,

$$\phi_{N-k} = 2\ell\pi - \phi_k \tag{6.91}$$

where ℓ is an integer. As we also noted in that section, it follows from conditions (6.89) and (6.90) that the frequency response H_k is uniquely specified by $(N+1)/2$ samples for N odd and by $(N/2+1)$ samples for N even.

If the filter is to be linear-phase, then h_n must also satisfy the symmetry requirement

$$h_n = h_{N-n-1} \tag{6.92}$$

with constant phase delay

$$\alpha = \frac{N-1}{2} \tag{6.93}$$

(see Section 6.4). Let us determine what effect condition (6.92) has on the filter phase response ϕ_k. We have

$$\begin{aligned} h_n &= \frac{1}{N}\sum_{k=0}^{N-1}[M_k e^{j\phi_k}]e^{jnk\Omega} \\ &= h^*_{N-n-1} \end{aligned} \tag{6.94}$$

because h_n is real. But

$$\begin{aligned} \text{DFT }[h^*_{N-n-1}] &= \sum_{n=0}^{N-1} h^*_{N-n-1}e^{-jnk\Omega} \\ &= \sum_{n=0}^{N-1}[h^*_{N-n-1}e^{j(N-n-1)k\Omega}]e^{-j(N-1)k\Omega} \\ &= \sum_{m=0}^{N-1}[h^*_m e^{jmk\Omega}]e^{-j(N-1)k\Omega}, \quad (m = N-n-1) \\ &= H_k^* e^{-j(N-1)k\Omega} \end{aligned} \tag{6.95}$$

It follows that

$$h_{N-n-1}^* = \frac{1}{N}\sum_{k=0}^{N-1}[H_k^* e^{-j(N-1)k\Omega}]e^{jnk\Omega} \tag{6.96}$$

$$= \frac{1}{N}\sum_{k=0}^{N-1}[M_k e^{-j\phi_k} e^{-j(N-1)k\Omega}]e^{jnk\Omega} \tag{6.97}$$

because

$$H_k^* = M_k^* e^{-j\phi_k} = M_k e^{-j\phi_k} \tag{6.98}$$

From (6.94) and (6.97), we get

$$\frac{1}{N}\sum_{k=0}^{N-1}[M_k e^{-j\phi_k} e^{-j(N-1)k\Omega}]e^{jnk\Omega} = \frac{1}{N}\sum_{k=0}^{N-1}[M_k e^{j\phi_k}]e^{jnk\Omega} \tag{6.99}$$

or

$$e^{-j\phi_k} e^{-j(N-1)k\Omega} = e^{j\phi_k} \tag{6.100}$$

Therefore,

$$-\phi_k - (N-1)k\Omega = \phi_k \tag{6.101}$$

or

$$\begin{aligned} 2\phi_k &= -2k\pi + k\Omega \\ &= -2k\pi + 2k\pi/N \end{aligned} \tag{6.102}$$

Hence we get

$$\phi_k = -\frac{N-1}{N}k\pi \tag{6.103}$$

$k = 1, 2, \ldots, N-1$, as the constraint on ϕ_k for linear phase. We note that if we replace k by $N-k$ in (6.103), we get

$$\begin{aligned} \phi_{N-k} &= -(N-1)\pi + \frac{N-1}{N}k\pi \\ &= -(N-1)\pi - \phi_k \end{aligned} \tag{6.104}$$

which is seen to constrain ϕ_k further when compared with the condition (6.91) that is required for a real sequence. Define

$$P = \begin{cases} \frac{N-1}{2}, & \text{for } N \text{ odd} \\ \frac{N}{2} - 1, & \text{for } N \text{ even} \end{cases} \tag{6.105}$$

From (6.94) we have the unit pulse response.

$$h_n = \frac{1}{N}\sum_{k=0}^{N-1}[M_k e^{j\phi_k}]e^{jnk\Omega} \tag{6.106}$$

The exponent is

$$\begin{aligned}\phi_k + nk\Omega &= -\frac{N-1}{N}k\pi + nk2\pi/N \\ &= -k\pi + \frac{2n+1}{N}k\pi\end{aligned} \tag{6.107}$$

to satisfy condition (6.103). Therefore, (6.106) becomes

$$\begin{aligned}h_n &= \frac{1}{N}\sum_{k=0}^{N-1}(-1)^k M_k e^{j\frac{2n+1}{N}k\pi} \\ &= \frac{1}{N}\left[M_0 + \sum_{k=1}^{P}(-1)^k M_k e^{j\frac{2n+1}{N}k\pi}\right. \\ &\qquad \left. + (-1)^{N-k} M_{N-k} e^{j\frac{2n+1}{N}(N-k)\pi}\right]\end{aligned} \tag{6.108}$$

Noting that

$$e^{j\frac{2n+1}{N}(N-k)\pi} = -e^{-j\frac{2n+1}{N}k\pi} \tag{6.109}$$

and considering the symmetry condition (6.90), we get, for (6.108),

$$h_n = \frac{1}{N}\left[M_0 + 2\sum_{k=1}^{P}(-1)^k M_k \cos\frac{(2n+1)k\pi}{N}\right] \tag{6.110}$$

Then, to satisfy conditions (6.90), (6.91), and (6.103)—that is, to obtain a filter with real unit pulse response and linear phase—we must have

1. *N odd*

$$\phi_k = -\frac{N-1}{N}k\pi, \quad k = 0, 1, 2, \ldots, N-1 \tag{6.111}$$

$$M_k = M_{N-k}, \quad k = 1, 2, \ldots, N-1 \tag{6.112}$$

resulting in the unit pulse response

$$h_n = \frac{1}{N}\left[M_0 + 2\sum_{k=1}^{P}(-1)^k M_k \cos\frac{(2n+1)k\pi}{N}\right] \tag{6.113}$$

2. *N even*

$$\phi_k = \begin{cases} -\dfrac{N-1}{N}k\pi, & k = 0, 1, \ldots, \dfrac{N}{2} - 1 \\ \pi - \dfrac{N-1}{N}k\pi, & k = \dfrac{N}{2} + 1, \ldots, N-1 \end{cases} \tag{6.114}$$

$$\begin{aligned}M_k &= M_{N-k}, \quad k = 1, 2, \ldots, N-1 \\ M_{N/2} &= 0\end{aligned} \tag{6.115}$$

resulting in the unit pulse response

$$h_n = \frac{1}{N}\left[M_0 + 2\sum_{k=1}^{P}(-1)^k M_k \cos\frac{(2n+1)k\pi}{N}\right] \tag{6.116}$$

Alternatively, in terms of specified frequency response samples, we must have

$$\begin{aligned} H_k &= M_k e^{-j(N-1)k\pi/N}, \quad k = 0, 1, \ldots, P \\ H_{N-k} &= M_k e^{j(N-1)k\pi/N}, \quad k = 1, 2, \ldots, P \end{aligned} \tag{6.117}$$

In addition, for N even,

$$H_{N/2} = 0 \tag{6.118}$$

Now we can derive the transfer function for the designed filter. The development follows that in Jong [7]. From (6.84),

$$\begin{aligned} H(z) &= \frac{1}{N}\sum_{k=0}^{N-1} H_k \frac{1 - z^{-N}}{1 - e^{jk\Omega}z^{-1}} \\ &= \frac{1 - z^{-N}}{N}\left[H_0 + \sum_{k=1}^{P}\frac{H_k}{1 - e^{jk\Omega}z^{-1}} + \sum_{k=1}^{P}\frac{H_{N-k}}{1 - e^{jk\Omega}z^{-1}}\right] \end{aligned} \tag{6.119}$$

Substitute for the desired frequency response samples from (6.117) and from (6.118) (if N is even) to get

$$\begin{aligned} H(z) = \frac{1 - z^{-N}}{N}\Bigg[&\frac{M_0}{1 - z^{-1}} + \sum_{k=1}^{P}\frac{M_k e^{-j(N-1)k\pi/N}}{1 - e^{jk\Omega}z^{-1}} \\ &+ \sum_{k=1}^{P}\frac{M_k e^{j(N-1)k\pi/N}}{1 - e^{j(N-k)\Omega}z^{-1}}\Bigg] \end{aligned} \tag{6.120}$$

We note in the second summation that $e^{j(N-k)\Omega} = e^{-jk\Omega}$. Combine the two summations to get

$$H(z) = \frac{1 - z^{-N}}{N}\left[\frac{M_0}{1 - z^{-1}} + \sum_{k=1}^{P}\frac{(-1)^k 2M_k \cos\left(\frac{k\pi}{N}\right)(1 - z^{-1})}{1 - 2\cos(2k\pi/N)z^{-1} + z^{-2}}\right] \tag{6.121}$$

The terms inside the square brackets can be realized as a parallel arrangement of first- and second-order recursive elements.

To get an expression for the continuous frequency response of the designed filter, set $z = e^{j\omega T}$ in (6.121). The first term becomes

$$\frac{1 - e^{-jN\omega T}}{N}\frac{M_0}{1 - e^{-j\omega T}} = \frac{M_0}{N}\frac{e^{-jN\omega T/2}(e^{jN\omega T/2} - e^{-jN\omega T/2})}{e^{-j\omega T/2}(e^{j\omega T/2} - e^{-j\omega T/2})}$$

$$= e^{-j(N-1)\omega T/2} \frac{\sin N\omega T/2}{\sin \omega T/2} \frac{M_0}{N} \tag{6.122}$$

The summation becomes

$$\frac{1 - e^{-jN\omega T}}{N} \sum_{k=1}^{P} \frac{(-1)^k 2M_k \cos \frac{k\pi}{N}(1 - e^{-j\omega T})}{1 - 2\cos(k\pi/N)e^{-j\omega T} + e^{-j2\omega t}}$$

$$= -\frac{4}{N} e^{-j(N+1)\omega T/2} \sin \frac{N\omega T}{2} \sin \frac{\omega T}{2} \sum_{k=1}^{P} \frac{(-1)^k M_k \cos(k\pi/N)}{e^{-j\omega T}[\cos \omega T - \cos(2k\pi/N)]}$$

$$= \frac{2}{N} e^{-j(N-1)\omega T/2} \left[\cos \frac{(N-1)\omega T}{2} - \cos \frac{(N+1)\omega T}{2}\right] \times$$

$$\sum_{k=1}^{P} \frac{(-1)^k M_k \cos(k\pi/N)}{\cos 2k\pi/N - \cos \omega T} \tag{6.123}$$

because

$$\sin \frac{N\omega T}{2} \sin \frac{\omega T}{2} = 1/2 [\cos(N-1)\omega T/2 - \cos(N+1)\omega T/2]$$

We combine (6.122) and (6.123) to get the frequency response.

$$H(j\omega T) = \frac{e^{\frac{-j(N-1)\omega T}{2}}}{N} \left\{ M_0 \frac{\sin \frac{N\omega T}{2}}{\sin \frac{\omega T}{2}} \right.$$

$$\left. + 2\left[\cos \frac{(N-1)\omega T}{2} - \cos \frac{(N+1)\omega T}{2}\right] \sum_{k=1}^{P} \frac{(-1)^k M_k \cos(k\pi/N)}{\cos 2k\pi/N - \cos \omega T} \right\} \tag{6.124}$$

For convenience in applying the results of this section, we will summarize the equations required for filter design.

6.8.5 Summary of Equations for Filter Design by the Frequency Sampling Method

For a finite pulse response filter with real pulse response and linear phase, the N given frequency response samples must satisfy the conditions

$$H_k = M_k e^{-j(N-1)k\pi/N}, \quad k = 0, 1, \ldots, P \tag{6.125}$$

$$H_{N-k} = M_k e^{j(N-1)k\pi/N}, \quad k = 0, 1, \ldots, P \tag{6.126}$$

where

$$M_{N-k} = M_k, k = 1, 2, \ldots, N-1 \tag{6.127}$$

$$P = \begin{cases} (N-1)/2, & \text{for } N \text{ odd} \\ N/2 - 1, & \text{for } N \text{ even} \end{cases} \tag{6.128}$$

and

$$H_{N/2} = 0, \quad \text{for } N \text{ even} \tag{6.129}$$

The unit pulse response of the designed filter is

$$h_n = \frac{1}{N}\left[M_0 + 2\sum_{k=1}^{P}(-1)^k M_k \cos\frac{(2n+1)k\pi}{N}\right] \tag{6.130}$$

The transfer function is

$$H(z) = \frac{1-z^{-N}}{N}\left[\frac{M_0}{1-z^{-1}} + \sum_{k=1}^{P}\frac{(-1)^k 2M_k \cos\left(\frac{k\pi}{N}\right)(1-z^{-1})}{1-2\cos(2k\pi/N)z^{-1}+z^{-2}}\right] \tag{6.131}$$

with continuous frequency response

$$H(j\omega T) = \frac{e^{-j(N-1)\omega T/2}}{N} \times \left\{R_0 + 2\left[\cos\frac{(N-1)\omega T}{2} - \cos\frac{(N+1)\omega T}{2}\right]\sum_{k=1}^{P} R_k\right\} \tag{6.132}$$

where

$$R_0 = \frac{M_0 \sin(N\omega T/2)}{\sin(\omega T/2)} \tag{6.133}$$

$$R_k = \frac{(-1)^k M_k \cos(k\pi/N)}{\cos(2\pi k/N) - \cos(\omega T)} \tag{6.134}$$

6.8.6 Example

In applying the frequency sampling method to the design of approximations to ideal lowpass, highpass, bandpass, and bandstop filters, we note that the stopband samples of the desired response will be zero, thereby reducing the number of terms in the transfer function (6.131) and the frequency response (6.132). An example will illustrate the procedure.

EXAMPLE 6.4

Using the frequency sampling method, design a lowpass digital filter to meet the following specifications:

Cutoff frequency $f_c = 8$ Hz

Gain approximately unity in the passband and zero in the stopband

Linear phase

Sampling frequency $f_s = 100$ Hz

Take $N = 25$.

Solution

First it is necessary to find the fundamental frequency Ω. We have

$$\Omega = \frac{2\pi}{NT} = \frac{\omega_s}{N} = \frac{2\pi(100)}{25} = 8\pi \text{ rad/s}$$

Because the cutoff frequency is

$$\omega_c = 16\pi$$

the desired magnitude response samples in the operational range are

$$M_k = \begin{cases} 1, & 0 \le k \le 2 \\ 0, & 3 \le k \le 12 \end{cases}$$

To meet the symmetry requirements, (6.127) implies that

$$M_k = \begin{cases} 1, & 0 \le k \le 2 \\ 0, & 3 \le k \le 22 \\ 1, & k = 23, 24 \end{cases}$$

From (6.125) and (6.126), the required frequency response samples are

$$H_k = \begin{cases} e^{-j24k\pi/25}, & 0 \le k \le 2 \\ 0, & 3 \le k \le 22 \\ e^{j24k\pi/25}, & k = 23, 24 \end{cases}$$

We obtain the filter transfer function from (6.131), noting that for this case, there are only two nonzero terms in the summation. The transfer function is

$$H(z) = \frac{1 - z^{-25}}{25}\left[\frac{1}{1 - z^{-1}} - \frac{1.9842(1 - z^{-1})}{1 - 1.9372z^{-1} + z^{-2}} + \frac{1.9372(1 - z^{-1})}{1 - 1.7526z^{-1} + z^{-2}}\right]$$

A block diagram for the filter is shown in Figure 6.17.

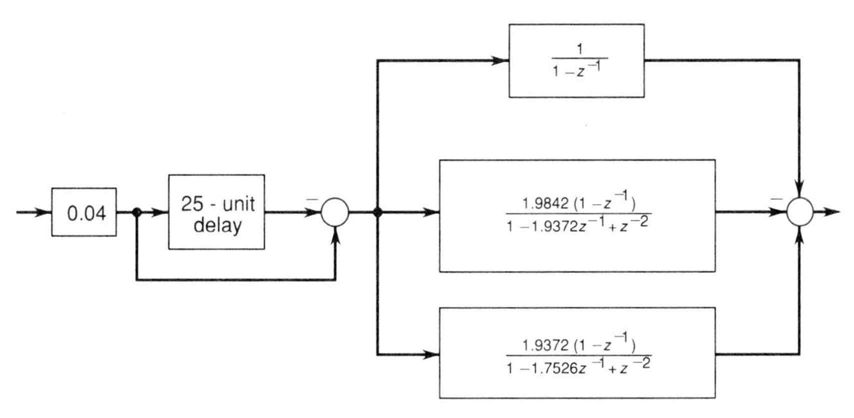

FIGURE 6.17 Filter arrangement for Example 6.4.

The continuous frequency response according to (6.132) becomes, for this example,

$$H(j\omega T) = \frac{e^{-j12\omega T}}{25}\left[\frac{\sin 12.5\omega T}{\sin 0.5\omega T} - 2(\cos 12\omega T - \cos 13\omega T) \times \left(\frac{0.9921}{0.9686 - \cos\omega T} - \frac{0.9686}{0.8763 - \cos\omega T}\right)\right]$$

The magnitude response for the filter is shown as the solid curve in Figure 6.18. The comblike nature of the response in the stopband is evident. Although the response at the frequency sample points (0, 4, 8, 12, ..., 44, 48 Hz) agrees with that specified, the oscillations between samples are such that the minimum attenuation in the stopband is only about 16 dB.

The situation can be improved if, instead of specifying that the response samples go directly from unity in the passband to zero in the stopband, we allow a transition region. For example, let

$$H_k = \begin{cases} e^{-j24k\pi/25}, & 0 \le k \le 2 \\ 0.5e^{-j24k\pi/25}, & k = 3 \\ 0, & 4 < k \le 21 \\ 0.5e^{j24k\pi/25}, & k = 22 \\ e^{j24k\pi/25}, & k = 23, 24 \end{cases}$$

This results in an *additional* term in the transfer function,

$$\frac{1 - z^{-25}}{25}\left[\frac{-0.9298(1 - z^{-1})}{1 - 1.4579z^{-1} + z^{-2}}\right]$$

and in the frequency response,

$$\frac{e^{-j12\omega T}}{25}\left[\frac{-0.9298(\cos 12\omega T - \cos 13\omega T)}{0.7290 - \cos\omega T}\right]$$

The result of adding the transition samples is shown as the dashed curve in Figure 6.18, where now the minimum attenuation in the stopband is about 30 dB. This improvement was obtained at the expense of a widened transition region in the continuous response. As we noted earlier, optimization methods can be used to determine the best values for the transition samples in order to maximize the minimum attenuation and reduce intersample oscillation in the continuous frequency response.

6.9 SUMMARY

The theory of the discrete Fourier transform for finite sequences is developed, and its properties are examined. The linear convolution of two finite sequences can be obtained by means of an appropriate circular convolution. Fast Fourier transform algorithms are described that can be used to reduce

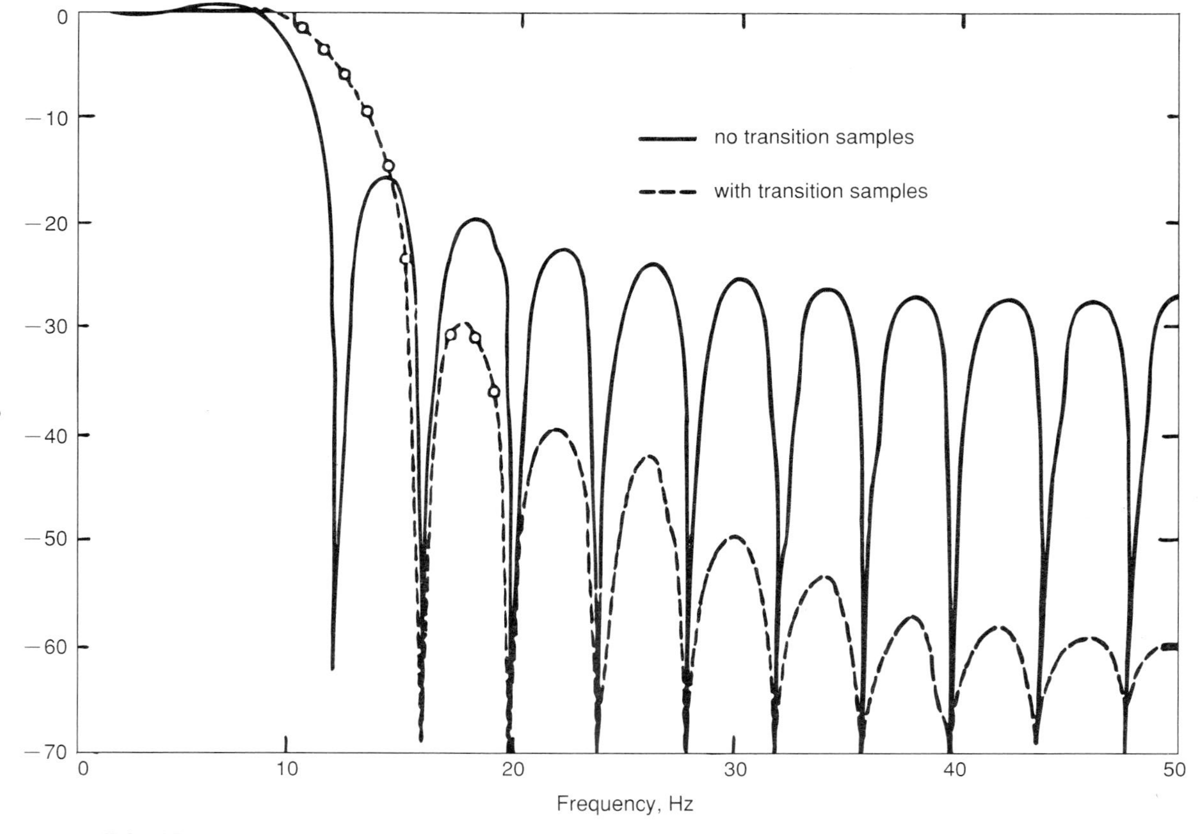

FIGURE 6.18 Magnitude responses for lowpass filter design from frequency sampling (Example 6.4).

the number of computations required to evaluate the DFT. The utility of the DFT for digital filtering is demonstrated. In particular, the frequency sampling method can be used to design linear-phase filters with finite real pulse responses the frequency responses of which agree with specified values at the sample instances.

REFERENCES FOR CHAPTER 6

1. B. Gold and C. M. Rader, *Digital Processing of Signals* (New York: McGraw-Hill, 1969).
2. A. V. Oppenheim and R. W. Schafer, *Digital Signal Processing* (Englewood Cliffs, NJ: Prentice-Hall, 1975).
3. L. R. Rabiner and B. Gold, *Theory and Application of Digital Signal Processing* (Englewood Cliffs, NJ: Prentice-Hall, 1975).
4. J. W. Cooley and J. W. Tukey, "An Algorithm for the Machine Computation of Complex Fourier Series," *Mathematics of Computation* 19 (April 1965): 297–301.
5. G. D. Bergland, "A Guided Tour of the Fast Fourier Transform," *IEEE Spectrum* (July 1969): 41–52.
6. J. W. Cooley, P. A. Lewis, and P. D. Welch, "Historical Notes on the Fast Fourier Transform," *IEEE Trans. Audio Electroacoustics* (June 1967): 76–79.
7. M. T. Jong, *Methods of Discrete Signal and System Analysis* (New York: McGraw-Hill, 1982), Chap. 8.
8. L. R. Rabiner, B. Gold, and C. A. McGonegal, "An Approach to the Approximation Problem for Nonrecursive Digital Filters," *IEEE Trans. Audio Electroacoustics* AV-18 (June 1970): 83–106.

EXERCISES FOR CHAPTER 6

1. If $X_k = X(k)$ is the DFT of a finite sequence $x_n = x(n), n = 0, 1, \ldots, N-1$, which may be complex in general, prove that

$$x_n^* = \text{DFT}\{X_k^*/N\}$$

where * denotes complex conjugate.

2. Prove that when we reverse the time sequence, it is equivalent to reversing the frequency sequence. That is, prove that

$$\text{DFT}\{x_{-n}\} = X_{-k} = X_{N-k}$$

3. Prove that if $x(n)$ is a real sequence, and

$$x(n) = -x(N-n)$$

then its DFT is purely imaginary. That is, prove that

$$Re[X(k)] = 0$$

4. Prove that, like the continuous Fourier transform, the DFT has a frequency selectivity property. That is, prove that

$$\text{DFT}\left\{\sum_q a_q e^{jqn\Omega}\right\} = Na_k, \quad k = 0, 1, \ldots, N-1$$

where q is integer.

5. A finite sequence $x(n)$ is shown in Figure P6.1. Sketch the sequence $x((-n))_5$.

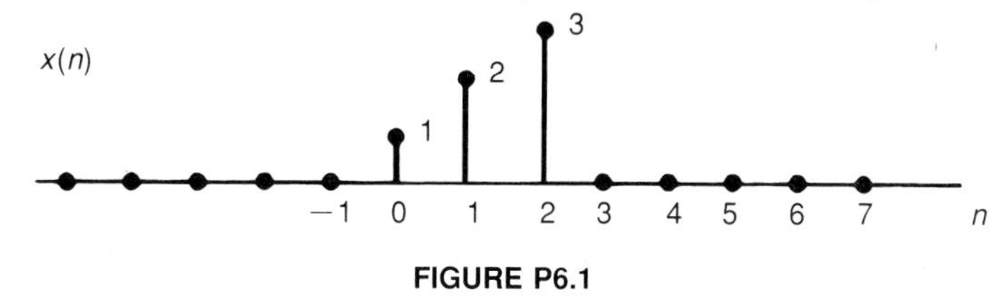

FIGURE P6.1

6. Compute the DFT of each of the following finite-length sequences considered to be of length N.

a) $x(n) = \delta(n)$

b) $x(n) = \delta(n - n_0), \quad 0 < n_0 < N - 1$

c) $x(n) = a^{(n-n_0)}, \quad 0 < a < 1, \quad 0 < n_0 < N - 1$

d) $x(n) = R_N(n)$

7. Find the DFT of the sequence $x(n) = \sin\frac{n\pi}{N}, \quad n = 0, 1, \ldots, N-1$

(*Hint*: Write the sinusoid in exponential form.)

8. Find the DFT $X(k), k = 0, 1, \ldots, N-1$, of the Hamming window sequence

$$x(n) = 0.54 + 0.42\cos\frac{\pi n}{N}, \quad n = 0, 1, \ldots, N-1$$

9. Find and sketch the DFT $X(k), k = 0, 1, \ldots, N-1$, of the sequence

$$x(n) = 0.42 + 0.5\cos\frac{2n\pi}{N} + 0.08\cos\frac{4n\pi}{N}, \quad n = 0, 1, \ldots, N-1$$

10. Compute the DFT of the square-wave sequence

$$x(n) = \begin{cases} 1, & 0 \le n \le \frac{N}{2} - 1 \\ -1, & N/2 \le n \le N - 1 \end{cases}$$

where N is even.

11. Find the DFT of the 8-point sequence

$$x(n) = \begin{cases} 0, & 0 \le n \le 4 \\ 1, & 5 \le n \le 6 \\ 0, & n = 7 \end{cases}$$

12. For the finite sequences

$$h(n) = \begin{cases} (1/2)^n, & 0 \le n \le 4 \\ 0, & \text{otherwise} \end{cases}$$

and

$$x(n) = \begin{cases} 1, & 0 \le n \le 2 \\ 0, & \text{otherwise} \end{cases}$$

find *graphically* the circular convolution

$$y(n) = x(n)\circledN h(n) = \sum_{m=0}^{N-1} h(m)x((n-m))$$

over one period, $n = 0, 1, \ldots, N-1$, where $N = 6$.

13. **a)** Find graphically the 7-point circular convolution of the two sequences shown in Figure P6.2.

b) Using an appropriate *circular* convolution, find graphically the *linear* convolution of the sequences. Compare your result with that for part a) to identify the aliasing that results from taking a circular convolution with too few points.

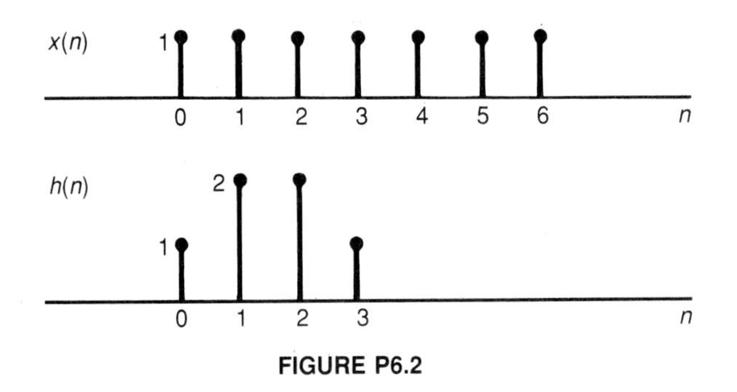

FIGURE P6.2

14. It is desired to use a power-of-2 FFT to compute samples of the spectrum of a continuous signal. If the signal is known to contain no frequency above 2 kHz, and it is necessary to distinguish between spectral peaks 40 Hz apart (that is, the frequency resolution is 40 Hz), find

a) The minimum sampling rate

b) The minimum number of samples

c) The minimum signal length

15. A 4-point complex sequence is given by $1 + j2, 3 + j4, 5 + j2, 2 + j3$.

a) Find the DFT by direct computation.

b) Use the FFT as described in Section 6.7.1 to find the DFT, and confirm that the result is the same as that obtained in part a).

16. Write your own program to implement a power-of-2 decimation-in-time FFT algorithm based on the description in Section 6.7.1. Use your program to find the DFT of the Hamming window sequence in Exercise 8 if $N = 16$. Compare your result with that obtained in Exercise 8.

17. Show that (6.86) follows from (6.85).

18. Show that (6.121) follows from (6.120).

19. Use the frequency sampling method to design a highpass digital filter to meet the following specifications:

Signals to be processed contain no frequencies above 50 Hz

Passband cutoff frequency – 40 Hz

Gain approximately unity in the passband and zero in the stopband

Linear phase

Take $N = 21$. Write the transfer function and plot the magnitude response for the resulting filter.

20. Design a bandpass filter by means of the frequency sampling method to meet the following specifications:

Passband between 55 and 65 Hz

Gain approximately unity in the passband and zero in the stopbands

Linear phase

Sampling frequency $f_s = 200$ Hz

Choose a suitable minimum value for N on the basis of the specifications. If the magnitude response is not satisfactory with regard to attenuation in the stopbands, allow the magnitude of one sample in the transition regions to lie between zero and 1. (*Note*: The interactive program FREQSAMP on the diskette based on (6.131) and (6.132) may be used to solve this and similar problems. But first you should try to get the results without benefit of the program.)

APPENDIX 6A

Diskette Program Example

The FORTRAN program DFT computes the discrete Fourier transform (DFT) or the inverse DFT of a sequence via a fast Fourier transform (FFT) algorithm. It is arranged for interactive operation.

The user has the choice of processing an arbitrary input sequence or a square-wave example. The input data must be written as a complex sequence with a specified format (if the sequence is real, zeros should be appended for the imaginary part). The input and output data are stored in a file DFT.DAT created by the program. In the example, the DFT of an exponential sequence is computed.

```
DFT

DO YOU WANT TO PROCESS AN ARBITRARY INPUT SEQUENCE,
(IS), OR A SQUARE WAVE EXAMPLE (SQ) (ENTER IS/SQ):
IS

INPUT DATA MUST BE WRITTEN AS A COMPLEX SEQUENCE
WITH FORMAT(5X,F15.6,5X,F15.6) FOR DIRECT DFT AND
FORMAT(40X,F15.6,5X,F15.6) FOR INVERSE DFT

ENTER LENGTH OF INPUT SEQUENCE (MUST BE POWER OF 2):
8
WHAT IS THE NAME OF THE FILE CONTAINING THE SEQUENCE?
DFTEX.DAT

DO YOU WANT DIRECT DFT (ENTER Y/N):
Y

OUTPUT DATA IN FILE DFT.DAT
Stop - Program terminated.
```

K	TIME SEQUENCE			FREQUENCY SEQUENCE		
1	1.000000	+ J	.000000	1.992188	+ J	.000000
2	.500000	+ J	.000000	1.186093	+ J	-.648695
3	.250000	+ J	.000000	.796875	+ J	-.398437
4	.125000	+ J	.000000	.688907	+ J	-.179945
5	.062500	+ J	.000000	.664062	+ J	.000000
6	.031250	+ J	.000000	.688907	+ J	.179945
7	.015625	+ J	.000000	.796875	+ J	.398437
8	.007813	+ J	.000000	1.186093	+ J	.648695

CHAPTER 7

BASIC CONCEPTS OF PROBABILITY THEORY AND RANDOM PROCESSES

■ 7.1 INTRODUCTION

It has been said that physical quantities are never known exactly and that there is no such thing as an exact science. Digital filter design is no exception. There are errors in measurements, errors in system models representing the filtering, and errors in implementing the finished designs. In fact, errors arise at every stage of a design. Fortunately, many of these errors are so small that they can be neglected. In other words, we can treat many signals as *deterministic* (known exactly and repeatable) and many models as representing the real-world situation to a high degree of approximation. Adopting this viewpoint makes it possible to handle very many design problems, even if a signal to be filtered is clearly nondeterministic, or *random* (a noisy radar signal, for example). In such cases we can often design a filter, using the methods discussed in Chapters 4 and 5, to remove the noise and pass only the desired signal. There often comes a stage, however, when random effects have to be taken into account. For example,

1. It may not be a straightforward task to separate the desired signal from the noise simply by knowing the frequency band of the noise.
2. We may want to extract some desired quantity from a sequence of noisy measurements by filtering when we know that the sequence is not repeatable.
3. When we come to implement the designed filter, we find that only discrete sets of amplitude values are feasible in a computer, and the errors are treated as noise as discussed in the next chapter.

It follows that if we limit ourselves to deterministic design and analysis, we may be severely restricting our scope and field of application. Perhaps the random effects can be neglected; if so, all is well. But the only way we can find out is by acquiring the necessary tools to analyze the situation. For that reason, some basic concepts in probability theory and random processes are discussed in this chapter. The field of probability theory is a vast one addressed in numerous good texts, such as [1] through [5]. Our treatment of the subject is a brief review of concepts we will need for the discussion of quantization effects in Chapter 8 and for estimation and the design of the discrete Kalman filter in Chapter 9. The reader who is already familiar with probability theory and random processes can turn directly to those later chapters.

7.2 RANDOM VARIABLES

Fundamental to the concept of probability is the notion of an experiment that has at least two possible outcomes. Associated with the experiment is a set of outcomes on the sample space S. Before performing the experiment, we do not know what outcome α will occur. With each outcome α_j is associated a number $P(\alpha_j)$, $0 \leq P(\alpha_j) \leq 1$. By the relative-frequency definition of probability, if we perform an experiment n times under identical circumstances, and the outcome or event A occurs in n_A of these trials, then the probability of event A is defined as

$$P(A) = \lim_{n \to \infty} \frac{n_A}{n} \tag{7.1}$$

An axiomatic approach to probability requires that the probability function satisfy certain axioms [1], and the theory is based on the results of this requirement. The axioms are

I.

$$P(A) \geq 0 \tag{7.2}$$

II.

$$P(S) = 1 \tag{7.3}$$

where S is the event space containing all possible outcomes of the experiment. It is also called the certain event.

III. If A and B are disjoint or mutually exclusive events (that is, if the occurrence of one precludes the occurrence of the other), then

$$P(A + B) = P(A) + P(B) \tag{7.4}$$

where $P(A + B)$ is the probability that either A or B will occur—that is, the probability of the *union* of A and B, written $A \cup B$ or $A + B$.

From these axioms, it is possible to draw many other conclusions. These can be stated in the form of basic theorems, given here without proof.

Theorem 1

For all events A and B such that A is a subset of B (that is, if A occurs, B must also occur), we have

$$P(A) \leq P(B) \tag{7.5}$$

Theorem 2

For all events A,

$$P(\bar{A}) = 1 - P(A) \tag{7.6}$$

where $\bar{A}$ is the event that A does not occur—that is, the *complement* of A. As a corollary,

$$P(A) \leq P(S) = 1 \tag{7.7}$$

Theorem 3

For all events A and B,

$$P(A + B) = P(A) + P(B) - P(AB) \tag{7.8}$$

where AB is the event that both A and B occur—that is, the intersection of A and B, often written $A \cap B$ or AB. If A and B are disjoint, they contain no common elements, so their intersection is the null or impossible event ϕ. Then $P(AB) = P(\phi) = 0$, in accordance with axiom III.

Theorem 4

For all events A and B such that A is a subset of B,

$$P(B - A) = P(B) - P(A) \tag{7.9}$$

where $P(B - A)$ is the probability of the event made up of all the elements of B that are not in A.

The conditional probability of an event A given that B has occurred is

$$P(A|B) = \frac{P(AB)}{P(B)} \tag{7.10}$$

If A and B are disjoint, then $P(AB) = 0$, and therefore

$$P(A|B) = 0 \tag{7.11}$$

That is, if B occurs, A cannot occur. If A is a subset of B, then $AB = A$ and

$$P(A|B) = \frac{P(A)}{P(B)} \tag{7.12}$$

If B is a subset of A, then $AB = B$ and

$$P(A|B) = 1 \tag{7.13}$$

Two events A and B are said to be independent if

$$P(AB) = P(A)P(B) \tag{7.14}$$

In that case,

$$P(A|B) = \frac{P(A)P(B)}{P(B)} = P(A) \tag{7.15}$$

That is, the occurrence of B does not affect that of A.

Theorem of total probability

Consider n mutually exclusive events B_i that decompose the sample space S. That is, $B_i \cap B_j = \phi$, $i \neq j$, and $\bigcup_{i=1}^{n} B_i = S$. Then, if A is an arbitrary event,

$$P(A) = P(AS) = \sum_{i=1}^{n} P(AB_i) = \sum_{i=1}^{n} P(A|B_i)P(B_i) \tag{7.16}$$

Bayes' theorem

For mutually exclusive events B_i, $i = 1, \ldots, n$, decomposing S, and an arbitrary event A, we have

$$P(B_j|A) = \frac{P(B_jA)}{P(A)} = \frac{P(A|B_j)P(B_j)}{\sum_{i=1}^{n} P(A|B_i)P(B_i)} \tag{7.17}$$

where the denominator of the last expression follows from (7.16).

The following example illustrates some of the preceding concepts, in particular the use of Bayes' theorem.

EXAMPLE 7.1

Either John or Matt is selected to shoot at a target. There is a probability of 0.6 that John will be selected. John and Matt have probabilities 0.7 and 0.5, respectively, of hitting the target. If the experiment of shooting once at the target results in a hit, what is the probability that John was selected?

Solution

Let J be the event that John was selected, and let M be the event that Matt was selected. Because one or other of the two—but not both—must be chosen, events J and M are mutually exclusive. Therefore, $P(M) = 1 - P(J) = 0.4$. Let H be the event of a hit on the target. From the problem statement, we can define the conditional probabilities $P(H|J) = 0.7$ and $P(H|M) = 0.5$. We require the probability that John was selected, given that a hit was made. That is, we need to find $P(J|H)$.

From Bayes' theorem (7.17), we have

$$P(J|H) = \frac{P(JH)}{P(H)} = \frac{P(H|J)P(J)}{P(H|J)P(J) + P(H|M)P(M)}$$

$$= \frac{(0.7)(0.6)}{(0.7)(0.6) + (0.5)(0.4)}$$

$$= 0.68$$

Repeated Independent Trials

Consider n identical independent trials, by which we mean that the outcome of any trial is independent of the outcome of any other trial. Then, if A is the event of success in any one trial and $P(A) = p$, we have

$$P\{k \text{ successes in } n \text{ trials}\} = \binom{n}{k} p^k q^{n-k} \tag{7.18}$$

where $q = 1 - p$ and $\binom{n}{k}$ is the number of combinations of n things taken k at a time. Expression (7.18) is called the binomial distribution. From (7.18),

$$P\{\textit{at least } k \text{ successes in } n \text{ trials}\} = \sum_{i=k}^{n} \binom{n}{i} p^i q^{n-i} \tag{7.19}$$

Prior to our taking a measurement or set of measurements subject to error, the situation may be regarded as an experiment the outcome of which is the actual set of measurements made. If another set of measurements were made, it is likely that the outcome would be different. It is convenient to associate a real number with every outcome of an experiment. For example, when flipping a coin, we can associate a 1 with the outcome "heads" and a 0 with the outcome "tails." This introduces the notion of a random variable.

Definition

A random variable $X = X(\alpha)$ is a real and single-valued *function* the domain of which is the sample space S and the range of which is a subset of the real numbers.

The random variable may be either continuous or discrete. When the experiment has been completed, X takes on a value of x. (Capital letters will be used for random variables, and the corresponding lower-case letters for the values they assume as a result of the experiment.) For a continuous random variable X, the probability distribution function—also called the cumulative probability—is defined as

$$F_X(x) = P\{X \leq x\} = \text{ probability that } X \leq x \tag{7.20}$$

and the probability density function as

$$f_X(x) = \frac{dF_X(x)}{dx} \tag{7.21}$$

if the derivative exists. Note that

$$f_X(x)dx = F_X(x + dx) - F_X(x) = P\{x < X \leq x + dx\} \tag{7.22}$$

It follows that

$$F_X(x) = \int_{-\infty}^{x} f_X(x')dx' \tag{7.23}$$

$$P\{x_1 < x \leq x_2\} = F_X(x_2) - F_X(x_1) = \int_{x_1}^{x_2} f_X(x')dx' \tag{7.24}$$

and

$$P\{-\infty \leq X \leq \infty\} = \int_{-\infty}^{\infty} f_X(x')dx' = 1 \tag{7.25}$$

Knowledge of $F_X(x)$ or $f_X(x)$ completely defines the random variable X. We note from the definitions that both $F_X(x)$ and $f_X(x)$ are non-negative.

If X is a discrete random variable that can take on any one of the discrete values x_i, $i = 1, 2, \ldots, n$ (n may be infinite) as the result of an experiment, then

$$F_X(x) = P\{X \leq x\} = \sum_i p_i \tag{7.26}$$

where i is such that $x_i \leq x$, and

$$p_i = P\{X = x_i\} \tag{7.27}$$

We can write the discrete distribution (7.27) as a probability density function if we make use of a delta function,

$$f_X(x) = \sum_{i=1}^{n} p_i \delta(x - x_i) \tag{7.28}$$

where

$$\delta(x - x_i) = \begin{cases} \infty, & x = x_i \\ 0, & \text{otherwise} \end{cases} \tag{7.29}$$

Then

$$\begin{aligned} \int_{-\infty}^{\infty} f_X(x)dx &= \int_{-\infty}^{\infty} \left[\sum_{i=1}^{n} p_i \delta(x - x_i) \right] dx \\ &= \sum_{i=1}^{n} p_i \int_{-\infty}^{\infty} \delta(x - x_i)dx = \sum_{i=1}^{n} p_i = 1 \end{aligned} \tag{7.30}$$

The expected value or mean value of a random variable X is defined as

$$\langle X \rangle = \int_{-\infty}^{\infty} x f_X(x)dx \tag{7.31}$$

We will use the symbol $\langle\ \rangle$ throughout to denote the expected value, or expectation, of the random variable within the brackets. It is easily shown that the expected-value operation is linear. The proof is left as an exercise. For brevity, $\langle X \rangle$ is often written μ_X. The variance of X, which is a measure of the spread about the mean, is defined as

$$\text{Var } X = \langle (X - \mu_X)^2 \rangle = \int_{-\infty}^{\infty} (x - \mu_X)^2 f_X(x)dx = \langle X^2 \rangle - \mu_X^2 \tag{7.32}$$

For brevity, Var X is often written σ_X^2, σ_X being the standard deviation. Higher-order moments are defined similarly.

The joint probability density function of two random variables X and Y is

$$f_{XY}(x, y)dx, dy = P\{x < X \le x + dx, y < Y \le y + dy\} \tag{7.33}$$

which is the probability of the intersection of the two events within the right-hand parentheses. Its integral over all x and y is equal to unity. The random variables X and Y are said to be independent if

$$f_{XY}(x, y) = f_X(x) f_Y(y) \tag{7.34}$$

The covariance of two random variables X and Y is defined as

$$\begin{aligned} \text{Cov }(XY) &= \langle (X - \mu_X)(Y - \mu_Y) \rangle \\ &= \int_{-\infty}^{\infty} \int_{-\infty}^{\infty} (x - \mu_X)(y - \mu_Y) f_{XY}(x, y)dx\, dy \end{aligned} \tag{7.35}$$

Upon expanding (7.35), we get

$$\text{Cov } XY = \langle (X-\mu_X)(Y-\mu_Y)\rangle = \langle XY\rangle - \mu_X\mu_Y \tag{7.36}$$

Thus, if X and Y are independent,

$$\text{Cov } XY = 0 \tag{7.37}$$

If (7.37) is true, X and Y are said to be uncorrelated. Therefore, independence implies uncorrelation, but the converse is not necessarily true (except for Gaussian-distributed random variables, which we will discuss shortly).

The conditional density function of X given that $Y = y$ has occurred is

$$f_X(x|Y=y) = \frac{f_{XY}(x,y)}{f_Y(y)} \tag{7.38}$$

By Bayes' rule for density functions, this can be written

$$f_X(x|Y=y) = \frac{f_Y(y|X=x)f_X(x)}{f_Y(y)} \tag{7.39}$$

Let us look at some density functions that arise often.

1. *Uniform distribution*

If a random variable X is equally likely to take on any value between two limits a and b and cannot assume any value outside that range, it is said to be uniformly distributed in the range $[a, b]$ with density function

$$f_X(x) = \begin{cases} \dfrac{1}{b-a}, & a \le x \le b \\ 0, & \text{otherwise} \end{cases} \tag{7.40}$$

EXAMPLE 7.2

The error in an instrument reading is uniformly distributed between -5 and 2 millivolts (mV). Find the mean, root mean square (rms), and standard deviation of the error.

Solution

Let the random variable E denote the error. Then the mean is

$$\mu_E = \int_a^b e f_E(e)de = \frac{1}{7}\int_{-5}^{2} e\,de = -1.5 \text{ mV}$$

The mean-square value is

$$\langle E^2\rangle = \int_{-\infty}^{\infty} e^2 f_E(e)de = \frac{1}{7}\int_{-5}^{2} e^2 de = 6.33 \text{ (mV)}^2$$

Therefore,

$$\text{rms value} = 2.52 \text{ mV}$$

The variance is given by

$$\sigma_E^2 = \langle E^2 \rangle - \mu_E^2 = 4.08 \ \ (\text{mV})^2$$

Therefore, the standard deviation is

$$\sigma_E = 2.02 \ \ \text{mV}$$

---

2. *Rayleigh distribution*

Random variables that cannot assume negative values sometimes have a Rayleigh distribution:

$$f_X(x) = \begin{cases} \dfrac{2x}{\propto} \exp(-x^2/\propto), & x \geq 0 \\ 0, & x < 0 \end{cases} \tag{7.41}$$

Both mean and variance depend on the positive parameter $\propto$.

3. *Binomial distribution*

This is the distribution for a discrete random variable as given in (7.18). It can be written as a probability density function, using delta functions.

$$f_X(x) = \sum_{k=0}^{n} \binom{n}{k} p^k q^{n-k} \delta(x - k) \tag{7.42}$$

4. *Gaussian distribution*

The most common of all probability density functions is the normal, or Gaussian, density, which is defined by

$$f_X(x) = \frac{1}{\sigma_X \sqrt{2\pi}} \exp[-(x - \mu_X)^2/(2\sigma_X^2)] \tag{7.43}$$

for a single random variable. The density function is completely defined by the mean μ_X and the variance σ_X^2. For brevity, it is often written $N(\mu_X, \sigma_X^2)$. With a change of variable $U = (X - \mu_X)/\sigma_X$, the integral of (7.43) with respect to x equals the integral of

$$f_U(u) = \frac{1}{\sqrt{2\pi}} \exp\,[-u^2/2] \tag{7.44}$$

with respect to u, which is the form assumed in tables for the Gaussian distribution.

For two random variables X and Y, the Gaussian joint probability density function is

$$f_{XY}(x, y) = \frac{1}{2\pi\sigma_X\sigma_Y\sqrt{1 - r^2}} \exp(-J/2) \tag{7.45}$$

where

$$J = \frac{1}{1 - r^2} \left[\frac{(x - \mu_X)^2}{\sigma_X^2} - \frac{2r(x - \mu_X)(y - \mu_Y)}{\sigma_X \sigma_Y} + \frac{(y - \mu_Y)^2}{\sigma_Y^2} \right] \tag{7.46}$$

and r is the correlation coefficient defined by

$$r = \frac{\text{Cov }(XY)}{\sigma_X \sigma_Y} \tag{7.47}$$

We note that with no correlation between X and Y (that is, $r = 0$), the joint density (7.45) reduces to the product of the *marginal* densities $f_X(x)$ and $f_Y(y)$, implying independence also.

A random vector is a vector (an $n \times 1$ column matrix) the components or elements of which are random variables. A joint density function can be defined for the n random variables. We will use boldface type for vectors in this book. For a Gaussian random vector $\mathbf{X}$ the transpose of which is the row vector

$$\mathbf{X}' = [X_1, X_2, \ldots, X_n] \tag{7.48}$$

the joint density function is

$$f_{\mathbf{X}}(\mathbf{x}) = \frac{1}{(2\pi)^{n/2}|S|^{1/2}} \exp[-\frac{1}{2}(\mathbf{x} - \mu_{\mathbf{X}})' S^{-1}(\mathbf{x} - \mu_{\mathbf{X}})] \tag{7.49}$$

where the $n \times n$ covariance matrix S is defined as

$$\begin{aligned} S &= \langle (\mathbf{X} - \mu_{\mathbf{X}})(\mathbf{X} - \mu_{\mathbf{X}})' \rangle \\ &= \begin{bmatrix} \langle (X_1 - \mu_1)^2 \rangle \cdots \langle (X_1 - \mu_1)(X_n - \mu_n) \rangle \\ \vdots \qquad\qquad \vdots \\ \langle (X_n - \mu_n)(X_1 - \mu_1) \rangle \cdots \langle (X_n - \mu_n)^2 \rangle \end{bmatrix} \end{aligned} \tag{7.50}$$

and $|S|$ is the determinant of S. Note that the covariance matrix is symmetric: $s_{ji} = s_{ij}$.

An expression identical to (7.39) can be written for the conditional density for a random vector, with random variables X and Y replaced by random vectors $\mathbf{X}$ and $\mathbf{Y}$.

EXAMPLE 7.3

Express the second-order Gaussian distribution (7.45) in the vector form (7.49).

Solution

Define the vector

$$\mathbf{X} = \begin{bmatrix} X \\ Y \end{bmatrix}$$

with mean vector

$$\mu_{\mathbf{X}} = \begin{bmatrix} \mu_X \\ \mu_Y \end{bmatrix}$$

Then the covariance matrix is

$$S = \begin{bmatrix} \sigma_X^2 & r\sigma_X\sigma_Y \\ r\sigma_X\sigma_Y & \sigma_Y^2 \end{bmatrix}$$

with determinant

$$|S| = \sigma_X^2\sigma_Y^2 - r^2\sigma_X^2\sigma_Y^2 = (1 - r^2)\sigma_X^2\sigma_Y^2$$

Therefore,

$$|S|^{1/2} = \sqrt{1 - r^2}\sigma_X\sigma_Y$$

and the inverse of S is

$$S^{-1} = \frac{1}{|S|}\begin{bmatrix} \sigma_Y^2 & -r\sigma_X\sigma_Y \\ -r\sigma_X\sigma_Y & \sigma_X^2 \end{bmatrix}$$

Inserting these quantities in (7.49) and expanding the resulting expression confirms that it is the same as (7.45).

One reason why the Gaussian distribution plays such an important role in probability theory is given by the *central limit theorem*, which states that under fairly general conditions, the sum of *n independent* random variables is a random variable the probability distribution of which approaches the Gaussian for large n, no matter what the distributions of the individual random variables. Often, n need not be very large before the density function of the sum takes on the characteristic Gaussian shape. The reader can confirm this by doing Exercise 14, which deals with the sum of three uniformly distributed independent random variables.

7.3 THE BEST MEAN-SQUARE ESTIMATE OF A RANDOM VARIABLE WHEN A MEASUREMENT HAS BEEN MADE

Consider a random variable X. Suppose that a measurement $Y = y$ has been made and that the conditional density function $f_X(x|Y = y)$ is known. What is the best estimate of $X = x$ in the sense of minimum mean-square error?

Let the estimate be

$$\hat{x} = g(y) \tag{7.51}$$

where g is some function of y to be determined. It may be nonlinear. The situation is depicted in Figure 7.1.

Prior to the measurement's being made, the estimate is a random variable $\hat{X}$. If $Y = y$ is measured, then $\hat{X} = \hat{x}$. Define the error in estimation as

$$E = X - \hat{x} \tag{7.52}$$

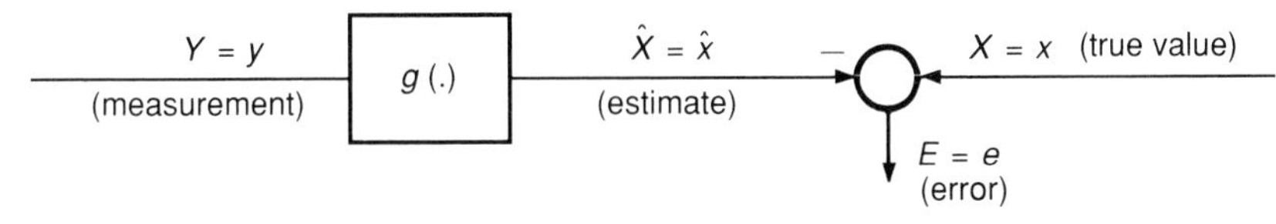

FIGURE 7.1 Estimate of random variable X.

The error E is a random variable because X is random. We want to select the estimate $\hat{x}$ so as to minimize the mean-square error, given that a particular measurement $Y = y$ has been made. The conditional mean-square error is

$$J = \langle E^2|Y = y\rangle = \langle (X - \hat{x})^2|Y = y\rangle = \int_{-\infty}^{\infty} (x - \hat{x})^2 f_X(x|Y = y)dx \tag{7.53}$$

Differentiate J with respect to $\hat{x}$ and set the result equal to zero to get the best, or optimal, value for $\hat{x}$:

$$\frac{\delta J}{\delta \hat{x}} = -2\int_{-\infty}^{\infty} (x - \hat{x}) f_X(x|Y = y)dx = 0 \tag{7.54}$$

Therefore,

$$\hat{x}\int_{-\infty}^{\infty} f_X(x|Y = y)dx = \int_{-\infty}^{\infty} x f_X(x|Y = y)dx \tag{7.55}$$

Because

$$\int_{-\infty}^{\infty} f_X(x|Y = y)dx = 1 \tag{7.56}$$

it follows that

$$\hat{x} = \langle X|Y = y\rangle = g(y) \tag{7.57}$$

Thus the best estimate of x is the conditional mean, or the expected value of X, given a measurement y. In general, this is a nonlinear function of y. However, if $f_X(x|Y = y)$ is Gaussian, then the best estimate is a *linear* function of the measurement. A similar result can be derived for a random vector $\mathbf{X}$.

7.4 FUNCTIONS OF RANDOM VARIABLES

A function of a random variable is itself a random variable. Let

$$Y = H(X) \tag{7.58}$$

Given $f_X(x)$, the probability density function of X, we want to find the probability density function of Y—that is, $f_Y(y)$. Suppose H is a single-valued function, either nondecreasing or nonincreasing as shown in Figure 7.2(a) and (b), respectively. From the figure, for the nondecreasing function,

$$P\{X \leq x\} = P\{Y \leq y\}$$

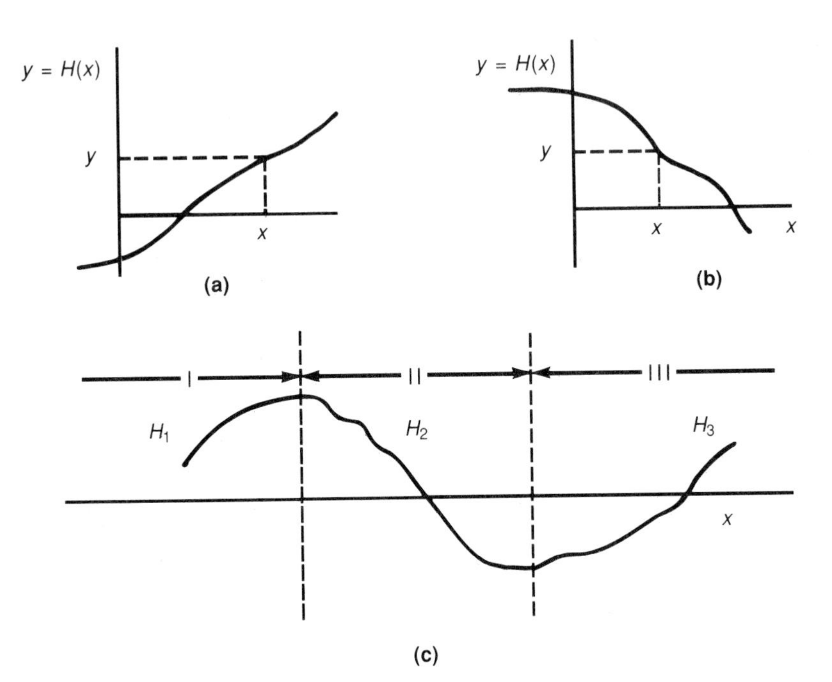

FIGURE 7.2 Obtaining the probability density function of a random variable. (a) Nondecreasing function. (b) Nonincreasing function. (c) Function not single-valued.

or

$$F_X(x) = F_Y(y)$$

which implies that

$$f_X(x)dx = f_Y(y)dy$$

Therefore,

$$f_Y(y) = f_X(x)\frac{dx}{dy} \tag{7.59}$$

For the nonincreasing function, we have

$$P\{X \leq x\} = P\{Y \geq y\}$$

or

$$F_X(x) = 1 - F_Y(y)$$

This implies that

$$f_X(x)dx = -f_Y(y)dy$$

Therefore,

$$f_Y(y) = -f_X(x)\frac{dx}{dy} \tag{7.60}$$

Because $dx/dy < 0$ in this case, $f_Y(y) \geq 0$. We can combine (7.59) and (7.60) to get

$$f_Y(y) = f_X(x)\left|\frac{dx}{dy}\right| \tag{7.61}$$

From (7.58),

$$X = H^{-1}(Y) = G(Y) \tag{7.62}$$

where G is the inverse of the single-valued function H. Then (7.61) can be written

$$f_Y(y) = f_X[G(y)]\left|\frac{dG(y)}{dy}\right| \tag{7.63}$$

If H is not single-valued, its domain is broken up into regions over which it is single-valued, as shown in Figure 7.2(c). Then, treating each single-valued function H_i as before, we get

$$f_Y(y) = \sum_{i=1}^{n} f_X[G_i(y)]\left|\frac{dG_i(y)}{dy}\right| \tag{7.64}$$

where $G_i = H_i^{-1}$ and $n = 3$ for the function depicted in Figure 7.2. Let us consider an example.

EXAMPLE 7.4

Let X be a random variable distributed $N[\mu, \sigma^2]$. If $Y = e^X$, $X \geq 0$, and $Y = e^{-X}$, $X < 0$, find the probability density function of Y.

Solution

In this case, $H_1(X) = e^{-X}$ and $H_2(X) = e^X$. Hence $G_1(Y) = -\log_e Y$ and $G_2(Y) = \log_e Y$. Also, $dG_1(y)/dy = -1/y$ and $dG_2(y)/dy = 1/y$. Therefore, from (7.64),

$$\begin{aligned} f_Y(y) &= f_X(-\log_e y)\left|-\frac{1}{y}\right| + f_X(\log_e y)\left|\frac{1}{y}\right| \\ &= \frac{1}{y}[f_X(-\log_e y) + f_X(\log_e y)] \\ &= \frac{1}{\sqrt{2\pi}\sigma y}\left\{\exp\left|-\frac{(-\log_e(y-\mu)^2}{2\sigma^2}\right| + \exp\left|-\frac{(\log_e(y-\mu)^2}{2\sigma^2}\right|\right\}, \quad y \geq 1 \\ &= 0, \quad y < 1 \end{aligned}$$

7.5 CHARACTERISTIC FUNCTIONS

The characteristic function of a random variable X is defined as

$$C_X(\omega) = \int_{-\infty}^{\infty} e^{j\omega x} f_X(x)dx \tag{7.65}$$

We see that this is the same as the Fourier transform of the probability density function as it was defined in Chapter 2 except that the exponent has a positive sign. Thus the characteristic function and the density function form a Fourier transform pair with

$$f_X(x) = \frac{1}{2\pi} \int_{-\infty}^{\infty} e^{-j\omega x} C_X(\omega)d\omega \tag{7.66}$$

We note from (7.65) that

$$C_X(\omega) = \langle e^{j\omega X} \rangle \tag{7.67}$$

This facilitates evaluation of the probability density of the sum of independent random variables in terms of the individual densities. For example, let $Z = X + Y$, where X and Y are independent random variables with densities $f_X(x)$ and $f_Y(y)$, respectively. Then the characteristic function of Z is

$$C_Z(\omega) = \langle e^{j\omega Z} \rangle = \langle e^{j\omega(X+Y)} \rangle = \langle e^{j\omega X} \rangle \langle e^{j\omega Y} \rangle \tag{7.68}$$

Hence

$$C_Z(\omega) = C_X(\omega)C_Y(\omega) \tag{7.69}$$

From the convolution theorem for Fourier transforms, $f_Z(z)$ is the linear convolution of $f_X(x)$ and $f_Y(y)$—that is,

$$\begin{aligned} f_Z(z) &= f_X(x) * f_Y(y) \\ &= \int_{-\infty}^{\infty} f_X(\sigma) f_Y(z - \sigma)d\sigma \end{aligned} \tag{7.70}$$

In general, if

$$Z = \sum_{i=1}^{N} X_i \tag{7.71}$$

and the X_i are independent, then

$$C_Z(\omega) = \prod_{i=1}^{N} C_{X_i}(\omega) \tag{7.72}$$

Therefore,

$$f_Z(z) = f_1(x_1) * f_2(x_2) * \cdots * f_N(x_N) \tag{7.73}$$

The characteristic function is also called the moment-generating function, because it can be used to get the moments of the random variable, as described below. We have

$$C_X(\omega) = \int_{-\infty}^{\infty} e^{j\omega x} f_X(x)dx \tag{7.74}$$

$$\left.\frac{dC_X(\omega)}{d\omega}\right|_{\omega=0} = j\left.\int_{-\infty}^{\infty} xe^{j\omega x} f_X(x)dx\right|_{\omega=0} = j\langle X\rangle \tag{7.75}$$

and

$$\left.\frac{d^2C_X(\omega)}{d\omega^2}\right|_{\omega=0} = j^2\langle X^2\rangle \tag{7.76}$$

In general,

$$\left.\frac{d^nC_X(\omega)}{d\omega^n}\right|_{\omega=0} = j^n\langle X^n\rangle \tag{7.77}$$

and the nth moment is given by

$$\langle X^n\rangle = (-j)^n \left.\frac{d^nC_X(\omega)}{d\omega^n}\right|_{\omega=0} \tag{7.78}$$

7.6 RANDOM PROCESSES

Loosely speaking, a random process is a finite or infinite set of random variables, ordered in time. A more precise definition is that a random process $X(t) = X(t, \alpha)$ is a *function* the domain of which is the sample space and the range of which is a set of real functions of time.

Each time function is called a realization of the random process or a random signal. The functions of time may be either continuous or discrete. For the applications in this text, discrete time functions, or sequences, and discrete random processes are of primary interest. Nevertheless, it is instructive to examine both continuous and discrete processes, and we shall do so. As a simple example of a discrete random process, consider the following. A sequence $X(t)$, $t = kT$, is generated by ten successive throws of a die, and the value of each sample is equal to that showing on the uppermost face after each throw. A second sequence is generated by a second set of ten throws, and so on. The collection of sequences so generated is called the *ensemble* of the random process $X(t)$, $t = kT$. These sequences might be as shown in Figure 7.3. The ensemble of a continuous random process $X(t)$ might be a collection of continuous waveforms $x(t)$ forming the output of a noise generator, as shown in Figure 7.4. At any time $t = t_1$ or $t = kT$, the value $x(t)$ is a realization of a random variable $X(t)$ with associated first-order cumulative probability distribution and probability density functions

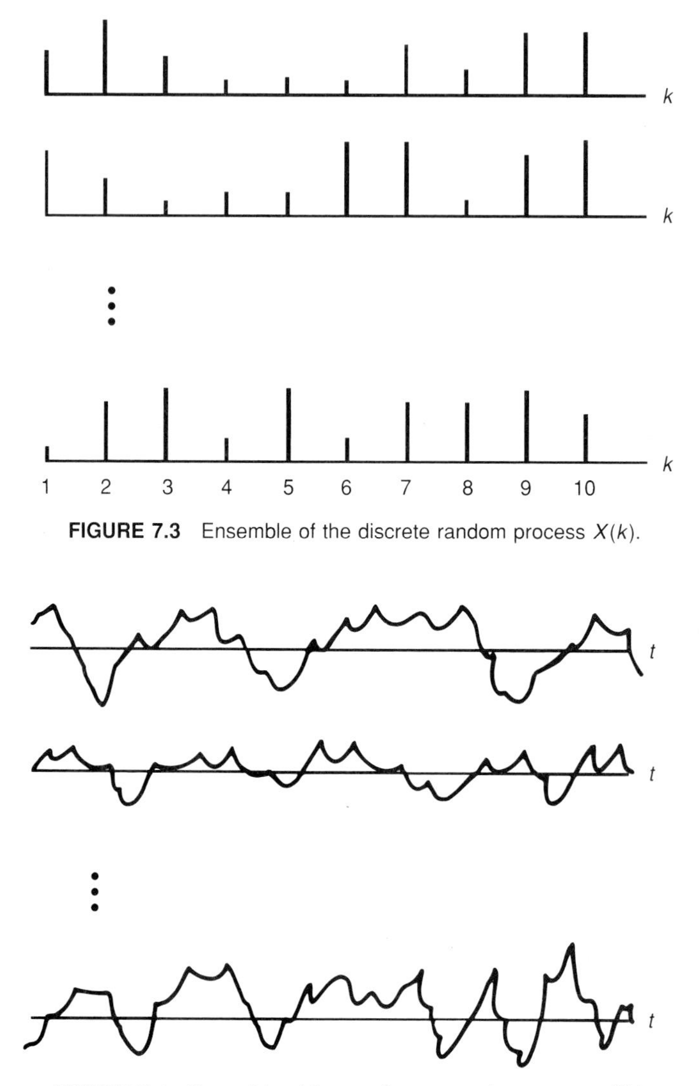

FIGURE 7.3 Ensemble of the discrete random process $X(k)$.

FIGURE 7.4 Ensemble of the continuous random process $X(t)$.

defined, respectively, as

$$F_1(x, t) = P\{X(t) \le x\} \tag{7.79}$$

and

$$f_1(x, t) = \frac{\partial F_1(x, t)}{\partial x} \tag{7.80}$$

In general, these are functions of time, as are the mean, variance, and other moments based on them.

1. *Mean*

$$\langle X(t)\rangle = \mu_X(t) = \int_{-\infty}^{\infty} x f_1(x,t)dx \tag{7.81}$$

2. *Variance*

$$\langle [X(t)-\mu_X(t)]^2\rangle = \sigma_X^2(t) = \int_{-\infty}^{\infty} [x-\mu_X(t)]^2 f_1(x,t)dx \tag{7.82}$$

3. kth *moment*

$$\langle X^k(t)\rangle = m_X^k(t) = \int_{-\infty}^{\infty} x^k f_1(x,t)dx \tag{7.83}$$

As we have noted, the time variable t may be either continuous or discrete. If at any time t, the random variable $X(t)$ is discrete also, then we must use (7.28) to convert the discrete distribution to $f_1(x,t)$ for use in the integrals (7.81) through (7.83). For the meaning of, say, $\langle X(t)\rangle$, we consider an ensemble of time functions making up the random process $X(t)$, as shown in Figure 7.5, select a particular time t, and average the values of all the realizations over the ensemble. We could do this at many different times and get an idea of how the mean varies with time. However, this would not tell us how the values of $X(t)$ at two different times t_1 and t_2 are related and, hence, the frequency content of the process. For this, we need the second-order or joint distribution function $F_2(x_1,x_2,t_1,t_2)$ between the two random variables $X(t_1)$ and $X(t_2)$ ($t_1 = kT$, $t_2 = mT$ for a discrete-time process). This function is

$$F_2(x_1,x_2,t_1,t_2) = P\{X(t_1)\le x_1, \quad X(t_2)\le x_2\} \tag{7.84}$$

The associated joint density function is

$$f_2(x_1,x_2,t_1,t_2) = \frac{\partial^2 F_2(x_1,x_2,t_1,t_2)}{\partial x_1 \partial x_2}$$

and

$$\begin{aligned} f_2(x_1,x_2,t_1,t_2)dx_1dx_2 = \ & P\{x_1 < X(t_1) \le x_1 + dx_1, \\ & x_2 < X(t_2) \le x_2 + dx_2\} \end{aligned} \tag{7.85}$$

The distribution functions F_1 and F_2, or the densities f_1 and f_2, are called the first- and second-order statistics of the process. A complete description of a random process would involve the joint density functions of n random variables, where n is very large or infinite. Clearly, achieving such a description is impractical. In practice, a probabilistic description of the process is restricted to the joint density function $f_2(x_1,x_2,t_1,t_2)$ of two random variables $X(t_1)$ and $X(t_2)$, where t_1 and t_2 are *any* two time instances in the domain of the random process $X(t)$. We note that if $t_1 = t_2$, the second-order density gives the first-order density.

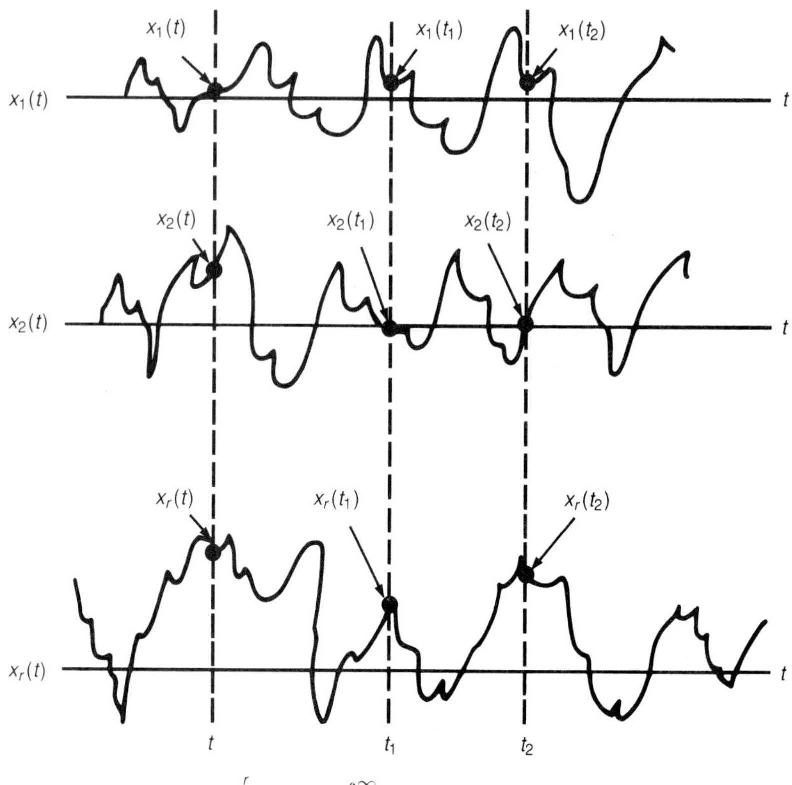

Mean $\mu_X(t) = \lim_{r\to\infty} \frac{1}{r} \sum_{i=1}^{r} x_i(t) = \int_{-\infty}^{\infty} x f_1(x,t)\,dx$

Autocorrelation function

$$\phi_X(t_1,t_2) = \lim_{r\to\infty} \frac{1}{r} \sum_{i=1}^{r} x_i(t_1)x_i(t_2) = \int_{-\infty}^{\infty}\int_{-\infty}^{\infty} x_1 x_2 f_2(x_1,x_2,t_1,t_2)\,dx_1\,dx_2$$

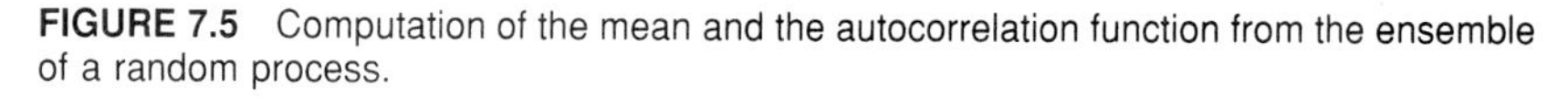

FIGURE 7.5 Computation of the mean and the autocorrelation function from the ensemble of a random process.

From the second-order density, we can get the autocorrelation function and the autocovariance function of the random process $X(t)$. The autocorrelation function is defined as

$$\begin{aligned}\phi_X(t_1,t_2) &= \langle X(t_1)X(t_2)\rangle \\ &= \int_{-\infty}^{\infty}\int_{-\infty}^{\infty} x_1 x_2 f_2(x_1,x_2,t_1,t_2)\,dx_1\,dx_2 \qquad (7.86)\end{aligned}$$

and is a measure of the correlation between, or interdependence of, $X(t_1)$ and $X(t_2)$. The physical significance of this operation is apparent in Figure 7.5.

If we subtract the means in (7.86) before taking the expected value, we have the autocovariance function

$$\gamma_X(t_1, t_2) = \langle [X(t_1) - \mu_1][X(t_2) - \mu_2] \rangle$$
$$= \int_{-\infty}^{\infty} \int_{-\infty}^{\infty} (x_1 - \mu_1)(x_2 - \mu_2) f_2(x_1, x_2, t_1, t_2) dx_1 dx_2 \quad (7.87)$$

where $\mu_1 = \mu_X(t_1)$ and $\mu_2 = \mu_X(t_2)$.

We can also define crosscorrelation and crosscovariance functions between *two* separate random processes $X(t)$ and $Y(t)$. Thus the crosscovariance function is

$$\gamma_{XY}(t_1, t_2) = \langle [X(t_1) - \mu_X][Y(t_2) - \mu_Y] \rangle$$
$$= \int_{-\infty}^{\infty} \int_{-\infty}^{\infty} (x - \mu_X)(y - \mu_Y) f_{XY}(x, y, t_1, t_2) dx\, dy \quad (7.88)$$

where

$$f_{XY}(x, y, t_1, t_2) = P\{x < X(t_1) < x + dx, \quad y < Y(t_2) \le y + dx\} \quad (7.89)$$

and $\mu_X = \mu_X(t_1)$, $\mu_Y = \mu_Y(t_2)$. The crosscorrelation function $\phi_{XY}(t_1, t_2)$ is defined in a similar manner to $\gamma_{XY}(t_1, t_2)$, but the means μ_X and μ_Y are not removed.

Both correlation functions and covariance functions are extensively used in the analysis of random processes. They are particularly useful when the process is *stationary*. A random process is said to be stationary if its statistics are independent of time. It is *strictly* stationary if

$$f_n(x_1, x_2, \ldots, x_n, t_1, \ldots, t_n) = f_n(x_1, \ldots, x_n, t_1 + \epsilon, \ldots, t_n + \epsilon) \quad (7.90)$$

for any n and ϵ. This would imply that

$$f_1(x, t) = f_1(x, t + \epsilon) = f_1(x) \quad (7.91)$$

independent of t. Also,

$$f_2(x_1, x_2, t_1, t_2) = f_2(x_1, x_2, t_1 + \epsilon, t_2 + \epsilon)$$
$$= f_2(x_1, x_2, \tau) \quad (7.92)$$

where $\tau = t_2 - t_1$, and so on for higher-order density functions. All other random processes are said to be nonstationary. It is difficult to prove strict-sense stationarity. If we limit ourselves to a second-order density function to describe the process, then the process is said to be weakly stationary, or stationary *in the wide sense*, if

1. $\langle X(t) \rangle = \mu_X$, independent of t
2. $\gamma_X(t_1, t_2) = \gamma_X(\tau)$ [or $\phi_X(t_1, t_2) = \phi_X(\tau)$] (7.93)

Thus, for a wide-sense stationary random process, the autocovariance and autocorrelation are functions of a single time variable τ:

$$\phi_X(\tau) = \langle X(t)X(t+\tau)\rangle \tag{7.94}$$

and

$$\gamma_X(\tau) = \langle [X(t) - \mu_X][X(t+\tau) - \mu_X]\rangle \tag{7.95}$$

Similarly, for the crosscorrelation and crosscovariance of two wide-sense stationary random processes $X(t)$ and $Y(t)$,

$$\phi_{XY}(\tau) = \langle X(t)Y(t+\tau)\rangle \tag{7.96}$$

and

$$\gamma_{XY}(\tau) = \langle [X(t) - \mu_X][Y(t+\tau) - \mu_Y]\rangle \tag{7.97}$$

In this text, when we mention stationary random processes, we will mean stationary in the wide sense as we have just defined it. For discrete-time processes, the time shift τ is a discrete time variable; that is, $\tau = \ell T$, where ℓ is integer.

We note that the analysis of nonstationary processes is difficult. Their correlation and covariance functions depend on two time variables t_1 and t_2 as given above. Hence we would need two-dimensional Fourier transforms to describe their frequency spectra. Very few analytic results are available for nonstationary processes and how they are operated upon by systems. In general, it is necessary to investigate them on a computer. Fortunately, in practice, many random processes are stationary. Others may be considered to be stationary if a time-varying trend is removed by filtering, as described in Chapter 5. We shall concentrate on stationary processes for the remainder of this chapter.

7.7 PROPERTIES OF AUTOCORRELATION AND AUTOCOVARIANCE FUNCTIONS FOR A STATIONARY RANDOM PROCESS *X*(*t*)

The autocorrelation and autocovariance functions have some useful properties.

1. *For zero time shift,*

$$\phi_X(0) = \langle X^2(t)\rangle \tag{7.98}$$

which is the mean-square value of the process. Also,

$$\gamma_X(0) = \langle (X(t) - \mu_X)^2\rangle \tag{7.99}$$

which is the variance σ_X^2 of $X(t)$. These relationships follow from the definitions (7.86) and (7.87). It is evident from this property that both the mean-square value and the variance are constant with respect to time.

2. *Both are even functions of* τ.
The proof is straightforward:

$$\phi_X(-\tau) = \langle X(t)X(t-\tau)\rangle$$

Let $u = t - \tau$ or $t = u + \tau$. Then

$$\phi_X(-\tau) = \langle X(u)X(u+\tau)\rangle = \phi_X(\tau) \tag{7.100}$$

A similar proof holds for $\gamma_X(\tau)$.

3. *Both functions achieve their maximum values at* $\tau = 0$.
To prove this, we note that for all values of t and τ

$$[X(t) \pm X(t+\tau)]^2 \geq 0$$

This implies that

$$\langle [X(t) \pm X(t+\tau)]^2\rangle \geq 0$$

Expand this, noting that the expected-value operation is linear.

$$\langle X^2(t)\rangle \pm 2\langle X(t)X(t+\tau)\rangle + \langle X^2(t+\tau)\rangle \geq 0$$

or

$$2\phi_X(0) \pm 2\phi_X(\tau) \geq 0$$

This means that

$$\phi_X(0) \geq \phi_X(\tau)$$

and

$$\phi_X(0) \geq -\phi_X(\tau)$$

Therefore,

$$\phi_X(0) \geq |\phi_X(\tau)| \tag{7.101}$$

A similar proof holds for $\gamma_X(\tau)$.

4. $\lim_{\tau\to\infty} \phi_X(\tau) = \mu_X^2$, *assuming that there are no periodicities in* $X(t)$.

Proof

As $\tau \to \infty$, the random variables $X(t)$ and $X(t+\tau)$ become statistically independent. Therefore,

$$\begin{aligned}\lim_{\tau\to\infty} \phi_X(\tau) &= \lim_{\tau\to\infty} \langle X(t)X(t+\tau)\rangle \\ &= \lim_{\tau\to\infty} \langle X(t)\rangle\langle X(t+\tau)\rangle \\ &= \mu_X^2 \end{aligned} \tag{7.102}$$

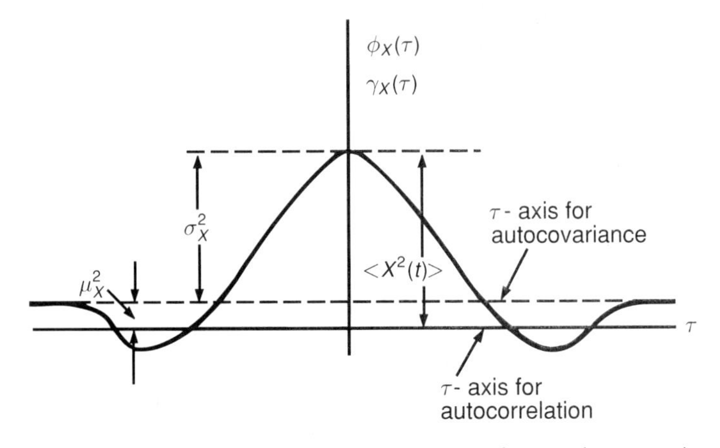

FIGURE 7.6 Typical autocorrelation and autocovariance for a stationary random process $X(t)$.

We can prove in the same manner that

$$\lim_{\tau \to \infty} \gamma_X(\tau) = 0 \tag{7.103}$$

In Figure 7.6, the foregoing properties are summarized for the two functions. This figure takes into account the fact that

$$\begin{aligned} \gamma_X(\tau) &= \langle [X(t) - \mu_X][X(t+\tau) - \mu_X] \rangle \\ &= \langle X(t)X(t+\tau) \rangle - \mu_X \langle X(t+\tau) \rangle - \mu_X \langle X(\tau) \rangle + \mu_X^2 \\ &= \phi_X(\tau) - \mu_X^2 \end{aligned} \tag{7.104}$$

The discrete-time autocorrelation and autocovariance functions have similar properties. The proof is left as an exercise.

7.8 ERGODIC PROCESSES

Sometimes, from knowledge of how the random process $X(t)$ is generated, it is possible to derive an analytic expression for it. Then we can compute its mean, variance, autocorrelation, and so on. For example, a sinewave generator might have errors in amplitude and phase, and its output might be a random process described by

$$X(t) = A \sin(\omega t + \Theta) \tag{7.105}$$

where amplitude A and phase Θ are random variables. In most cases, however, no analytic expression is available. Instead, we may have a record, or sequence of data, generated by an unknown or vaguely known random process and may wish to extract meaningful information about its statistics or averages. If the process can be assumed to be stationary, we can make use of the *ergodic hypothesis:*

The Ergodic Hypothesis

A stationary random process $X(t)$ is said to be ergodic in the most general sense if all its statistics can be determined from a *single* realization $x_k(t)$ of the process, provided the realization is long enough.

It follows that if a process is ergodic, time averages equal ensemble averages in the limit as the length of the record becomes infinite. For example, with a continuous record of length L, we have

$$\textit{Mean:}\quad \mu_X = \lim_{L\to\infty} \frac{1}{L}\int_0^L x_k(t)dt \tag{7.106}$$

$$\textit{Variance:}\quad \sigma_X^2 = \lim_{L\to\infty} \frac{1}{L}\int_0^L [x_k(t)-\mu_X]^2 dt \tag{7.107}$$

$$\textit{Autocovariance:}\quad \gamma_X(\tau) = \lim_{L\to\infty} \frac{1}{L}\int_0^L [x_k(t)-\mu_X][x_k(t+\tau)-\mu_X]dt \tag{7.108}$$

If the record is discrete with N samples x_i, the corresponding quantities are

$$\textit{Mean:}\quad \mu_X = \lim_{N\to\infty} \frac{1}{N}\sum_{i=1}^{N} x_i \tag{7.109}$$

$$\textit{Variance:}\quad \sigma_X^2 = \lim_{N\to\infty} \frac{1}{N}\sum_{i=1}^{N} (x_i-\mu_X)^2 \tag{7.110}$$

$$\textit{Autocovariance:}\quad \gamma_X(\ell) = \lim_{N\to\infty} \frac{1}{N}\sum_{i=1}^{N} (x_i-\mu_X)(x_{i+\ell}-\mu_X) \tag{7.111}$$

Of course, in practice we do not have records of infinite length, so the averages we can compute differ from the true values. These averages are called sample mean, sample variance, and so on, and they are designated by different letters. For instance, for a discrete time record,

$$\textit{Sample mean:}\quad \bar{x} = \frac{1}{N}\sum_{i=1}^{N} x_i \tag{7.112}$$

$$\textit{Sample variance:}\quad s_X^2 = \frac{1}{N}\sum_{i=1}^{N} (x_i-\bar{x})^2 \tag{7.113}$$

$$\textit{Sample autocovariance:}\quad r_X(\ell) = \frac{1}{N}\sum_{i=1}^{N-\ell} (x_i-\bar{x})(x_{i+\ell}-\bar{x}) \tag{7.114}$$

where $\ell = 0, 1, \ldots, N-1$.

In general, it is not possible to prove that a process is ergodic; the contention is merely a hypothesis. However, making this assumption—justified or not—often gives useful results if the underlying process is stationary. The sample quantities in (7.112) through (7.114) can be shown to be within specified confidence intervals of the true values, but this is outside the scope of our discussion.

7.9 SPECTRAL DENSITIES

A random process is a collection or ensemble of time functions, continuous or discrete. Each time function $x_k(t)$ has a frequency content that can be determined by taking the Fourier transform of the time function. However, the frequency content of any single realization is not very useful, because there is no guarantee that the particular realization will recur. Instead, we need a frequency characterization of the entire process based on the average behavior of the realizations. This is provided by the *power spectral density*.

According to the Wiener–Khintchine theorem, the power spectral density $\Gamma_X(\omega)$ is the Fourier transform of the autocovariance function. Sometimes it is taken as the Fourier transform of the autocorrelation function, but this gives rise to a delta function at the origin if the process has a nonzero mean. Therefore, for a continuous process, we will take the first definition,

$$\begin{aligned} \Gamma_X(\omega) &= \int_{-\infty}^{\infty} \gamma_X(\tau) e^{-j\omega\tau} d\tau \\ &= 2\int_{0}^{\infty} \gamma_X(\tau) \cos\omega\tau d\tau \end{aligned} \tag{7.115}$$

because γ_X is an even function. Therefore, the power spectral density is a real function of frequency. In other words, phase information is lost in the averaging process over the ensemble. The power spectral density is non-negative. Proof of this is left as an exercise.

It follows that

$$\gamma_X(\tau) = \frac{1}{2\pi}\int_{-\infty}^{\infty} \Gamma_X(\omega) e^{j\omega\tau} d\omega \tag{7.116}$$

If $\tau = 0$,

$$\gamma_X(0) = \sigma_X^2 = \frac{1}{2\pi}\int_{-\infty}^{\infty} \Gamma_X(\omega) d\omega \tag{7.117}$$

(that is, the variance equals a constant times the area under the power spectral density curve).

For a discrete process $X(t) = X(kT)$, the power spectral density is the Fourier transform of the autocovariance sequence $\gamma_X(\tau) = \gamma_X(lT)$, or the z-transform of the sequence evaluated on the unit circle—that is, at $z = e^{j\omega T}$

We have

$$\Gamma_X(z)\big|_{z=e^{j\omega T}} = \Gamma_X(\omega T) = \sum_{\ell=-\infty}^{\infty} \gamma_X(\ell T)e^{-j\ell\omega T}$$

$$= 2\sum_{\ell=0}^{\infty} \gamma_X(\ell T)\cos \ell\omega T \qquad (7.118)$$

because γ_X is an even function. As with the continuous case, $\Gamma_X(\omega T)$ is real, and all phase information is lost. Take the inverse transform of $\Gamma_X(z)$,

$$\gamma_X(\ell T) = \frac{1}{2\pi j}\oint_C \Gamma_X(z)z^{\ell-1}dz \qquad (7.119)$$

where C lies in the region of convergence for $\Gamma_X(z)$. It follows from (7.119) that the variance of $X(kT)$ is

$$\sigma_X^2 = \gamma_X(0) = \frac{1}{2\pi j}\oint_C \Gamma_X(z)z^{-1}\,dx \qquad (7.120)$$

A random process that is particularly useful is a stationary purely random process, or *white noise*. For the continuous case, this process has a flat power spectral density at all frequencies, as shown in Figure 7.7(a). This is impractical: It would require an infinite variance for the process (recall that the variance is proportional to the area under the power spectral density curve). In other words, it would require samples arbitrarily close together to be independent. Its autocovariance is a delta function. Thus, if the spectral density is

$$\Gamma_X(\omega) = A \qquad (7.121)$$

where A is a constant, then

$$\Gamma_X(\tau) = A\delta(\tau) \qquad (7.122)$$

In practice, we use what is called bandlimited white noise, where the spectral density is assumed to be constant out to a certain frequency ω_0, as shown in Figure 7.7(b). Then

$$\Gamma_X(\omega) = \begin{cases} A, & |\omega| < \omega_0 \\ 0, & \text{otherwise} \end{cases} \qquad (7.123)$$

The autocovariance function is the inverse Fourier transform of this.

$$\gamma_X(\tau) = \frac{A}{\pi}\frac{\sin\omega_0\tau}{\tau} \qquad (7.124)$$

It is depicted in Figure 7.7(b). This function allows some correlation between adjacent values of $X(t)$. We may make the cutoff frequency ω_0 as large as we like so that the bandlimited white noise approaches the ideal. We can use the ideal white noise of Figure 7.7(a) in mathematical computations and can use the bandlimited white noise when we want to approximate the ideal in practical situations.

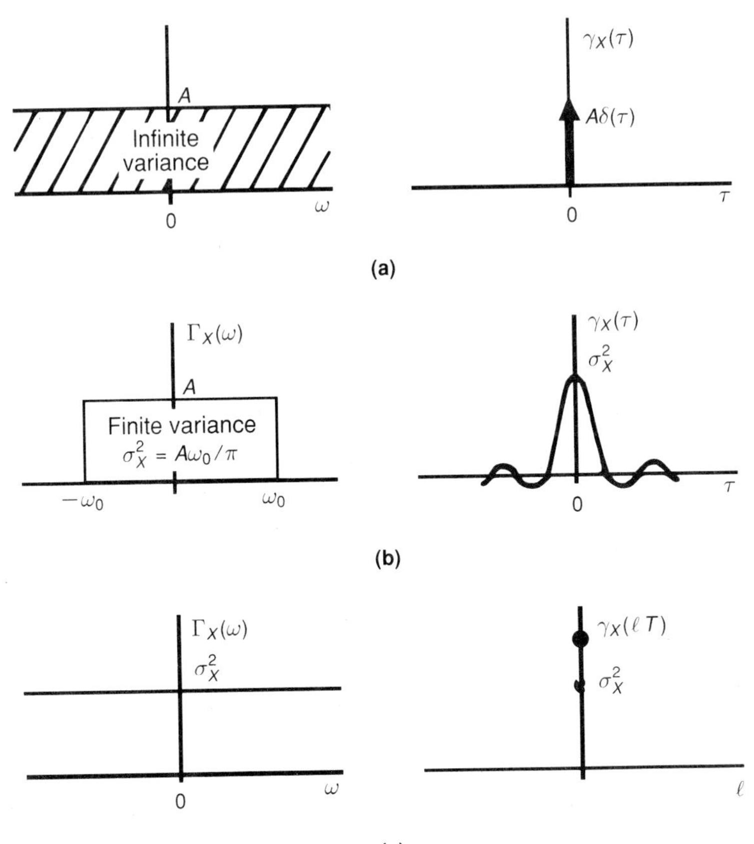

FIGURE 7.7 Power spectral density and autocovariance functions for continuous and discrete white noise processes. (a) White noise (continuous process). (b) Bandlimited white noise (continuous process). (c) White noise (discrete process).

No such difficulty arises for discrete random processes or sequences, wherein the samples are separated by a time step $T > 0$, and, accordingly, an infinite variance is not required between adjacent samples. In this case, if the white-noise process has zero mean and variance σ_X^2, then

$$\gamma_X(\tau) = \gamma_X[(i-j)T] = \gamma_X(\ell T) = \sigma_X^2 \delta_{\ell,0} \tag{7.125}$$

where i and j are integers, $\ell = i - j$, and

$$\delta_{\ell,0} = \begin{cases} 1, & \text{if } \ell = 0 \\ 0, & \text{if } \ell \neq 0 \end{cases} \tag{7.126}$$

From (7.118), the spectral density is

$$\Gamma_X(\omega) = \sum_{\ell=-\infty}^{\infty} \gamma_X(\ell T) e^{-j\ell\omega T}$$

$$= \sum_{\ell=-\infty}^{\infty} \sigma_X^2 \delta_{\ell,0} e^{-j\ell\omega T} = \sigma_X^2 \tag{7.127}$$

which is a constant, independent of frequency. This is shown in Figure 7.7(c). We note that for a discrete process, the variance is *not* proportional to the area under the spectral density curve but is given by the integral (7.120).

For white noise, whether continuous or discrete, the first-order probability density function can be Gaussian, uniform, or any other valid density. It is called Gaussian white noise if the first-order density function is Gaussian. What makes it white noise is the independence of samples, regardless of how close they are.

The cross-spectral density is the Fourier transform of the crosscovariance function. That is, it is

$$\Gamma_{XY}(j\omega) = \int_{-\infty}^{\infty} \gamma_{XY}(\tau) e^{-j\omega\tau} d\tau \tag{7.128}$$

for a continuous process. For a discrete process, it is

$$\Gamma_{XY}(j\omega T) = \sum_{\ell=-\infty}^{\infty} \gamma_{XY}(\ell T) e^{-j\ell\omega T} \tag{7.129}$$

The cross-spectral density, in general, has both magnitude and phase.

We shall now use the concepts of covariance functions and spectral densities to analyze how a linear system operates on a random process.

7.10 RANDOM PROCESS APPLIED TO A LINEAR SYSTEM

Suppose that a random process $X(t)$ is applied as input to a linear time-invariant system. By saying that the process is applied, we mean that at a given trial, a particular realization or time function $x_i(t)$ is applied, resulting in an output $y_i(t)$ related to the input through the convolution integral (for a continuous system)

$$y_i(t) = \int_{-\infty}^{\infty} x_i(t_1) h(t - t_1) dt_1 \tag{7.130}$$

where $h(t)$ is the impulse response or weighting function of the system. Taking another trial, with a second realization $x_j(t)$, results in a second output $y_j(t)$ related to x_j through the convolution integral. We can repeat the trials as often as desired, thereby generating an ensemble of outputs $y_i(t)$, or a random process $Y(t)$. We might never get the same input or output twice, so there is little to be gained by looking at the characteristics of any one input or output. Instead, we look at the *average* behavior as exemplified by the covariance functions and spectral densities of the input and output processes $X(t)$ and $Y(t)$, respectively. Because every input and corresponding output realization are related by the convolution integral, we can express the relationship between input and output process as

$$Y(t) = \int_{-\infty}^{\infty} h(t_1)X(t - t_1)dt_1 \tag{7.131}$$

Take expected values of both sides of (7.131).

$$\langle Y(t)\rangle = \int_{-\infty}^{\infty} h(t_1)\langle X(t - t_1)\rangle dt_1$$

or

$$\mu_Y = \mu_X \int_{-\infty}^{\infty} h(t_1)dt_1 \tag{7.132}$$

This relationship indicates how the mean value of the stationary random process is transformed by the linear system. In the following development, for both continuous and discrete processes, we assume that the input process has zero mean μ_X. It follows from (7.132) that $\mu_Y = 0$ also. This simplifies the expressions and does not change the principles involved. If $\mu_X \neq 0$, then we can subtract the mean from both input and output process—for example,

$$Y(t) - \mu_Y = \int_{-\infty}^{\infty} h(t_1)[X(t - t_1) - \mu_X]dt_1$$

and proceed from there.

Let us now determine how the input and output autocovariances are related. From (7.131),

$$Y(t + \tau) = \int_{-\infty}^{\infty} h(t_2)X(t + \tau - t_2)\, dt_2 \tag{7.133}$$

Therefore,

$$Y(t)Y(t + \tau) = \int_{-\infty}^{\infty} dt_1 h(t_1)X(t - t_1) \int_{-\infty}^{\infty} dt_2 h(t_2)X(t + \tau - t_2)$$

Take expected values of both sides.

$$\begin{aligned}
\gamma_Y(\tau) &= \langle Y(t)Y(t + \tau)\rangle \\
&= \int_{-\infty}^{\infty} dt_1 h(t_1) \int_{-\infty}^{\infty} dt_2 h(t_2)\langle X(t - t_1)X(t + \tau - t_2)\rangle \\
&= \int_{-\infty}^{\infty} dt_1 h(t_1) \int_{-\infty}^{\infty} dt_2 h(t_2)\gamma_X(\tau + t_1 - t_2)
\end{aligned} \tag{7.134}$$

If the input is white noise with $\gamma_X(\tau) = A\delta(\tau)$, then (7.134) becomes

$$\gamma_Y(\tau) = A \int_{-\infty}^{\infty} h(t_1)h(\tau + t_1)dt_1 \tag{7.135}$$

The lower limit of integration can be set to zero because h is realizable.

Expression (7.134) seems to be a rather complex relationship between input and output covariances. We can obtain a much simpler relationship between input and output spectral densities by taking the Fourier transform of both sides of (7.134). To simplify notation, let $s = j\omega$. Then

$$\begin{aligned}\Gamma_Y(s) &= \int_{-\infty}^{\infty} \gamma_Y(\tau)e^{-s\tau}\,d\tau \\ &= \int_{-\infty}^{\infty} d\tau e^{-s\tau}\left[\int_{-\infty}^{\infty} dt_1 h(t_1)\int_{-\infty}^{\infty} dt_2 h(t_2)\gamma_X(\tau + t_1 - t_2)\right] \quad (7.136)\end{aligned}$$

from (7.134). Change the order of integration, assuming uniform convergence of the infinite-limit integrals. Then

$$\begin{aligned}\Gamma_Y(s) &= \int_{-\infty}^{\infty} dt_1 h(t_1)\int_{-\infty}^{\infty} dt_2 h(t_2)\int_{-\infty}^{\infty} d\tau\gamma_X(\tau + t_1 - t_2)e^{-s\tau} \\ &= \int_{-\infty}^{\infty} dt_1 h(t_1)e^{st_1}\int_{-\infty}^{\infty} dt_2 h(t_2)e^{-st_2}\,\times \\ &\qquad \int_{-\infty}^{\infty} d\tau\gamma_X(\tau + t_1 - t_2)e^{-s(\tau+t_1-t_2)} \quad (7.137)\end{aligned}$$

In the last integral, change the variable of integration to $t_3 = \tau + t_1 - t_2$. Note that t_1 and t_2 are constant with respect to this integral, so $d\tau = dt_3$. Then

$$\begin{aligned}\Gamma_Y(s) &= \int_{-\infty}^{\infty} dt_1 h(t_1)e^{st_1}\int_{-\infty}^{\infty} dt_2 h(t_2)e^{-st_2}\int_{-\infty}^{\infty} dt_3\gamma_X(t_3)e^{-st_3} \\ &= H(-s)H(s)\Gamma_X(s) \quad (7.138)\end{aligned}$$

Because $s = j\omega$,

$$\Gamma_Y(\omega) = H(-j\omega)H(j\omega)\Gamma_X(\omega) = |H(j\omega)|^2\Gamma_X(\omega) \quad (7.139)$$

This is a very useful relationship between input and output spectral densities. If $X(t)$ is white noise with spectral density $\Gamma_X(\omega) = A$, then

$$\Gamma_Y(\omega) = |H(j\omega)|^2 A \quad (7.140)$$

so we can generate processes with different spectral densities and autocovariances by our selection of the transfer function $H(s)$.

EXAMPLE 7.5

A white-noise process $X(t)$ with $\Gamma_X(\omega) = A$ is applied to a linear system the transfer function of which is $H(s) = a/(s + a)$. Find the power spectral density, autocovariance, and variance of the output.

Solution

From (7.140), the output spectral density is

$$\Gamma_Y(\omega) = \frac{a^2 A}{\omega^2 + a^2}$$

The autocovariance is the inverse Fourier transform of this, namely

$$\gamma_Y(\tau) = Ce^{-a|\tau|}$$

where $C = aA/2$. This is called an exponential covariance function. The variance is

$$\sigma_Y^2 = \gamma_Y(0) = C$$

For a discrete process input to a discrete system, we have a similar development. Here we use the convolution summation to relate input and output processes:

$$Y(n) = \sum_{m=-\infty}^{\infty} h(m)X(n-m) \tag{7.141}$$

where $Y(n) = Y(nT)$, $X(m-n) = X[(m-n)T]$, and $h(m) = h(mT)$ is the unit pulse response or weighting function. Taking expected values of both sides of (7.141) yields

$$\langle Y(n)\rangle = \sum_{m=-\infty}^{\infty} h(m)\langle X(n-m)\rangle$$

or

$$\mu_Y = \mu_X \sum_{m=-\infty}^{\infty} h(m) \tag{7.142}$$

Now we relate input and output covariances

$$Y(n+\ell) = \sum_{p=-\infty}^{\infty} h(p)X(n+\ell-p) \tag{7.143}$$

and

$$\begin{aligned}\gamma_Y(\ell) &= \langle Y(n)Y(n+\ell)\rangle \\ &= \sum_{m=-\infty}^{\infty} h(m) \sum_{p=-\infty}^{\infty} h(p)\langle X(n-m)X(n+\ell-p)\rangle \\ &= \sum_{m=-\infty}^{\infty} h(m) \sum_{p=-\infty}^{\infty} h(p)\gamma_X(\ell+m-p)\end{aligned} \tag{7.144}$$

This gives the relationship between input and output autocovariances. If the input is discrete white noise with $\gamma_X(\ell) = \sigma_X^2\delta_{\ell,0}$, then $\gamma_X(\ell+m-p) = \sigma_X^2\delta_{p,\ell+m}$, and (7.144) becomes

$$\gamma_Y(\ell) = \sigma_X^2 \sum_{m=-\infty}^{\infty} h(m)h(\ell+m) \tag{7.145}$$

where the lower limit may be taken as zero because h is realizable. To get the spectral relationships, take the z-transform of (7.144).

$$\Gamma_Y(z) = \sum_{\ell=-\infty}^{\infty} \gamma_Y(\ell) z^{-\ell}$$

$$= \sum_{\ell=-\infty}^{\infty} \left[\sum_{m=-\infty}^{\infty} h(m) \sum_{p=-\infty}^{\infty} h(p) \gamma_X(\ell + m - p) \right] z^{-\ell} \tag{7.146}$$

Change the order of summation.

$$\Gamma_Y(z) = \sum_{m=-\infty}^{\infty} h(m) \sum_{p=-\infty}^{\infty} h(p) \sum_{\ell=-\infty}^{\infty} \gamma_X(\ell + m - p) z^{-\ell}$$

$$= \sum_{m=-\infty}^{\infty} h(m) z^{m} \sum_{p=-\infty}^{\infty} h(p) z^{-p} \sum_{\ell=-\infty}^{\infty} \gamma_X(\ell + m - p) z^{-(\ell+m-p)}$$

$$= H(z^{-1}) H(z) \Gamma_X(z) \tag{7.147}$$

In the second-to-last step, we set index $k = \ell + m - p$. Then, to get power spectral densities, replace z by $e^{j\omega T}$ in (7.147), obtaining

$$\Gamma_Y(\omega T) = |H(j\omega T)|^2 \Gamma_X(\omega T) \tag{7.148}$$

which is similar to (7.139) for the continuous case. Again, we can shape the output spectrum by passing a discrete white noise process with $\Gamma_X(\omega) = A$ through an arbitrary stable transfer function $H(z)$, obtaining

$$\Gamma_Y(\omega T) = |H(j\omega T)|^2 A \tag{7.149}$$

Using the same procedures, we can derive expressions for the cross-spectral density between input and output processes in terms of the spectral density of the input process, and the system function. These expressions are

$$\Gamma_{XY}(s) = H(s) \Gamma_X(s) \tag{7.150}$$

or

$$\Gamma_{XY}(j\omega) = H(j\omega) \Gamma_X(\omega) \tag{7.151}$$

for the continuous case and, for the discrete case,

$$\Gamma_{XY}(z) = H(z) \Gamma_X(z) \tag{7.152}$$

or

$$\Gamma_{XY}(j\omega T) = H(j\omega T) \Gamma_X(\omega T) \tag{7.153}$$

Proof of these relationships is left as an exercise. Expressions (7.151) and (7.153) indicate that for a white noise input, the cross-spectral density is proportional to the system frequency response. We can use this fact to identify the frequency response of an unknown system, or "black box," by applying white noise as input and computing the cross-spectral densities of the input and output processes.

7.11 THE MARKOV PROPERTY

Let us now examine some of the foregoing concepts in terms of the Markov property [6]. In Section 7.4, we stated that preparing a complete description of a general random process $X(t)$ would require knowledge of the joint probability density function of all the random variables the process comprises. More specifically, it would require knowledge of all possible probability density functions $f_N(x(t_0), \ldots, x(t_N))$ for all t_0, t_1, ... , t_N (not necessarily evenly spaced) in the interval $[t_0, t_f]$ and for any integer N between 1 and infinity. This information being impossible to obtain, it is comforting to find that most random processes, in practice, are well approximated by Markov processes of low order.

Definition

A random process $X(\tau)$ is said to be a Markov process if, for all $t_k > t_{k-1} > \cdots > t_1 > t_0$ in the interval $[t_0, t_f]$, the conditional probability density function

$$f[x(t_k)|x(t_{k-1}) \cdots x(t_1)x(t_0)] = f[(x(t_k)|x(t_{k-1})] \tag{7.154}$$

It follows that the random process is completely specified if we have the density functions $f[x(t_k)|x(t_{k-1})]$ and $f[x(t_{k-1})]$ or, equivalently, $f[x(t_k), x(t_{k-1})]$, for all t_k and t_{k-1} in the interval $[t_0,\ t_f]$. We note that this is the information we need to define the autocorrelation and autocovariance functions, $\phi_x(t_k,\ t_{k-1})$ and $\gamma_x(t_k,\ t_{k-1})$, respectively.

Definition

A Markov process is said to be stationary if $f[x(t_k)|x(t_{k-1})]$ and $f[x(t_{k-1})]$ are independent of k for all t_k and t_{k-1} in $[t_0,\ t_f]$.

This implies that $f[x(t_k), x(t_{k-1})]$ is independent of k and depends only on the difference $|t_k - t_{k-1}|$. If

$$f[x(t_k)|x(t_{k-1})] = f[x(t_k)] \tag{7.155}$$

then the process is purely random, or white noise. It is nonstationary unless $f[x(t_k)]$ is independent of k.

The foregoing definitions apply equally to continuous and to discrete Markov processes. Let us call a realization of the latter a Markov sequence. For this, the conditional density function (7.154) becomes

$$f(x_k|x_{k-1}, \ldots, x_1, x_0) = f(x_k|x_{k-1}) \tag{7.156}$$

That is, all knowledge regarding the previous values of the sequence samples is contained in x_{k-1}.

The reader can generate a simple Markov sequence as follows:

1. Throw a die or spin a wheel marked in such a way as to indicate a score when the wheel stops. Note the score.
2. Repeat the experiment as often as desired, each time noting both the score obtained and the *cumulative* score.

If the trials are independent of each other, the cumulative score will be a Markov sequence. After the kth trial, the cumulative score will be

$$y_k = y_{k-1} + v_k \tag{7.157}$$

where v_k is the score resulting from the kth trial. Equation (7.157) has the familiar form of a first-order recursive filter with coefficient $A = 1$. However, the input sequence v_k is a realization of a random process the samples of which are independent; that is, v_k is discrete white noise. In general, we can generate a Markov sequence by passing a white-noise sequence v_k through a first-order recursive filter, as shown in Figure 7.8. Here, for generality, we show a time-varying feedback gain A_k. This would give a sequence y_k belonging to a nonstationary Markov process Y_k. Conversely, we can consider a given Markov process as having resulted from white noise passed through a first-order continuous filter or a recursive digital filter.

A Markov process is sometimes called a first-order autoregressive process. Although the samples of the white-noise input sequence are mutually independent, the output sequence has correlated samples because of the filtering. This agrees with our findings in Section 7.8—for example, (7.145) and (7.148) with a white-noise input. White noise can be passed through higher-order recursive filters to generate higher-order autoregressive processes. By writing such systems in state-variable form, as shown in the next section, we can treat these cases as *vector* Markov processes.

A Markov process with the added restriction that $f[x(t_k)]$ and $f[x(t_k)|x(t_{k-1})]$ are Gaussian for all t_k and t_{k-1} in $[t_0, t_f]$ is called a Gauss–Markov process. Therefore, the first- and second-order probability density functions of a Gauss–Markov process $X(\tau)$ can be determined from the two *deterministic* time functions

$$\mu_X(t) = \langle X(t) \rangle \tag{7.158}$$

$$\sigma_X^2(t) = \langle [X(t) - \mu_X(t)]^2 \rangle \tag{7.159}$$

These will be constants if the process is stationary. It can be shown that linear transformations of a Gaussian random process are Gaussian [1]. It follows that we can always represent a Gauss–Markov random process as the output of a first-order continuous filter or recursive digital filter the input to which is Gaussian white noise and the initial state of which is Gaussian. We shall use this fact later in Chapter 9 in connection with the discrete Kalman filter.

EXAMPLE 7.6

A first-order discrete autoregressive process $X(k)$ is generated by the difference equation

$$X_k = AX_{k-1} + V_k, \quad k = 0, 1, 2, \ldots$$

where V_k is discrete white noise with autocovariance function $\gamma V(\ell) = \sigma_V^2 \delta(\ell)$, and A is a constant with $|A| < 1$. Find the normalized autocovariance function $\rho_X(\ell) = \gamma X(\ell)/\sigma_X^2$.

Solution

The given difference equation has the solution

$$X_k = \sum_{i=0}^{\infty} h_i V_{k-i}$$

if zero initial conditions are assumed. From Example 1.2, the unit pulse response h_i is equal to A^i.

The autocovariance function of the process X_k is

$$\begin{aligned}
\gamma X(\ell) &= \langle X_k X_{k+\ell} \rangle \\
&= \left\langle \sum_{i=0}^{\infty} h_i V_{k-i} \sum_{j=0}^{\infty} h_j V_{k+\ell-j} \right\rangle \\
&= \sum_{i=0}^{\infty} h_i \sum_{j=0}^{\infty} h_j \langle V_{k-i} V_{k+l-j} \rangle \\
&= \sum_{i=0}^{\infty} h_i \sum_{j=0}^{\infty} h_j \gamma V(\ell + i - j) \\
&= \sum_{i=0}^{\infty} h_i \sum_{j=0}^{\infty} h_j \sigma_V^2 \delta(\ell + i - j) \\
&= \sigma_V^2 \sum_{i=0}^{\infty} h_i h_{i+\ell}
\end{aligned}$$

Therefore,

$$\sigma_X^2 = \gamma X(0) = \sigma_V^2 \sum_{i=0}^{\infty} h_i^2$$

Thus the normalized autocovariance function of X_k is

$$\rho_X(\ell) = \gamma X(\ell)/\gamma X(0) = \frac{\sum_{i=0}^{\infty} h_i h_{i+\ell}}{\sum_{i=0}^{\infty} h_i^2} = A^\ell$$

The normalized autocovariance function for a Markov process, as derived in this example, can be used to find the correlation *length* of the process. The correlation length ℓ_c will be defined as the number of samples for $\rho_X(\ell)$ to fall to 0.368 (or e^{-1}) of that sample's value (unity) at $\ell = 0$. The correlation length is a measure of the interdependence between the samples of a random process. Thus $A^{\ell c} = e^{-1}$, or

$$\ell_c = -\frac{1}{\log_e A}$$

If it is desired to measure the correlation length in time or distance, simply multiply ℓ_c by the time or distance between samples.

It is sometimes necessary to generate a Markov process that has a desired standard deviation and correlation length. This is readily done, as shown in the following example.

EXAMPLE 7.7

Generate a Markov process by passing white noise through a first-order recursive filter, as shown in Figure 7.8, with A constant. The process $X(k)$ is to have a standard deviation σ_X of 40 ft and a correlation length of 1 second. Find the value of coefficient A and the standard deviation of the white-noise process $V(k)$. The computer time step is 0.1 second.

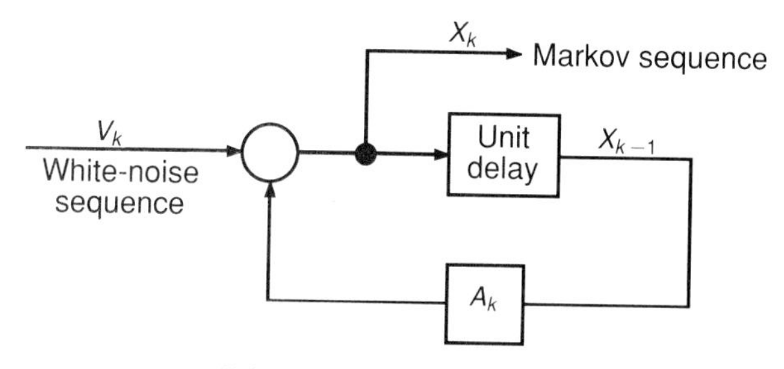

FIGURE 7.8 Markov process.

Solution

The number of samples in the correlation length is $\ell_c = 1.0/0.1 = 10$. Because $A^{\ell_c} = e^{-1}$, it follows that

$$A = (0.368)^{1/\ell_c} = 0.905$$

From (7.145), with appropriate change of notation, and setting $\ell = 0$, we have

$$\sigma_X^2 = \gamma_X(0) = \sigma_V^2 \sum_{m=0}^{\infty} h^2(m) = \sigma_V^2/(1 - A^2)$$

because $h(m) = A^m$. It follows that

$$\sigma_V = \sigma_X(1 - A^2)^{1/2} = 17.03$$

is the value for the standard deviation of the white noise.

7.12 VECTOR RANDOM PROCESSES AND THE MARKOV PROPERTY

Often it is convenient, for simplicity of notation, to consider two or more random processes that have similar roles to be components of a *vector* random process. Some examples:

1. Magnitude and direction of wind at a fixed location as a function of time
2. Amplitude and phase of a radio wave at a receiver as a function of time
3. Position and velocity of an aircraft in geographic x, y, z coordinates as a function of time

In the first two examples, the vector random process has two components; that is, it has dimensionality of two. In the third example, the dimensionality is six. We will use a boldface capital letter for a vector random process. At any particular time t ($t = kT$ for a discrete process), it is a vector random variable. Thus, for an n-dimensional vector,

$$\mathbf{X}(t) = \begin{bmatrix} X_1(t) \\ X_2(t) \\ \vdots \\ X_n(t) \end{bmatrix} \tag{7.160}$$

When an operation is indicated for a vector process, it means that the operation is applied to each of the component processes. For example,

$$\langle \mathbf{X}(t) \rangle = \begin{bmatrix} \langle X_1(t) \rangle \\ \langle X_2(t) \rangle \\ \vdots \\ \langle X_n(t) \rangle \end{bmatrix} \qquad \text{Var } \mathbf{X}(t) = \begin{bmatrix} \text{Var } X_1(t) \\ \text{Var } X_2(t) \\ \vdots \\ \text{Var } X_n(t) \end{bmatrix} \tag{7.161}$$

$$\gamma_{\mathbf{X}}(t) = \begin{bmatrix} \gamma_{X_1}(t) \\ \vdots \\ \gamma_{X_n}(t) \end{bmatrix} \qquad \mu_{\mathbf{X}}(t) = \begin{bmatrix} \mu_{X_1}(t) \\ \vdots \\ \mu_{X_n}(t) \end{bmatrix}$$

If $\mathbf{X}(t)$ is a stationary vector random process, then each component process $X_1(t), \ldots, X_n(t)$ of $\mathbf{X}(t)$ is stationary. If $\mathbf{X}(t)$ is also Gaussian, then the first-order vector probability function of $\mathbf{X}(t)$ has the form (7.49) with covariance matrix S given by (7.50).

Suppose we have a random process $Y(k)$ generated by passing discrete white noise $V(k)$ through a second-order recursive filter.

$$Y(k) = A_1 Y(k-1) + A_2 Y(k-2) + V(k) \tag{7.162}$$

Expressed as in (7.162), $Y(k)$ is not a Markov process. However, we can convert (7.162) into a *vector* Markov form by defining

$$\begin{aligned} X_1(k) &= Y(k-1) \\ X_2(k) &= Y(k-2) \end{aligned} \tag{7.163}$$

Then we can write (7.162) as the two first-order equations

$$\begin{aligned} X_1(k+1) &= A_1 X_1(k) + A_2 X_2(k) + V(k) \\ X_2(k+1) &= X_1(k) \end{aligned}$$

or

$$\begin{bmatrix} X_1(k+1) \\ X_2(k+1) \end{bmatrix} = \begin{bmatrix} A_1 & A_2 \\ 1 & 0 \end{bmatrix} \begin{bmatrix} X_1(k) \\ X_2(k) \end{bmatrix} + \begin{bmatrix} 1 \\ 0 \end{bmatrix} V(k) \tag{7.164}$$

Define the random vector process

$$\mathbf{X}(k) = \begin{bmatrix} X_1(k) \\ X_2(k) \end{bmatrix} \tag{7.165}$$

Then (7.164) becomes

$$\mathbf{X}_1(k+1) = A\mathbf{X}(k) + BV(k) \tag{7.166}$$

where

$$A = \begin{bmatrix} A_1 & A_2 \\ 1 & 0 \end{bmatrix}; \qquad B = \begin{bmatrix} 1 \\ 0 \end{bmatrix}$$

Equation (7.166) is in the Markov form, and $\mathbf{X}(k)$ is said to be a vector Markov process. We can generalize this procedure by exciting an nth-order recursive filter with a discrete white noise process and defining an nth-order vector process $\mathbf{X}(k)$ that satisfies a vector equation of the form (7.166). Thus the study of Markov processes has widespread application.

7.13 SUMMARY

The axioms on which probability theory is based are reviewed, and a foundation is laid for the treatment and manipulation of random variables. This leads to a discussion of random processes, or time-varying random variables, and how to describe them. We concentrate on the correlation and spectral characteristics of stationary random processes, including ergodic processes, and we derive the properties of the output process that results from the application of a stationary process as input to a linear time-invariant system. The Markov property is discussed for scalar and vector processes.

REFERENCES FOR CHAPTER 7

1. A. Papoulis, *Probability, Random Variables and Stochastic Processes* (New York: McGraw-Hill, 1965).
2. E. Parzen, *Stochastic Processes* (San Francisco: Holden-Day, 1962).
3. W. Feller, *An Introduction to Probability Theory and Its Applications* Vol. I (New York: Wiley, 1968).
4. G. M. Jenkins and D. G. Watts, *Spectral Analysis and Its Applications* (San Francisco: Holden-Day, 1968).
5. A. W. Drake, *Fundamentals of Applied Probability Theory* (New York: McGraw-Hill, 1967).
6. A. E. Bryson and Y. C. Ho, *Applied Optimal Control* (Waltham, MA: Blaisdell, 1969), Chap. 11.

EXERCISES FOR CHAPTER 7

1. A pair of dice is thrown, and the sum of the numbers on the uppermost faces is noted. What is the sample space for this experiment? If only the larger of the two numbers is noted (it must be larger), what is the sample space?

2. State a reasonable range and domain for each of the following random variables:
   a) The weight of a male high school student selected at random
   b) The number of throws of a single die before a five results
   c) The cumulative sum resulting from three flips of a coin if 1 corresponds to "heads" and 0 to "tails"

3. Two players alternately flip a coin up to a maximum of 4 times each. The player who gets a head first wins. Define the three possible outcomes of this experiment as a) first player wins, b) second player wins, and c) no one wins. What is the probability of each outcome?

4. Three marksmen, A, B, and C, have probabilities of 0.6, 0.7, and 0.9, respectively, of hitting a distant target.

a) If each fires two shots, what is the probability of at least one hit?

b) If all three simultaneously fire one shot, and one hit is noted, what is the probability that it was fired by marksman A?

c) One of the marksmen, chosen at random, fires at the target. What is the probability that the target is hit?

5. A population consists of 40% males and 60% females. A certain physical trait is known to occur in 10% of males and in 1% of females. A person is selected at random and is found to have the trait. What is the probability that this person is male?

6. Prove that the random variable X with density function

$$f_X(x) = \frac{1}{\sigma_X \sqrt{2\pi}} \exp\left[-(x - \mu_X)^2/(2\sigma_X^2)\right]$$

has mean μ_X and variance σ_X^2.

7. From the joint density function $f_{XY}(x, y)$ in (7.45), find the *marginal* density function $f_Y(y)$ by integrating over all x. Then find the conditional density function $f_X(x|Y = y)$.

8. Which of the following are valid probability density functions?

a)

$$f_X(x) = \begin{cases} ae^{-bx}, & x \geq 0 \\ 0, & x < 0 \end{cases}$$

where a and b are positive

b)

$$f_X(x) = \begin{cases} axe^{-bx}, & x \geq 0 \\ 0, & x < 0 \end{cases}$$

where a and b are positive

c)

$$f_{XY}(x, y) = \begin{cases} ae^{(x+y)}, & x < 0 \text{ and } y < 0 \\ 0, & \text{otherwise} \end{cases}$$

9. Prove that taking an expected value is a linear operation.

10. Find the characteristic function that corresponds to each of the following probability density functions.

a) Uniform

$$f_X(x) = \begin{cases} \dfrac{1}{a}, & -a/2 \leq x \leq a/2 \\ 0, & \text{otherwise} \end{cases}$$

b) Gaussian

$$f_X(x) = \frac{1}{\sigma_X\sqrt{2\pi}} \exp\left[-(x-\mu_X)^2/(2\sigma_X^2)\right]$$

c) Rayleigh

$$f_X(x) = \begin{cases} \dfrac{2x}{\alpha} \exp(-x^2/\alpha), & x \geq 0 \\ 0, & x < 0 \end{cases}$$

d) Exponential

$$f_X(x) = \begin{cases} ae^{-ax}, & x \geq 0 \\ 0, & x < 0 \end{cases}$$

11. Prove that the sum of two independent Gaussian random variables is itself Gaussian.

12. Using the characteristic function found in Exercise 10 for the Gaussian-distributed random variable, find the third, fourth, and fifth moments, if μ_X is zero.

13. A measurement $Y = y$ is made of a random variable X. If X and Y are jointly Gaussian according to (7.45), what is the best estimate (in the sense of minimum mean-square error) of X based on y. Is the estimate $\hat{x}$ a linear or a nonlinear function of y?

14. Find the probability density function of the sum of three independent random variables X_1, X_2, and X_3, each of which is uniformly distributed between $-\frac{1}{2}$ and $\frac{1}{2}$.

15. Let random variable X have the probability density function

$$f_X(x) = \begin{cases} e^{-x}, & x \geq 0 \\ 0, & x < 0 \end{cases}$$

Find the probability density of
a) X^2
b) $\sqrt{X}$
c) $\log_e X$

16. Three independent observations x_1, x_2, and x_3 are made of a random variable X for which the probability density function is $f_X(x)$ and the cumulative distribution function is $F_X(x)$. Find the probability that $X_1 > X_2 > X_3$, where X_1, X_2, and X_3 are random variables describing the three observations (before they are made).

17. A random process is described by

$$X(t) = A\sin(\omega t + \Theta)$$

where amplitude A is a random variable with density function $f_A(a)$, and phase Θ is uniformly distributed from 0 to 2π. Determine whether $X(t)$ is a stationary process.

18. A random process

$$X(t) = a \sin(\omega t + \Theta)$$

has random phase uniformly distributed from 0 to 2π. Determine whether the process is stationary.

19. Prove that the autocorrelation and the autocovariance for a stationary discrete random process have properties analogous to those given in Section 7.7.

20. Show that the power spectral density of a stationary random process $X(t)$ is even and non-negative.

21. A stationary random process is applied to a linear time-invariant system. Prove that the output process is also stationary.

22. Confirm the relationships (7.150) and (7.152) for cross-spectral densities.

23. The normalized autocovariance of a random process is defined as the ratio of autocovariance to variance. That is, $\rho(\tau) = \gamma(\tau)/\gamma(0)$, where $-1 \leq \rho \leq 1$ and $\tau = \ell T$ for a discrete process. If discrete white noise $X(k)$ is applied as input to a first-order recursive filter $y_k = Ay_{k-1} + x_k$, show that the normalized autocovariance of the output is $\rho(\ell) = A^{|\ell|}$.

24. The input to a linear time-invariant system is a stationary random process $X(t)$ the autocovariance function of which is

$$\gamma_X(\tau) = Ae^{-b|\tau|} \cos \omega_0 \tau$$

where A, b, and ω_0 are real positive constants. Find the power spectrum of the output process $Y(t)$ if the system transfer function is $a/(s+a)$.

25. Given a stationary random process $X(t)$ with autocorrelation function $\phi_X(\tau)$, find the autocorrelation function of the derivative $\dot{X}(t)$.

26. Discrete white noise with variance σ_X^2 is applied as input to

a) A first-order recursive filter $y_k = Ay_{k-1} + x_k$

b) A second-order recursive filter $y_k = A_1 y_{k-1} + A_2 y_{k-1} + x_k$

Find the output power spectral density in both cases.

27. A random process has the autocovariance function

$$\gamma_Y(\tau) = Ae^{-b|\tau|} \cos \omega_0 \tau$$

where A, b, and ω_0 are real positive constants. It is desired to generate the process by passing white noise through a continuous filter. What is the transfer function of the required filter?

28. Suppose that discrete white noise is input to a filter represented by the following difference equation:

$$y_k = \sum_{i=1}^{n} A_i y_{k-i} + \sum_{i=0}^{m} B_i v_{k-i}$$

If $A_i = 0$, $i = 1, n$, the output process is called a *moving-average* (MA) process. If $B_i = 0$, $i = 1, m$, it is called an *autoregressive* (AR) process. And if at least some of the A and B coefficients are nonzero, it is called an *autoregressive moving-average* (ARMA) process. Consider the case where $n = 3$ and $m = 2$, and write the resulting ARMA process as a vector Markov process.

CHAPTER 8

QUANTIZATION EFFECTS IN DIGITAL FILTERS

8.1 INTRODUCTION

The theory of digital filter design up to this point has been based on the assumption that the signals and parameters being dealt with can take on any finite amplitude or value. In fact, however, because of the finite-word-length limitation on any digital device, only discrete values of amplitude are permissible; that is, the amplitude is quantized. Taking these discrete values into account in the filter equations would lead to nonlinear equations that would be impossible to handle rigorously in general. For general-purpose computers, with long word lengths or many bits available to represent a number, quantization effects may not be significant. In general, the effect increases as the word length decreases. We need mathematical models that will enable us to predict the effect of finite word length on filter performance—and so ascertain whether a problem exists. Several such models are discussed in this chapter. Sometimes, the model merely puts bounds on the minimum register length needed, but that in itself is a worthwhile result. A particularly useful technique is based on the assumption that the amount of quantization is small compared with the signal and parameter values—that is, that there is "fine" quantization. Then the quantization error can be treated as noise and the problem becomes a linear one.

The main types of quantization errors that arise in digital filtering are due to quantization of

1. Input signals
2. Coefficients in the difference equations
3. The products needed for the iterations

For example, with the first-order digital filter

$$y_k = Ay_{k-1} + x_k \tag{8.1}$$

error type 1 refers to quantization of input x_k, type 2 occurs in the representation of parameter A, and type 3 deals with formation of the product Ay_{k-1} needed for each iteration. These error effects will be considered separately.

In what follows, fixed-point arithmetic is assumed. Because of its speed and the low-cost hardware associated with it, fixed-point arithmetic is often preferred in smaller computers and dedicated devices operating in real time. Because of the shorter registers available in such cases, quantization errors are more likely to be a problem. A similar analysis can be performed for floating-point arithmetic.

Before examining the effects of quantization, we will discuss the binary representation of numbers, fixed-point arithmetic, and floating-point arithmetic.

8.2 NUMBER REPRESENTATION

We can represent a number V to any desired accuracy by the finite series

$$V = \sum_{i=n_1}^{n_2} c_i r^i \tag{8.2}$$

where r, which is called the radix of the representation, is the number of different symbols used, and the coefficients c_i are integers with

$$0 \le c_i < r \tag{8.3}$$

The part of the series (8.2) with negative powers of r is separated from the remainder of the series by what is called the radix point.

If $r = 10$, we have the decimal representation with ten different symbols 0 through 9, and the radix point is the decimal point. For example,

$$\begin{aligned} 60.295 &= \sum_{i=-3}^{2} c_i 10^i \\ &= (6 \times 10^1) + (2 \times 10^{-1}) + (9 \times 10^{-2}) + (5 \times 10^{-3}) \end{aligned}$$

With $r = 2$, we have the binary representation with the two symbols 0 and 1, which are said to be complements of each other. This system is extensively used in the implementation of digital hardware [4]. The radix point is then called the binary point. For example, the binary number,

$$101.1101 = (1 \times 2^2) + (1 \times 2^0) + (1 \times 2^{-1}) + (1 \times 2^{-2}) + (1 \times 2^{-4})$$

has the decimal value 5.8125. To convert from decimal to binary, we divide the part to the left of the decimal point repeatedly by 2 and arrange the *remainder* in reverse order (right to left). We multiply the part to the right of the decimal point repeatedly by 2, each time removing the integer part, and write the remainder in order (left to right).

EXAMPLE 8.1

Solution

Convert the decimal number 627.625 to binary form.

Integer Part	Remainder	Fractional Part	Remainder
$627 \div 2 = 313$	1	$0.625 \times 2 = 1.250$	1 (Binary number ↓)
$313 \div 2 = 156$	1	$0.250 \times 2 = 0.500$	0
$156 \div 2 = 78$	0	$0.500 \times 2 = 1.000$	1
$78 \div 2 = 39$	0	$0.000 \times 2 = 0.000$	0
$39 \div 2 = 19$	1		
$19 \div 2 = 9$	1		
$9 \div 2 = 4$	1 ↑		
$4 \div 2 = 2$	0		
$2 \div 2 = 1$	0		
$1 \div 2 = 0$	1 (Binary number)		

Therefore,

$$627.625_{\text{decimal}} = 1001110011.101_{\text{binary}}$$

Binary numbers are manipulated in a similar manner to decimal numbers:

Addition

$0 + 0 = 0$
$0 + 1 = 1$
$1 + 0 = 1$
$1 + 1 = 0$ and carry 1

Subtraction

$0 - 0 = 0$
$1 - 0 = 1$
$0 - 1 = 1$ and borrow 1
$1 - 1 = 0$

Multiplication

$0 \times 0 = 1 \times 0 = 0 \times 1 = 0$
$1 \times 1 = 1$

Division

$1 \div 1 = 1$
$0 \div 1 = 0$
Division by 0 is not defined.

Other number systems often used in digital computers are the octal ($r = 8$) and the hexadecimal ($r = 16$). In both cases the radix is a power of 2, so these numbers can be converted readily to binary. We will concentrate on the latter system.

8.3 MACHINE ARITHMETIC

8.3.1 Introduction

The basic element in a digital computer is the two-state device. A one is associated with one state, a zero with the other. The device is said to contain one *bit* of information. A number N of such devices can be assembled to form a register containing N bits of information—an N-bit number or word. Clearly, the computer is ideally suited for manipulation of binary numbers.

Implementation of a first-order recursive filter, as shown in Figure 8.1, illustrates some of the important operations involved. The latest-computed output y_{k-1} is stored in the output register in the form of an N-bit number. It is then multiplied by the N-bit number representing the coefficient A, which was stored in the coefficient register. The product Ay_{k-1} (after being rounded off to an N-bit number) is added to the current input x_k, (also an N-bit number) to form the current input y_k, which, in turn, is stored for multiplication by A in the next iteration. The whole procedure starts with an initial value y_{-1} in the output register. This value may be zero or nonzero. Higher-order filters can be implemented in a similar manner.

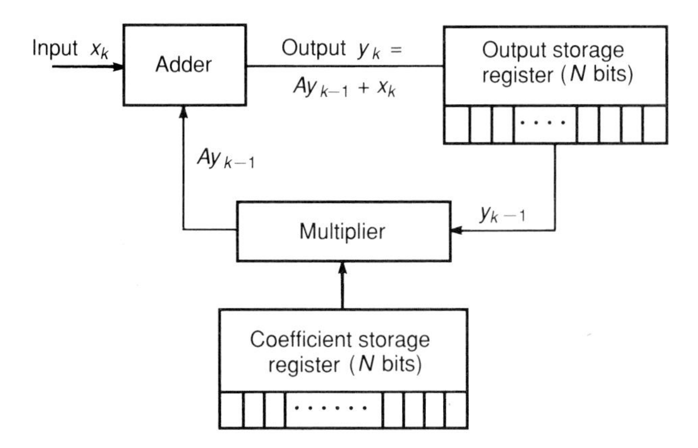

FIGURE 8.1 Implementation of a first-order recursive filter.

There are several different methods of carrying out the arithmetic operations involved. The most commonly used are: fixed-point arithmetic and floating-point arithmetic. In fixed-point arithmetic, the binary point has a fixed location in the register. In floating-point arithmetic, it does not. We will consider each representation in turn.

8.3.2 Fixed-Point Arithmetic

With the restrictions discussed below, fixed-point arithmetic has the form (8.2). It is suited for manipulating integer or fractional numbers, because the products of integers are integers, and the products of fractions are fractions. However, if it is necessary to round off the products, as in the case of recursive filters, it is better to limit the fixed-point representation to fractional numbers, because we cannot reduce the number of bits representing an integer. This is not too much of a restriction; we can scale the numbers so that we are manipulating fractions. The leftmost bit (called the most significant bit, or MSB) is usually reserved for the sign of the number. This bit is called the sign bit, and a zero indicates a positive number, a one a negative number. The rest of the bits give the magnitude of the number. The rightmost bit is called the least significant bit (LSB).

The manner in which negative numbers are represented gives three different forms for fixed-point arithmetic:

1. The sign-magnitude form
2. The one's-complement form
3. The two's-complement form

With the sign-magnitude form, a 6-bit representation of a number and its negative would be, for example,

$$0.4375 = 0.01110$$

$$-0.4375 = 1.01110$$

With the one's-complement form, the positive number is represented as in sign-magnitude notation, but its negative is obtained by complementing all the bits of the positive number. For our example,

$$0.4375 = 0.01110$$

$$-0.4375 = 1.10001$$

This is the same as subtracting the magnitude from $2 - 2^{-(N-1)}$, where N is the number of bits:

$$\begin{aligned} 2 - 2^{-(N-1)} &= 10.00000 - 0.00001 \\ &= 1.11111 \end{aligned}$$

Now subtract $0.4375 = 0.01110$.

$$\begin{array}{r} 1.11111 \\ -0.01110 \\ \hline 1.10001 \end{array} \quad \text{as before}$$

With the two's-complement form, the positive number is represented as before, but its negative is obtained by complementing all the bits of the positive number and adding a 1 to the least significant bit. Thus

$$\begin{aligned} 0.4375 &= 0.01110 \\ -0.4375 &= 1.10001 + 0.00001 \\ &= 1.10010 \end{aligned}$$

This is the same as subtracting the magnitude of the number from 2:

$$\begin{array}{rr} 2.0 = & 10.00000 \\ - & 0.01110 \\ \hline & 1.10010 \end{array} \quad \text{as before}$$

We note that $-7/16 = -0.4375$ is being represented by the binary equivalent of $2 - \frac{7}{16} = 1\frac{9}{16} = 1.5625$, which is the complement with respect to 2 of the number $\frac{7}{16}$. Hence the term *two's complement*. Thus the magnitude of the negative number ($\frac{7}{16}$ in this case) is given by

$$2 - 1 - \sum_{i=1}^{N-1} c_i 2^{-i} = 1 - \sum_{i=1}^{N-1} c_i 2^{-i} \tag{8.4}$$

or, in binary,

$$1.00000 - 1.10010 = 0.01110$$

In Table 8.1 we compare the three number systems for a 3-bit register.

TABLE 8.1 Different Number Representations for 3-bit Word Length

Binary Number	Decimal Equivalent Using		
	Sign-magnitude	One's-complement	Two's-complement
0.11	3/4	3/4	3/4
0.10	2/4	2/4	2/4
0.01	1/4	1/4	1/4
0.00	0	0	0
1.00	−0	−3/4	−4/4 = −1
1.01	−1/4	−2/4	−3/4
1.10	−2/4	−1/4	−2/4
1.11	−3/4	−0	−1/4

We note from the table that there are two representations for zero in sign-magnitude and one's-complement but no -1 point. The two's-complement form has one zero and can represent decimal numbers from -1 to $1 - 2^{-2}$, or, in general, from -1 to $1 - 2^{-(N-1)}$ for an N-bit register. The two's-complement system is often used in digital-filter implementations because it is convenient for addition and subtraction.

EXAMPLE 8.2

Using a 4-bit two's-complement representation, a) subtract 0.625 from 0.250 and b) subtract 0.250 from 0.625.

Solution

a)

Decimal	*Two's-complement*
0.250	0.010 } add
−0.625	1.011 }
−0.375	1.101 = −0.375

b)

Decimal	*Two's-complement*
0.625	0.101 } add
−0.250	1.110 }
0.375	0.011 = 0.375

(With two's-complement we neglect the carry bit at the leftmost, or most significant, position.)

Addition and subtraction by the one's-complement system are similar, but any carry bit at the most significant position is brought around to the rightmost, or least significant, position. Sign-magnitude addition and subtraction are more complex. Accordingly, this system is used more for multiplication, which is simply performed by multiplying the magnitude bits and adjusting the sign of the product (multiplying positive by positive or negative by negative yields positive, and multiplying negative by positive yields negative).

EXAMPLE 8.3

Multiply 0.625 by 0.250 using sign-magnitude representation.

Solution

Decimal	*Sign-magnitude*
0.625	0.101
0.250	0.010
0000	000
3125	101
1250	000
0.156250	0.001010 = 0.156250

Multiplication by one's-complement and two's-complement arithmetic is more difficult and may require special hardware or algorithms.

We note that because we are multiplying fractional numbers, there is no problem of overflow in multiplication for any of the three arithmetics. However, overflow can occur with addition if the sum of the fractions being added is greater than one. If overflow occurs in an intermediate step of an addition, there will be no overflow in the final sum, provided that the absolute value of the latter is less than one. An example will illustrate this.

EXAMPLE 8.4

Add $0.3125 + 0.7500 + (-0.6250)$ using 5-bit one's-complement arithmetic.

Solution

Decimal		*One's-complement*
0.3125		0.0101
+0.7500		0.1100
1.0625		1.0001 ← (incorrect; MSB = 1 implies negative)
−0.6250		1.0101
0.4375	correct result →	0.0111 ← end around carry

The overflow in addition that can occur with fixed-point arithmetic (as a result of the small dynamic range of numbers) is a drawback. Floating-point arithmetic, which we will discuss next, does not have this disadvantage.

8.3.3 Floating-Point Arithmetic

With binary floating-point arithmetic, a number V is represented as

$$V = M \times 2^E \tag{8.5}$$

where M, called the *mantissa*, lies in the range

$$\frac{1}{2} \leq |M| < 1 \tag{8.6}$$

and E, called the *exponent*, is a finite integer. Thus, in floating-point notation,

$$0.01101 = 0.1101 \times 2^{-1}$$

and

$$10111.001 = 0.10111001 \times 2^{5}$$

This representation requires storage of the two fixed-point numbers M and E. Usually, about three-quarters of the register bits are devoted to M and one-quarter to E. Figure 8.2 shows the register layout for the two floating-point numbers noted.

Multiplication of two floating-point numbers is straightforward; multiply the mantissas and add the exponents. Thus if

$$V_1 = M_1 \times 2^{E_1} \tag{8.7}$$

and

$$V_2 = M_2 \times 2^{E_2} \tag{8.8}$$

then

$$V_1 V_2 = M_1 M_2 \times 2^{(E_1+E_2)} \tag{8.9}$$

For division, divide the mantissas and subtract the exponents. Thus

$$V_1/V_2 = (M_1/M_2) \times 2^{(E_1-E_2)} \tag{8.10}$$

Let us consider an example.

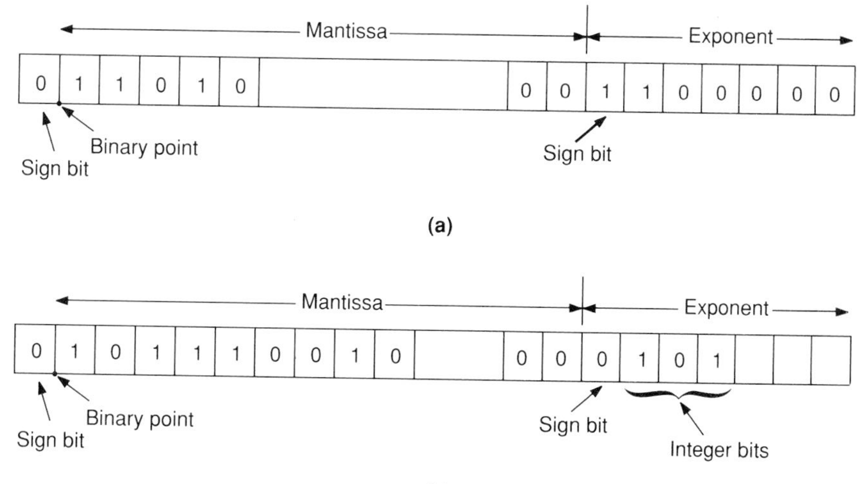

FIGURE 8.2 Register arrangement for two floating-point numbers. (a) Number $0.11010 \times 2^{-1} = 0.11010 \times 2^{1.100000}$ (b) Number $0.1011101 \times 2^{5} = 0.1011101 \times 2^{0101.000}$

EXAMPLE 8.5

Multiply the binary floating-point versions of the decimal numbers 2.75 and 0.0625.

Solution

Decimal	*Binary*	Floating point
2.75	$10.11 = 0.1011 \times 2^{2}$	$= 0.1011 \times 2^{010.0}$
$\times 0.0625$	$0.0001 = 0.1000 \times 2^{-3}$	$= 0.1000 \times 2^{111.0}$
$= 0.171875$		$0.01011 \times 2^{1001.0}$
		$= 0.1011 \times 2^{10010.0}$
		$= 0.6875 \times 2^{-2}$ (decimal)
		$= 0.171875$

Addition and subtraction of two floating-point numbers is slightly more difficult: Adjust the smaller of the numbers so that it has the same exponent as the other, and add or subtract the mantissas. Then adjust the sum or difference, if necessary, so that its mantissa satisfies (8.6). Consider an example.

EXAMPLE 8.6

Add the floating-point versions of the decimal numbers 2.75 and 0.0625

Solution

Decimal	*Binary*
2.75	$0.1011 \times 2^{010.0}$
$+0.0625$	$0.1000 \times 2^{111.0}$
2.8125	

The number $0.1000 \times 2^{111.0} = 0.000001 \times 2^{010.0}$. Now we can add the two numbers.

$$
\begin{array}{l}
\;\;\, 0.101100 \times 2^{010.0} \\
+0.000001 \times 2^{010.0} \\
\hline
= 0.101101 \times 2^{010.0} \\
= 2.8125 \text{ (decimal)}
\end{array}
$$

8.3.4 A Comparison of Fixed-Point and Floating-Point Arithmetics

Fixed-point arithmetic with two's-complement representation of negative numbers is often used with small computers or special-purpose hardware for real-time processing, because it offers fast operation and relatively economical implementation. Its drawbacks include a small dynamic range (the range of numbers that can be represented) and low accuracy if small numbers are truncated to fit the register. Round-off errors exist only for multiplication, but the fact that the register may overflow with addition may lead to incorrect results. No overflow is possible with multiplication.

Operation with floating-point arithmetic is slower and more expensive than operation with fixed-point arithmetic, because operations are performed on two fixed-point numbers—the mantissa and the exponent. This also results in costlier hardware. As a result, floating-point is used more with larger, general-purpose computers for processing stored data (data not processed in real time). The main advantage that floating-point arithmetic offers over fixed-point arithmetic is the increased dynamic range. We note that round-off errors can occur with both addition and multiplication of floating-point numbers.

In view of the large number of bits available to represent numbers in general-purpose computers, it is safe to say, in general, that quantization errors in such cases have little impact on signal processing. We will put more emphasis, then, on quantization errors associated with fixed-point arithmetic because, as we have noted, it is used more with small computers and hardware having relatively short registers. It should be observed, however, that an increasing number of microprocessors and digital-signal-processing chips *are* supporting floating-point arithmetic in hardware (on the chip itself).

8.4 QUANTIZATION OF INPUT SIGNAL

8.4.1 Introduction

Many of the important concepts involved in determining the effect of finite word length in filter design can be introduced by considering the quantization of the input signal to the filter. If a signal with amplitude range R is to be represented by an N-bit word, then the number of values, or quantization levels, that can be represented is 2^N. For example, with fixed-point arithmetic, $(N-1)$ bits give 2^{N-1} magnitude values, and the sign bit gives the additional information of whether the value is positive or negative, resulting in 2^N different levels. The difference between adjacent levels, or the quantization step, in terms of the true range of the signal, is

$$q_0 = \frac{\text{range of signal}}{\text{number of quantization levels}} = \frac{R}{2^N} \tag{8.11}$$

With fixed-point representation of fractional numbers, if the range of the signal exceeds ± 1, it is necessary to scale the signal. The actual quantization step of the scaled numbers in the register is reduced accordingly.

EXAMPLE 8.7

Determine the quantization levels of a continuous signal of range ±20 V after it has been sampled and then has been processed in an A/D converter with a 4-bit capacity.

Solution

From (8.11), the quantization step for the unscaled signal is

$$q_0' = \frac{40}{2^4} = 2.5 \text{ V}$$

The quantization step for the signal, scaled to be in the range ±1, is

$$q_0 = \frac{2}{2^4} = 0.125 \text{ V}$$

which is $2^{-(N-1)} = 2^{-3}$, or the value corresponding to a one in the rightmost (least significant) bit in the register.

The $2^4 = 16$ possible levels of the signal might be represented in decimal and binary form as shown in Table 8.2. Negative values have a one's-complement representation.

If we add an offset of $q_0/2$ to positive signal values and subtract $q_0/2$ from negative values, we avoid the duplication of zero inherent in the one's-complement representation and reduce the quantization errors that can occur at the highest and lowest signal values. This arrangement is shown in Table 8.3.

Figure 8.3 shows the continuous signal, the sampled unquantized signal, and the quantized signal, where the quantization is effected by rounding off to the nearest level.

TABLE 8.2 Example of Quantization Levels Using the One's-Complement System

Decimal		Binary	Decimal		Binary
Unscaled	Scaled		Unscaled	Scaled	
0	0	0.000	−0	−0	1.111
2.50	0.125	0.001	−2.50	−0.125	1.110
5.00	0.250	0.010	−5.00	−0.250	1.101
7.50	0.375	0.011	−7.50	−0.375	1.100
10.00	0.500	0.100	−10.00	−0.500	1.011
12.50	0.625	0.101	−12.50	−0.625	1.010
15.00	0.750	0.110	−15.00	−0.750	1.001
17.50	0.875	0.111	−17.50	−0.875	1.000

TABLE 8.3 Quantization Levels with Offset

Decimal		Binary	Decimal		Binary
Unscaled	Scaled		Unscaled	Scaled	
1.25	0.0625	0.000	−1.25	−0.0625	1.111
3.75	0.1875	0.001	−3.75	−0.1875	1.110
6.25	0.3125	0.010	−6.25	−0.3125	1.101
8.75	0.4375	0.011	−8.75	−0.4375	1.100
11.25	0.5625	0.100	−11.25	−0.5625	1.011
13.75	0.6875	0.101	−13.75	−0.6875	1.010
16.25	0.8125	0.110	−16.25	−0.8125	1.001
18.75	0.9375	0.111	−18.75	−0.9375	1.000

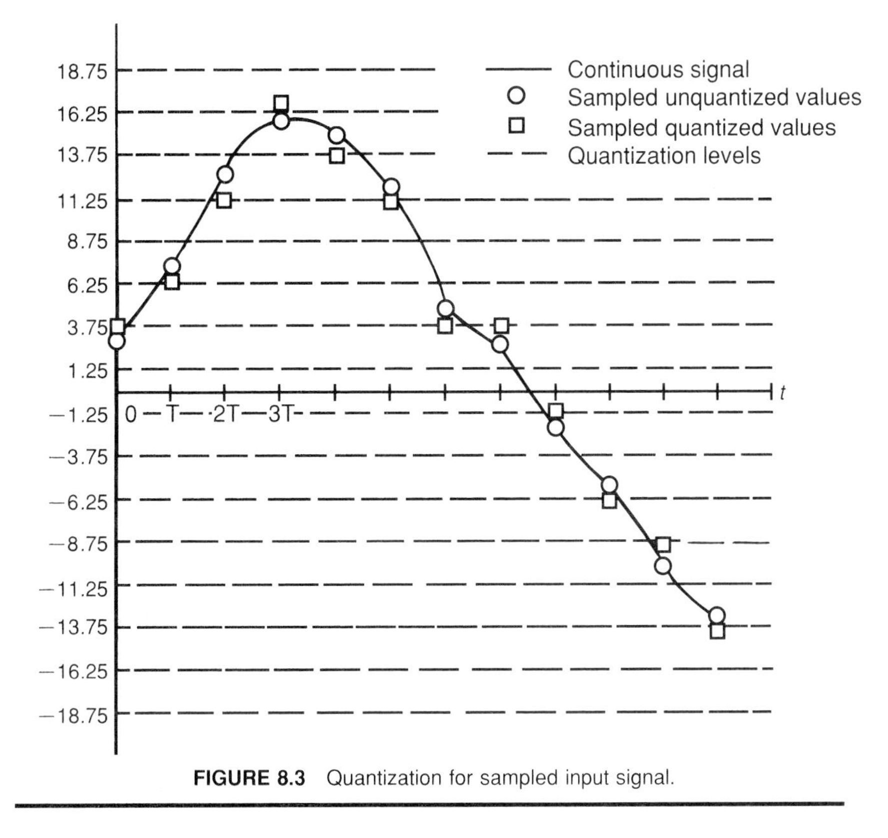

FIGURE 8.3 Quantization for sampled input signal.

8.4.2 Quantization Methods

Three common methods of quantization are

1. *Rounding*. As shown in Figure 8.3, the signal value is approximated by the nearest quantization level.

2. *Truncation.* The signal value is approximated by the highest quantization level that is not greater than the signal.
3. *Sign-magnitude truncation.* This is the same as truncation for positive signals, but negative signal values are approximated by the nearest quantization level that is not lower than the signal.

These descriptions apply to quantization in fixed-point arithmetic. The two truncation methods result from the different treatments of negative numbers with sign-magnitude, one's-complement, and two's-complement representations. Quantization to a floating-point number is discussed below. These methods are also applicable to the quantization of filter parameter values and the results of multiplication.

With rounding, if the value lies halfway between two levels, the logic can be arranged so that the number is approximated either by the nearest higher level or by the nearest lower level, or a random choice between the two levels can be made. At any time instant kT, let the error due to quantization be

$$e(kT) = x(kT) - x_A(kT) \tag{8.12}$$

where x and x_A are the quantized and unquantized values, respectively. It is evident that for rounding,

$$-\frac{q_0}{2} \leq e(kT) \leq \frac{q_0}{2} \tag{8.13}$$

The nonlinear relationship between x and x_A is shown in Figure 8.4. For fixed-point numbers, the error satisfies (8.13), regardless of whether sign-magnitude, one's-complement, or two's-complement is used for negative numbers. This is because rounding is independent of the sign of the number; it depends on the number's magnitude only.

In floating-point arithmetic, only the mantissa is affected by quantization. If

$$x_A = M_A \times 2^E \tag{8.14}$$

and

$$x = M \times 2^E \tag{8.15}$$

then

$$e = x - x_A = (M - M_A)2^E \tag{8.16}$$

But for rounding,

$$-q_0/2 \leq (M - M_A) \leq q_0/2 \tag{8.17}$$

Then, from (8.16),

$$-2^E q_0/2 \leq e \leq 2^E q_0/2 \tag{8.18}$$

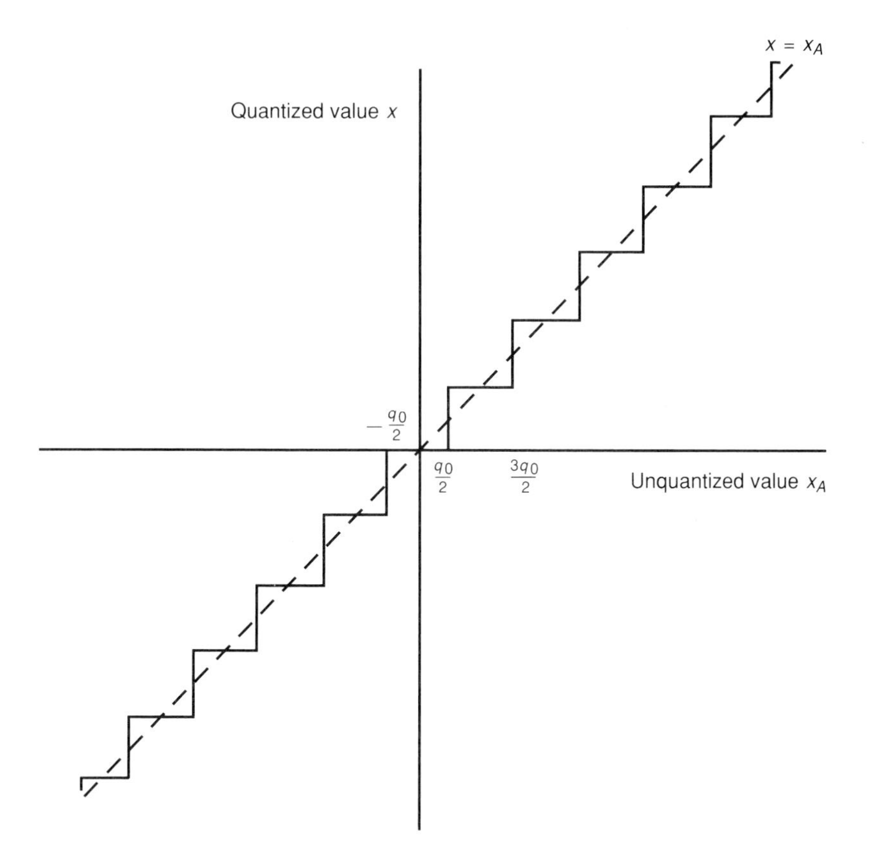

FIGURE 8.4 Relationship between quantized and unquantized values for rounding.

which gives the *absolute* error in the floating-point number due to quantization of the mantissa. Define the *relative* error as ϵ, where

$$x = x_A(1 + \epsilon) \tag{8.19}$$

Then

$$e = \epsilon x_A \tag{8.20}$$

and (8.18) can be written as

$$-2^E q_0/2 \leq \epsilon x_A \leq 2^E q_0/2 \tag{8.21}$$

or

$$-2^E q_0/2 \leq \epsilon M_A 2^E \leq 2^E q_0/2 \tag{8.22}$$

which implies that

$$-q_0/2 \leq \epsilon M_A \leq q_0/2 \tag{8.23}$$

The mantissa satisfies

$$\frac{1}{2} \leq M_A < 1 \tag{8.24}$$

If $M_A = \frac{1}{2}$, we get the maximum range of the relative error, from (8.23), as

$$-q_0 \leq \epsilon \leq q_0 \tag{8.25}$$

for rounding.

If the quantization method is truncation, the number is approximated in fixed-point arithmetic by the nearest level that does not exceed it (Figure 8.5). In this case, we see that the error $e = x - x_A$ is negative or zero and that

$$-q_0 < e \leq 0 \tag{8.26}$$

This applies to all positive numbers in sign-magnitude, one's-complement, or two's-complement representations. The situation must be examined further, however, for negative numbers. Consider first the two's-complement representation. From (8.4), the *magnitude* of the negative number with N_1 bits is

$$A_1 = 1 - \sum_{i=1}^{N_1-1} c_i 2^{-i} \tag{8.27}$$

If we truncate the number to N bits, the magnitude becomes

$$A = 1 - \sum_{i=1}^{N-1} c_i 2^{-i} \tag{8.28}$$

The change in magnitude of the negative number that results from truncation is

$$\begin{aligned} A - A_1 &= \sum_{i=1}^{N_1-1} c_i 2^{-i} - \sum_{i=1}^{N-1} c_i 2^{-i} \\ &= \sum_{i=N}^{N_1-1} c_i 2^{-i} \\ &\geq 0 \end{aligned} \tag{8.29}$$

Because the magnitude increases with truncation, the negative number in two's-complement becomes more negative. The maximum value of the magnitude error is given when all the c_i are unity in (8.29). In that case,

$$\begin{aligned} A - A_1 &= 2^{-(N-1)} - 2^{-(N_1-1)} \\ &< q_0 \end{aligned} \tag{8.30}$$

because $q_0 = 2^{-(N-1)}$. Therefore, the range of error with negative numbers

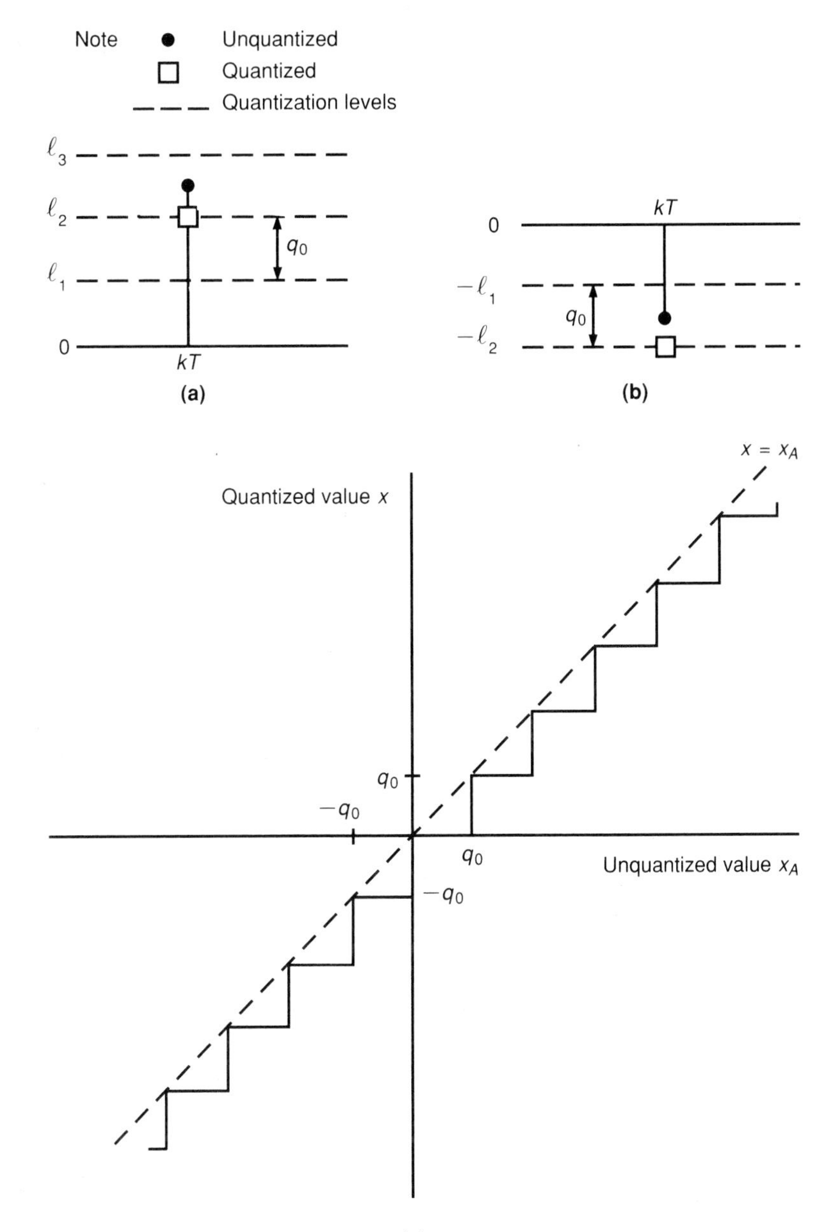

(c)

FIGURE 8.5 Relationship between quantized and unquantized values for truncation. (a) Positive number. (b) Negative number. (c) Truncation characteristic.

in the two's-complement system is

$$-q_0 < e \leq 0 \tag{8.31}$$

as depicted in Figure 8.5. This is the same as for positive numbers; compare (8.26).

With one's-complement, the magnitude of the negative number with N_1 bits is

$$A_1 = 1 - \sum_{i=1}^{N_1-1} c_i 2^{-i} - 2^{-(N_1-1)} \tag{8.32}$$

With truncation to N bits, the magnitude becomes

$$A = 1 - \sum_{i=1}^{N-1} c_i 2^{-i} - 2^{-(N-1)} \tag{8.33}$$

The change in magnitude due to truncation is

$$\begin{aligned} A - A_1 &= \sum_{i=1}^{N_1-1} c_i 2^{-i} - \sum_{i=1}^{N-1} c_i 2^{-i} + 2^{-(N_1-1)} - 2^{-(N-1)} \\ &= \sum_{i=N}^{N_1-1} c_i 2^{-i} - (2^{-(N-1)} - 2^{-(N_1-1)}) \\ &< 0 \end{aligned} \tag{8.34}$$

Therefore, the magnitude of the negative number decreases with truncation—that is, the negative numbers become less negative. The situation is illustrated in Figure 8.6, which shows sign-magnitude truncation as we have defined it. Therefore, the range of error for negative numbers in one's-complement is

$$0 \leq e < q_0 \tag{8.35}$$

if we truncate an unquantized number.

In sign-magnitude representation, the bits representing the magnitude of a negative number are the same as for the corresponding positive number. Only the sign bit is different. Hence when we truncate a negative number, the magnitude decreases and the truncated value is at the nearest quantitative level that is not less than the number, which implies sign-magnitude truncation, as shown in Figure 8.6.

Finally, we consider truncation of the mantissa of a floating-point number. From (8.16),

$$e = x - x_A = (M - M_A)2^E \tag{8.36}$$

With two's-complement representation of the mantissa,

$$-q_0 < M - M_A \leq 0 \tag{8.37}$$

or

$$-2^E q_0 < e \leq 0 \tag{8.38}$$

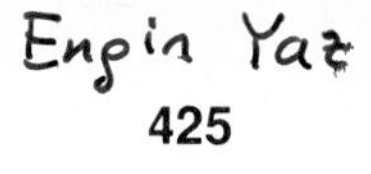

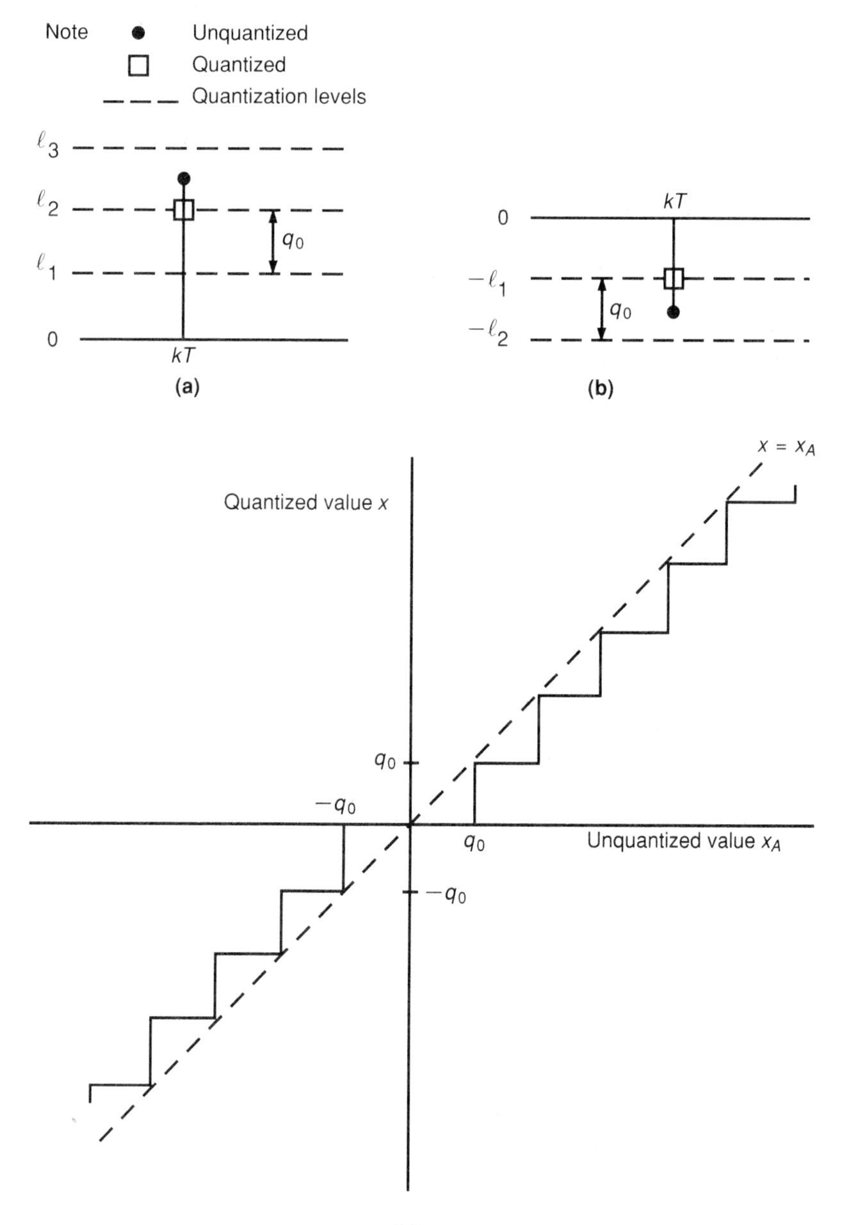

FIGURE 8.6 Relationship between quantized and unquantized values for sign-magnitude truncation. (a) Positive number. (b) Negative number. (c) Sign-magnitude truncation characteristic.

from (8.36). Because $e = \epsilon x_A$ from (8.20), we have

$$-2^E q_0 < \epsilon x_A \leq 0 \tag{8.39}$$

or

$$-2^E q_0 < \epsilon M_A 2^E \leq 0 \tag{8.40}$$

which implies that

$$-q_0 < \epsilon M_A \leq 0 \tag{8.41}$$

If $M_A = \frac{1}{2}$, we get the maximum range of the relative error ϵ as

$$2q_0 < \epsilon \leq 0 \tag{8.42}$$

If $M_A = -\frac{1}{2}$, the relative error range is

$$0 \leq \epsilon < 2q_0 \tag{8.43}$$

With a one's-complement representation, the error for truncation of positive values of the mantissa is

$$-q_0 < M - M_A \leq 0 \tag{8.44}$$

or

$$-2^E q_0 < e \leq 0 \tag{8.45}$$

With

$$e = \epsilon x_A = \epsilon M_A 2^E \tag{8.46}$$

and $M_A = \frac{1}{2}$, we get the maximum range of the relative error for positive M_A as

$$-2q_0 < \epsilon \leq 0 \tag{8.47}$$

For negative mantissa values, the error is

$$0 \leq M - M_A < q_0 \tag{8.48}$$

or

$$0 \leq e < 2^E q_0 \tag{8.49}$$

With (8.46) and $M_A = -\frac{1}{2}$, the maximum range of the relative error for negative M_A is

$$-2q_0 < \epsilon \leq 0 \tag{8.50}$$

which is the same as for positive M_A in (8.47). This holds also for floating-point numbers wherein the mantissa has a fixed-point sign-magnitude representation. The proof is left as an exercise.

The error ranges for quantization to fixed-point and to floating-point numbers are summarized in Table 8.4.

8.4.3 Statistical Model for Fine Quantization

We saw in the last section that the result of quantizing an input signal $x_A(kT)$ is to introduce a rounding error, or a truncation error, that can vary over a

TABLE 8.4 Quantization Error Ranges

Type of Quantization	Type of Arithmetic	Fixed-point Number Error Range	Floating-Point Number Relative-Error Range
Rounding	Sign-magnitude, one's-complement, two's-complement	$-q_0/2 \leq e \leq q_0/2$	$-q_0 \leq \epsilon \leq q_0$
Truncation	Two's-complement	$-q_0 < e \leq 0$	$-2q_0 < \epsilon \leq 0, M_A > 0$ $0 \leq \epsilon < 2q_0, M_A < 0$
Sign-magnitude truncation	One's-complement, sign-magnitude	$-q_0 < e \leq 0, x_A > 0$ $0 \leq e < q_0, x_A < 0$	$-2q_0 < \epsilon \leq 0$

Notes:
1. $e = x - x_A = \epsilon x_A$, where x_A is the unquantized number, x is the quantized number, and ϵ is the relative error.
2. q_0 is the quantization step.
3. $x_A = M_A 2^E$ in floating-point arithmetic.

finite range. For an arbitrarily varying signal with fine quantization (such as $N \geq 8$ bits), there is no reason to assume that any particular error value is more likely to occur than another within the range. Thus we can assume that the error may be treated as a uniformly distributed random variable $E(kT)$ that can take on values $e(kT)$ within the error range. (See Chapter 7 for a review of probability concepts.) The probability density function for rounding is shown in Figure 8.7. We will confine the discussion to fixed-point numbers. From the figure, it is clear that the expected value of $E(kT)$ is zero and the variance is

$$\sigma_E^2 = \int_{-\infty}^{\infty} e^2 f_E(e)de = q_0^2/12 \tag{8.51}$$

Suppose the unquantized signal $x_A(kT)$ is itself a realization of a stationary random process $X_A(kT)$ with variance σ_X^2. If the quantization error is regarded as noise, then the signal-to-noise ratio (SNR) is

$$\frac{\sigma_X^2}{\sigma_E^2} = \frac{12\sigma_X^2}{q_0^2} = \frac{12 \times 2^{2N}}{R^2}\sigma_X^2 \tag{8.52}$$

where R is the range of x_A. In decibels, the SNR is

$$10\log_{10}\left(\frac{\sigma_X^2}{\sigma_E^2}\right) = 10\log_{10}12 + 20N\log_{10}2 - 20\log_{10}R + 10\log_{10}\sigma_X^2 \tag{8.53}$$

This indicates that the SNR increases by approximately 6 dB for each bit added to the register length.

Probability densities for truncation errors and sign-magnitude truncation errors are shown in Figure 8.8. If the quantization method is truncation, the mean and variance of the error are $-q_0/2$ and $q_0^2/12$, respectively. For sign-magnitude truncation, the mean is $-q_0/2$ and the variance $q_0^2/12$ if $x_A(kT)$ is known to be positive; they are $q_0/2$ and $q_0^2/12$, respectively, if $x_A(kT)$ is known to be negative. If the sign of $x_A(kT)$ is unknown, as is often the case, then the error mean is zero and the variance $q_0^2/3$. Proving these facts is left as an exercise.

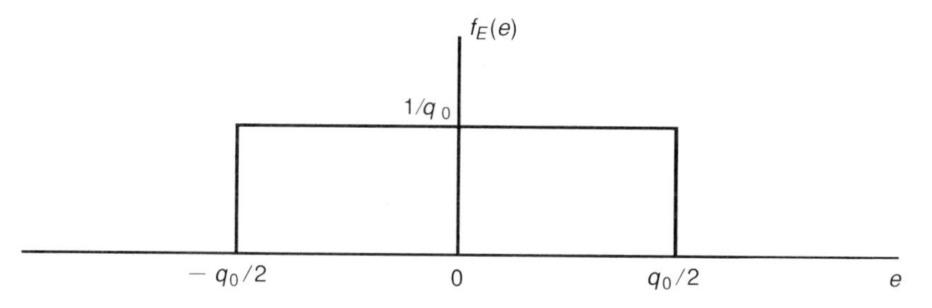

FIGURE 8.7 Probability density function for quantization error due to rounding.

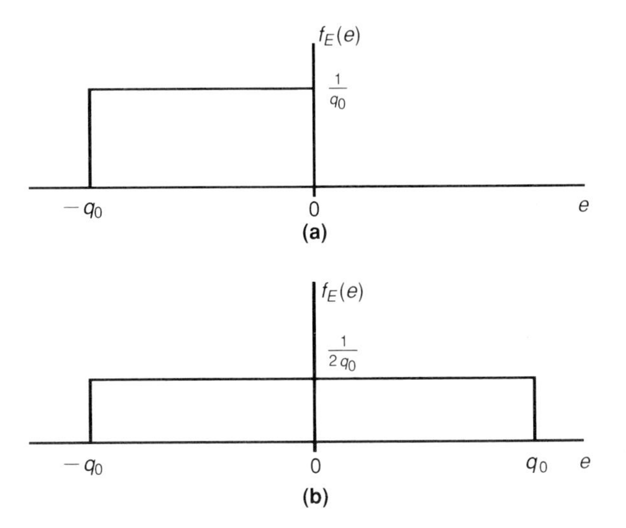

FIGURE 8.8 Probability density functions for (a) quantization error that is due to truncation and (b) quantization error that is due to sign-magnitude truncation if the sign of the unquantized signal is unknown.

In many practical situations, rounding is preferred to other methods of quantizing the signal for three reasons:

1. The error signal is independent of the type of arithmetic.
2. The mean of the error signal is zero.
3. No other quantization method gives a lower variance.

The effect of quantization of the input signal is that of noise superimposed on the original unquantized signal, as shown in Figure 8.9.

If the digital filter is linear, then the effect of the noise alone on the output $y(kT)$ can be computed; that is, the variance of the output due to quantization error $e(kT)$ can be obtained. For signal fluctuations such that many quantization levels are traversed from one sample to the next, it is reasonable to assume that error $e(kT)$ at any sampling time kT is independent of $e(mT)$ at any other sampling time mT. Bennett [1] has shown that this assumption holds for most signals encountered in practice, provided that the quantization is fine enough. Clearly, it does not hold if $x_A(kT)$ is a constant, for example.

Let the digital filter have weighting function $h(kT)$, and let $f(kT)$ be the output noise due to quantization of the input, as shown in Figure 8.9(b). Then, writing f_k for $f(kT)$, h_n for $h(nT)$, and e_{k-n} for $e[(k-n)T]$, we get

$$f_k = \sum_{n=0}^{k} h_n e_{k-n} \tag{8.54}$$

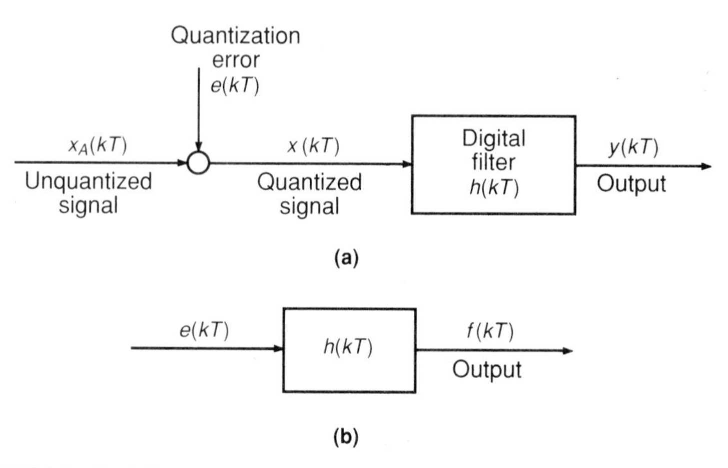

FIGURE 8.9 Model for error due to fine quantization of input signal. (a) Error model. (b) Output due to quantization noise.

Considering the error sequence e_k to be a realization of a random process E_k, we find that the output sequence f_k is a realization of a random process F_k, so (8.54) can be written

$$F_k = \sum_{n=0}^{k} h_n E_{k-n} \tag{8.55}$$

Consider the variance of any term $h_n E_{k-n}$ in the sum (8.55)

$$\text{Var}\,(h_n E_{k-n}) = h_n^2 \,\text{Var}\, E_{k-n} = \frac{q_0^2}{12} h_n^2 \tag{8.56}$$

Here we are assuming that rounding is the quantization method. If another method were used, with nonzero mean and with a different variance, (8.56) would be modified accordingly.

Because the variance of the sum of independent random variables is the sum of their variances, it follows that, if the quantization errors are assumed to be independent at different sampling instances, then the variance of the output F_k is

$$\sigma_F^2(k) = \frac{q_0^2}{12} \sum_{n=0}^{k} h_n^2 \tag{8.57}$$

The variance builds up to a finite steady-state value, provided that the filter is stable—that is, its poles lie within the unit circle in the z-plane. The steady-state variance is

$$\lim_{k\to\infty} \sigma_F^2(k) = \sigma_{Fss}^2 = \frac{q_0^2}{12} \sum_{n=0}^{\infty} h_n^2 \tag{8.58}$$

An alternative expression can be obtained for the steady-state variance of the output in terms of the filter transfer function $H(z)$. It is

$$\sigma_{Fss}^2 = \frac{q_0^2}{12} \sum \begin{array}{c} \text{residues of } H(z)H(z^{-1})z^{-1} \text{ at poles} \\ \text{of same within the unit circle} \end{array} \tag{8.59}$$

This expression is easier to evaluate than (8.58) in general and is derived as follows:

$$H(z) = \sum_{n=0}^{\infty} h_n z^{-n} \tag{8.60}$$

and

$$H(z^{-1}) = \sum_{m=0}^{\infty} h_m z^m \tag{8.61}$$

Therefore,

$$H(z)H(z^{-1}) = \sum_{n=0}^{\infty} \sum_{m=0}^{\infty} h_n h_m z^{m-n} \tag{8.62}$$

Multiply both sides by z^{-1} and integrate with respect to z about a closed contour C in the z-plane.

$$\oint_C H(z)H(z^{-1})z^{-1}\,dz = \oint_C \sum_{n=0}^{\infty} \sum_{m=0}^{\infty} h_n h_m z^{m-n-1}\,dz \tag{8.63}$$

If C lies within the regions of convergence for both (8.60) and (8.61), then the order of summation and the order of integration on the right-hand side of (8.63) can be interchanged. Note that the unit circle lies in the intersection of the regions of convergence for $H(z)$ and $H(z^{-1})$, provided that $H(z)$ is stable. Hence the unit circle is a logical choice for C. Then

$$\oint_C H(z)H(z^{-1})z^{-1}\,dz = \sum_{n=0}^{\infty} \sum_{m=0}^{\infty} h_n h_m \oint_C z^{m-n-1}\,dz \tag{8.64}$$

Because C encloses the origin of the z-plane, then by a theorem due to Cauchy,

$$\oint_C z^{m-n-1}\,dz = 2\pi j \delta_{mn} \tag{8.65}$$

where

$$\delta_{mn} = \begin{cases} 1 & \text{if } m = n \\ 0 & \text{if } m \neq n \end{cases} \tag{8.66}$$

Therefore,

$$\oint_C H(z)H(z^{-1})z^{-1}\,dz = 2\pi j \sum_{n=0}^{\infty} \sum_{m=0}^{\infty} h_n h_m \delta_{mn} = 2\pi j \sum_{n=0}^{\infty} h_n^2 \tag{8.67}$$

Hence

$$\sum_{n=0}^{\infty} h_n^2 = \frac{1}{2\pi j} \oint_C H(z)H(z^{-1})z^{-1}\,dz \tag{8.68}$$

Equation (8.59) follows from (8.58) and (8.68).

To illustrate the use of expressions (8.58) and (8.59), consider the following example.

EXAMPLE 8.8

Use (8.58) and (8.59) to find the steady-state variance of the noise in the output that is due to quantization of the input for the first-order filter

$$y_k = Ay_{k-1} + x_k \tag{8.69}$$

with $|A| < 1$.

Solution

The pulse response is $A^k u(k)$. From (8.57), the variance of the output noise is

$$\sigma_F^2(k) = \frac{q_0^2}{12} \sum_{n=0}^{k} A^{2n} = \frac{q_0^2}{12} \frac{1 - A^{2(k+1)}}{1 - A^2} \tag{8.70}$$

The steady-state variance as k approaches infinity is

$$a_{Fss}^2 = \frac{q_0^2}{12(1 - A^2)} \tag{8.71}$$

To use (8.59), note that

$$H(z) = \frac{z}{z - A} \tag{8.72}$$

with a pole at $z = A$, and

$$H(z^{-1}) = \frac{1}{1 - Az} \tag{8.73}$$

with a pole at $z = 1/A$, which is outside the unit circle. Then, by (8.59),

$$\sigma_{Fss}^2 = \frac{q_0^2}{12} \sum \left(\text{residues of } \frac{z}{z - A} \cdot \frac{1}{1 - Az} z^{-1} \text{ at } z = A \right) = \frac{q_0^2}{12(1 - A^2)} \tag{8.74}$$

as before. For higher-order systems, it would be easier to use the residue method (8.59), rather than (8.58), because of the complexity of the pulse response.

If the pole of the foregoing filter is very close to the unit circle (that is, if $A = 1 - \epsilon$, where $\epsilon \ll 1$), then the steady-state variance of the output noise is

$$\sigma_{Fss}^2 = \frac{q_0^2}{12\epsilon(2-\epsilon)} \approx \frac{q_0^2}{24\epsilon} \tag{8.75}$$

Thus for this case, the noise variance is inversely proportional to the distance of the pole from the unit circle and directly proportional to the gain of the filter at zero frequency—that is, $1/\epsilon$.

8.5 ERRORS CAUSED BY INEXACT VALUES OF FILTER PARAMETERS

8.5.1 Effect on Stability

To ensure stability of a recursive digital filter, all its poles must lie within the unit circle in the z-plane. In many cases, it is desirable for a pole or a pair of poles to lie close to the unit circle. Then, if the quantization step is so coarse that the actual representation of the poles falls on or outside the unit circle, the implemented filter becomes undamped or unstable.

Consider, for example, the first-order filter

$$y_k = Ay_{k-1} + x_k \tag{8.76}$$

and let N be the number of significant bits available to represent the coefficient A, the magnitude of which may vary, in general, from zero to unity for a stable filter ($-1 < A < 1$). Then the step size between quantized values is, from (8.11),

$$q_0 = 2^{-(N-1)} \tag{8.77}$$

Thus if $\epsilon = 1 - A$ is the distance of the pole to the unit circle, the smallest value of ϵ that can be represented accurately is $2^{-(N-1)}$ if truncation is the quantization method. It follows that, for stability,

$$N \geq -\frac{\log_{10}\epsilon}{\log_{10}2} + 1 = -\frac{\log_{10}(1-A)}{\log_{10}2} + 1 \tag{8.78}$$

With rounding,

$$N \geq -\frac{\log_{10}(1-A)}{\log_{10}2} \tag{8.79}$$

for stability. Consider an example.

EXAMPLE 8.9

a) Let $A = e^{-aT}$, where $a = 1$ rad/s and $T = 10^{-3}$ seconds. If truncation is the quantization method, find the minimum number of bits required to represent A so as not to result in instability.

b) Say only 9 bits are available and $T = 10^{-3}$ seconds as before. Find a.

Solution

a)

$$1 - A = 1 - e^{-aT} \approx aT$$

Therefore,

$$N \geq -\frac{\log_{10} aT}{\log_{10} 2} + 1 = 11 \text{ bits}$$

b) We have

$$9 = -\frac{\log_{10}(10^{-3}a)}{0.3} + 1$$

which requires

$$a = 4 \text{ rad/s}$$

For higher-order filters, a pole location depends, in general, on more than one coefficient. Consider the second-order filter

$$y_k = A_1 y_{k-1} - A_2 y_{k-2} + x_k \tag{8.80}$$

with transfer function

$$H(z) = \frac{z^2}{z^2 - A_1 z + A_2} \tag{8.81}$$

If the poles are complex at $z = re^{\pm j\theta}$ with $\theta = \omega_r T$, then

$$\begin{aligned} r &= \sqrt{A_2} \\ \cos\theta &= A_1/(2\sqrt{A_2}) \end{aligned} \tag{8.82}$$

Let ΔA_1 and ΔA_2 be quantization errors in the coefficients A_1 and A_2. The corresponding errors in the pole locations are Δr and $\Delta\theta$ (depicted in Figure 8.10 for positive errors in r and θ). Assume that ΔA_1 and ΔA_2 are relatively small so that we can neglect higher powers of ΔA_1 and ΔA_2. The changes in pole position due to changes in the nominal values of the coefficients are

$$\Delta r = \frac{\partial r}{\partial A_1}\Delta A_1 + \frac{\partial r}{\partial A_2}\Delta A_2 = \frac{1}{2r}\Delta A_2 \tag{8.83}$$

and

$$\Delta\theta = \frac{\partial\theta}{\partial A_1}\Delta A_1 + \frac{\partial\theta}{\partial A_2}\Delta A_2 = \frac{-1}{2r\sin\theta}\Delta A_1 + \frac{1}{2r^2\tan\theta}\Delta A_2 \tag{8.84}$$

where the sensitivity coefficients $\partial r/\partial A_1$ and so on are derived from (8.82). From (8.84) it is evident that large errors in θ (and hence in ω_r) will result when θ is small. Therefore, narrow-band lowpass filters are very sensitive

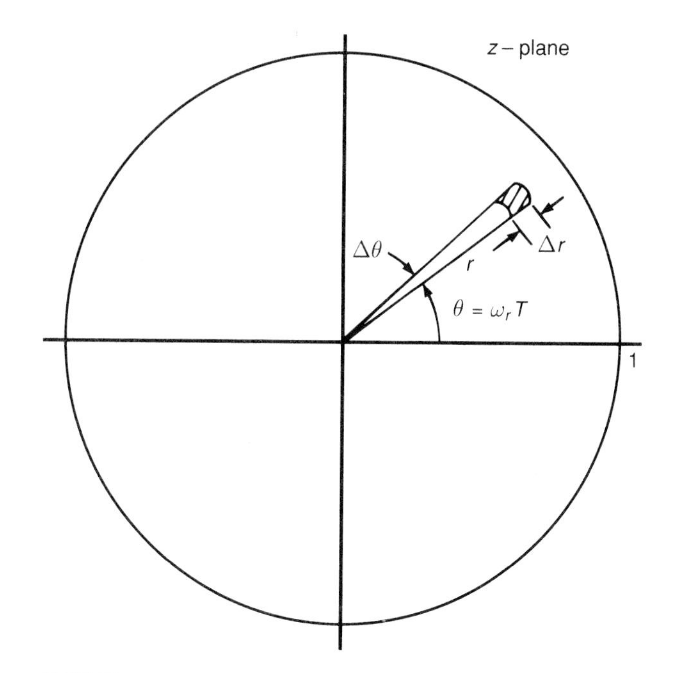

FIGURE 8.10 Error in pole location due to quantized coefficients for second-order filter.

to quantization effects. As shown in [2], this sensitivity can be eliminated by means of a coupled arrangement of the second-order filter (8.80):

$$\begin{aligned} y_k &= Ay_{k-1} + Bu_{k-1} + Lx_k \\ u_k &= Cy_{k-1} + Du_{k-1} \end{aligned} \tag{8.85}$$

where

$$\begin{aligned} A &= D = r\cos\theta \\ B &= -C = r\sin\theta \end{aligned} \tag{8.86}$$

However, this coupled form requires more computation and would be used only if quantization were a problem.

General Expressions for Direct or Canonic Form ($n > 1$) We will now derive general expressions for the change in pole locations that is due to changes in the coefficients of the nth-order difference equation with $n > 1$:

$$y_k = \sum_{i=1}^{n} A_1 y_{k-1} + \sum_{i=0}^{t} B_i x_{k-i} \tag{8.87}$$

First, for simplicity, consider the case where all the poles are real [3]. Then the transfer function can be written

$$H(z) = \frac{\sum\limits_{i=0}^{t} B_i z^{-i}}{1 - \sum\limits_{i=1}^{n} A_i z^{-1}} = \frac{\sum\limits_{i=0}^{t} B_i z^{-i}}{\prod\limits_{i=1}^{n}(1 - p_i z^{-1})} \tag{8.88}$$

where p_i, $i = 1, n$, are the poles. Equating the denominators of (8.88) gives

$$1 - \sum_{i=1}^{n} A_i z^{-i} = \prod_{i=1}^{n}(1 - p_i z^{-1}) \tag{8.89}$$

Take partial derivatives of both sides with respect to a particular coefficient A_ℓ, $1 \leq \ell \leq n$.

$$\begin{aligned} -z^{-\ell} &= \frac{\partial}{\partial A_\ell} \prod_{i=1}^{n}(1 - p_i z^{-1}) \\ &= -z^{-1}\frac{\partial p_1}{\partial A_\ell} \prod_{i=2}^{n}(1 - p_i z^{-1}) \cdots - z^{-1}\frac{\partial p_m}{\partial A_\ell} \prod_{\substack{i=1 \\ i \neq m}}^{n}(1 - p_i z^{-1}) \cdots \\ &\quad - z^{-1}\frac{\partial p_n}{\partial A_\ell} \prod_{i=1}^{n-1}(1 - p_i z^{-1}) \end{aligned} \tag{8.90}$$

Here it is assumed that for $n > 2$, the filter is realized in nth-order direct or canonic form, as opposed to a series or parallel arrangement of first- and second-order filters. Thus a change in the value of a particular coefficient can affect all the poles.

At the mth pole, $z = p_m$. Substituting for z in (8.90) gives

$$-p_m^{-\ell} = p_m^{-1}\frac{\partial p_m}{\partial A_\ell} \prod_{\substack{i=1 \\ i \neq m}}^{n}(1 - p_i p_m^{-1}) \tag{8.91}$$

Note that all other products on the right-hand side of (8.91) are zero, because $(1 - p^m z^{-1})$ is a factor. Solving for the sensitivity coefficient gives

$$\frac{\partial p_m}{\partial A_\ell} = \frac{p_m^{-\ell+1}}{\prod\limits_{\substack{i=1 \\ i \neq m}}^{n} 1 - p_i p_m^{-1}} \tag{8.92}$$

$$= \frac{p_m^{n-\ell}}{\prod\limits_{\substack{i=1 \\ i \neq m}}^{n} (p_m - p_i)} \tag{8.93}$$

The total change in the location of the mth pole due to errors in the n coefficients $A_1, \ldots, A_n$, is given by the total derivative

$$\Delta p_m = \sum_{\ell=1}^{n} \frac{\partial p_m}{\partial A_\ell} \Delta A_\ell \tag{8.94}$$

Examination of (8.93) reveals that the magnitude of the denominator of $\partial p_m / \partial A_\ell$ is proportional to the *lengths* of the vectors from the pole p_m to all other poles p_i. Hence if p_m is very close to another pole, then p_m and that other pole will be very sensitive to changes in the coefficients.

EXAMPLE 8.10

A second-order digital filter has real poles at p_1 and p_2. It is implemented in direct form.

a) From the general expression (8.93), write an expression for the change in location of these poles that is due to changes in the coefficients of the corresponding difference equation.
b) If the nominal values of p_1 and p_2 are 0.98 and 0.94, respectively, what is the minimum number of bits required to ensure that the filter cannot go unstable as a result of coefficient quantization? Assume rounding.

Solution

a) From (8.93),

$$\frac{\partial p_m}{\partial A_\ell} = \frac{p_m^{n-\ell}}{\prod\limits_{\substack{i=1 \\ i \neq m}}^{n} (p_m - p_i)}, \qquad \ell = 1, 2, \text{ and } m = 1, 2$$

Therefore,

$$\frac{\partial p_1}{\partial A_1} = \frac{p_1}{p_1 - p_2} \qquad \frac{\partial p_1}{\partial A_2} = \frac{1}{p_1 - p_2}$$

$$\frac{\partial p_2}{\partial A_1} = \frac{p_2}{p_2 - p_1} \qquad \frac{\partial p_2}{\partial A_2} = \frac{1}{p_2 - p_1}$$

The total variation in pole position is

$$\Delta p_m = \sum_{\ell=1}^{2} \frac{\partial p_m}{\partial A_\ell} \Delta A_\ell$$

so

$$\Delta p_1 = \frac{\partial p_1}{\partial A_1} \Delta A_1 + \frac{\partial p_1}{\partial A_2} \Delta A_2 = \frac{1}{p_1 - p_2} [p_1 \Delta A_1 + \Delta A_2]$$

and

$$\Delta p_2 = \frac{\partial p_2}{\partial A_1} \Delta A_1 + \frac{\partial p_2}{\partial A_2} \Delta A_2 = \frac{1}{p_2 - p_1} [p_2 \Delta A_1 + \Delta A_2]$$

b) We need to find ΔA_1 and ΔA_2. The denominator of the filter transfer function has the form

$$(z - p_1)(z - p_2) = z^2 - A_1 z + A_2$$

where $A_1 = p_1 + p_2$ and $A_2 = p_1 p_2$. For stability, $-2 < A_1 < 2$ and $-1 < A_2 < 1$. For fixed-point arithmetic, the A_1 coefficient can be scaled to give a fractional number, although for filter coefficients, the binary point is often moved to the right to accommodate coefficients with magnitudes greater than unity. In either case (see Example 8.7), we can compute the quantization step from the range, and number of bits N. Thus, for A_2,

$$q_0 = 4/2^N$$

and for rounding,

$$\Delta A_1 = q_0/2 = 2/2^N$$

We could choose the same quantization step for the A_2 register, in which case we would need $N - 1$ bits, because the range of A_2 is half that of A_1. Alternatively, we could take N bits for both registers and let the quantization step for A_2 be $2/2^N = q_0/2$ so that, for rounding,

$$\Delta A_2 = q_0/4 = 1/2^N$$

Let us make the latter choice. Then, from the foregoing expressions for change in pole position,

$$\Delta p_1 = \frac{1}{0.98 - 0.94}[(0.98)2 + 1.0]/2^N = 74/2^N$$

and

$$\Delta p_2 = \frac{1}{0.94 - 0.98}[(0.94)2 + 1.0]/2^N = -72/2^N$$

Pole p_1, being closer to the unit circle, is more likely to cause the filter to go unstable if it is inexactly represented. For stability, we must have

$$1 - p_1 = 0.02 > \Delta p_1 = 74/2^N$$

or

$$2^N > 3700$$

This implies that $N = 12$ bits is the minimum register length.

To complete this analysis, it is necessary to generalize the nth-order filter to include complex poles so that, instead of (8.89), the following expression arises:

$$1 - \sum_{i=1}^{n} A_i z^{-i} = \prod_{i=1}^{q}(1 - p_i z^{-1}) \prod_{k=1}^{s}[1 - 2r_k(\cos\theta_k)z^{-1} + r_k^2 z^{-2}] \quad (8.95)$$

where $s = (n - q)/2$ with q simple poles and s pairs of complex poles. Differentiate (8.95) with respect to a particular coefficient A_ℓ, $1 \leq \ell \leq n$, and solve for the sensitivity coefficients $\partial p_m/\partial A_\ell$, $1 \leq m \leq q$, and $\partial r_g/\partial A_\ell$ and $\partial \theta_g/\partial A_\ell$, $1 \leq g \leq s$. After some algebra, the following results are obtained:

For the simple poles p_m, $1 \leq m \leq q$,

$$\frac{\partial p_m}{\partial A_\ell} = \frac{p_m^{-\ell+1}}{\prod_{\substack{i=1 \\ i \neq m}}^{q} (1 - p_i p_m^{-1}) \prod_{k=1}^{s} [1 - 2r_k(\cos\theta_k)p_m^{-1} + r_k^2 p_m^{-2}]} \tag{8.96}$$

For the complex poles $r_g \exp(\pm\theta_g)$, $1 \leq g \leq s$,

$$\frac{\partial r_g}{\partial A_\ell} = \frac{-r_g^{-\ell+1} \sin[(\ell - 1)\theta_g]}{2C_g \sin\theta_g} \tag{8.97}$$

$$\frac{\partial \theta_g}{\partial A_\ell} = \frac{r_g^{-\ell}\{\sin[(\ell - 2)\theta_g] - \cos\theta_g \sin[(\ell - 1)\theta_g]\}}{2C_g \sin^2\theta_g} \tag{8.98}$$

where

$$C_g = \prod_{i=1}^{q}(1 - p_i z^{-1}) \prod_{\substack{k=1 \\ k \neq g}}^{n}(1 - 2r_k \cos\theta_k z^{-1} + r_k^2 z^{-2})\Big|_{z = r_g e^{j\theta_g}} \tag{8.99}$$

The total derivatives are

$$\Delta p_m = \sum_{\ell=1}^{n} \frac{\partial p_m}{\partial A_\ell} \Delta A_\ell, \quad m = 1, \ldots, q \tag{8.100}$$

$$\Delta r_g = \sum_{\ell=1}^{n} \frac{\partial r_g}{\partial A_\ell} \Delta A_\ell, \quad g = 1, \ldots, s \tag{8.101}$$

$$\Delta \theta_g = \sum_{\ell=1}^{n} \frac{\partial \theta_g}{\partial A_\ell} \Delta A_\ell, \quad g = 1, \ldots, s \tag{8.102}$$

Again, it is evident that if the poles are tightly clustered, as for a narrow lowpass or a narrow bandpass digital filter (Figure 8.11), the poles of the direct-form realization are sensitive to quantization errors in the coefficients. Furthermore, the greater the number of clustered poles, the greater the sensitivity.

Cascade and Parallel Forms: On the other hand, the cascade and parallel forms realize each pair of complex poles separately. Thus the error in a given pole is independent of its distance from the other poles of the system: We get $\partial r/\partial A_l$ and $\partial\theta/\partial A_l$ separately for each pair of poles. Therefore, in general, the parallel and cascade forms are preferred over the direct form

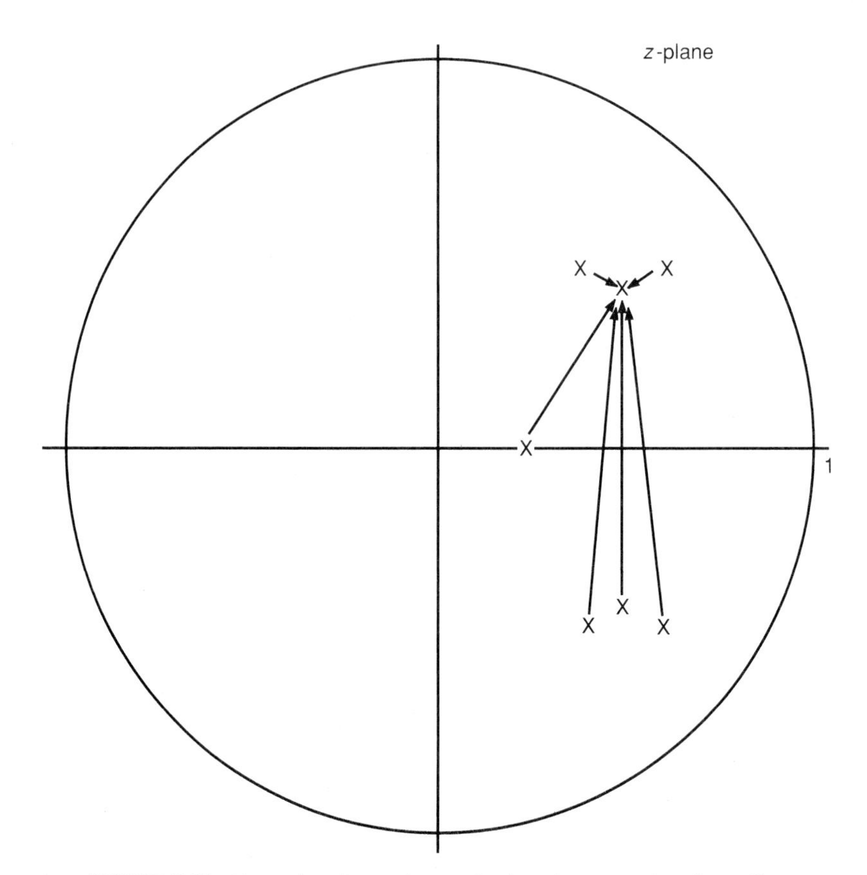

FIGURE 8.11 Vector lengths to clustered poles of a narrow-bandpass filter.

from the standpoint of parameter quantization. This is particularly true for narrow-bandpass frequency-selective filters with tightly-clustered poles and zeros.

The effect of coefficient changes in zero locations for nonrecursive filters can be examined in a similar manner. Here also, although instability is not a problem, a cascade form is preferred to a direct form.

8.5.2 Errors in Frequency Response Due to Coefficient Quantization

We have seen that when a digital filter is designed with frequency response $H_0(j\omega T)$, it is usually an approximation to an ideal or desired filter with frequency response $H_d(j\omega T)$. When the designed filter is implemented as a difference-equation algorithm, the finite number of bits available to represent the coefficients results in a filter the frequency response $H_q(j\omega T)$ of which differs, in general, from both the ideal response and the designed response. The specifications to which $H_0(j\omega T)$ was designed allow some tolerance or error about $H_d(j\omega T)$, as illustrated in Figure 8.12 for the magnitude response of a typical lowpass design.

Ideally, we would like to know the minimum number of bits required to represent the filter coefficients, before the implemented frequency response lies outside the prescribed error bounds. We will concentrate on the magnitude response. With reference to Figure 8.12, we must have the following conditions satisfied:

$$\begin{aligned} |M_d(\omega T) - M_q(\omega T)| &< \delta_p, \quad \omega \leq \omega_c \\ |M_d(\omega T) - M_q(\omega T)| &< \delta_s, \quad \omega \geq \omega_r \end{aligned} \tag{8.103}$$

where M_d and M_q are the desired and the implemented magnitude responses, respectively. The error tolerances δ_p and δ_s might be specified, for example, as the maximum allowable passband ripple and the maximum allowable stopband ripple (related to the minimum allowable stopband attenuation), respectively, for the ideal lowpass prototype. To find the minimum number of bits required to satisfy conditions (8.103), we start with $M_q(\omega T)$ computed with a small number of bits—say 4—representing the coefficients and check whether conditions (8.103) are satisfied over the frequency ranges specified. If not, we increase the number of bits by one and repeat the process until the conditions are met. To obtain this exact result could entail much computation.

A statistical method for obtaining a conservative estimate of the number of bits required was suggested by Avenhaus [5] (and modified by Crochiere [6]). Avenhaus proposed that the coefficient quantization errors be treated as random variables, as the errors were for the input quantization case. Let

$$M_q(\omega T) = M_0(\omega T) + \Delta M(\omega T) \tag{8.104}$$

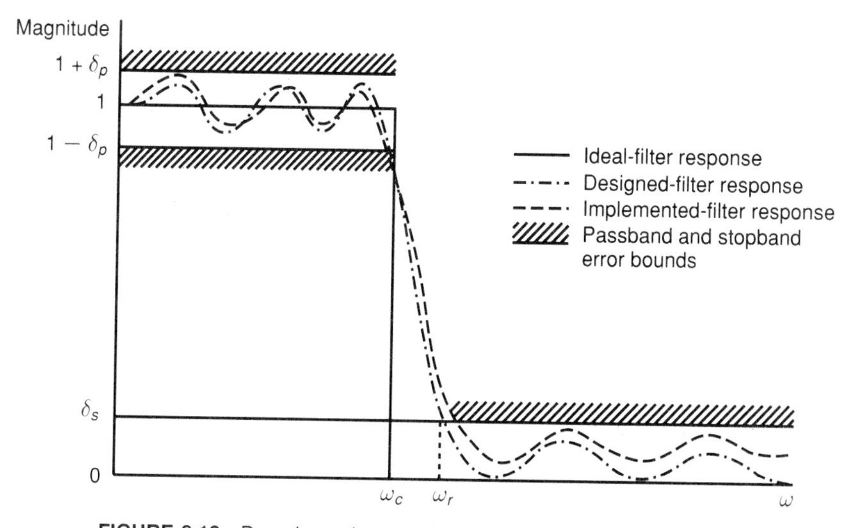

FIGURE 8.12 Bounds on the magnitude response of a lowpass filter.

where M_0 is the designed magnitude response and ΔM is the error due to quantization. For brevity, we will omit the frequency argument ωT and emphasize that both M_q and M are functions of the filter coefficients c_i, $i = 1, 2, \ldots, r$, by writing

$$M = M(c_1, c_2, \ldots, c_r) \tag{8.105}$$

where M is the functional form for M_0 and M_q in terms of the coefficients. Expand M_q in a Taylor series about the nominal or designed value M_0.

$$M_q = M_0 + \left.\frac{\partial M}{\partial c_1}\right|_0 \Delta c_1 + \cdots + \left.\frac{\partial M}{\partial c_r}\right|_0 \Delta c_r + \text{(higher-order terms)} \tag{8.106}$$

where the partial derivatives are evaluated at the designed values. If we assume that the coefficient errors Δc_i are small enough for us to neglect the higher-order terms and use (8.104), we get

$$\Delta M \approx \sum_{i=1}^{r} \left.\frac{\partial M}{\partial c_i}\right|_0 \Delta c_i = \sum_{i=1}^{r} S_i \Delta c_i \tag{8.107}$$

The terms $S_i \approx \partial M / \partial c_i$ indicate the sensitivity of the magnitude response with respect to the coefficients. As we have noted, the coefficient errors Δc_i will be treated as random variables with uniform distributions, just as for the input quantization case. For rounding, the mean is

$$\mu_{\Delta c_i} = 0, \quad i = 1, \ldots, r \tag{8.108}$$

and the variance is

$$\sigma^2_{\Delta c_i} = q_0^2/12, \quad i = 1, \ldots, r \tag{8.109}$$

Because ΔM is the weighted sum of random variables by (8.107), it is also a random variable, with mean

$$\mu_{\Delta M} = \left\langle \sum_{i=1}^{r} S_i \Delta c_i \right\rangle = \sum_{i=1}^{r} S_i \langle \Delta c_i \rangle = 0 \tag{8.110}$$

and variance

$$\sigma^2_{\Delta M} = \text{Var}\left[\sum_{i=1}^{r} S_i \Delta c_i\right] = \sum_{i=1}^{r} \text{Var}\,[S_i \Delta c_i] \tag{8.111}$$

under the assumption that the coefficient errors Δc_i are independent. It follows that

$$\sigma^2_{\Delta M} = \sum_{i=1}^{r} S_i^2 \,\text{Var}\,(\Delta c_i) = \frac{q_0^2}{12} S^2 \tag{8.112}$$

where

$$S^2 = \sum_{i=1}^{r} S_i^2 \tag{8.113}$$

By the central limit theorem (Chapter 7), because ΔM is the sum of identically distributed random variables, its distribution approaches the Gaussian. Then the probability density function of ΔM is approximately

$$f(\Delta M) = \frac{1}{\sigma_{\Delta M}\sqrt{2\pi}} \exp\left(-\frac{\Delta M^2}{2\sigma^2_{\Delta M}}\right) \tag{8.114}$$

Therefore, the probability that ΔM lies in the range

$$|\Delta M| \le \Delta M_{\max}$$

is

$$p = P\{-\Delta M_{\max} < \Delta M \le \Delta M_{\max}\} = \int_{-\Delta M_{\max}}^{\Delta M_{\max}} f(\Delta M)\, d(\Delta M) \tag{8.115}$$

where $f(\Delta M)$ is given by (8.114). Define

$$t = \Delta M/\sigma_{\Delta M} \tag{8.116}$$

Then, with this change of variable, (8.115) can be written in the standard form

$$\begin{aligned} p &= \frac{1}{\sqrt{2\pi}} \int_{-t_m}^{t_m} \exp(-t^2/2)\, dt \\ &= \frac{2}{\sqrt{2\pi}} \int_{0}^{t_m} \exp(-t^2/2)\, dt \end{aligned} \tag{8.117}$$

because the integrand is an even function. For any value of $t_m = \Delta M_{\max}/\sigma_{\Delta M}$, the probability p can be evaluated from the Gaussian error function

$$erf(t_1) = \int_{t_1}^{\infty} \exp(-t^2/2)\, dt \tag{8.118}$$

tabulated in many handbooks of mathematical tables. Conversely, if p is specified, we can obtain the required t_m from the tables. For example, $t_m = 2.58$ if $p = 0.99$, and $t_m = 1.96$ if $p = 0.95$.

From Figure 8.12, we deduce the bounds on ΔM for a lowpass filter as

$$\Delta M_{\max}(\omega T) = \begin{cases} \delta_p - |M_0(\omega T) - M_d(\omega T)|, & \omega \le \omega_c \\ \delta_s - |M_0(\omega T) - M_d(\omega T)|, & \omega \ge \omega_r \end{cases} \tag{8.119}$$

Similar bounds can be derived for highpass, bandpass, and bandstop filters.

EXAMPLE 8.11

It is desired to implement a designed filter with N bits for the coefficients. The range of coefficient values is ± 2. What bound must be imposed on the coefficient quantization step q_0 so that we have 99% confidence that the error in the magnitude response is less than the permitted error bound $\Delta M_{\max} = 0.05$, if the sum of the squares of the sensitivity coefficients is 9.42? What is the value of N?

Solution

From the error-function tables, we find there is a 0.005 probability that a Gaussian random variable in standard form exceeds $t_1 = 2.58$ and a 0.005 probability that it is less than $-t_1 = -2.58$. Hence there is a 0.99 probability that it lies within the bounds ± 2.58, or, as is generally stated in the context of statistics, we are 99% confident that it lies within these bounds. In this case,

$$t_1 = t_m = \Delta M_{\max}/\sigma_{\Delta M} = \sqrt{12}\Delta M_{\max}/(q_0 S) \tag{8.120}$$

from (8.112). This gives the bound on the quantization step.

$$\begin{aligned} q_0 &\leq \frac{\sqrt{12}\Delta M_{\max}}{t_m S} \\ &= \frac{\sqrt{12}\ 0.05}{(2.58)\sqrt{9.42}} = 0.02 \end{aligned} \tag{8.121}$$

We note that even when q_0 satisfies these bounds, there is a 1% probability that the error in the magnitude response will exceed its bounds. However, this risk is tolerable.

Because the number of bits required and the quantization step are related by

$$q_0 = R.2^{-N} \tag{8.122}$$

where R is the range of the coefficients, it follows that

$$\begin{aligned} N &\geq \log_2 R + \log_2 \frac{1}{q_0} \\ &= \log_2 R + \log_2 \frac{t_m S}{\sqrt{12}\Delta M_{\max}} \\ &= 7.51 \end{aligned} \tag{8.123}$$

Therefore, the minimum number of bits required to ensure the desired accuracy is $N = 8$.

If we have an explicit expression for the designed magnitude response $M(\omega T) = M_0(\omega T)$ in terms of the filter coefficients, we can evaluate the sensitivity coefficients $\partial M/\partial c_i$ directly and hence compute S. Often, however, such an expression is difficult to derive for higher-order filters, and we work instead with the designed frequency response $H(j\omega T) = H_0(j\omega T)$. We have

$$H = Me^{j\phi} = M\cos\phi + jM\sin\phi \tag{8.124}$$

where we (temporarily) omit the frequency argument for brevity. Then

$$\frac{\partial H}{\partial c_i} = \frac{\partial M}{\partial c_i}\cos\phi - M\sin\phi\frac{\partial \phi}{\partial c_i} + j\left[\frac{\partial M}{\partial c_i}\sin\phi + M\cos\phi\frac{\partial \phi}{\partial c_i}\right]$$

$$= Re\left[\frac{\partial H}{\partial c_i}\right] + j\,Im\left[\frac{\partial H}{\partial c_i}\right] \tag{8.125}$$

where $Re[\quad]$ and $Im[\quad]$ denote the real and imaginary parts, respectively. To get the sensitivity coefficients, we multiply the real part by $\cos\phi$ and the imaginary part by $\sin\phi$ and add the results.

$$\frac{\partial M}{\partial c_i} = \cos\phi\; Re\left[\frac{\partial H}{\partial c_i}\right] + \sin\phi\; Im\left[\frac{\partial H}{\partial c_i}\right], \quad i = 1, 2, \ldots, r \tag{8.126}$$

In the same way, we can evaluate the sensitivity coefficients for the phase as follows:

$$\frac{\partial \phi}{\partial c_i} = \left\{\cos\phi\; Im\left[\frac{\partial H}{\partial c_i}\right] - \sin\phi\; Re\left[\frac{\partial H}{\partial c_i}\right]\right\}/M, \quad i = 1, 2, \ldots, r \tag{8.127}$$

This, however, requires an explicit expression for $M(\omega T)$.

The validity of Avenhaus' assumption that the coefficient errors can be treated as random variables may be subject to debate. The design coefficients are known in advance; so are the errors, if a register of a particular length is chosen. Where, then, is the randomness? One might argue that the designed coefficients are only one set of many sets that could be chosen and, also, that the register length is but one of many available. The purist might refute that contention, but the fact remains that the Avenhaus method provides a conservative bound on register length that agrees well with experimental data.

8.6 ERROR CAUSED BY QUANTIZATION OF PRODUCTS

Multiplication of a number N bits long by another N-bit number results in a product that is a $2N$-bit number. If all $2N$ bits were saved and used, no quantization operation would be associated with the formation of a product. In practice, however, because the number would grow in length by N bits at each iteration, it is necessary each time to reduce the product to, say, N bits by some suitable round-off rule. (The term *round-off* is used whether the quantization method is rounding or truncation.)

A quantization is therefore associated with the formation of a product. The effect of this quantization on the filter depends on the filter arrangement. If it is assumed that the round-off error is independent from one iteration to the next, then the statistical model for fine quantization can be used (see Section 8.4), but noise sources must be introduced at various points in the system after a multiplication, instead of being simply added to the input signal. In effect, the multiplier is modeled as an infinite-precision operation followed by the additive noise source so that the overall result equals some quantization level.

Consider, for example, the second-order recursive filter

$$y_k = A_1 y_{k-1} + A_2 y_{k-2} + B_0 x_k + B_1 x_{k-1} \tag{8.128}$$

There are four multiplications involved if A_1, A_2, B_0, and B_1 are non-unity. Associated with each multiplication is round-off noise $e_i(kT)$, $i = 0, 3$. Consider first a direct realization of (8.128), as shown in Figure 8.13.

The separate noise sources may be replaced by a single resultant:

$$e(kT) = \sum_{i=0}^{3} e_i(kT) \tag{8.129}$$

This acts through the main summing junction as shown in Figure 8.13. Note that with this direct-form arrangement of the filter, the noise passes through the filter poles only; that is, the zeros have no effect on the noise. Thus the amplification or attenuation of the round-off noise is different, in general, from that of the input noise.

If all the quantization steps are the same (q_0), then for rounding,

$$\text{Var } E_i(kT) = \frac{q_0^2}{12}, \quad i = 0, 1, 2, 3 \tag{8.130}$$

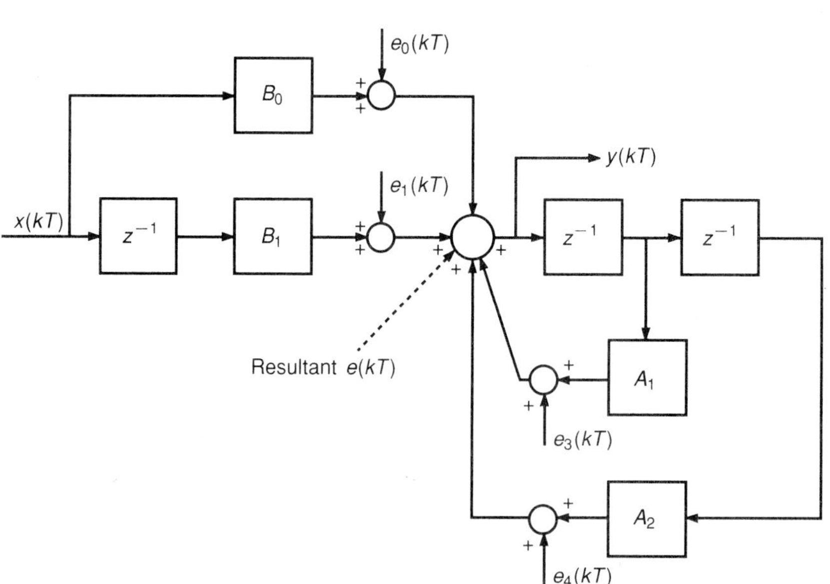

FIGURE 8.13 Multiplication round-off noise for the direct realization of a second-order recursive filter.

Given that the quantization errors are mutually independent, the variance of the resultant is the sum of the variances of the different components:

$$\text{Var } E(kT) = \sum_{i=0}^{3} \text{Var } E_i(kT) = \frac{q_0^2}{3} \tag{8.131}$$

The portion of the filter through which the round-off noise passes is shown in Figure 8.14. The output $f(kT)$ due to the noise forms part of the quantized output $y(kT)$.

The transfer function relating the transforms of $e(kT)$ and $f(kT)$ is

$$\frac{F(z)}{E(z)} = H'(z) = \frac{1}{1 - A_i z^{-1} - A_2 z^{-2}} \tag{8.132}$$

This is, of course, different from the transfer function $H(z)$ of the complete filter, which includes the zeros:

$$H(z) = \frac{B_0 + B_1 z^{-1}}{1 - A_1 z^{-1} - A_2 z^{-2}} \tag{8.133}$$

The steady-state variance of the output due to round-off noise is

$$\sigma^s_{Fss} = \frac{q_0^2}{3} \sum \begin{array}{c}\text{residues of } H'(z)H'(z^{-1})z^{-1} \text{ at poles}\\ \text{of same within unit circle}\end{array} \tag{8.134}$$

where $H'(z)$ is given in (8.132).

Now consider a canonic arrangement of filter (8.128). In this case, the round-off errors can be represented as noise sources positioned as shown in Figure 8.15.

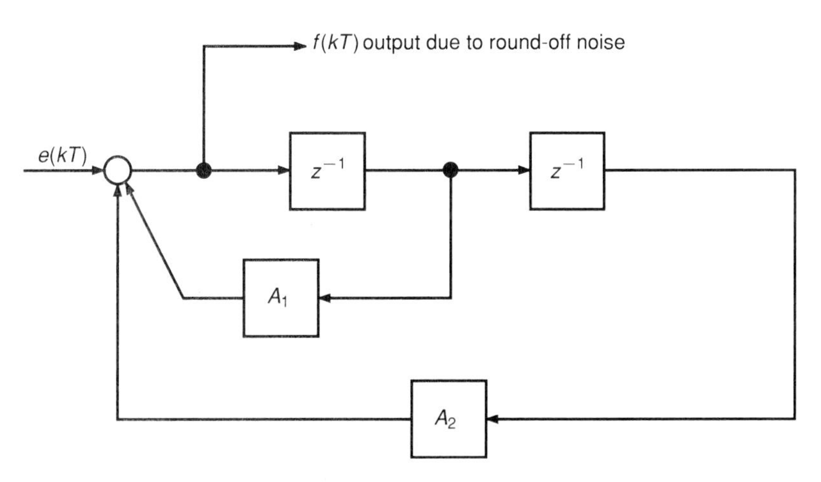

FIGURE 8.14 Portion of a second-order recursive filter affected by round-off noise for direct realization.

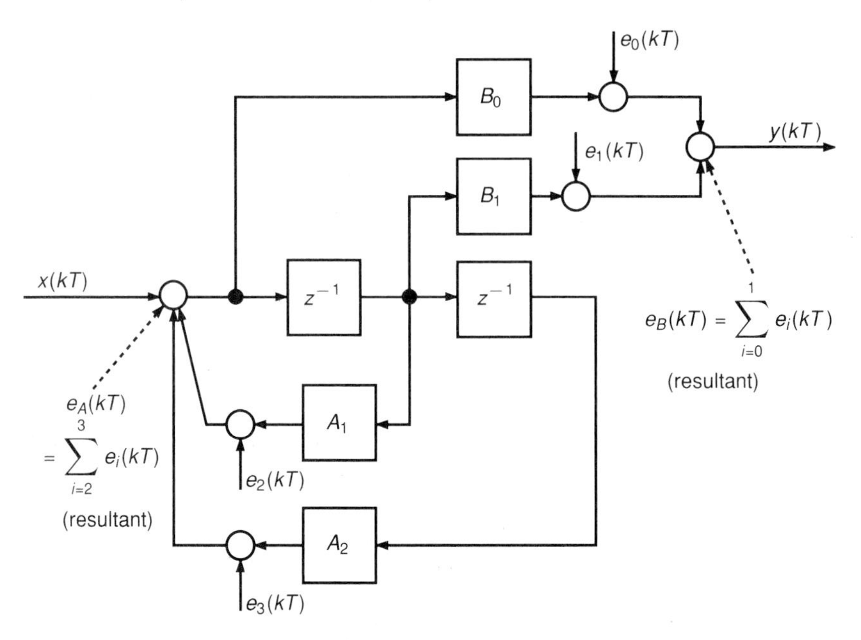

FIGURE 8.15 Multiplication round-off noise for a canonic representation of a second-order recursive filter.

Note that

$$e_A(kT) = \sum_{i=2}^{3} e_i(kT) \tag{8.135}$$

with variance $q_0^2/6$ passes through the entire filter, whereas

$$e_B(kT) = \sum_{i=0}^{1} e_i(kT) \tag{8.136}$$

with variance $q_0^2/6$ is simply a noise added to the output. In this case, the steady state variance of the output due to round-off noise is

$$\sigma_{fss} = \frac{q_0^2}{6}\left[1 + \sum \begin{matrix}\text{residues of } H(z)H(z^{-1})z^{-1} \text{ at poles} \\ \text{of same within unit circle}\end{matrix}\right] \tag{8.137}$$

where $H(z)$ is given in (8.133). Without inserting numerical values for the coefficients, it is not possible to determine whether the direct or the canonic form yields less output noise due to quantization error.

EXAMPLE 8.12

Compare the steady-state variance of output noise due to arithmetic round-off for the filter

$$H(z) = \frac{B_0 + B_1 z^{-1}}{1 - 2r\cos\theta z^{-1} + r^2 z^{-2}}$$

implemented in a) direct form and in b) canonic form, if $r = 0.9$, $\theta = \pi/4$, $B_0 = 1.1$, $B_1 = 0.3$, and the quantization step is q_0.

Solution

a) *Direct form*

From Figures 8.13 and 8.14, the steady-state variance of output noise is

$$\sigma_{Fss}^2 = \frac{q_0^2}{3} \sum \text{residues of } H'(z)H'(z^{-1})z^{-1} \text{ at poles of } H'(z)$$

$$= \frac{q_0^2}{3} \frac{1+r^2}{1-r^2} \frac{1}{r^4 - 2r^2 \cos 2\theta + 1}$$

$$= 1.92 q_0^2$$

b) *Canonic form*

From Figure 8.15,

$$\sigma_{fss}^2 = q_0^2/6 + q_0^2/6 \sum \text{residues of } H(z)H(z^{-1})z^{-1} \text{ at poles of } H(z)$$

$$= \frac{q_0^2}{6} \left[1 + \frac{(B_0^2 + B_1^2)(1+r^2) - 4B_0B_1 r \cos\theta}{(r^4 - 2r^2 \cos 2\theta + 1)(1-r^2)} \right]$$

$$= 1.07 q_0^2$$

Thus the canonic form gives less output noise for the given parameter values. Note that the values of B_0 and B_1 do not affect the noise variance in the direct form.

8.7 SCALING

In Section 8.3.4 we saw that although no round-off errors can occur with the addition of fixed-point numbers, the resulting sum may overflow—that is, exceed the dynamic range of the register. For a nonrecursive filter, this effect is felt in the distortion of output $y(k)$, which is the weighted sum of the current input and the previous inputs. The effect is much more serious for a recursive filter, because the errors are fed back and rapidly render the filter useless. In both cases, scaling is necessary to reduce the magnitudes of the signals at certain points in the filter and thereby prevent overflow under normal operating conditions.

Consider first the nonrecursive case with unit pulse response $h(k)$, $k = 0, 1, \ldots, N-1$ and with output related to input $x(k)$ by the convolution summation

$$y(k) = \sum_{i=0}^{N-1} h(i)x(k-i) \tag{8.138}$$

Take absolute values of both sides.

$$|y(k)| = \left| \sum_{i=0}^{N-1} h(i)x(k-i) \right| \leq \sum_{i=0}^{N-1} |h(i)|\,|x(k-i)| \tag{8.139}$$

Suppose that the input x is bounded and $|x(k)| \leq 1$. Then

$$|y(k)| \leq \sum_{i=0}^{N-1} |h(i)| \tag{8.140}$$

The right side of this expression is called the ℓ_1 norm of h. That is,

$$\|h\|_1 = \sum_{i=0}^{N-1} |h(i)| \tag{8.141}$$

If we divide each of the unit-pulse-response samples by the scaling factor $\|h\|_1$, we are assured that $|y(k)| \leq 1$, and overflow is avoided in the scaled filter $h'(k) = h(k)/\|h\|_1$, $k = 0, 1, \ldots, N-1$. However, scaling with the ℓ_1 norm is conservative, and other scaling measures are often used.

The ℓ_2 norm of h is defined as

$$\|h\|_2 = \left[\sum_{i=0}^{N-1} h^2(i) \right]^{\frac{1}{2}} \tag{8.142}$$

It is not difficult to show that $\|h\|_2 \leq \|h\|_1$. Other norms often used for scaling purposes are based on the frequency response $H(j\omega T)$. From the inverse z-transform, replacing z by $e^{j\omega T}$ in (3.68), we can write the inverse Fourier transform of $H(j\omega T)$ as

$$h(k) = \frac{1}{2\pi} \int_{-\pi}^{\pi} H(j\omega T) e^{jk\omega T}\, d(\omega T) \tag{8.143}$$

Define the L_p norm ($p \geq 1$) of a Fourier transform $H(j\omega T)$ as

$$\|H\|_p = \left[\frac{1}{2\pi} \int_{-\pi}^{\pi} |H(j\omega T)|^p\, d(\omega T) \right]^{1/p} \tag{8.144}$$

If the integral is finite, then the limit as p approaches infinity exists, and the L_∞ norm of H is

$$\|H\|_\infty = \max_{-\pi \leq \omega T \leq \pi} |H(j\omega T)| \tag{8.145}$$

(See [14].) This is the peak value of $|H(j\omega T)|$ over all frequencies. The L_∞ norm is sometimes called the Chebyshev norm and is designated $\|H\|_c$. Again $\|H\|_c \leq \|h\|_1$. This norm is perhaps the easiest to use. It involves plotting the magnitude response $M(\omega T) = |H(j\omega T)|$ and selecting its maximum value.

Similar norms can be defined for scaling recursive filters. The upper limits of the summations for the ℓ_1 and ℓ_2 norms now become infinite because of the infinite number of unit-pulse-response samples $h(k)$. However, for a stable filter, $h(k)$ approaches zero as k increases, convergence being more rapid the closer the filter poles are to the origin of the z-plane. Hence a practical estimate of the ℓ_1 and ℓ_2 norms can be obtained experimentally by applying a unit pulse input to the unscaled filter and truncating the summations as $h(k)$ becomes negligibly small.

The recursive filter may be implemented to have several summing points. The output of each summer must be selected to avoid overflow. Thus there will be several unit pulse responses $h_r(k)$ and corresponding transfer functions $H_r(z)$ relating the input $x(k)$ and the intermediate signals $y_r(k)$. A simple example is shown in the canonic representation of the second-order filter in Figure 8.16. Here

$$H(z) = \frac{B_0 + B_1 z^{-1} + B_2 z^{-2}}{1 - A_1 z^{-1} - A_2 z^{-2}} \tag{8.146}$$

$$H_1(z) = \frac{1}{1 - A_1 z^{-1} - A_2 z^{-2}} \tag{8.147}$$

There are two unit pulse responses involved, $h(k)$ and $h_1(k)$. We can define ℓ_1 and ℓ_2 norms with respect to h and h_1 or can define the L_∞ norm with respect to H and H_1 as described above. Suppose we select the ℓ_2 norm for scaling. Then we divide the *numerator* coefficients of $H(z)$ by the larger of $\|h\|_2$ and $\|h_1\|_2$ to scale the filter, obtaining the scaled transfer function $H'(z)$. If $H(z)$ is, for example, the first section of a two-section fourth-order filter connected in cascade, then it is necessary to scale the second section also [15]. This is done by computing the transfer functions and associated pulse responses from the section-2 input (which is y in Figure 8.16) to the output of the summers where overflows can occur, as we did

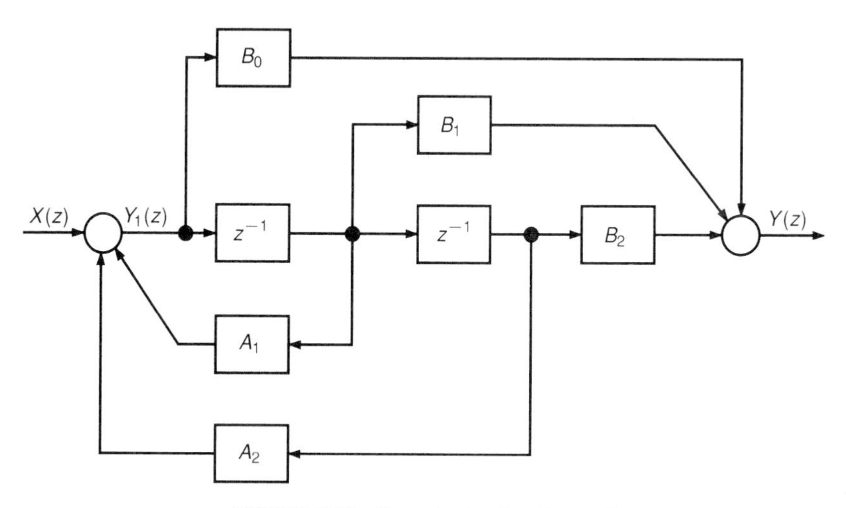

FIGURE 8.16 Canonic structure for scaling.

for section 1. Then we multiply these transfer functions by the scaled transfer function $H'(z)$ for section 1 to get the transfer function from section-1 input x to the corresponding variable in section 2. From the corresponding unit pulse responses, the ℓ_2 norm (and the ℓ_1 norm) can be computed. We divide the numerator coefficients of section 2 by the ℓ_2 norm to get the scaled section-2 transfer function. Additional cascaded sections are handled similarly.

8.8 COARSE QUANTIZATION

The preceding analyses of quantization effects have been based on the assumption of fine quantization, permitting linearization of what is essentially a nonlinear problem. In some applications, for economic reasons, it is desirable to reduce the number of bits as much as possible. There are cases where the resulting quantization is so coarse that the fine quantization model yields erroneous results or otherwise fails to predict the actual behavior of the implemented filter. The assumption of a signal fluctuating over many quantization steps in a sampling period is violated if the quantization is coarse and/or the signal is constant (including zero).

Ideally, the designer would like to know what effect coarse quantization has on filter performance so that he or she can make an intelligent decision about the minimum number of bits to use in cases where this is a consideration. Unfortunately, however, coarse quantization poses a difficult, nonlinear system problem, and there are few general results comparable to the linear case. The problem has been studied extensively. Although Blackman [7] is generally credited with being the first investigator to detect the existence of "deadbands" in the filter output due to coarse quantization, Monroe, in an earlier book [8], gave an example of this phenomenon for a first-order filter.

For a stable linear filter, the output should approach zero in the steady state if there is zero input, and it should approach a constant value if the input is a constant. With coarse quantization, however, an output can occur even with zero input if there is a nonzero initial condition on one of the registers. This output can be a fixed value lying within a deadband, or it can take the form of a periodic oscillation, or limit cycle, the magnitude and frequency of which are independent of the initial conditions or applied input. Whether a fixed output value or a limit cycle occurs depends on how the quantization is effected. Three simple examples will illustrate the phenomena of deadband and limit cycle.

EXAMPLE 8.13

Consider a first-order recursive filter where the input x_k is added to the product Ay_{k-1} before quantization. Let the output be rounded off to the nearest integer. If the unquantized output lies exactly halfway between integers, the logic is arranged so that the output is quantized to the next higher integer for positive integers and to the next lower integer for negative integers. The system is shown in Figure 8.17. For $A = 0.9$ and initial condition $y_{-1} = 92$, find the steady state output to a constant input $x_k = 10$.

The unquantized filter has the equation

$$y_k = Ay_{k-1} + x_k \tag{8.148}$$

According to the final-value theorem (Chapter 3), the "correct" steady state output is

$$\begin{aligned}\lim_{k\to\infty} y_k &= \lim_{z\to 1}(1 - z^{-1})Y(z) \\ &= \lim_{z\to 1}(1 - z^{-1})\frac{1}{1 - 0.9z^{-1}}\frac{10}{1 - z^{-1}} \\ &= 100\end{aligned} \tag{8.149}$$

With quantization, the filter equation becomes

$$y'_k = Q[Ay'_{k-1} + x_k] \tag{8.150}$$

where y'_k is the quantized output and $Q[\quad]$ denotes the quantization operation as shown in Figure 8.17. We can obtain the steady-state value of y'_k by an iterative method as follows. From (8.150),

$$\begin{aligned}y'_0 &= Q[Ay'_{-1} + x_0] = Q[(0.9)92 + 10] = Q[92.8] = 93 \\ y'_1 &= Q[Ay'_0 + x_1] = Q[(0.9)93 + 10] = Q[93.7] = 94 \\ y'_2 &= Q[Ay'_1 + x_2] = Q[(0.9)94 + 10] = Q[94.6] = 95 \\ y'_3 &= Q[Ay'_2 + x_3] = Q[(0.9)95 + 10] = Q[95.5] = 96 \\ y'_4 &= Q[Ay'_3 + x_4] = Q[(0.9)96 + 10] = Q[96.4] = 96\end{aligned}$$

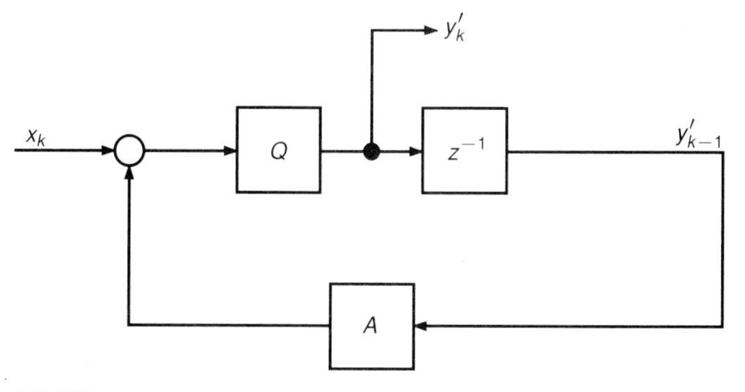

FIGURE 8.17 First-order filter with coarse quantization for Example 8.13.

So we get a steady-state output of 96 instead of the correct value of 100. For an initial condition of 105 or more, the steady-state output is 105. For initial conditions within the interval [96, 105], the output remains at the initial value. Thus there is a deadband of [96, 105] about the correct value of 100. The slight asymmetry is due to the logic in rounding upward for values halfway between positive integers.

This example of a deadband is generalized in [2] to the case of an nth-order recursive filter implemented in direct form with constant input. It does, however, require the assumption that all feedback signals are summed together and added to the input before quantization. It is thus a special case. Briefly, the deadband bounds are obtained as follows. Let

$$y_k = \sum_{i=1}^{n} A_i y_{k-i} + \sum_{i=0}^{m} B_i x_{k-i} \tag{8.151}$$

be the equation for infinite-precision arithmetic. For a constant input $x_k = \overline{X}$, the correct steady state value for the output is given by the final-value theorem as

$$Y_c = \left(\sum_{i=0}^{m} B_i \overline{X}\right) \Big/ \left(1 - \sum_{i=1}^{n} A_i\right) \tag{8.152}$$

When quantization is taken into account, (8.151) is replaced by

$$y'_k = Q\left[\sum_{i=1}^{n} A_i y'_{k-i} + \sum_{i=0}^{m} B_i x_{k-i}\right] = \sum_{i=1}^{n} A_i y'_{k-i} + \sum_{i=0}^{m} B_i x_{k-i} + e_k \tag{8.153}$$

where y'_k is the quantized output and e_k is the quantization error. In the limit as k approaches infinity, y_k approaches Y', the steady-state quantized output, and (8.153) becomes

$$Y' = \left(\sum_{i=1}^{n} A_i\right) Y' + \left(\sum_{i=0}^{m} B_i\right) \overline{X} + \lim_{k\to\infty} e_k \tag{8.154}$$

If we eliminate the input term by using (8.152), then (8.154) can be written as

$$(Y' - Y_c)\left(1 - \sum_{i=1}^{n} A_i\right) = \lim_{k\to\infty} e_k \tag{8.155}$$

For rounding, the magnitude of the error term e_k is less than $q_0/2$. Hence a bound on the magnitude of the error between the quantized output and the correct output Y_c is

$$|Y' - Y_c| \le \frac{q_0}{2\left|1 - \sum_{i=1}^{n} A_i\right|} \tag{8.156}$$

Note that this result is independent of the constant input $\overline{X}$. For Example 8.13, (8.156) gives an error bound of ± 5. Because $Y_c = 100$ in this example, the deadband bounds are [95, 105], which compare well with the bounds of [96, 105] found by iteration. For a second- or higher-order filter where the products are individually quantized *before* being summed, the bounds given by (8.156) are not applicable.

The following example illustrates further nonlinear behavior that results from coarse quantization.

EXAMPLE 8.14

Consider the same filter and quantization method as in Example 8.13, but now let $A = -0.9$, let the initial condition be 10, and let the input $x_k = 0$ for all k. Find the steady-state output.

Solution

From Chapter 3, we know that such a filter with a pole on the negative z-axis inside the unit circle will have a converging oscillatory response. With zero input and no quantization, the steady-state output is zero. Let us derive by iteration the steady-state output with quantization, as we did in the last example.

$$y'_0 = Q[Ay'_{-1}] = Q[(-0.9)10] = Q[-9.0] = -9$$
$$y'_1 = Q[Ay'_0] = Q[(-0.9)(-9)] = Q[8.1] = 8$$
$$y'_2 = Q[Ay'_1] = Q[(-0.9)8] = Q[-7.2] = -7$$
$$y'_3 = Q[Ay'_2] = Q[(-0.9)(-7)] = Q[6.3] = 6$$
$$y'_4 = Q[Ay'_3] = Q[(-0.9)6] = Q[-5.4] = -5$$
$$y'_5 = Q[Ay'_4] = Q[(-0.9)(-5)] = Q[4.5] = 5$$
$$y'_6 = Q[Ay'_5] = Q[(-0.9)5] = Q[-4.5] = -5$$

We see that the steady-state quantized output settles into a periodic oscillation that is independent of the initial condition. This is a limit cycle. It is stable in this case, because the response converges to it. The reader can easily confirm that if the initial condition is less than 5, the response will converge to 5.

Second-order systems can exhibit similar behavior, as the following example demonstrates.

EXAMPLE 8.15

Let the linear filter equation be

$$y_k = A_1 y_{k-1} + A_2 y_{k-2} \tag{8.157}$$

with given initial conditions y_{-1} and y_{-2} and zero input. With the multiplications rounded off individually before summing, the quantized system becomes

$$y'_k = Q[A_1 y'_{k-1}] + Q[A_2 y'_{k-2}] \tag{8.158}$$

Find the steady-state output from the given initial conditions.

Solution

Equation (8.158) can be written as two coupled first-order equations in the form

$$\begin{aligned} x_1(k+1) &= x_2(k) \\ x_2(k+1) &= Q[A_2 x_1(k)] + Q[A_1 x_2(k)] \end{aligned} \tag{8.159}$$

where $x_1(k) = y'_{k-2}$ and $x_2(k) = y'_{k-1}$.

Let the products be rounded to the nearest integer, values halfway between quantization levels being rounded to the higher level. Equation (8.159) can be used to solve recursively for x_1 and x_2 for specific values of coefficients and initial conditions. A phase-plane plot of $x_2(k)$ versus $x_1(k)$ is shown in Figure 8.18. This plot is drawn for $A_1 = 1.65$, $A_2 = -0.81$, $x_1(0) = y_{-2} = 10$, and $x_2(0) = y_{-1} = 10$. For clarity, the discrete points of the trajectory are joined. The trajectory for the unquantized or linear version, (8.157), is also shown for comparison. Note that the latter converges smoothly to rest at the origin, as predicted by theory. However, the quantized system settles into a closed periodic trajectory about the origin. This is a limit cycle, and for the given parameter values, the maximum amplitude or bound for the output $x_2(k)$ is 2 units, as shown. The period of oscillation is $10T$, where T is the sampling period. The limit cycle is independent of initial conditions. It is left as an exercise to show that if the products were summed before being quantized, then, for the given conditions, the output $x_2(k)$ would settle to a constant value of 2 units with no oscillation.

To have a nonzero periodic output for zero input is particularly bothersome for speech processing, for example. Hence the existence, bounds, and period of the limit cycle are matters of particular interest to the designer, who must assess the gravity of the problem and decide how to alleviate it, if necessary by finer quantization. One can always solve a system such as (8.159) and determine whether a limit cycle exists, but this approach is somewhat tedious in general. It would be desirable to have a method whereby, on examination of the system equations, one could test for the existence of a limit cycle and, if it existed, predict its bounds and period. References [9] through [13], among others, deal with this problem.

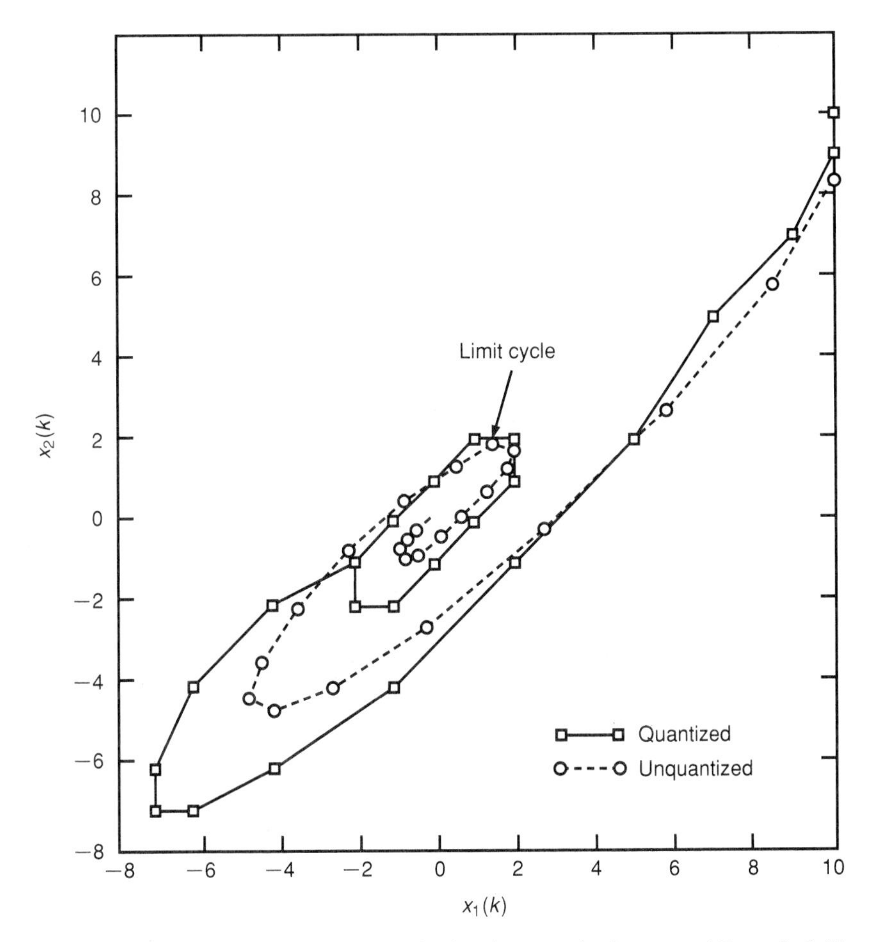

FIGURE 8.18 Phase-plane plot for quantized and unquantized system of Example 8.15.

In [9], Jackson treats the existence of deadbands and limit cycles in first- and second-order systems with fixed-point arithmetic. This is of practical importance for higher-order filters, because, as we noted earlier in this chapter, many of them are realized as series or parallel arrangements of first- and second-order filters. Jackson's approach is, in effect, a linearization of what is essentially a nonlinear problem. He postulates that if a deadband or sustained oscillation in the form of a limit cycle exists, the "effective" poles of the filter lie on the unit circle in the z-plane.

Take, for example, a first-order filter as in Example 8.13, described by (8.150). Then, for rounding,

$$|Q[Ay'_{k-1}] - Ay'_{k-1}| \leq q_0/2 \tag{8.160}$$

where q_0 is the quantization step. For values of k that bring the quantized output y'_k into the deadband, we have

$$|Q[Ay'_{k-1}]| = |y'_{k-1}| \tag{8.161}$$

implying that the effective values of A are ± 1 in such cases. The inequality in (8.160) is strengthened by

$$|Q[Ay'_{k-1}]| - |Ay'_{k-1}| \leq q_0/2 \tag{8.162}$$

Therefore, from (8.161), (8.162) becomes

$$|y'_{k-1}| - |Ay'_{k-1}| \leq q_0/2 \tag{8.163}$$

This gives the bounds on the deadband as

$$|y'_{k-1}| \leq \frac{q_0}{2(1-|A|)} \tag{8.164}$$

Note that with zero input, (8.164) gives the same bounds as (8.156) for positive A but different bounds for negative A. We see that (8.164) accurately predicts the limit cycle bounds for Example 8.14, whereas (8.156) does not. We conclude that (8.156) holds only for positive values of coefficients A_i.

The second-order case will now be considered. Take the system of (8.158), for example, which is repeated here for convenience:

$$y'_k = Q[A_1 y'_{k-1}] + Q[A_2 y'_{k-2}] \tag{8.165}$$

For the unquantized or linear situation, the poles of this filter are located at

$$z = r \exp(\pm j\theta) \tag{8.166}$$

where

$$r = \sqrt{-A_2} \tag{8.167}$$

$$\theta = \cos^{-1}(A_1/2r) \tag{8.168}$$

For the coefficient values of Example 8.15, the poles lie at $0.9 \exp(\pm j23.56°)$. From intuition, one might expect a limit cycle, if it exists, to have a frequency closely related to the resonant frequency of the poles. The latter is

$$\omega_r = \frac{23.56}{180}\frac{\pi}{T} = \frac{\omega_s}{15.28} \tag{8.169}$$

However, the true value of the limit cycle frequency is $\omega_s/10$, as we obtained it by iteration, so the intuitive approach gives a sizable error. If we assume that the effective poles lie on the unit circle, then substituting $r = 1$ into (8.168) gives $\theta = 34.41°$ for the angular position of the effective poles. This yields an effective resonant frequency of

$$\omega_r^1 = \frac{\omega_s}{10.46} \tag{8.170}$$

which agrees well with the true limit cycle frequency of $\omega_s/10$.

For the limit cycle bounds, proceed as for the first-order case.

$$|Q[A_2 y'_{k-2}] - A_2 y'_{k-2}| \leq q_0/2 \tag{8.171}$$

so that if the effective $-A_2 = r^2 = 1$, we have

$$|Q[A_2 y'_{k-2}]| = |y'_{k-2}| \tag{8.172}$$

for values of k within the limit cycle. The bounds are thus

$$|y'_{k-2}| \leq \frac{q_0}{2(1 - |A_2|)} \tag{8.173}$$

This gives bounds of ± 2.63 for Example 8.15, compared with the true bounds of ± 2 found by iteration, so the prediction is reasonable.

Reference [9] includes diagrams indicating the relative values of A_1 and A_2 for which limit cycles can exist. In addition, for the special cases wherein the effective poles are at ± 1, corresponding to limit cycle frequencies of 0 and $\omega_s/2$, the limit cycle bounds are derived as follows:

$$|y'_{k-2}| \leq \frac{q_0}{1 - |A_1| - A_2} \tag{8.174}$$

Although Jackson's bounds are derived on a heuristic basis, they are found to be exceeded only in exceptional cases. More conservative bounds are derived by Long and Trick in [12].

If a higher-order filter is made up of second-order sections in parallel, the foregoing theory applies. If the sections are in series, however, only the first section can be considered to have zero input, and its output limit cycle serves as the input to the next section and so on. Jackson postulates that the succeeding sections may display their own limit cycle behavior if the input limit cycle amplitude is small or, alternatively, may filter the output of the previous section if that amplitude is large.

If higher-order filters are realized by other than first- and second-order sections, the limit cycle behavior can become very complex, and no comprehensive theory is available. Very conservative bounds on limit cycle amplitudes for sections of arbitrary order are derived by Sandberg and Kaiser in [11]. Kaneko in [13] treats the case of limit cycles in digital filters utilizing floating-point arithmetic and derives some general results.

Limit cycles can occur with fine quantization also, as we will see in the next section. Claasen and Kristiansson [16] examined the question of stability in second-order recursive filters with overflow nonlinearities, and they obtained necessary and sufficient conditions for the absence of overflow phenomena. Claasen and others [17] analyzed quantization noise resulting from magnitude truncation in digital filters excited by white noise.

8.9 LIMIT CYCLE OSCILLATIONS DUE TO OVERFLOW

8.9.1 Introduction

As noted in Section 8.4, if we know the range R of a number to be stored and the length (N bits) of the register, we can represent the number in

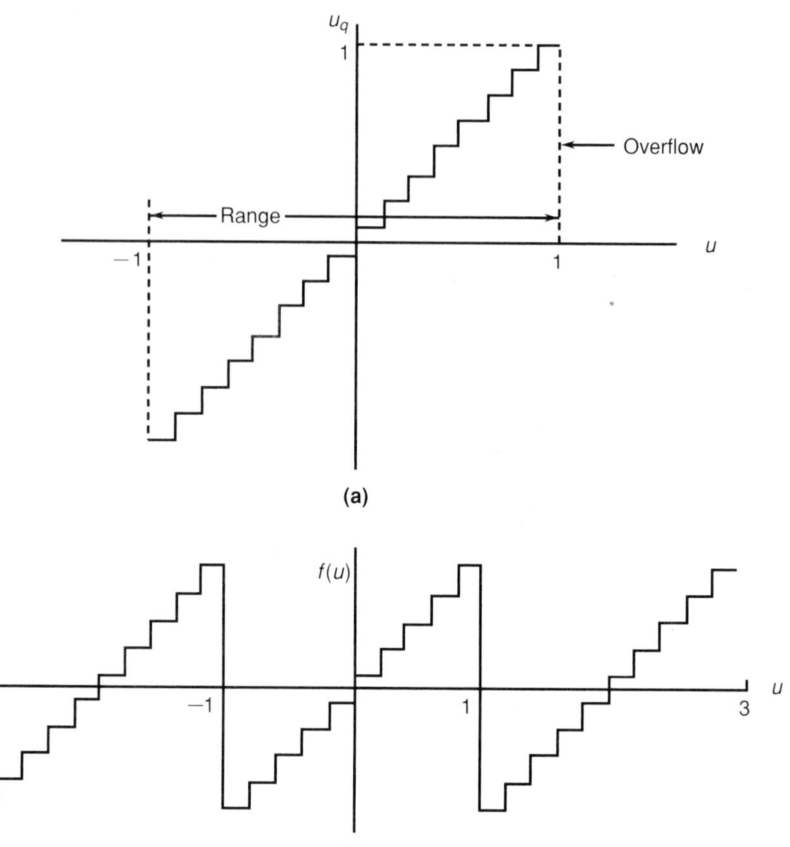

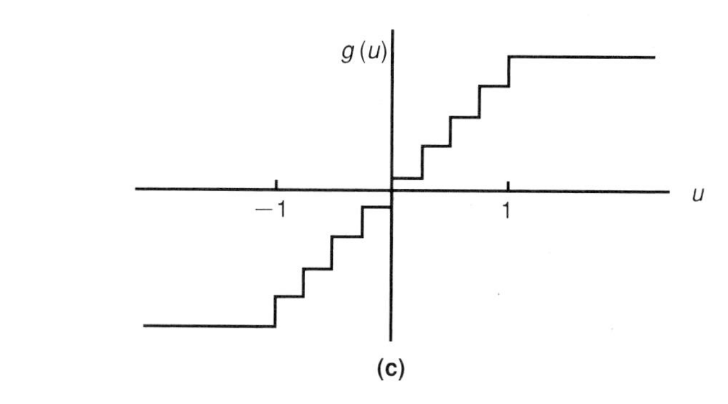

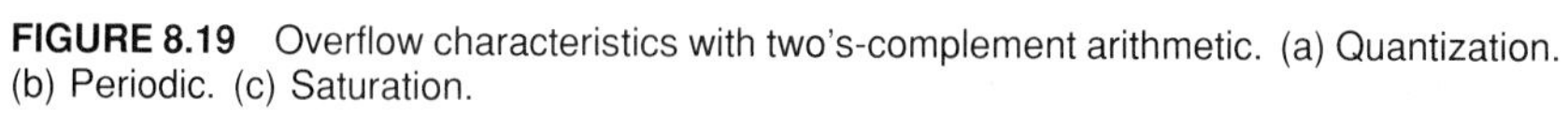

FIGURE 8.19 Overflow characteristics with two's-complement arithmetic. (a) Quantization. (b) Periodic. (c) Saturation.

quantized form, as shown, for example, in Figure 8.4. If, however, a positive or negative number resulting from an arithmetic operation lies outside the register range, an overflow occurs. This can cause large errors in filtering data. We will treat an important case that gives rise to overflow oscillations in a filter with zero input, following the procedure of Roberts and Mullis [18].

The largest positive number that can be represented in a 4-bit register is 0.111. Suppose that two's-complement arithmetic is used. If a slightly larger number occurs, the register overflows, adding 0.001 to 0.111, which results in a large *negative* number 1.001. Similarly, if a large negative number lies just outside the negative range of the register, it results in a large positive number being stored in the register. Thus the error can be as large as R, the range of the register. When scaled, $R = 2$, as shown in Figure 8.19(a). If nothing is done about the overflow, it has a periodic characteristic $f(u)$ with period 2, as depicted in Figure 8.19(b), for a number u.

If instead of permitting the overflow to go unchecked, we set an overflowed number to the largest (or smallest) number that can be represented, we have a saturation characteristic, as shown in Figure 8.19(c). This approach results in a smaller overflow error, in general, than that given by the unchecked overflow, but it requires more logic to implement.

With fine quantization, it is convenient for purposes of analysis to ignore the quantization effects and concentrate on the overflow problem. The function $f(u)$ in Figure 8.19(b) then becomes the sawtooth wave shown in Figure 8.20. The function $f(u)$ has the following properties:

1. $|f(u)| \leq 1$ (8.175)
2. $f(u) = u - N$ for $N - 1 < u < N + 1$, $N = 0, \pm 2, \pm 4, \pm 6, \ldots$

8.9.2 Digital Filters in State-Variable Form with Application to Overflow Oscillations

In order to investigate the effect of overflow oscillations on a digital filter, it is convenient to express the filter in state-variable form. This involves replacing the nth-order difference equation describing the filter by n first-order equations. Suppose the transfer function is given by

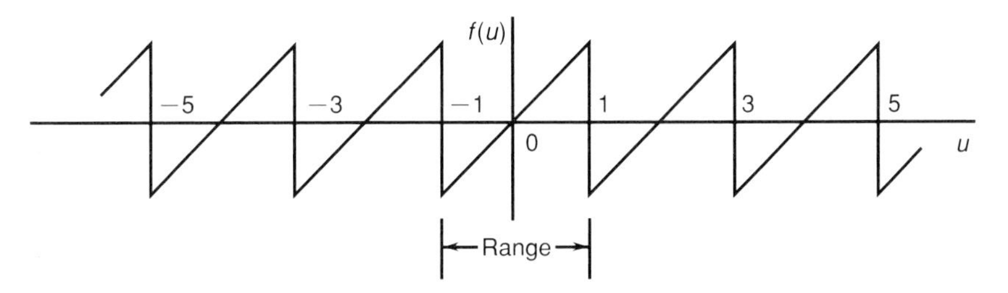

FIGURE 8.20 Overflow characteristic, ignoring quantization.

$$H(z) = \frac{\sum_{i=0}^{m} B_i z^{-i}}{1 - \sum_{i=1}^{n} A_i z^{-i}} \tag{8.176}$$

Express the filter in canonic form (see Section 3.12), as shown in Figure 8.21. Label the output of the n delay units from left to right as $x_1, x_2, \ldots, x_n$. These are the state variables. Then we can write the n first-order equations for the filter as

$$\begin{aligned} x_1(k+1) &= A_1 x_1(k) + A_2 x_2(k) + \cdots + A_n x_n(k) + u(k) \\ x_2(k+1) &= x_1(k) \\ x_3(k+1) &= x_2(k) \\ &\vdots \\ x_n(k+1) &= x_{n-1}(k) \end{aligned} \tag{8.177}$$

The filter output becomes

$$y(k) = B_0[A_1 x_1(k) + \cdots + A_n x_n(k)] + B_1 x_1(k) + \cdots B_m x_m(k) \tag{8.178}$$

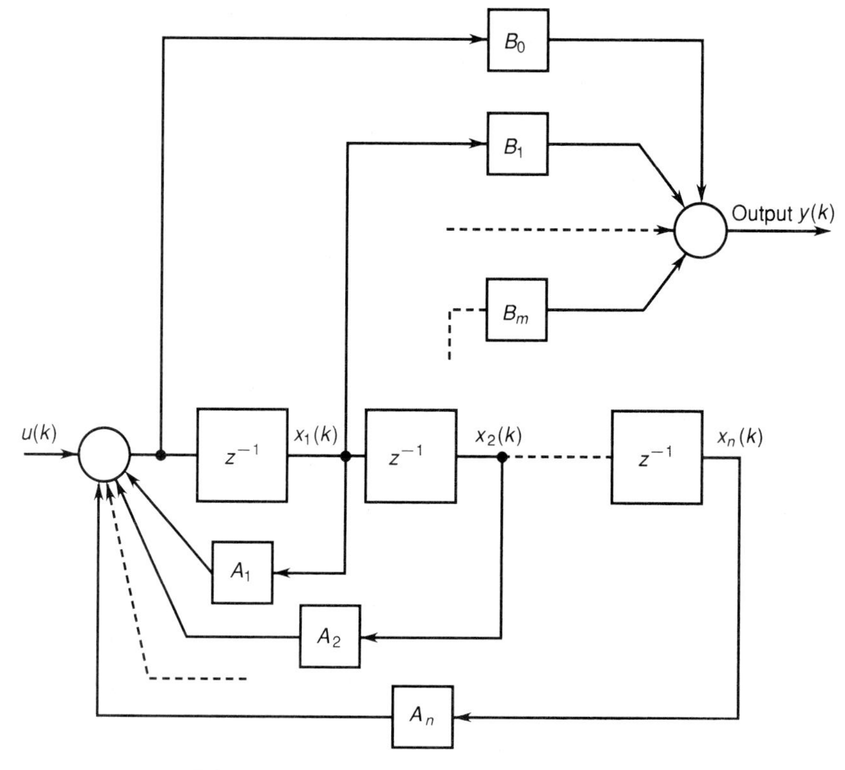

FIGURE 8.21 Definition of state variables, $m < n$.

Define the n-dimensional state vector or column matrix

$$\mathbf{x} = [x_1, x_2, \ldots, x_n]'$$

where the prime denotes the transpose of the matrix. Then we can write (8.177) and (8.178), respectively, as

$$\mathbf{x}(k+1) = A\mathbf{x}(k) + Bu(k) \tag{8.179}$$

$$y(k) = C\mathbf{x}(k) \tag{8.180}$$

where

$$A = \begin{bmatrix} A_1 & A_2 & A_3 & \ldots & A_{n-1} & A_n \\ 1 & 0 & 0 & \ldots & 0 & 0 \\ 0 & 1 & 0 & \ldots & 0 & 0 \\ \vdots & \vdots & \vdots & & \vdots & \vdots \\ 0 & 0 & 0 & \ldots & 1 & 0 \end{bmatrix} \qquad B = \begin{bmatrix} 1 \\ 0 \\ \vdots \\ 0 \end{bmatrix} \tag{8.181}$$

$$C = [(B_1 + B_0 A_1), (B_2 + B_0 A_2), \ldots, (B_m + B_0 A_m), B_0 A_{m+1}, \ldots, B_0 A_n]$$

The state-variable representation is not unique. Another set of n variables could be chosen. However, if the filter transfer function is given, this method of defining the variables is one of the simplest.

Now let us return to the overflow problem. Suppose there is zero input $u(k)$ so that (8.179) becomes

$$\mathbf{x}(k+1) = A\mathbf{x}(k) \tag{8.182}$$

Given a stable filter, with no restriction on register length, the state $x(k)$ would converge to zero from any initial condition in the absence of an input. With a finite register, the situation may be different. Let each component of the state vector $x(k)$ be represented by two's-complement arithmetic and lie within the register range $[-1, 1]$. In other words, we assume that the state vector $x(k)$ lies in the n-dimensional "hypercube" centered at the origin with a side length of 2 units. For a second-order filter, this becomes a square as shown in Figure 8.22. (If we took quantization into account, the square would have a rectangular grid with grid lines representing the quantization levels. For simplicity, we will ignore quantization effects.) When we transfer $x(k)$ according to (8.182), the resulting $x(k+1)$ may fall outside the square, as shown at point b in the figure, indicating an overflow, or inside the square, as shown at point d, with no overflow. In general, the occurrence of an overflow depends on the location of $x(k)$ within the square, and also on the matrix A.

If an overflow occurs and is unchecked, then the overflow function f defined in (8.175) operates on $A\mathbf{x}(k)$ to bring the state x back into the register range at point e in Figure 8.22. The same process can be repeated for all the points in the square. The trajectories starting at some points eventually come to rest at the origin. However, trajectories starting at other points in the square may converge not to the origin but to a sustained oscillation,

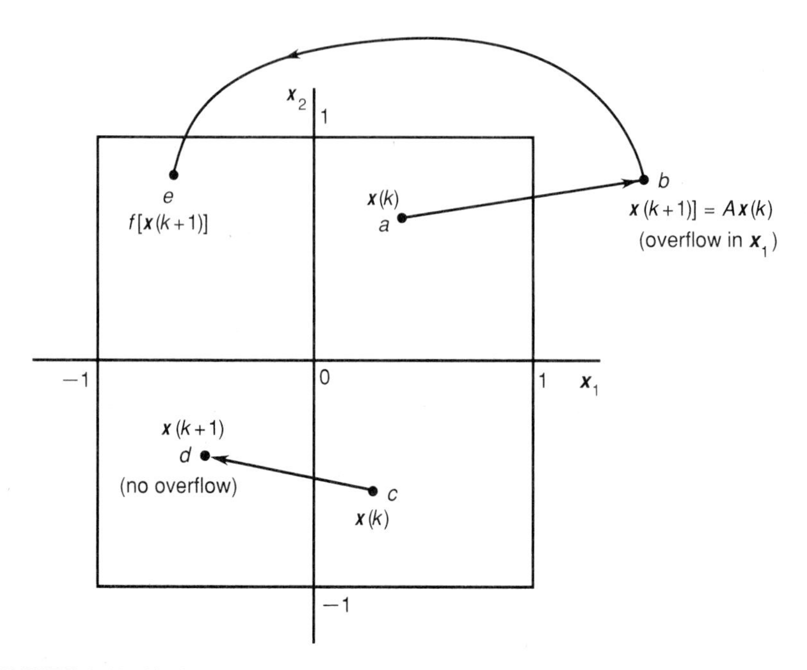

FIGURE 8.22 Unchecked overflow for second-order filter with two's-complement arithmetic.

or limit cycle, between two or more points in the square. The following example illustrates this phenomenon.

EXAMPLE 8.16

Consider a second-order recursive digital filter with poles at $0.75 \pm j0.5$. Find the region in the state space grid $[-1 \leq x_1 \leq 1,\ -1 \leq x_2 \leq 1]$ for which the initial state converges to the origin for zero input. Find the fixed points in the state space grid to which the initial state converges after an overflow.

Solution

The denominator of the filter transfer function is $z^2 - 1.5z + 0.8125$. From this, the system matrix can be written as

$$A = \begin{bmatrix} 1.5 & -0.8125 \\ 1 & 0 \end{bmatrix}$$

If we multiply certain states on the perimeter of the $[-1, 1]$ square by the matrix A, as shown in Figure 8.23, we see that some values overflow even on this first iteration. For example, the point $(1, -1)$, when multiplied by A, gives $(2.3125, -1)$ for $\mathbf{x}(k+1) = A\mathbf{x}$. When the function (8.175) is applied to the overflow value of x_1, the state becomes $(0.3125, -1)$, as shown. It would be tedious to iterate manually from even one initial state to see whether the trajectory starting there converges to the origin. Instead, we use a computer program that includes the overflow function (8.175) to check a grid of starting points in the square, thereby defining the region of points that converge. The program OVERFLOW is on the diskette. When it was applied to the current problem, the results shown in Figures 8.24 and 8.25 were obtained. Figure 8.24 indicates the regions of the square (unshaded) that contain starting points for trajectories that converge to the origin. The shaded regions contain starting points for trajectories that settle into sustained oscillations, or limit cycles, because of overflow.

There are three distinct limit cycles for this filter, as shown in Figure 8.25. The oscillation between points a and b has a period of two time steps. The limit cycles cde and fgh have periods of three time steps. The coordinates of the points on which the oscillations are sustained are

$$\begin{aligned}
a, b &: \quad (-0.6038, 0.6038), (0.6038, -0.6038) \\
c, d, e &: \quad (0.8596, -0.7952), (-0.0645, 0.8596), (-0.7952, -0.0645) \\
f, g, h &: \quad (0.7952, -0.0645), (-0.8596, 0.7952), (-0.0645, -0.8596)
\end{aligned}$$

These results were obtained via the computer program OVERFLOW. It is easy to use a hand calculation to verify that the limit cycles do exist. Take limit cycle ab, for example. Multiply the coordinates of point a by matrix A and apply the overflow function (8.175) to the result. Then

$$f\begin{bmatrix} 1.5 & -0.8125 \\ 1 & 0 \end{bmatrix}\begin{bmatrix} -0.6038 \\ 0.6038 \end{bmatrix} = f\begin{bmatrix} -1.3962 \\ -0.6038 \end{bmatrix} = \begin{bmatrix} 0.6038 \\ -0.6038 \end{bmatrix}$$

which is point b. Similarly for point b,

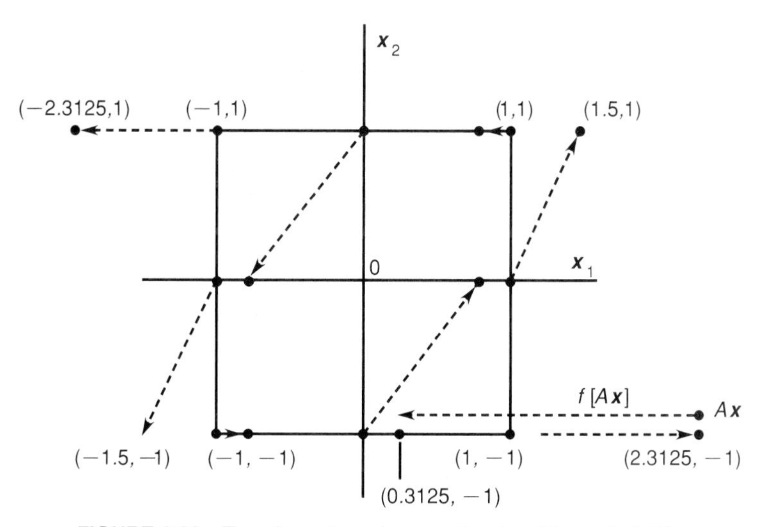

FIGURE 8.23 Transformation after one iteration (Example 8.16).

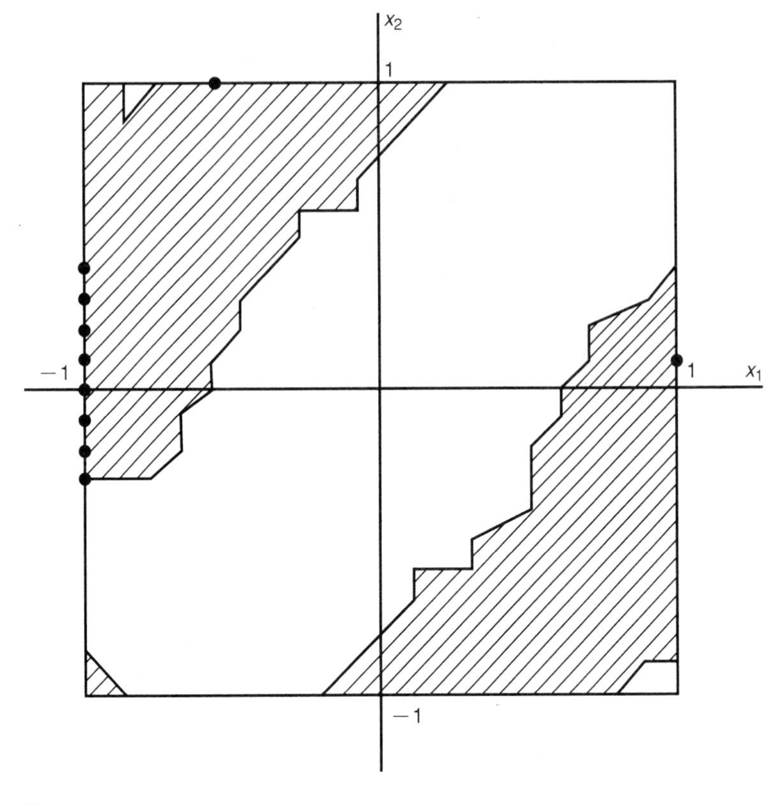

Note

Shaded areas:	Trajectories starting there converge to limit cycles.
Unshaded areas:	Trajectories starting there converge to rest at origin.
• • • • •:	Isolated starting points for trajectories converging to origin.

FIGURE 8.24 Partition of the $[-1;1]$ square (Example 8.16).

$$f\begin{bmatrix} 1.5 & -0.8125 \\ 1 & 0 \end{bmatrix}\begin{bmatrix} 0.6038 \\ -0.6038 \end{bmatrix} = f\begin{bmatrix} 1.3962 \\ 0.6038 \end{bmatrix} = \begin{bmatrix} -0.6038 \\ 0.6038 \end{bmatrix}$$

which is point a. We have confirmed that the trajectory oscillates between points a and b, as indicated by the computer result.

It is evident that limit cycles of the magnitude indicated by Example 8.16 would have a calamitous effect on filter operation and must be prevented. One way (mentioned already) to prevent them is to use a saturation characteristic, but this involves considerable extra logic to check for overflow after each iteration. A simpler solution, which may not, however, be feasible in many cases, is to increase the range of numbers that can be represented so that when they are scaled, there is no overflow.

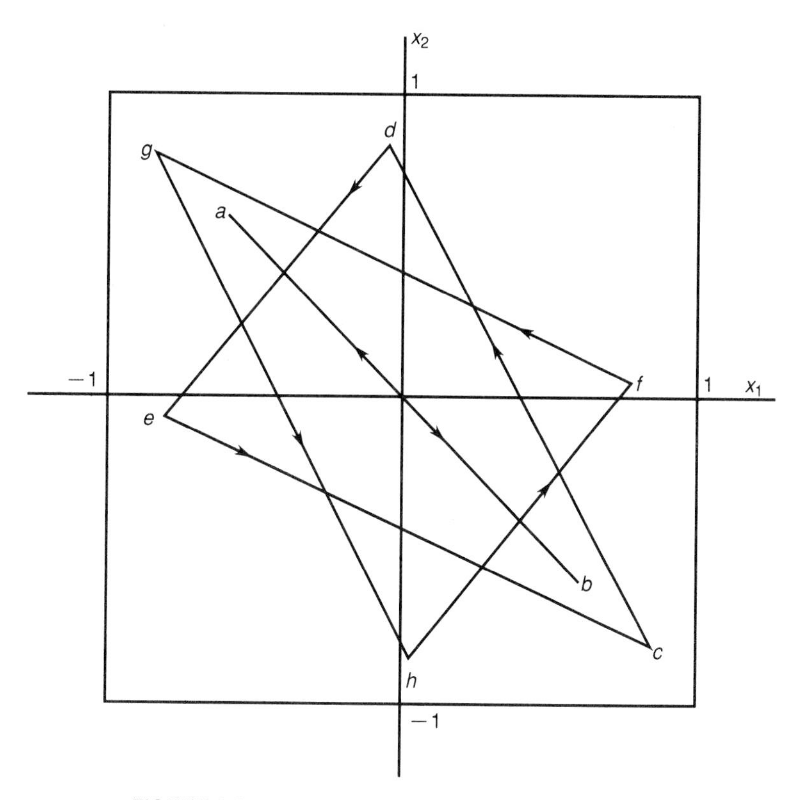

FIGURE 8.25 Limit cycles due to overflow (Example 8.16).

A third remedy, suggested in [18], is based on the following reasoning. A filter defined by a given difference equation can be implemented in many different ways, with different system matrices A, depending on the definition of state variables. If we can find a particular matrix A that, when multiplying the state vector x lying in the $[1, -1]$ hypercube, does not increase the length of x, then there is no danger of an overflow. Defining the length of a vector $\mathbf{x}$ by

$$|\mathbf{x}| = \left(\sum_{i=1}^{n} x_i^2 \right)^{\frac{1}{2}}$$

we can write the norm of the matrix A as follows:

$$|A| = \max_{\text{all } x, x \neq 0} \frac{|A\mathbf{x}|}{|\mathbf{x}|}$$

If $|A| < 1$, then all vectors $\mathbf{x}$ decrease in length when multiplied by A. In [18], it is shown that so-called *normal* matrices ($A'A = AA'$) have the property that $|A| < 1$, which ensures that no overflow will occur in $\mathbf{x}$. See [18] for further details.

8.10 SUMMARY

After a discussion of representing numbers in binary notation and of machine arithmetic, methods for the analysis of the effects of finite word length on digital-filter design are considered. If fine quantization is assumed, then the quantization errors can be treated as noise, and a stochastic model can be used to get the variance of the output noise due to quantization of the input or to round-off of the multiplications involved in the filtering. A different model is used to find the error in pole location for a recursive filter due to errors in the coefficients of the difference equation defining the filter. This model may be used to check the effect on filter stability, or on frequency response, of the number of bits used to implement the coefficients.

For coarse quantization, where a stochastic model is not valid, the problem becomes a nonlinear one and there are few general results. For first- and second-order filters, deadbands and limit cycles may arise in the filter output, and it is possible to predict their bounds under certain assumptions. Limit cycles can also arise in the case of fine quantization if there is arithmetic overflow.

REFERENCES FOR CHAPTER 8

1. W. R. Bennett, "Spectra of Quantized Signals," *Bell System Technical Journal* 27 (July 1948): 446–472.
2. B. Gold and C. M. Rader, *Digital Processing of Signals* (New York: McGraw-Hill, 1969).
3. J. F. Kaiser, "Some Practical Consideration in the Realizations of Linear Digital Filters," *Proc. 3rd Annual Allerton Conf.*, October 1965.
4. A. V. Oppenheim and R. W. Schafer, *Digital Signal Processing* (Englewood Cliffs, NJ: Prentice-Hall, 1975).
5. E. Avenhaus, "On the Design of Digital Filters with Coefficients of Limited Word Length," *IEEE Trans. Audioelectroacoustics*, AU–20 (August 1972): 206–212.
6. R. E. Crochiere, "A New Statistical Approach to the Coefficients Word Length Problem for Digital Filters," *IEEE Trans. Circuits Systems*, CAS–22 (March 1975): 190–196.
7. R. B. Blackman, *Linear Data-Smoothing in Theory and Practice* (Reading, MA: Addison-Wesley, 1965).
8. A. J. Monroe, *Digital Processes for Sampled Data Systems* (New York: Wiley, 1962).
9. L. B. Jackson, "Analysis of Limit Cycles Due to Multiplication Rounding in Recursive Digital (Sub)Filters," *Proc. 7th Annual Allerton Conf. Circuit System Theory*, 1969: 69–79.

10. S. R. Parker and S. F. Hess, "Limit-Cycle Oscillations in Digital Filters," *IEEE Trans. on Circuit Theory*, CT–18, no. 6 (November 1971) 687–697.
11. I. W. Sandberg and J. F. Kaiser, "A Bound on Limit Cycles in Fixed-Point Implementation of Digital Filters," *IEEE Trans. Audioelectroacoustics.* AU–20, (June 1972): 110–112.
12. J. L. Long and T. N. Trick, "An Absolute Bound on Limit Cycles Due to Roundoff Errors in Digital Filters," *IEEE Trans. Audioelectroacoustics*, AU–21, (February 1973): 27–30.
13. T. Kaneko, "Limit-Cycle Oscillations in Floating-Point Digital Filters," *IEEE Trans. Audioelectroacoustics*, AU–21, (April 1973): 100–106.
14. L. B. Jackson, "On the Interaction of Roundoff Noise and Dynamic Range in Digital Filters," *Bell Sys. Tech. J.* 49 (February 1970): 159–184.
15. T. W. Parks and C. S. Burrus, *Digital Filter Design* (New York: Wiley, 1987).
16. T. A. C. M. Claasen and L. O. G. Kristiansson, "Necessary and Sufficient Conditions for the Absence of Overflow Phenomena in a Second-Order Recursive Digital Filter," *IEEE Trans. Acoust., Speech, Signal Processing*, ASSP–23 (December 1975): 509–515.
17. T. A. C. M. Claasen, W. F. G. Mecklenbräuker, and J. B. H. Peck: "Quantization Noise Analysis for Fixed-Point Digital Filters Using Magnitude Truncation for Quantization," *IEEE Trans. Circuits Syst.*, CAS–22 (November 1975): 887–895.
18. R. A. Roberts and C. T. Mullis, *Digital Signal Processing* (Reading, MA: Addison-Wesley, 1987).

EXERCISES FOR CHAPTER 8

1. Identify the various types of quantization errors that can occur in a digital filter represented by the difference equation

$$y_k = A_1 y_{k-1} + A_2 y_{k-2} + A_3 y_{k-3} + B_0 x_k + B_1 x_{k-1} + B_2 x_{k-2}$$

2. Express the following binary numbers in decimal form.
a) 1101.011 b) 10.0101

3. Write the following decimal numbers in binary form.
a) 514.0 b) 15.21875 c) −101.3625 d) −0.109375

4. Write the binary numbers in Exercise 2 as floating-point numbers.

5. Write the sign-magnitude, the one's-complement, and the two's-complement representations of the binary form of the following decimal numbers.
a) −0.53125 b) −0.140625

6. Express the decimal values 0.78125 and 0.1875 in one's-complement binary representation, and subtract the latter number from the former.

7. Express the decimal values 0.78125 and 0.1875 in floating-point binary representation, and find their product.

8. Add the floating-point binary representations of the decimal numbers 432.0 and 0.15625.

9. Suppose we truncate the mantissa of a floating-point number, where the mantissa has a sign-magnitude fixed-point representation. Prove that the maximum range of the relative error is given by (8.50) for both positive and negative values of the mantissa.

10. Find the mean and variance of the quantization error treated as a random variable for

a) Truncation

b) Sign-magnitude truncation for the following three cases:

i) Input signal known to be positive

ii) Input signal known to be negative

iii) Sign of input signal unknown

Assume fine quantization with step q_0.

11. Show that the variance of output noise due to input quantization is approximately $q_0^2/(48\epsilon \sin^2\theta)$ for a digital filter with transfer function $H(z) = 1/(1 - 2r(\cos\theta)z^{-1} + r^2z^{-2})$. Assume $\epsilon = 1 - r \ll 1$ and that q_0 is the quantization step.

12. A digital filter with transfer function

$$H(z) = \frac{z(z - 0.4050)}{(z^2 - 1.2728z + 0.8100)}$$

is used to process signals the amplitude range of which is ± 100 V. An eight-bit A/D converter is used. Assuming fine quantization, find the standard deviation of the output noise due to input quantization.

13. The transfer function

$$H(z) = \frac{2z(z^2 - 0.9546z + 0.4050)}{(z - 0.6364)(z^2 - 1.2728z + 0.8100)}$$

results from a parallel arrangement of a first-order filter $H_1(z) = 1/(1 - Az^{-1})$ and a second-order filter $H_2(z) = 1/(1 - 2r(\cos\theta)z^{-1} + r^2z^{-2})$, where $A = r\cos\theta$, $r = 0.9$, and $\theta = 45°$. What is the standard deviation of output noise due to quantization of the input, for the parallel arrangement of the two filters?

14. A third-order digital filter has real poles at p_1, p_2, and p_3. It is implemented in direct form.

a) From the general expressions (8.93) and (8.94), write an expression for the bounds on the change in location of these poles due to changes in the coefficients of the corresponding difference equation.

b) If $p_1 = 0.60$, $p_2 = 0.95$, and $p_3 = 0.98$, what is the minimum number of bits required to ensure that the filter cannot go unstable as a result of coefficient quantization? Assume rounding.

15. Compute the sensitivity coefficients $\partial M(\omega T)/\partial c_i$ for the filter the transfer function of which is

$$H(z) = \frac{z}{z^2 - A_1 z - A_2}$$

16. The coefficients h_n, $n = 0, 1, \ldots, N-1$, of a *nonrecursive* filter are quantized with quantization step q_0. That is, the actual output is given by

$$y_k = \sum_{n=0}^{N-1} (h_n + e_n)x_{k-n}$$

where h_n are the designed or nominal values of coefficients, and e_n are the corresponding errors due to quantization. If the e_n are considered to be realizations of mutually independent random variables E_n, uniformly distributed with zero mean, show that the frequency response error over all frequencies has a standard deviation $\leq (N/3)^{\frac{1}{2}} q_0/2$ about the nominal frequency response.

17. Draw block diagrams for the direct and canonic forms of the filter

$$y_k = Ay_{k-1} + x_k + Bx_{k-1}$$

where $|A| < 1$ and $B \neq 1$. On these diagrams, mark the noise inputs due to quantization of each product (assume fine quantization). Compute and compare the steady-state variances of the output due to this noise in both cases if $A = 0.8$ and $B = 0.5$. Take q_0 as the quantization step.

18. The coefficients A, B, C, and D are non-unity in the filter shown in Figure P8.1. Derive an expression for the steady-state variance of the noise in the output y_k due to rounding off of the multiplications if the quantization step is q_0. Assume fine quantization and rounding.

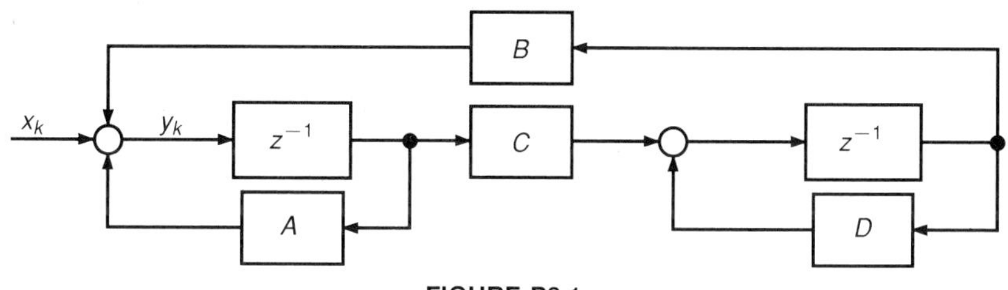

FIGURE P8.1

19. Repeat Exercise 18 for the filter shown in Figure P8.2 if the coefficients are non-unity.

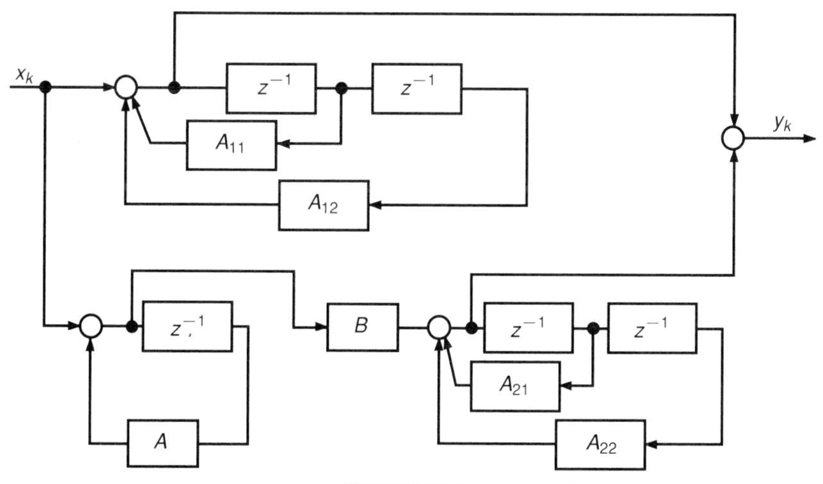

FIGURE P8.2

20. For the first-order filter of Figure 8.16, let $A = 0.95$, let $y_{-1} = 10$, and let the input be zero. If the quantization is the same as for Example 8.13, compare the actual output due to quantization with the standard deviation of output noise that would result from a "fine"-quantization model.

21. Take the second-order filter of Example 8.15, but assume that the products are summed before round-off. Show that the output converges to a fixed value rather than to an oscillation.

CHAPTER 9

DIGITAL FILTERING APPLIED TO ESTIMATION: THE DISCRETE KALMAN FILTER

9.1 INTRODUCTION

When measurement data are taken, the true value of the quantity or signal being measured is often corrupted by noise. If it were known that the noise had, for example, high-frequency components only, and that the signal was of low frequency, then applying simple lowpass filtering to the measurement would suffice to separate signal from noise. In general, however, the situation is more complex; the noise frequencies may overlap those of the signal, or it may be desired to estimate, from the noisy measurement, some quantity that is not being directly measured or observed. The latter requirement often arises in optimal-control problems where the control law is a function of all the states of the system, many of which are not measured [1]. In cases where there is some knowledge of the noise statistics and of the process generating the signal, it is possible to use a powerful and effective technique known as Kalman filtering [2] to estimate the signal—or, more generally, the state of the generated process—from a sequence of measurements. The estimate is optimal, in the sense of minimizing the mean-square-error, subject to the constraint of using a linear filter.

In this chapter, before dealing with the Kalman filter, we treat some simple examples to introduce the basic concepts of estimation. The alpha–beta tracker is discussed briefly. The alpha–beta tracker has a form similar to that of a second-order Kalman filter, but it has constant gains and so can be analyzed in the frequency domain, via the z-transform. Because of the time-varying gains and the general nonstationary characteristics of the Kalman filter, it is designed and analyzed in the time domain.

Concepts in probability theory, which are used in the estimation problem, were discussed in Chapter 7. Central to these is the notion of conditional probability, because we are interested in estimating some quantity given that a measurement, or set of measurements, has been made. In deriving the Kalman filter, the Markov property is used to describe the process the state of which we are trying to estimate from the measurements.

Some elementary knowledge of state variables and manipulation of matrices is needed to appreciate the formulation and solution of the Kalman filtering problem. The state-variable form is simply a concise shorthand notation for describing a dynamic system of the order n, where n is any finite integer. It enables the analyst to avoid the proliferation of detail associated with the analysis of higher-order systems—and so to concentrate on the important points of the design. A review of state variables and the discretization of continuous linear systems is presented in Appendix 9A. The differentiation of a quadratic form with respect to a vector and a matrix is described in Appendix 9B. The results are used to get the optimal estimator or filter.

9.2 A BASIC ESTIMATION PROBLEM

Let us first consider a simple example that illustrates some important concepts in estimation. Suppose that it is desired to estimate the value of an unknown constant x from a series of inaccurate measurements $y_i, i = 1, \ldots, k$. We can write

$$y_i = x + v_i, \quad i = 1, \ldots k \tag{9.1}$$

In this case, v_i is the error in the ith measurement. For the moment, no assumptions are made regarding the statistics of the error.

What is the best estimate of x on the basis of the k measurements? Before attempting to answer that question, we note that the word *best* is meaningful, in this context, only if it is defined with respect to some criterion or performance index. For the present example, take the best estimate $\hat{x}$ as the one that minimizes the sum of the squares of the errors between the measurements and the estimate. To find $\hat{x}$, we let

$$J = \sum_{i=1}^{k} v_i^2 \tag{9.2}$$

and, for the purpose of minimization, x in (9.1) is permitted to vary, being equal to different estimates of the unknown constant. Thus J is the performance index, and the estimate $\hat{x}$ that minimizes J is called the least-squares estimate.

From (9.1), (9.2) may be written

$$J = \sum_{i=1}^{k} (y_i - x)^2 \tag{9.3}$$

Differentiate J with respect to x,

$$\frac{\partial J}{\partial x} = -2 \sum_{i=1}^{k} (y_i - x) \tag{9.4}$$

and set the result equal to zero,

$$\sum_{i=1}^{k} (y_i - x) = 0 \tag{9.5}$$

to obtain

$$\hat{x} = \frac{1}{k} \sum_{i=1}^{k} y_i \tag{9.6}$$

where $\hat{x}$ is the value of x that satisfies (9.5). The least-squares estimate given by (9.6) is simply the mean of the measurements, which is a reasonable result.

The operation defined by (9.6) is in a nonrecursive form, as discussed in Chapter 5, with measurements y_i serving as the input. It is necessary to store k measurements before the single output $\hat{x}$ is obtained. As the number of measurements changes, so does $\hat{x}$. To emphasize that $\hat{x}$ is a function of k, rewrite (9.6) as

$$\hat{x}_k = \frac{1}{k} \sum_{i=1}^{k} y_i \tag{9.7}$$

A recursive estimate of $\hat{x}_k$ can be obtained as follows. If we have $(k + 1)$ measurements, then

$$\begin{aligned}
\hat{x}_{k+1} &= \frac{1}{k+1} \sum_{i=1}^{k+1} y_i \\
&= \frac{k}{k+1} \left(\frac{1}{k} \sum_{i=1}^{k} y_i \right) + \frac{1}{k+1} y_{k+1} \\
&= \frac{k}{k+1} \hat{x}_k + \frac{1}{k+1} y_{k+1} \\
&= \left(1 - \frac{1}{k+1} \right) \hat{x}_k + \frac{1}{k+1} y_{k+1} \\
&= \hat{x}_k + \frac{1}{k+1} (y_{k+1} - \hat{x}_k)
\end{aligned} \tag{9.8}$$

It follows that

$$\hat{x}_k = \hat{x}_{k-1} + \frac{1}{k}(y_k - \hat{x}_{k-1}) \tag{9.9}$$

This has the form of a recursive filter with input y_k and output $\hat{x}_k$. The quantity $(y_k - \hat{x}_{k-1})$ is called the *measurement residual* and is multiplied by a time-varying gain $1/k$. As k increases, it is evident from (9.9) that less and less weight is given to the current measurement y_k. In the limit as $1/k$ approaches zero, the measurement is disregarded; that is, it is considered that the current measurement contributes no new information. A block diagram of the filter (9.9) is given in Figure 9.1.

The recursive form of (9.9) requires less storage than the nonrecursive estimate of (9.7). In the absence of additional information, the initial condition for (9.9) may be taken as $\hat{x}_{k-1} = 0$. More complex estimators, including the Kalman filter, have the general recursive form of the simple system shown in Figure 9.1, with a time-varying gain, or gain matrix, multiplying the measurement residual.

To introduce some other concepts that will be useful later, consider the foregoing type of problem in vector form. Suppose

$$\mathbf{y} = H\mathbf{x} + \mathbf{v} \tag{9.10}$$

where the measurement vector $\mathbf{y}$ is an $m \times 1$ column matrix, the vector $\mathbf{x}$ is an $n \times 1$ column matrix of unknown constants to be estimated, and the measurement noise or measurement-inaccuracy vector $\mathbf{v}$ is $m \times 1$. Boldface type will be used for vector quantities. The coefficient matrix H is $m \times n$; that is, has m rows and n columns. It is desired to find the estimate $\hat{\mathbf{x}}$ of vector $\mathbf{x}$ that minimizes the error squared:

$$J = \mathbf{v}'\mathbf{v} = (y - H\mathbf{x})'(y - H\mathbf{x}) \tag{9.11}$$

where $\mathbf{v}'$ denotes the transpose of $\mathbf{v}$. The performance index J is in a quadratic form. In order to get the least-squares estimate of vector $\mathbf{x}$, differentiate J with respect to $\mathbf{x}$ and set the result equal to zero. Differentiation of a quadratic form with respect to a vector or matrix is reviewed in Appendix 9B. The results given therein will be used in this problem.

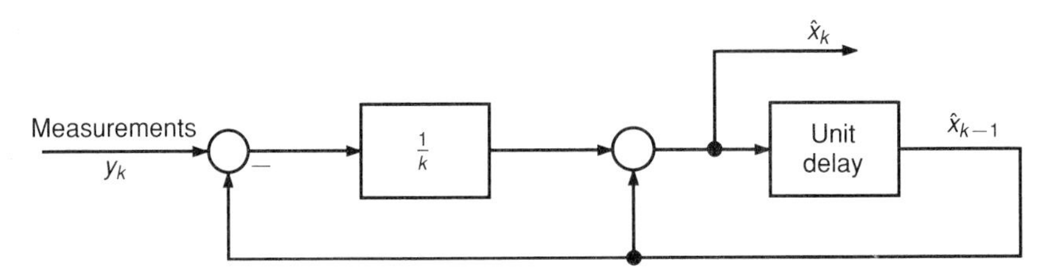

FIGURE 9.1 Recursive filter to give the least-squares estimate of an unknown constant x on the basis of a series of inaccurate measurements.

First expand J into a more suitable form for differentiation.

$$J = \mathbf{y}'\mathbf{y} - \mathbf{y}'H\mathbf{x} - \mathbf{x}'H'\mathbf{y} + \mathbf{x}'H'H\mathbf{x} \tag{9.12}$$

Using Equations (9B.23), (9B.24), and (9B.25) of Appendix 9B, we find that the derivative with respect to $\mathbf{x}$ is

$$\nabla_{\mathbf{X}} J = -H'\mathbf{y} - H'\mathbf{y} + 2H'H\mathbf{x} \tag{9.13}$$

Set the derivative equal to zero to obtain

$$\hat{\mathbf{x}} = (H'H)^{-1}H'\mathbf{y} \tag{9.14}$$

for the estimate. The term $(H'H)^{-1}$ is the inverse of $H'H$.

In general, $\hat{\mathbf{x}}$ given by (9.14) is a vector estimate of the unknown constant vector $\mathbf{x}$. If x is simply a scalar quantity, as in (9.1), we can still use result (9.14) by writing (9.1) as

$$\begin{bmatrix} y_1 \\ \vdots \\ y_k \end{bmatrix} = \begin{bmatrix} 1 \\ \vdots \\ 1 \end{bmatrix} x + \begin{bmatrix} v_1 \\ \vdots \\ v_k \end{bmatrix} \tag{9.15}$$

or

$$\mathbf{y} = H\mathbf{x} + \mathbf{v} \tag{9.16}$$

which is the form (9.10) with $m = k$, $n = 1$, and $H' = [1, 1, \ldots, 1]$.

Then, from (9.14),

$$\hat{x} = \left\{ [1, 1, \ldots, 1] \begin{bmatrix} 1 \\ \vdots \\ 1 \end{bmatrix} \right\}^{-1} [1, 1, \ldots, 1] \begin{bmatrix} y_1 \\ \vdots \\ y_k \end{bmatrix} \tag{9.17}$$

$$= \frac{1}{k} \sum_{i=1}^{k} y_i \tag{9.18}$$

which is the nonrecursive form obtained in (9.6).

We will now introduce some probabilistic concepts into the simple estimation problem just treated. (See Chapter 7 for a review of probability.) Suppose that for each measurement y_i in (9.1), the noise or inaccuracy v_i is a realization of a random variable V_i. In this context, as in Chapter 7, a capital letter will be used for a random variable or vector, whereas the corresponding lower-case letter will denote the value that the random variable takes on as a result of a measurement or experiment. Let V_i have zero mean and variance σ_V^2 for all i. Assume that the noise samples are independent. Because V_i is a random variable, the measurement, *in advance* of being made, must also be considered a random variable. That is, (9.1) can be written in terms of random variables Y_i and V_i as

$$Y_i = x + V_i, \quad i = 1, \ldots, k \tag{9.19}$$

The least-squares estimate (9.6) is the result of a given set of measurements. If other sets of measurements were made, other values would probably be obtained for $\hat{x}$. Thus the estimate $\hat{x}$ based on a given set of measurements may be considered a realization of a random variable $\hat{X}$ that is often called an *estimator*. If the *mean* of an estimator is equal to the quantity being estimated (x in this case), the estimator is said to be unbiased. Consider the estimation error random variable defined by

$$E = x - \hat{X} \tag{9.20}$$

The expected value or mean value of E, denoted by $\langle E \rangle$, is

$$\begin{aligned}
\langle E \rangle &= \langle x - \hat{X} \rangle \\
&= \left\langle x - \frac{1}{k}\sum_{i=1}^{k} Y_i \right\rangle \\
&= \left\langle x - \frac{1}{k}\sum_{i=1}^{k} (x + V_i) \right\rangle \\
&= \left\langle -\frac{1}{k}\sum_{i=1}^{k} V_i \right\rangle \\
&= -\frac{1}{k}\sum_{i=1}^{k} \langle V_i \rangle \\
&= 0
\end{aligned} \tag{9.21}$$

because $\langle V_i \rangle = 0$. It follows from (9.20) that

$$\langle x - \hat{X} \rangle = 0 \tag{9.22}$$

or

$$\langle \hat{X} \rangle = x \tag{9.23}$$

Hence

$$\hat{X} = \frac{1}{k}\sum_{i=1}^{k} Y_i \tag{9.24}$$

is an unbiased estimator of x.

The variance of the error E—that is, its spread about its expected value (zero)—is

$$\begin{aligned}
\text{Var}\ (E) &= \langle E^2 \rangle \\
&= \langle (x - \hat{X})^2 \rangle \\
&= \left\langle \left[x - \frac{1}{k}\sum_{i=1}^{k} Y_i \right]^2 \right\rangle
\end{aligned}$$

$$= \left\langle \frac{1}{k^2} \sum_{i=1}^{k} V_i^2 \right\rangle$$

$$= \frac{1}{k^2} \sum_{i=1}^{k} \langle V_i^2 \rangle$$

$$= \frac{\sigma_V^2}{k} \tag{9.25}$$

In the second-to-last step, we used the fact that the noise samples are independent to eliminate the cross-coupling terms. Thus we can make the variance of the estimation error as small as we want by increasing k, the number of measurements.

9.3 POSITION AND RATE ESTIMATION: THE ALPHA–BETA TRACKER

Consider an application wherein we wish to estimate the range x and the range rate s of a target on the basis of a sequence $y_i, i = 1, 2, \ldots$, of radar measurements. The method we shall use is instructive because it is, in some ways, similar to that used for deriving the Kalman filter and because it results in a similar filter structure.

Typically, in radar range measurements, a pulse is transmitted at time t. Ideally, the pulse is reflected off the target and is received back at the radar after the elapsed time Δt_c corresponding to the range from radar to target. That is,

$$x = \frac{c}{2} \Delta t_c \tag{9.26}$$

where c is the velocity of light. In practice, however, because of errors caused by imperfect reflection, atmospheric effects, and receiver noise, Δt_c (and hence x) cannot be measured exactly. Instead, an elapsed time Δt is noted, and the range measurement is

$$y = \frac{c}{2} \Delta t$$

$$= \frac{c}{2} (\Delta t_c + \epsilon)$$

$$= x + v \tag{9.27}$$

where $\epsilon = \Delta t - \Delta t_c$, $x = c\Delta t_c/2$ and $v = c\epsilon/2$. Equation (9.27) has the same form as the example in the previous section, where we saw that even when the true range x is constant, we can improve the accuracy of the estimate $\hat{x}$ by taking a sequence of measurements. In this application, we have assumed a target range rate s to be estimated, so the range x is time-varying. Denote x_k as the current range (range at time k) and s_k as the current range rate. We seek a *recursive* filter that will give estimates of x_k and s_k on the basis of previous estimates of same. As a guide, we have the form (9.9) for the recursive estimate of a constant. Let

$\hat{x}_k =$ estimate of x_k based on all measurements up to and including y_k

$\hat{s}_k =$ estimate of s_k based on all measurements up to and including y_k

$\hat{x}_k^* =$ estimate of x_k based on all measurements up to and including y_{k-1}.

Estimate $\hat{x}_k^*$ is called the *a priori* or *predicted* estimate of x_k; $\hat{x}_k$ and $\hat{s}_k$ are called the *a posteriori* or *updated* estimates of x_k and s_k, respectively.

Suppose that after the $(k-1)$st measurement, the estimates $\hat{x}_{k-1}$ and $\hat{s}_{k-1}$ have been made. Before measuring y_k, we want to predict the range x_k. Different predictions could be made. A reasonable one is

$$\hat{x}_k^* = \hat{x}_{k-1} + T\hat{s}_{k-1} \tag{9.28}$$

where T is the interval between pulses. Now, y_k is measured. If y_k differs from $\hat{x}_k^*$, a correction based on the residual $(y_k - \hat{x}_k^*)$ is made so that we get $\hat{x}_k$:

$$\hat{x}_k = \hat{x}_k^* + \alpha(y_k - \hat{x}_k^*). \tag{9.29}$$

where α is a positive constant. Similarly, the range rate estimate is updated by an increment

$$\Delta\hat{s}_k = \frac{1}{T}(y_k - \hat{x}_k^*) \tag{9.30}$$

so that

$$\hat{s}_k = \hat{s}_{k-1} + \frac{\beta}{T}(y_k - \hat{x}_k^*) \tag{9.31}$$

where β is a positive constant. The set of equations

$$\hat{x}_k = \hat{x}_k^* + \alpha(y_k - \hat{x}_k^*) \tag{9.32}$$

$$\hat{s}_k = \hat{s}_{k-1} + \frac{\beta}{T}(y_k - \hat{x}_k^*) \tag{9.33}$$

$$\hat{x}_k^* = \hat{x}_{k-1} + T\hat{s}_{k-1} \tag{9.34}$$

defines what is called the alpha–beta tracker. This is an effective way of processing range data. Note the resemblance of (9.32) to (9.9). In contrast to (9.9), however, the gain α is constant, as is β/T in (9.33). A block diagram of the α–β tracker is given in Figure 9.2. As we will soon see, the tracker is similar in form to a second-order Kalman filter for range and range rate estimation, but the latter has time-varying gains.

Considerable study has been devoted to determining the best values for parameters α and β (see, for example, [3] and [4]). Because the gains are constant, z-transforms can be taken of (9.32), (9.33), and (9.34) to obtain

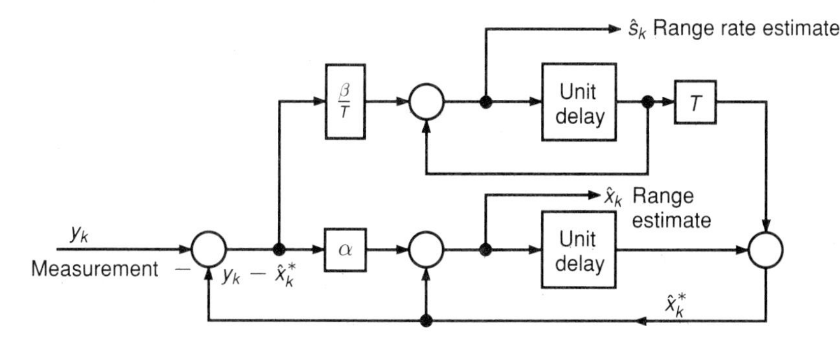

FIGURE 9.2 Alpha–beta tracker.

the transfer functions

$$H_1(z) = \frac{\hat{X}(z)}{Y(z)} = \frac{z(\alpha z + \beta - \alpha)}{z^2 + (\alpha + \beta - 2)z + (1 - \alpha)} \tag{9.35}$$

$$H_2(z) = \frac{\hat{X}^*(z)}{Y(z)} = \frac{z(\alpha + \beta) - \alpha}{z^2 + (\alpha + \beta - 2)z + (1 - \alpha)} \tag{9.36}$$

$$H_3(z) = \frac{\hat{S}(z)}{Y(z)} = \frac{\beta}{T} \frac{z(z-1)}{z^2 + (\alpha + \beta - 2)z + (1 - \alpha)} \tag{9.37}$$

The characteristic equation is

$$z^2 - (\alpha + \beta - 2)z + 1 - \alpha = 0 \tag{9.38}$$

The choice of values for α and β must reflect a trade-off between the conflicting requirements of low sensitivity to noisy measurements and a fast maneuver-following capability, as discussed in [4]. One possible choice is based on requiring that the tracker response be critically damped. From (9.38), this is ensured if

$$\alpha = 2\sqrt{\beta} - \beta \tag{9.39}$$

which gives a double pole at

$$z = p = 1 - \sqrt{\beta} \tag{9.40}$$

For stability, $0 < \beta < 4$.

Suppose that the noise variance is σ_V^2 (a constant), and assume that the noise samples v_k are mutually independent from measurement to measurement. From (9.27), the noise is additive. Furthermore, the system is linear. Therefore, we can determine the effect of the noise on the estimate $\hat{x}$ by considering the noise to act alone through the transfer function $H_1(z)$ given by (9.35). In Chapter 8 we derived an expression (8.58) relating the output noise variance to the input noise variance. In this case, it becomes

$$\sigma_F^2 = \sigma_V^2 \sum_{n=0}^{\infty} h_1^2(n) \tag{9.41}$$

where $h_1(n)$ is the pulse response corresponding to $H_1(z)$; that is, $h_1(n) = \mathcal{F}^{-1}\{H_1(z)\}$. The quantity $\sum_{n=0}^{\infty} h_1^2(n)$ is called the *variance ratio*. If it is less than unity, it attentuates the noise; if it is greater than unity, it amplifies the noise. As shown in Chapter 8, the variance ratio can be derived from the transfer function, because

$$\sum_{n=0}^{\infty} h_1^2(n) = \sum \text{ residues of } H_1(z)H_1(z^{-1})z^{-1} \text{ at poles of } H_1(z) \tag{9.42}$$

For the double pole at $z = p$, where p is given by (9.40), the residues are evaluated as

$$\sum_{n=0}^{\infty} h_1^2(n) = \frac{d}{dz}[(z-p)^2 H_1(z)H_1(z^{-1})z^{-1}]|_{z=p} \tag{9.43}$$

$$= \frac{(1+4p+5p^2)(1-p)}{(1+p)^3} \tag{9.44}$$

Thus the variance ratio is a function of p, which, in turn, is a function of β by (9.40). The ratio is tabulated in Table 9.1 and plotted versus gain β in Figure 9.3.

The closer β is to one, the closer the double pole is to the origin, and the faster the response. However, this makes the tracker more sensitive to

TABLE 9.1 Variance Ratio versus Gain and Pole Location for Alpha–Beta Tracker

Gain β	Double Pole $p = 1 - \beta$	Variance Ratio $\sum_{n=0}^{\infty} h_1^2(n)$
0.00	1.00	0.00
0.01	0.90	0.13
0.10	0.68	0.40
0.20	0.55	0.57
0.40	0.37	0.78
0.60	0.23	0.91
0.80	0.11	0.98
1.00	0.00	1.00
1.20	−0.10	0.98
1.40	−0.18	0.94
1.60	−0.26	0.93
1.80	−0.34	1.02
2.00	−0.41	1.41
2.20	−0.48	2.52
2.40	−0.55	5.26
2.60	−0.61	11.79

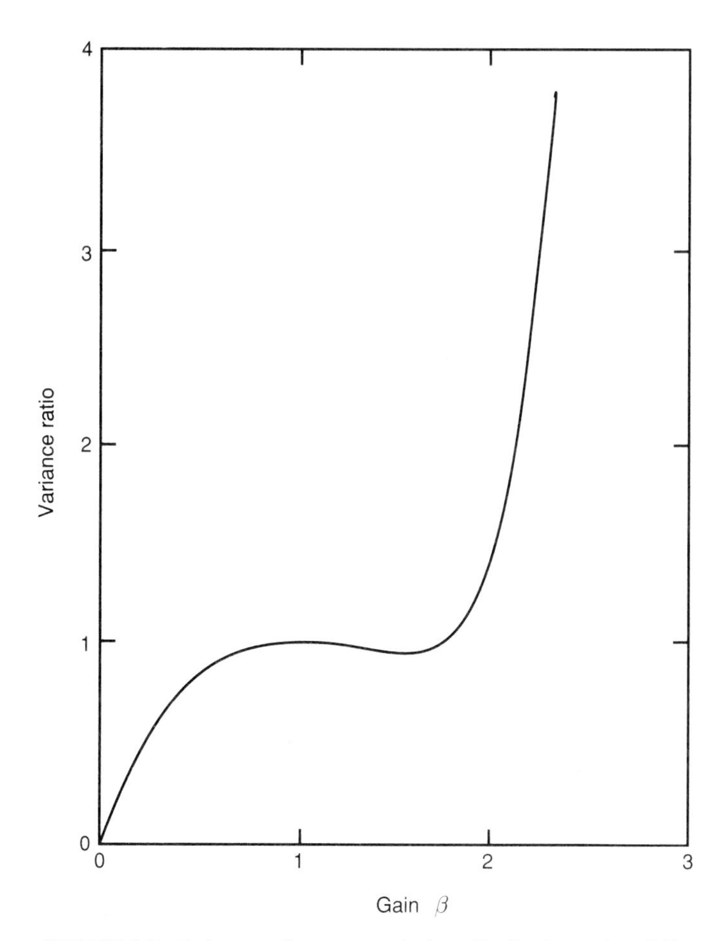

FIGURE 9.3 Variance ratio versus gain for critically damped α–β filter.

noise, as Table 9.1 suggests. A reasonable compromise between speed of response and noise sensitivity for this case would seem to be a value of β between 0.2 and 0.4. For further details, consult [4], which contains a good analysis of the problem.

9.4 STATIC ESTIMATION PROBLEM FOR GAUSSIAN-DISTRIBUTED RANDOM VECTORS

As a further preliminary step before we derive the discrete Kalman filter, consider the problem of estimating a random vector $\mathbf{X}$ on the basis of an observation or measurement vector $\mathbf{Y} = \mathbf{y}$, where

$$\mathbf{Y} = C\mathbf{X} + \mathbf{V} \tag{9.45}$$

Here $\mathbf{Y}$ and $\mathbf{V}$ are $m \times 1$, $\mathbf{X}$ is $n \times 1$, and matrix C is $m \times n$. The situation is depicted diagrammatically in Figure 9.4.

We shall make the following assumptions:

1. The random vector or "state" $\mathbf{X}$ is independent of the measurement-noise vector $\mathbf{V}$.
2. State $\mathbf{X}$ is Gaussian.

$$f_{\mathbf{x}}(\mathbf{x}) = \frac{1}{(2\pi)^{n/2}|S|^{1/2}} \exp\left[-\frac{1}{2}(\mathbf{x}-\bar{\mathbf{x}})' S^{-1}(\mathbf{x}-\bar{\mathbf{x}})\right] \tag{9.46}$$

where

$$\langle \mathbf{X} \rangle = \bar{\mathbf{x}} \tag{9.47}$$

$$\langle (\mathbf{X}-\bar{\mathbf{x}})(\mathbf{X}-\bar{\mathbf{x}})' \rangle = S \tag{9.48}$$

The matrix S is called the covariance matrix (see Chapter 7). Note that the mean $\bar{\mathbf{x}}$ could be a previous estimate of the state [1].

3. Measurement-noise vector $\mathbf{V}$ is Gaussian.

$$f_{\mathbf{v}}(\mathbf{v}) = \frac{1}{(2\pi)^{m/2}|R|^{1/2}} \exp\left(-\frac{1}{2}\mathbf{v}' R^{-1}\mathbf{v}\right) \tag{9.49}$$

where

$$\langle \mathbf{V} \rangle = 0 \tag{9.50}$$

$$\langle \mathbf{V}\mathbf{V}' \rangle = R \tag{9.51}$$

If the mean-square error $\langle (\mathbf{X}-\hat{\mathbf{X}})'(\mathbf{X}-\hat{\mathbf{X}}) \rangle$ is taken as the performance index, then the best estimate $\hat{\mathbf{x}}$ of $\mathbf{X}$ based on observation $\mathbf{y}$ is given by the conditional mean $\langle \mathbf{X}|\mathbf{Y}=\mathbf{y} \rangle$ (see Chapter 7). In order to compute the conditional mean, we need the conditional density function $f_{\mathbf{x}}(\mathbf{x}|\mathbf{Y}=\mathbf{y})$. This is obtained as follows:

By Bayes' rule,

$$f_{\mathbf{x}}(\mathbf{x}|\mathbf{Y}=\mathbf{y}) = \frac{f_{\mathbf{Y}}(\mathbf{y}|\mathbf{X}=\mathbf{x}) f_{\mathbf{x}}(\mathbf{x})}{f_{\mathbf{Y}}(\mathbf{y})} \tag{9.52}$$

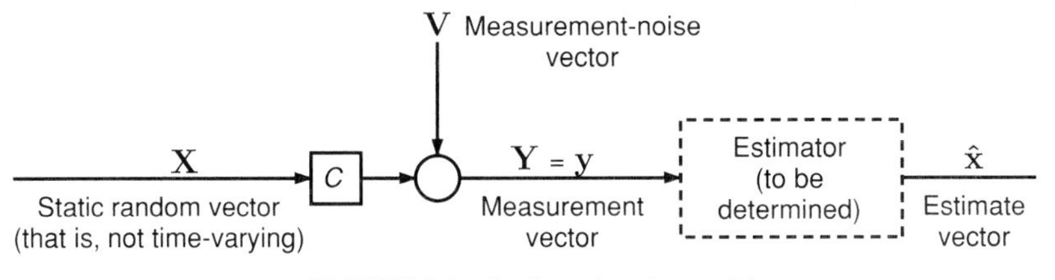

FIGURE 9.4 Static estimation problem.

The denominator of the term on the right side of (9.52) is constant because $\mathbf{y}$ is measured. The distribution of $f_{\mathbf{Y}}(\mathbf{y}|\mathbf{X} = \mathbf{x})$ is simply the distribution of the measurement noise, given by (9.49). So we can write

$$f_{\mathbf{Y}}(\mathbf{y}|\mathbf{X} = x) = \frac{1}{(2\pi)^{m/2}|R|^{1/2}} \exp\left[-\frac{1}{2}(\mathbf{y} - C\mathbf{x})' R^{-1} (\mathbf{y} - C\mathbf{x})\right] \quad (9.53)$$

Therefore, (9.52) becomes

$$f_{\mathbf{x}}(\mathbf{x}|\mathbf{Y} = \mathbf{y}) = \text{constant} \cdot \exp\left[-\frac{1}{2}J\right] \quad (9.54)$$

where

$$J = (\mathbf{x} - \bar{\mathbf{x}})' S^{-1} (\mathbf{x} - \bar{\mathbf{x}}) + (\mathbf{y} - C\mathbf{x})' R^{-1} (\mathbf{y} - C\mathbf{x}) \quad (9.55)$$

Because (9.54) is Gaussian,

$$\hat{\mathbf{x}} = \langle \mathbf{X}|\mathbf{Y} = \mathbf{y} \rangle \quad (9.56)$$

is given by the value of $\mathbf{x}$ that maximizes $f_{\mathbf{x}}(\mathbf{x}|\mathbf{Y} = \mathbf{y})$ or minimizes J. This is shown conceptually in Figure 9.5, although, of course, $\mathbf{X}$ is n-dimensional. Therefore, for a maximum,

$$\nabla_{\mathbf{x}} J = 0 \quad (9.57)$$

Using the relationships given in Appendix 9B for the differentiation of a quadratic form gives

$$\nabla_{\mathbf{x}} J = S^{-1}(\hat{\mathbf{x}} - \bar{\mathbf{x}}) - C' R^{-1} (\mathbf{y} - C\bar{\mathbf{x}}) = 0$$

or

$$S^{-1}(\hat{\mathbf{x}} - \bar{\mathbf{x}}) = C' R^{-1} (\mathbf{y} - C\hat{\mathbf{x}}) \quad (9.58)$$

Here the estimate $\hat{\mathbf{x}}$ is the value of $\mathbf{x}$ satisfying (9.57). Hence

$$\begin{aligned}(S^{-1} + C'R^{-1}C)\hat{\mathbf{x}} &= S^{-1}\bar{\mathbf{x}} + C'R^{-1}\mathbf{y} \\ &= (S^{-1} + C'R^{-1}C)\bar{\mathbf{x}} + C'R^{-1}(\mathbf{y} - C\bar{\mathbf{x}}) \quad (9.59)\end{aligned}$$

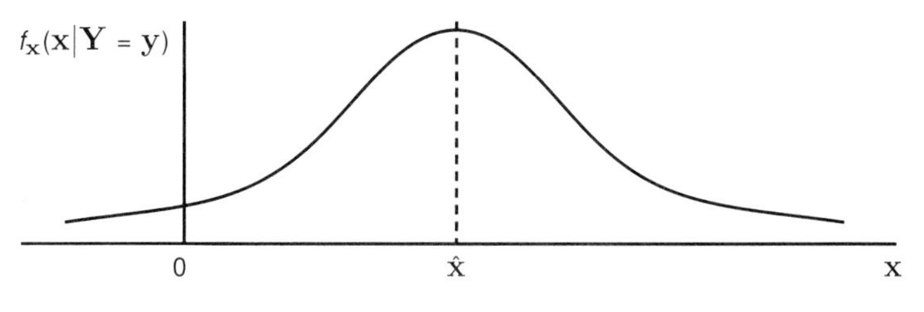

FIGURE 9.5 Conditional mean (conceptual).

Therefore, the best estimate is

$$\hat{\mathbf{x}} = \bar{\mathbf{x}} + G(\mathbf{y} - C\bar{\mathbf{x}}) \tag{9.60}$$

where the gain matrix is given by

$$G = (S^{-1} + C'R^{-1}C)^{-1}C'R^{-1} \tag{9.61}$$

Note that (9.60) gives the same form for the estimate as was obtained in (9.9) for the recursive estimate of a constant. This form will be used in the derivation of the Kalman filter in the next section.

The gain matrix is a function of the state and noise covariance matrices S and R and indicates how much weight should be placed on the measurement $\mathbf{y}$. It can be shown that the covariance matrix P of the error in the estimate is

$$\begin{aligned} P &= \langle [\mathbf{x} - \hat{\mathbf{x}}][\mathbf{x} - \hat{\mathbf{x}}]' \rangle \\ &= S^{-1} + C'R^{-1}C \end{aligned} \tag{9.62}$$

which is the quantity in the parentheses in (9.61). The proof is left as an exercise.

Further discussion of this class of problem is given in [1]. Even when $\mathbf{X}$ and $\mathbf{V}$ are not Gaussian but have covariance matrices S and R, a weighted-least-squares estimate of $\mathbf{X}$ can be obtained by minimizing the quadratic form (9.55).

9.5 DERIVATION OF THE DISCRETE KALMAN FILTER

Consider the broad class of problems wherein it is desired to estimate the time-varying state of a dynamic system or process from a sequence of measurements containing noise. Let $\mathbf{X}_k$ be the unknown state vector process at time $t = t_k = kT$, where T is the interval between measurements. Assume that the dynamic system is defined by a linear difference equation (see Appendix 9A),

$$\mathbf{X}_{k+1} = \phi_k \mathbf{X}_k + \Gamma_k \mathbf{W}_k + D_k \mathbf{u}_k \tag{9.63}$$

where $\mathbf{W}_k$ is a purely random vector sequence (discrete white noise) acting as a disturbance, and $\mathbf{u}_k$ is a deterministic input vector. The coefficient matrices $\phi_k (= \phi(t_{k+1}, t_k))$, $\Gamma_k [= \Gamma(t_{k+1}, t_k)]$, and $D_k [= h(t_{k+1}, t_k)]$ are assumed to be known, as is $\mathbf{u}_k$. The initial state $\mathbf{X}_0$ is assumed to be random.

Because of the random input $\mathbf{W}_k$ and the form of (9.63), if $\mathbf{u}_k = 0$, the state $\mathbf{X}_k$ is a Markov sequence as defined in Chapter 7. If $\mathbf{W}_k$ is Gaussian, then $\mathbf{X}_k$ is a Gauss–Markov sequence. The Gaussian assumption does not affect the derivation of the Kalman filter. However, if $\mathbf{X}_k$ and the measurement noise are Gaussian, then the filter derived is not only the best *linear* filter in the sense of minimizing the mean-square error but also the best *filter*, linear or nonlinear, in that sense.

Before proceeding, we note the following points:

1. The dynamic process may be continuous and described by a differential equation of the form

$$\dot{\mathbf{X}} = A\mathbf{X} + B_1\mathbf{W}(t) + B\mathbf{u}(t) \tag{9.64}$$

discussed in Appendix 9A. If the measurements are discrete, it is necessary to discretize (9.64) as shown in the appendix to put it in the form (9.63).

2. Not all the components of the disturbance function are necessarily white noise. Suppose, for example, that one disturbance component has the correlation function

$$\gamma_Y(\tau) = \sigma_W^2 e^{-b|\tau|} \tag{9.65}$$

As shown in Chapter 7, this can be regarded as the output of a first-order filter with transfer function

$$H(s) = \frac{\sigma_W\sqrt{2b}}{s+b} \tag{9.66}$$

and white-noise input $W(t)$. The corresponding differential equation is

$$\dot{Y} = -bY + \sigma_W\sqrt{2b}W(t) \tag{9.67}$$

The state vector $\mathbf{X}(t)$ can be expanded to include the additional component $Y(t)$, and (9.64) modified accordingly. This procedure is known as *augmenting* the state vector. The modified system is then discretized to get the form (9.63).

The observations are assumed to be linearly related to the state. That is,

$$\mathbf{Y}_k = C_k\mathbf{X}_k + \mathbf{V}_k \tag{9.68}$$

with C_k known. The measurement noise $\mathbf{V}_k$ is a purely random sequence.

With regard to dimensionality for the vector sequences involved, take $\mathbf{x}_k$: $n\times 1$, $\mathbf{y}_k$: $m\times 1$, $\mathbf{W}_k$: $p\times 1$, and $\mathbf{u}_k$: $r\times 1$. For example, the components of the $\mathbf{X}_k$ vector could be position, speed, acceleration, and angle of attack of a vehicle, giving $n = 4$. The random-disturbance vector $\mathbf{W}_k$ might allow for wind gusts and variation in thrust, resulting in $p = 2$. Perhaps in that example, measurements of position and speed only are made, so $m = 2$. The dimensions of the matrices ϕ_k, Γ_k, D_k, and C_k are made compatible with those of the vectors.

Furthermore, it is assumed that the means and covariances of the random sequences and variables involved are known. Specifically,

1. *Initial state*

$$\langle\mathbf{X}_0\rangle = \bar{\mathbf{x}}_0 \tag{9.69}$$

$$\langle(\mathbf{X}_0 - \bar{\mathbf{x}}_0)(\mathbf{X}_0 - \bar{\mathbf{x}}_0)'\rangle = S_0 \tag{9.70}$$

2. *Random disturbance*

$$\langle \mathbf{W}_k \rangle = 0 \tag{9.71}$$

Note: If the mean of $\mathbf{W}_k$ is not zero, include it as part of the deterministic input $\mathbf{u}_k$.

$$\langle \mathbf{W}_k \mathbf{W}'_\ell \rangle = \begin{cases} Q_k, \ell = k \\ 0, \ell \neq k \end{cases} = Q_k \delta_{k,\ell} \tag{9.72}$$

This means that there may be correlation between the disturbance components at any particular time k, but no correlation between disturbances at different times.

3. *Measurement noise*

$$\langle \mathbf{V}_k \rangle = 0 \tag{9.73}$$

$$\langle \mathbf{V}_k \mathbf{V}'_\ell \rangle = R_k \delta_{k,\ell} \tag{9.74}$$

4.

$$\langle \mathbf{X}_0 \mathbf{W}'_k \rangle = 0, \text{ for all } k \tag{9.75}$$

$$\langle \mathbf{X}_0 \mathbf{V}'_k \rangle = 0, \text{ for all } k \tag{9.76}$$

$$\langle \mathbf{W}_k \mathbf{V}'_\ell \rangle = 0, \text{ for all } k \text{ and } \ell \tag{9.77}$$

The last three relationships imply that there is no correlation between the initial state, the random disturbance, and the measurement noise. The situation is illustrated diagrammatically in Figure 9.6.

The problem is to estimate *sequentially* the state vector $\mathbf{X}_k$. After observations $\mathbf{Y} = \mathbf{y}_1, \mathbf{Y}_2 = \mathbf{y}_2, \ldots, \mathbf{Y}_k = \mathbf{y}_k$ have been made, we want to obtain the best linear estimate of $\mathbf{X}_k$ (that is, $\hat{\mathbf{X}}_k = \hat{\mathbf{x}}_k$) based on these observations. By *best estimate* is meant that the mean-square error,

$$\langle \mathbf{E}'_k \mathbf{E}_k \rangle = \langle (\mathbf{X}_k - \hat{\mathbf{X}}_k)'(\mathbf{X}_k - \hat{\mathbf{X}}_k) \rangle \tag{9.78}$$

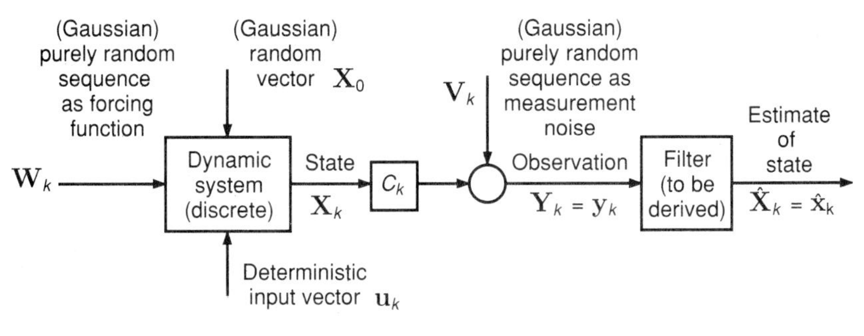

FIGURE 9.6 State estimation problem.

is minimized. The estimate is sequential, or recursive, in the sense that we use the best estimate of the *previous* value of the state ($\hat{\mathbf{x}}_{k-1}$) in order to compute $\hat{\mathbf{x}}_k$. Let $\hat{\mathbf{x}}_k$ be the estimate of $\mathbf{x}_k$ based on $\mathbf{y}_1, \mathbf{y}_2, \ldots, \mathbf{y}_k$. This is the updated estimate of $\mathbf{x}_k$. Let $\hat{\mathbf{x}}_k^*$ be the best estimate of $\mathbf{x}_k$ based on $\mathbf{y}_1, \mathbf{y}_2, \ldots, \mathbf{y}_{k-1}$. This is the predicted estimate of $\mathbf{x}_k$.

Using the static estimation problem of the previous section as a guide, we have

$$\hat{\mathbf{x}}_k = \hat{\mathbf{x}}_k^* + G_k(\mathbf{y}_k - C_k\hat{\mathbf{x}}_k^*) \tag{9.79}$$

where G_k is a time-varying gain matrix to be determined so as to minimize the mean-square error (9.78). It is an indication of how much weight should be placed on the current measurement $\mathbf{y}_k$.

How do we find $\hat{\mathbf{x}}_k^*$? We can predict $\hat{\mathbf{x}}_k^*$ from $\hat{\mathbf{x}}_{k-1}$ by a dynamic equation of the form (9.63)

$$\mathbf{x}_k^* = \phi_{k-1}\hat{\mathbf{x}}_{k-1} + \mathbf{h}_{k-1} \tag{9.80}$$

We omit a term in $\mathbf{W}_{k-1}$ because the random disturbance is unpredictable. In (9.80),

$$\mathbf{h}_{k-1} = D_{k-1}\mathbf{u}_{k-1}$$

By proving that it minimizes the mean square prediction error, we shall confirm that (9.80) is, in fact, reasonable. The best estimate of $\mathbf{x}_k$, given observations $\mathbf{y}_1, \mathbf{y}_2, \ldots, \mathbf{y}_{k-1}$, is

$$\begin{aligned}
\hat{\mathbf{x}}_k^* &= \langle \mathbf{X}_k | \mathbf{y}_1, \ldots, \mathbf{y}_{k-1} \rangle \\
&= \langle (\phi_{k-1}\mathbf{X}_{k-1} + \Gamma_{k-1}\mathbf{W}_{k-1} + \mathbf{h}_{k-1}) | \mathbf{y}_1, \ldots, \mathbf{y}_{k-1} \rangle \\
&= \phi_{k-1}\langle \mathbf{X}_{k-1} | \mathbf{y}_1, \ldots, \mathbf{y}_{k-1} \rangle + \Gamma_{k-1}\langle \mathbf{W}_{k-1} | \mathbf{y}_1, \ldots, \mathbf{y}_{k-1} \rangle + \mathbf{h}_{k-1} \\
&= \phi_{k-1}\hat{\mathbf{x}}_{k-1} + \Gamma_{k-1}\langle \mathbf{W}_{k-1} \rangle + \mathbf{h}_{k-1}
\end{aligned} \tag{9.81}$$

Because $\langle \mathbf{W}_{k-1} \rangle = 0$, (9.81) becomes (9.80), confirming the suitability of the latter expression.

The minimization procedure is straightforward. Express the mean-square error (9.78) in terms of G_k via (9.79). Then differentiate the mean-square error with respect to G_k and set the result equal to zero in order to find the minimizing value of G_k.

The estimation-error random vector is

$$\begin{aligned}
\mathbf{E}_k &= \mathbf{X}_k - \hat{\mathbf{X}}_k \\
&= \mathbf{X}_k - \hat{\mathbf{X}}_k^* - G_k(\mathbf{Y}_k - C_k\hat{\mathbf{X}}_k^*)
\end{aligned} \tag{9.82}$$

using (9.79). From (9.68),

$$\begin{aligned}
\mathbf{E}_k &= \mathbf{X}_k - \hat{\mathbf{X}}_k^* - G_k(C_k\mathbf{X}_k + \mathbf{V}_k - C_k\hat{\mathbf{X}}_k^*) \\
&= (I - G_kC_k)\mathbf{E}_k^* - G_k\mathbf{V}_k
\end{aligned} \tag{9.83}$$

where

$$\mathbf{E}_k^* \triangleq \mathbf{X}_k - \hat{\mathbf{X}}_k^* \tag{9.84}$$

is the *a priori* error in estimation of the error in the predicted estimate, and I is the identity matrix. Then the mean-square error is

$$\langle \mathbf{E}_k' \mathbf{E}_k \rangle = \langle [(I - G_k C_k)\mathbf{E}_k^* - G_k \mathbf{V}_k]'[(I - G_k C_k)\mathbf{E}_k^* - G_k \mathbf{V}_k] \rangle \tag{9.85}$$

Choose G_k so as to minimize this expression. First, rearrange (9.85) so as to exhibit G_k more explicitly for differentiation.

$$\begin{aligned}\langle \mathbf{E}_k{}' \mathbf{E}_k \rangle &= \langle \mathbf{E}_k^{*\prime} \mathbf{E}_k^* \rangle + \langle (C_k \mathbf{E}_k^* - \mathbf{V}_k)' G_k' G_k (C_k \mathbf{E}_k^* - \mathbf{V}_k) \rangle \\ &\quad - \langle \mathbf{E}_k^{*\prime} G_k (C_k \mathbf{E}_k^* - \mathbf{V}_k) \rangle - \langle (C_k \mathbf{E}_k^* - \mathbf{V}_k)' G_k' \mathbf{E}_k^* \rangle\end{aligned} \tag{9.86}$$

From the expressions (9B.26), (9B.27), and (9B.28) in Appendix 9B for differentiation of a quadratic form with respect to a matrix, we get

$$\nabla_{G_k}(\langle \mathbf{E}_k' \mathbf{E}_k \rangle) = 2G_k \langle (C_k \mathbf{E}_k^* - \mathbf{V}_k)(C_k \mathbf{E}_k^* - \mathbf{V}_k)' \rangle - 2\langle \mathbf{E}_k^* (C_k \mathbf{E}_k^* - \mathbf{V}_k)' \rangle = 0 \tag{9.87}$$

for an extremum. Therefore,

$$\begin{aligned}G_k[C_k \langle \mathbf{E}_k^* \mathbf{E}_k^{*\prime} \rangle C_k' - C_k \langle \mathbf{E}_k^* \mathbf{V}_k' \rangle - \langle \mathbf{V}_k \mathbf{E}_k^{*\prime} \rangle C_k' + \langle \mathbf{V}_k \mathbf{V}_k' \rangle] \\ - \langle \mathbf{E}_k^* \mathbf{E}_k^{*\prime} \rangle C_k' + \langle \mathbf{E}_k^* \mathbf{V}_k' \rangle = 0\end{aligned} \tag{9.88}$$

The three terms involving the correlation of $\mathbf{E}_k^*$ with $\mathbf{V}_k$ are zero, because $\mathbf{E}_k^*$ is based on $\mathbf{Y}_{k-1}$ and prior measurements. Define

$$P_k^* = \langle \mathbf{E}_k^* \mathbf{E}_k^{*\prime} \rangle \tag{9.89}$$

as the covariance matrix of the error in the predicted estimate. Then (9.88) becomes

$$G_k[C_k P_k^* C_k' + R_k] = P_k^* C_k' \tag{9.90}$$

or

$$G_k = P_k^* C_k' [C_k P_k^* C_k' + R_k]^{-1} \tag{9.91}$$

Thus if P_k^* is known, G_k can be computed, so the predicted estimate $\mathbf{x}_k^*$ can be used to give the filtered estimate $\hat{\mathbf{x}}_k$ by means of (9.79).

To facilitate the evaluation of P_k^*, we will express it in terms of P_{k-1}, where

$$P_k = \langle \mathbf{E}_k \mathbf{E}_k' \rangle \tag{9.92}$$

is the covariance matrix of the *a posteriori* error in estimation or the error in the updated estimate (the error the mean-square value of which was minimized). First, $\mathbf{E}_k^*$ can be written in terms of $\mathbf{E}_{k-1}$.

$$\begin{aligned}\mathbf{E}_k^* &= \mathbf{X}_k - \mathbf{X}_k^* \\ &= \phi_{k-1}\mathbf{X}_{k-1} + \Gamma_{k-1}\mathbf{W}_{k-1} + \mathbf{h}_{k-1} - \phi_{k-1}\hat{\mathbf{X}}_{k-1} - \mathbf{h}_{k-1} \\ &= \phi_{k-1}\mathbf{E}_{k-1} + \Gamma_{k-1}\mathbf{W}_{k-1}\end{aligned} \tag{9.93}$$

Therefore,

$$\begin{aligned}P_k^* &= \langle \mathbf{E}_k^* \mathbf{E}_k^{*\prime} \rangle \\ &= \langle [\phi_{k-1}\mathbf{E}_{k-1} + \Gamma_{k-1}\mathbf{W}_{k-1}][\phi_{k-1}\mathbf{E}_{k-1} + \Gamma_{k-1}\mathbf{W}_{k-1}]' \rangle \\ &= \phi_{k-1}\langle \mathbf{E}_{k-1}\mathbf{E}_{k-1}' \rangle \phi_{k-1}' + \phi_{k-1}\langle \mathbf{E}_{k-1}\mathbf{W}_{k-1}' \rangle \Gamma_{k-1}' \\ &\quad + \Gamma_{k-1}\langle \mathbf{W}_{k-1}\mathbf{E}_{k-1}' \rangle \phi_{k-1}' + \Gamma_{k-1}\langle \mathbf{W}_{k-1}\mathbf{W}_{k-1}' \rangle \Gamma_{k-1}' \end{aligned} \tag{9.94}$$

The two terms involving the correlation of $\mathbf{E}_{k-1}$ and $\mathbf{W}_{k-1}$ are zero, because $\mathbf{X}_{k-1}$ is uncorrelated with $\mathbf{W}_{k-1}$ ($\mathbf{X}_{k-1}$ depends on $\mathbf{W}_{k-2}$). Thus (9.94) becomes

$$P_k^* = \phi_{k-1}P_{k-1}\phi_{k-1}' + \Gamma_{k-1}Q_{k-1}\Gamma_{k-1}' \tag{9.95}$$

To complete the set of equations required for solving the problem, express P_k in terms of P_k' as follows:

$$\begin{aligned}P_k &= \langle \mathbf{E}_k \mathbf{E}_k' \rangle \\ &= \langle [(I - G_kC_k)\mathbf{E}_k^* - G_k\mathbf{V}_k][(I - G_kC_k)\mathbf{E}_k^* - G_k\mathbf{V}_k]' \rangle \end{aligned} \tag{9.96}$$

from (9.83). Expand this expression to get

$$\begin{aligned}P_k = (I - G_kC_k)P_k^*(I - G_kC_k)' + G_k\langle \mathbf{V}_k\mathbf{E}_k^{*\prime} \rangle (I - G_kC_k)' \\ - (I - G_kC_k)\langle \mathbf{E}_k^*\mathbf{V}_k' \rangle G_k' + G_k\langle \mathbf{V}_k\mathbf{V}_k' \rangle G_k' \end{aligned} \tag{9.97}$$

The two middle terms on the right-hand side are zero for the reason already stated. Hence

$$P_k = (I - G_kC_k)P_k^*(I - G_kC_k)' + G_kR_kG_k' \tag{9.98}$$

This equation is valid for any G_k as well as for the optimal G_k.

A more concise expression for P_k in terms of P_k^* can be found. From (9.98),

$$P_k = P_k^* - G_kC_kP_k^* - P_k^*C_k'G_k' + G_k[C_kP_k^*C_k' + R_k]G_k' \tag{9.99}$$

But

$$G_k[C_kP_k^*C_k' + R_k] = P_k^*C_k'$$

from the optimal value of G_k in (9.90), so

$$P_k = (I - G_kC_k)P_k^* \tag{9.100}$$

To start the iterative process, set P_0 equal to S_0, the covariance or uncertainty matrix for the initial state of X_0. The value is inserted in (9.95) to give P_1^*. For the initial estimate of the state, set $\hat{\mathbf{x}}_0$ equal to $\bar{\mathbf{x}}_0$, the mean value of the initial state.

Equations (9.79), (9.80), (9.91), (9.95), and (9.100) are the solution to the estimation problem and define the discrete Kalman filter [2], [5]. A summary of the main equations and underlying assumptions is contained in the next section.

9.6 SUMMARY OF EQUATIONS FOR THE DISCRETE KALMAN FILTER

Best linear estimate of current state $\mathbf{x}_k$ based on all observations up to and including $\mathbf{y}_k$:

$$\hat{\mathbf{x}}_k = \hat{\mathbf{x}}_k^* + G_k[\mathbf{y}_k - C_k\hat{\mathbf{x}}_k^*], \quad k = 1, 2, \ldots \tag{9.101}$$

Predicted estimate of $\mathbf{x}_k$ based on all observations up to and including $\mathbf{y}_{k-1}$:

$$\hat{\mathbf{x}}_k^* = \phi_{k-1}\hat{\mathbf{x}}_{k-1} + \mathbf{h}_{k-1} \tag{9.102}$$

Weighting or gain matrix:

$$G_k = P_k^* C_k'(C_k P_k^* C_k' + R_k)^{-1} \tag{9.103}$$

Covariance matrix of the error in the predicted estimate of the current state $\mathbf{x}_k$:

$$P_k^* = \phi_{k-1}P_{k-1}\phi_{k-1}' + \Gamma_{k-1}Q_{k-1}\Gamma_{k-1}' \tag{9.104}$$

Covariance matrix of the error in the updated or filtered estimate of $\mathbf{x}_k$:

$$P_k = (I - G_k C_k)P_k^* \tag{9.105}$$

Initial conditions (both $\bar{\mathbf{x}}_0$ and S_0 are given):

$$\hat{\mathbf{x}}_0 = \langle \mathbf{X}_0 \rangle = \bar{\mathbf{x}}_0 \tag{9.106}$$

$$P_0 = \langle (\mathbf{X}_0 - \bar{\mathbf{x}}_0)(\mathbf{X}_0 - \bar{\mathbf{x}}_0)' \rangle = S_0 \tag{9.107}$$

Assumptions:

1. *Dynamic system*

$$\mathbf{X}_{k+1} = \phi_k \mathbf{X}_k + \Gamma_k \mathbf{W}_k + \mathbf{h}_k \tag{9.108}$$

2. *Observation vector*

$$\mathbf{Y}_k = C_k \mathbf{X}_k + \mathbf{V}_k \tag{9.109}$$

3. *Statistics of initial state (known)*

$$\langle \mathbf{X}_0 \rangle = \bar{\mathbf{x}}_0 \tag{9.110}$$

$$\langle (\mathbf{X}_0 - \bar{\mathbf{x}}_0)(\mathbf{X}_0 - \bar{\mathbf{x}}_0)' \rangle = S_0 \tag{9.111}$$

4. *Statistics of random forcing function (known)*

$$\langle \mathbf{W}_k \rangle = 0 \quad \text{for all } k \tag{9.112}$$

$$\langle \mathbf{W}_k \mathbf{W}_\ell' \rangle = Q_k \delta_{k,\ell} \quad \text{for all } k, \ell \tag{9.113}$$

5. *Statistics of measurement noise (known)*

$$\langle \mathbf{V}_k \rangle = 0 \quad \text{for all } k \tag{9.114}$$

$$\langle \mathbf{V}_k \mathbf{V}'_\ell \rangle = R_k \delta_{k,\ell} \quad \text{for all } k, \ell \tag{9.115}$$

6. *Mutual independence of initial state, random forcing function, and measurement noise*

$$\langle \mathbf{X}_0 \mathbf{W}'_k \rangle = 0 \tag{9.116}$$

$$\langle \mathbf{X}_0 \mathbf{V}'_k \rangle = 0 \tag{9.117}$$

$$\langle \mathbf{V}_k \mathbf{W}'_\ell \rangle = 0 \quad \text{for all } k, \ell \tag{9.118}$$

Notes:

1. $$P_k^* = \langle \mathbf{E}_k^* \mathbf{E}_k^{*\prime} \rangle$$

   with the error in the predicted estimate defined as

$$\mathbf{E}_k^* = \mathbf{X}_k - \hat{\mathbf{X}}_k^*$$

2. $$P_k = \langle \mathbf{E}_k \mathbf{E}'_k \rangle$$

   with the error in the updated estimate defined as

$$\mathbf{E}_k = \mathbf{X}_k - \hat{\mathbf{X}}_k$$

A general schematic of the discrete Kalman filter is given in Figure 9.7. The Roman numerals I, II, and III over certain blocks indicate the order in which the equations are solved. Note from the upper part of the diagram that the filter contains a model of the dynamic system, excluding the random forcing function $\mathbf{W}_k$.

By manipulation of the equations, it is possible to obtain slightly different filter arrangements. For example, write (9.103) as

$$G_k(C_k P_k^* C'_k + R_k) = P_k^* C'_k$$

or

$$G_k R_k + (I - G_k C_k) P_k^* C'_k = P_k C'_k \tag{9.119}$$

from (9.105). Therefore, an alternative expression for the gain matrix is

$$G_k = P_k C'_k R_k^{-1} \tag{9.120}$$

If sufficient computer storage were available, it might be feasible to compute P_k and hence G_k in advance of the measurements being made, because (9.115) and (9.116) are independent of the measurements $\mathbf{y}_k$. Also, because of the recursive nature of the estimation process, it can be done in real time, that is, as the measurements are received.

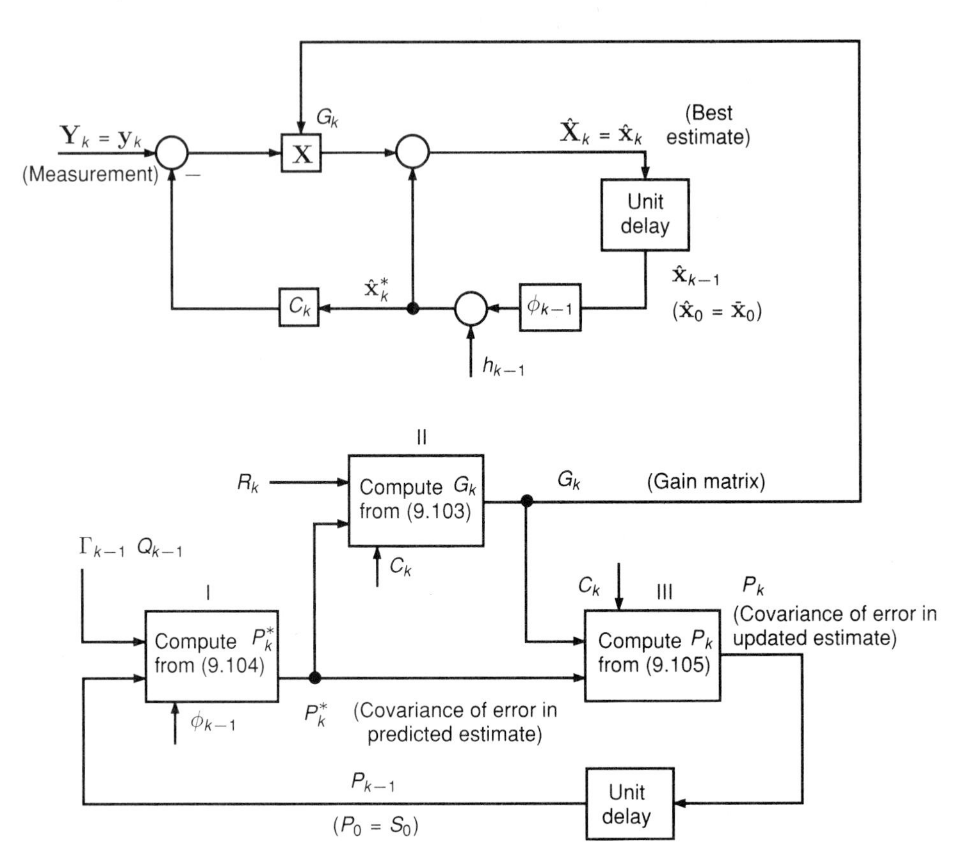

FIGURE 9.7 General form of discrete Kalman filter.

9.7 STEADY-STATE SOLUTION FOR DISCRETE KALMAN FILTER

If the matrices ϕ_k, Γ_k, and C_k are constant, and if $\mathbf{W}_k$ and $\mathbf{V}_k$ are stationary random processes (that is, Q_k and R_k are constant matrices) and the filter is stable, then the filter will reach a steady state so that P_k and P_k^* and hence G_k become constant matrices P, P^*, and G, respectively, as k approaches infinity.

Insert constant values in (9.105) to get

$$P = (I - GC)P^* \tag{9.121}$$

From (9.121), this may be written

$$P = (I - PC'R^{-1}C)P^* \tag{9.122}$$

Use (9.104) to eliminate P^*. Then

$$\begin{aligned} P &= (I - PC'R^{-1}C)(\phi P\phi' + \Gamma Q\Gamma') \\ &= \phi P\phi' + \Gamma Q\Gamma' - PC'R^{-1}C\phi P\phi' - PC'R^{-1}C\Gamma Q\Gamma' \end{aligned} \tag{9.123}$$

This is a matrix second-order nonlinear *algebraic* equation of the Riccati form to be solved for the steady-state value P. In general, it must be solved by computer. Once P is found, the steady-state value of the gain matrix can be obtained from

$$G = PC'R^{-1} \tag{9.124}$$

9.8 APPLICATION OF THE KALMAN FILTER EQUATIONS

With the exception of the first-order, or scalar, case, it is not possible to obtain an analytic solution for the Kalman filter equations. For higher-order systems ($n > 1$), the equations must be solved iteratively, as indicated in Figure 9.7.

Three examples are treated here. The first-order problem is solved completely. The second-order example is formulated for solution by iteration. The third example deals with error propagation and correction in a navigational system.

EXAMPLE 9.1

In this simple example, all quantities are scalars. It illustrates how to set up the problem for solution, the use of the filter equations, and the significance of the covariance equation. Suppose that a sequence of measurements is made of a discrete random process X_k the autocovariance of which is $\gamma_X(\ell) = \sigma^2 A^{|\ell|}/(1 - A^2)$. The measurements are noisy, and the measurement-noise sequence v_k is a realization of the noise random process V_k, which has mean zero and variance σ_V^2. Thus

$$y_k = x_k + v_k \tag{9.125}$$

We want to make a sequential estimate $\hat{x}_k$ of x_k, on the basis of the measurements, such that the mean-square error $\langle E^2 \rangle = \langle (X - \hat{X})^2 \rangle$ is minimized.

Solution

In order to cast the problem in the proper format for Kalman filtering, we must have the state x as the output of a dynamic system, and, if there is a random input, it should be white noise. From (7.135) we note that if discrete white noise W_k with variance σ^2 is input to a first-order recursive filter with coefficient A, then the output process X_k has autocovariance $\gamma_X(\ell) = \sigma^2 A^{|\ell|}/(1 - A^2)$. This gives us the model for the required dynamic system:

$$X_{k+1} = \phi X_k + W_k \tag{9.126}$$

where $\phi = A$ is a constant and

$$\langle W_k \rangle = 0$$

$$\langle W_k W_\ell \rangle = q\delta_{k,\ell}$$

with $q = \sigma^2$. Furthermore, we assume that the initial state X_0 is a random variable with

$$\langle X_0 \rangle = \bar{x}_0$$

$$\langle (X_0 - \bar{x}_0)^2 \rangle = s_0$$

For the measurement-noise process V,

$$\langle V \rangle = 0$$

$$\langle V_k V_\ell \rangle = r\delta_{k,\ell}$$

where $r = \sigma_V^2$. Also, we assume that X_0, W_k, and V_k are mutually independent.

Now that the problem is in the form for Kalman filtering, we can use the filter equations in Section 9.6. In using these equations, we have from (9.125) and (9.126) that $C_k = 1, \Gamma_k = 1, \phi_k = \phi, \mathbf{h}_k = 0, Q_k = q$, and $R_k = r$, all scalars and constant. The covariance equation becomes an equation for the error variance in this scalar case.

From (9.121), the time-varying gain is

$$g_k = p_k/r \tag{9.127}$$

Equation (9.104) becomes

$$p_k^* = \phi^2 p_{k-1} + q \tag{9.128}$$

and (9.105) gives

$$p_k = (1 - g_k)p_k^* \tag{9.129}$$

Substitute (9.127) and (9.128) into (9.129) to get

$$p_k = \left(1 - \frac{p_k}{r}\right)(\phi^2 p_{k-1} + q) \tag{9.130}$$

This is a difference equation for p_k that may be written

$$\frac{p_k}{r} = \frac{\phi^2 p_{k-1} + q}{\phi^2 p_{k-1} + q + r} \tag{9.131}$$

with initial condition $p_0 = s_0$. Note that (9.131) also gives g_k.

To get the steady-state value of p_k and g_k, either let $p_k = p_{k-1} = p$ in (9.131) and solve the resulting algebraic equation for p, or get the same

TABLE 9.2 Estimation-Error Variance and Filter Gains for Example 9.1.

k	Error Variance p_k	Gain $g_k = p_k/r$
0	12.00	2.40
1	3.53	0.71
2	3.32	0.66
3	3.31	0.66
4	3.31	0.66

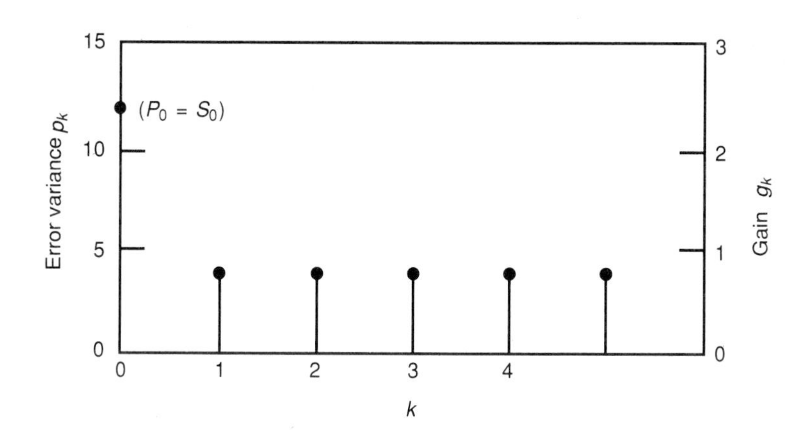

FIGURE 9.8 Estimation-error variance and filter gains for first-order example.

equation for p by solving (9.124) for this example. The result is

$$g = \frac{p}{r} = \frac{1}{2\phi^2}\left[-\left(1 - \phi^2 + \frac{q}{r}\right) \pm \sqrt{\left(1 - \phi^2 + \frac{q}{r}\right) + \frac{4q\phi^2}{r}}\right] \tag{9.132}$$

Table 9.2 and Figure 9.8 show how the error variance p_k and gain g_k converge rapidly to their constant steady-state values for $r = 5, q = 9, \phi = 0.5$, and $s_0 = 12$. The convergence to the steady-state values is equally rapid for other sets of parameter values.

In the next example, the dynamic system is second order and continuous. The measurements are discrete, so it is necessary to discretize the dynamic model used in the filter. The problem is formulated for solution by iteration.

EXAMPLE 9.2

A vehicle of mass m is acted on by a constant force u and a purely random disturbing force $W(t)$. In addition, a viscous drag force proportional to speed acts on the vehicle. The position and speed of the vehicle cannot be observed directly. However, a sequence of discrete measurements of position are made in the presence of noise. Assume that the statistics of the vehicle's initial state, the disturbing function, and the measurement noise are known. It is desired to outline the discrete Kalman filter for estimating the position and speed of the vehicle.

Solution

Let x_1 be the vehicle position, and let $x_2 = \dot{x}_1 = \dfrac{dx_1}{dt}$ be the speed. Then

$$\ddot{x}_1(t) = \frac{1}{m}[u + w(t) + a\dot{x}_1(t)] \tag{9.133}$$

where $\ddot{x}_1(t) = \dfrac{d^2x}{dt^2}$, and $w(t)$ is a particular realization or sample function of the random process $W(t)$. The system is shown in block diagram form in Figure 9.9.

In state-variable form (see Appendix 9A) [6], (9.133) can be written

$$\begin{bmatrix} \dot{x}_1 \\ \dot{x}_2 \end{bmatrix} = \begin{bmatrix} 0 & 1 \\ 0 & -\frac{a}{m} \end{bmatrix} \begin{bmatrix} x_1 \\ x_2 \end{bmatrix} + \begin{bmatrix} 0 \\ \frac{1}{m} \end{bmatrix} u + \begin{bmatrix} 0 \\ \frac{1}{m} \end{bmatrix} w(t) \tag{9.134}$$

or

$$\dot{\mathbf{x}} = A\mathbf{x} + Bu + \Gamma w(t) \tag{9.135}$$

where

$$\mathbf{x} = \begin{bmatrix} x_1 \\ x_2 \end{bmatrix}$$

$$A = \begin{bmatrix} 0 & 1 \\ 0 & -\frac{a}{m} \end{bmatrix}$$

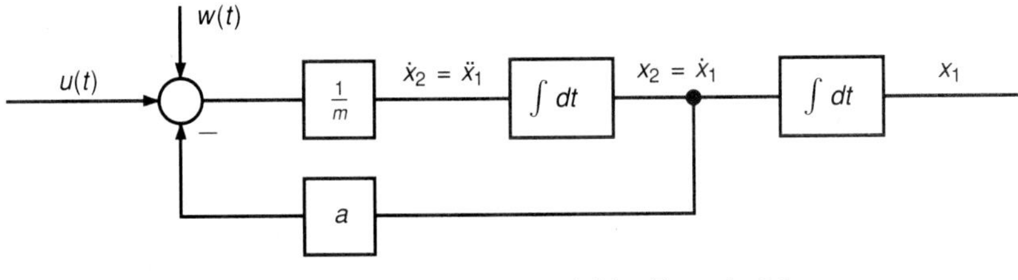

FIGURE 9.9 Dynamic model for Example 9.2.

$$B = \begin{bmatrix} 0 \\ \frac{1}{m} \end{bmatrix}$$

$$\Gamma = \begin{bmatrix} 0 \\ \frac{1}{m} \end{bmatrix}$$

Using results from Appendix 9A, we obtain as follows a discrete model of the dynamic system, which is required for inclusion in the filter. The solution of (9.135) with $w(t) = 0$ is

$$\mathbf{x}(t) = \phi(t, t_0)\mathbf{x}(t_0) + \int_{t_0}^{t} \phi(t, \varphi)B(\varphi)u(\varphi)d\varphi \tag{9.136}$$

where $\phi(t, t_0)$ is the state transition matrix of the homogeneous system

$$\dot{\mathbf{x}} = A\mathbf{x} \tag{9.137}$$

Because A is constant, the transition matrix $\phi(t, t_0) = \phi(t - t_0)$ can be computed as

$$\phi(t) = \mathcal{L}^{-1}\{[sI - A]^{-1}\} \tag{9.138}$$

and results in

$$\phi(t - t_0) = \begin{bmatrix} 1 & \frac{m}{a}(1 - \exp[-\frac{a}{m}(t - t_0)]) \\ 0 & \exp\left[\frac{a}{m}(t - t_0)\right] \end{bmatrix} \tag{9.139}$$

If $t_0 = t_k$ and $t = t_{k+1}$, (9.136) becomes

$$\mathbf{x}(t_{k+1}) = \phi(t_{k+1}, t_k)\mathbf{x}(t_k) + \int_{t_k}^{t_{k+1}} \phi(t_{k+1}, \varphi)B(\varphi)u\,d\varphi \tag{9.140}$$

where u is assumed constant, and $t_k \leq \varphi < t_{k+1}$. In the notation of Section 9.5, the discrete form of the dynamic model is

$$\mathbf{x}_{k+1} = \phi_k\mathbf{x}_k + h_k + \Gamma_k\mathbf{W}_k \tag{9.141}$$

Take ϕ_{ij} as the elements of ϕ in (9.139). Then

$$\begin{aligned}
\mathbf{h}_k &= \int_{t_k}^{t_{k+1}} \phi(t_{k+1}, \varphi)B(\varphi)u\;d\varphi \\
&= \int_{t_k}^{t_{k+1}} \begin{bmatrix} \phi_{11} & \phi_{12} \\ \phi_{21} & \phi_{22} \end{bmatrix} \begin{bmatrix} 0 \\ \frac{1}{m} \end{bmatrix} d\varphi \cdot u \\
&= \int_{t_k}^{t_{k+1}} \begin{bmatrix} \phi_{12}/m \\ \phi_{22}/m \end{bmatrix} d\varphi \cdot u \\
&= \begin{bmatrix} \frac{u}{a}(t_{k+1} - t_k) - \frac{mu}{a^2}\left(1 - \exp\left[-\frac{a}{m}(t_{k+1} - t_k)\right]\right) \\ \frac{u}{a}\left(1 - \exp\left[-\frac{a}{m}(t_{k+1} - t_k)\right]\right) \end{bmatrix}
\end{aligned} \tag{9.142}$$

If $t_{k+1} - t_k = T$, the interval between measurements, then

$$h_k = \begin{bmatrix} \frac{uT}{a} - \frac{mu}{a^2}[1 - \exp\left(-\frac{aT}{m}\right)] \\ \frac{u}{a}[1 - \exp\left(-\frac{aT}{m}\right)] \end{bmatrix} \tag{9.143}$$

which is constant for constant T, over one period.

If $v(t_k)$ is the noise on the measurement $y(t_k)$, then

$$\begin{aligned} y(t_k) &= x_1(t_k) + v(t_k) \\ &= [1 \quad 0] \begin{bmatrix} x_1(t_k) \\ x_2(t_k) \end{bmatrix} + v(t_k) \end{aligned} \tag{9.144}$$

or

$$y_k = C\mathbf{x}_k + v_k \tag{9.145}$$

where $\mathbf{x}_k = \mathbf{x}(t_k), y_k = y(t_k), v = v(t_k)$, and

$$C = [1 \quad 0] \tag{9.146}$$

It is assumed that the measurement noise is white with

$$\langle V_k \rangle = 0; \quad \langle V_k V_\ell \rangle = r\delta_{k,\ell} \tag{9.147}$$

It is assumed that the random forcing function W_k is a white-noise process with

$$\begin{aligned} \langle W_k \rangle &= 0 \\ \langle W_k W_\ell \rangle &= q\delta_{k,\ell} \end{aligned} \tag{9.148}$$

The stage is now set for us to write the filter equations using (9.104), (9.103), and (9.105). For (9.104),

$$\Gamma Q \Gamma' = q \begin{bmatrix} 0 \\ \frac{1}{m} \end{bmatrix} [0 \quad \tfrac{1}{m}] = \begin{bmatrix} 0 & 0 \\ 0 & \frac{q}{m^2} \end{bmatrix} \tag{9.149}$$

Hence, in component form, (9.104) becomes, for this problem (omitting the time variables for brevity),

$$p_{11}^* = \phi_{11}(p_{11}\phi_{11} + p_{12}\phi_{12}) + \phi_{12}(p_{21}\phi_{11} + p_{22}\phi_{12}) \tag{9.150}$$

$$p_{12}^* = p_{21}^* = \phi_{11}(p_{11}\phi_{21} + p_{12}\phi_{22}) + \phi_{12}(p_{21}\phi_{21} + p_{22}\phi_{22}) \tag{9.151}$$

$$p_{22}^* = \phi_{21}(p_{11}\phi_{21} + p_{12}\phi_{22}) + \phi_{22}(p_{21}\phi_{21} + p_{22}\phi_{22}) + \frac{q}{m^2} \tag{9.152}$$

where the components of ϕ are obtained from (9.139) as

$$\phi_{11} \stackrel{\Delta}{=} \phi_{11}(T) = 1 \tag{9.153}$$

$$\phi_{12} \stackrel{\Delta}{=} \phi_{12}(T) = \frac{m}{a}\left[1 - \exp\left(-\frac{aT}{m}\right)\right] \tag{9.154}$$

$$\phi_{21} \stackrel{\Delta}{=} \phi_{21}(T) = 0 \tag{9.155}$$

$$\phi_{22} \stackrel{\Delta}{=} \phi_{22}(T) = \exp\left(-\frac{aT}{m}\right) \tag{9.156}$$

with $T = t_{k+1} - t_k$. As noted in Section (9.5), the initial condition for (9.104) is $P_0 = S_0$, where

$$\langle(\mathbf{X}_0 - \bar{\mathbf{x}}_0)(\mathbf{X}_0 - \bar{\mathbf{x}}_0)'\rangle = S_0 \tag{9.157}$$

$$\langle\mathbf{X}_0\rangle = \bar{\mathbf{x}}_0 \tag{9.158}$$

From (9.103), the gain matrix is

$$G_k = \begin{bmatrix} g_1(t_k) \\ g_2(t_k) \end{bmatrix} = \begin{bmatrix} p_{11}^*(t_k)/(p_{11}^*(t_k) + r_1) \\ p_{12}^*(t_k)/(p_{11}^*(t_k) + r_1) \end{bmatrix} \tag{9.159}$$

The P_k matrix is obtained from (9.105) as

$$P_k = \begin{bmatrix} (1 - g_1(t_k))p_{11}^*(t_k) & (1 - g_1(t_k))p_{12}^*(t_k) \\ (1 - g_1(t_k))p_{12}^*(t_k) & -g_2(t_k)p_{12}^*(t_k) + p_{22}^*(t_k) \end{bmatrix} \tag{9.160}$$

The state estimate is

$$\hat{\mathbf{x}}_k = \hat{\mathbf{x}}_k^* + G_k[y_k - C\hat{\mathbf{x}}_k^*] \tag{9.161}$$

or, in scalar form,

$$\hat{x}_1(t_k) = \hat{x}_1^*(t_k) + g_1(t_k)[y(t_k) - \hat{x}_1^*(t_k)] \tag{9.162}$$

$$\hat{x}_2(t_k) = \hat{x}_2^*(t_k) + g_2(t_k)[y(t_k) - \hat{x}_1^*(t_k)] \tag{9.163}$$

The expected value of the initial state

$$\langle\mathbf{X}_0\rangle = \bar{\mathbf{x}}_0 \tag{9.164}$$

is assumed known. This is taken as the initial value for the estimate—that is,

$$\hat{x}_1(t_0) = \bar{x}_1(t_0) \tag{9.165}$$

$$\hat{x}_2(t_0) = \bar{x}_2(t_0) \tag{9.166}$$

Also,

$$\hat{\mathbf{x}}_k^* = \phi_k\hat{\mathbf{x}}_k + \mathbf{h}_k \tag{9.167}$$

or, in scalar form,

$$\hat{x}_1^*(t_k) = \hat{x}_1(t_{k-1}) + \phi_{12}\hat{x}_2(t_{k-1}) + h_1 \tag{9.168}$$

$$\hat{x}_2^*(t_k) = \phi_{22}\hat{x}_2(t_{k-1}) + h_2 \tag{9.169}$$

where h_1 and h_2 are given by (9.143), ϕ_{12} by (9.154), and ϕ_{22} by (9.156). A schematic of the filter is shown in Figure 9.10.

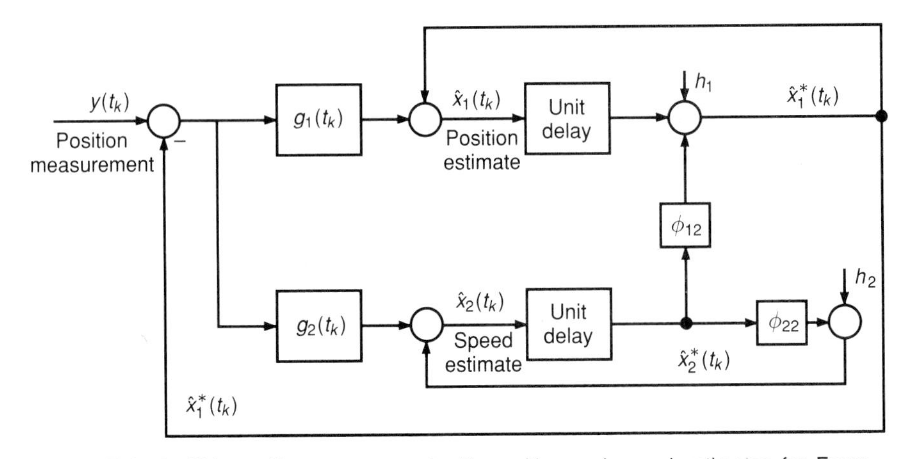

FIGURE 9.10 Kalman filter arrangement with position and speed estimates for Example 9.2.

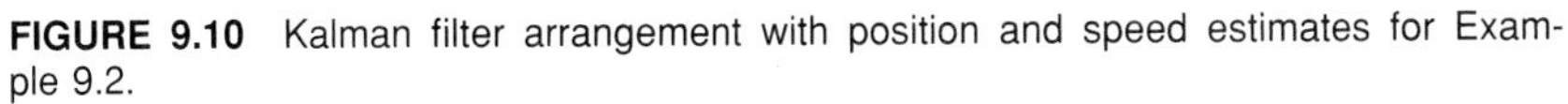

The filter equations outlined in Example 9.2 can be programmed to yield a recursive estimate of position and speed. However, if many such filtering problems are to be expected, it is more efficient to write a general Kalman filter program based on the equations given in Section 9.6 and then insert the parameters for the specific problem. The computer used should have subroutines to carry out matrix operations such as addition, subtraction, multiplication, inverse, and transpose. Kalman filters for estimating more than 50 state variables have been used in the aerospace industry. Filter orders up to 20 are common.

The next and final example deals with the use of a Kalman filter as an aid to inertial navigation, the science in which it had its earliest applications.

EXAMPLE 9.3

First we will provide some background on the navigational problem. An inertial reference unit (IRU) with stable platform is used to navigate an airborne vehicle. On the platform are mounted three gyroscopes ("gyros") and two accelerometers. The gyros measure angular rates, and their output signals are used to torque the platform so that, ideally, it is maintained locally level—that is, parallel to the local tangent plane to the earth's surface. In addition, the platform is turned constantly so that, again ideally, one of the fixed accelerometers always points north and the other always points east [see Figure 9.11(a)]. We integrate the accelerometer outputs to get the lateral velocity and position (latitude and longitude) of the vehicle relative to the earth. The vertical position and velocity are provided by an altimeter, which is generally more accurate than the IRU.

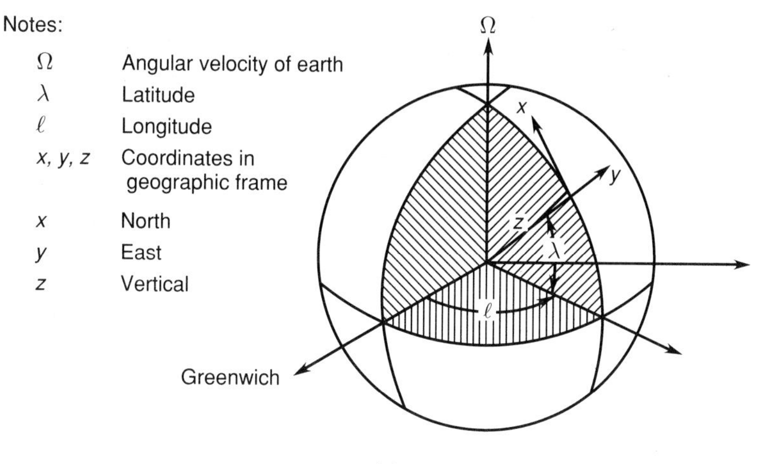

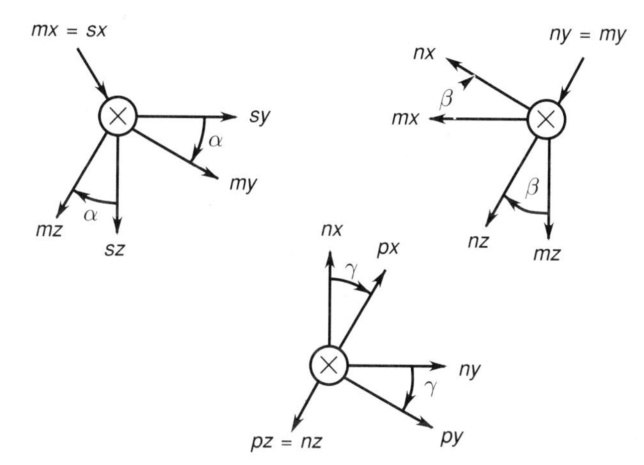

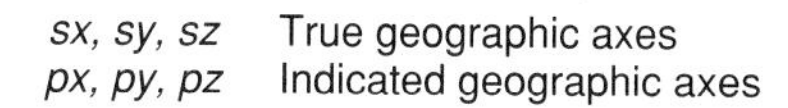

FIGURE 9.11 Conventions for inertial navigational problem. (a) Locally level or geographic coordinate frame. (b) Definition of error angles α; β, and γ.

Although in theory, the platform is supposed to maintain the orientation we have described, in practice, as a result of gyro errors or "drifts," the actual platform orientation differs from the ideal by small error angles α, β, and γ, as shown in Figure 9.11(b). Additional errors in the indicated vehicle position and velocity are caused by accelerometer biases, instrument misalignment on the platform, and errors in the initialization of the IRU when the vehicle is launched. Because of the integrating effects, the navigational errors grow with time as shown, for example, in Figures 9.12 and 9.13. In order to correct the false position and velocity readings given by the IRU, it is necessary to send position updates to the vehicle from a ground station. These updates, which themselves contain noise, serve as input measurements to a suitable Kalman filter installed in the vehicle. On the basis of the measurements, the filter estimates the state of the dynamic process, which in this case is the error system. The estimates of the error states are used to correct the IRU. The problem is to define the error system and the inputs so that we can use the filter equations of Section 9.6—and so determine the minimum number of updates necessary to reduce the navigational errors to manageable bounds.

Solution

To simplify the navigational equations of motion, we will assume that the vehicle is traveling at a constant speed along a circle of constant latitude. The derivation of the navigational-system equations in geographic coordinates is given in several standard texts (such as [8] and [9]). The relevant equations for a vehicle moving at constant ground speed are:

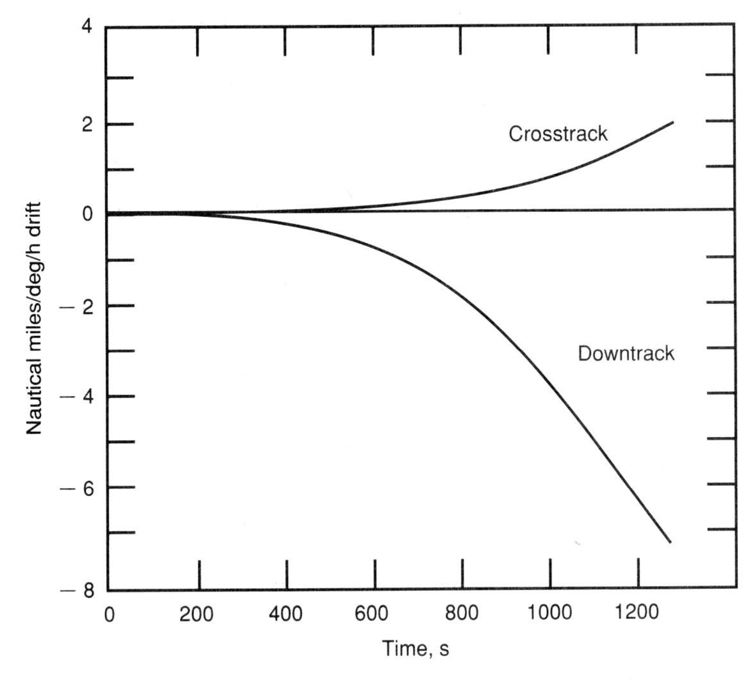

FIGURE 9.12 Position errors due to north gyro drift.

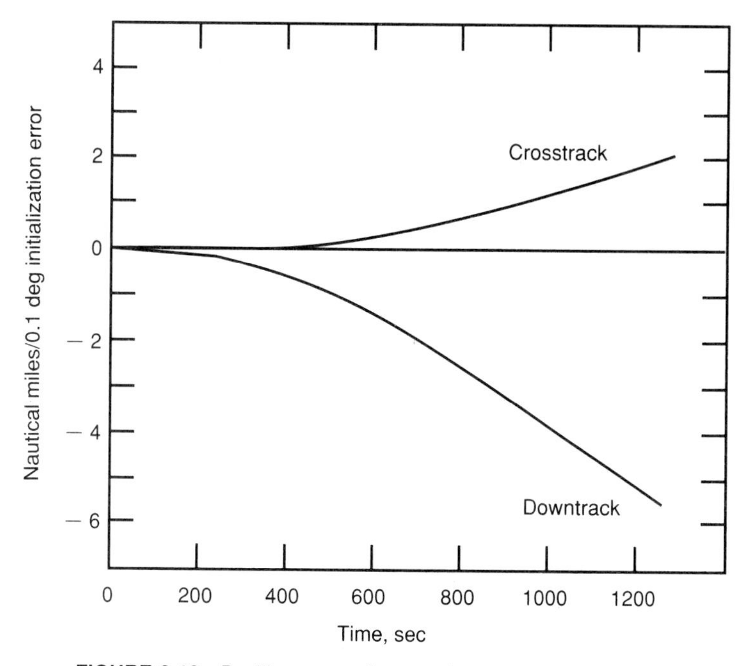

FIGURE 9.13 Position errors due to attitude initialization error α_0.

$$\begin{aligned}
\dot{\omega}_n^I &= (2\Omega\omega_e \sin\lambda - 2\Omega\omega_e^I \sin\lambda^I) + (\omega_e^2 \tan\lambda - \omega_e^I \tan\lambda^I) \\
&\quad - \gamma(\omega_n\omega_e \tan\lambda + 2\Omega\omega_n \sin\lambda) \\
&\quad - \beta(2\Omega\omega_e \cos\lambda + \omega_e^2 - g/R) && (9.170) \\
\dot{\omega}_e^I &= (\omega_n^I\omega_e^I \tan\lambda^I - \omega_n\omega_e \tan\lambda) + (2\Omega\omega_n^I \sin\lambda^I - 2\Omega\omega_n \sin\lambda) \\
&\quad - \gamma(2\Omega\omega_e \sin\lambda + \omega_e^2 \tan\lambda) \\
&\quad + \alpha(2\Omega\omega_e \cos\lambda + \omega_e^2 - g/R) && (9.171) \\
\dot{\lambda}^I &= \omega_n^I && (9.172) \\
\dot{\alpha} &= \Omega(\cos\lambda^I - \cos\lambda) + \omega_e^I - \omega_e + \gamma\omega_n \\
&\quad - \beta(\Omega \sin\lambda + \omega_e \tan\lambda) + U_x && (9.173) \\
\dot{\beta} &= -\omega_n^I + \omega_n + \alpha(\Omega \sin\lambda + \omega_e \tan\lambda) \\
&\quad + \gamma(\Omega \cos\lambda + \omega_e) + U_y && (9.174) \\
\dot{\gamma} &= -\Omega(\sin\lambda^I - \sin\lambda) - (\omega_e^I \tan\lambda^I - \omega_e \tan\lambda) \\
&\quad - \beta(\Omega \cos\lambda + \omega_e) - \alpha\omega_n + U_z && (9.175) \\
\dot{\ell}^I &= \omega_e^I \sec\lambda^I && (9.176)
\end{aligned}$$

In these equations, the variables with superscript I are those computed by the navigational system; the same variables with no superscript are the true values. In addition to those variables already defined in Figure 9.11, w_n is the rate of change of geographic latitude λ (the vector ω_n is in the negative-y, or west, direction), and $\omega_e = \dot{\ell}\cos\lambda$ (the vector ω_e is in the positive-x, or north, direction). The inputs U_x, U_y, and U_z are drift rates of the north, east, and vertical gyros, respectively. Furthermore, g is gravitional acceleration, and R is the distance of the vehicle from the center of the earth.

Define the navigational system errors as follows:

$$\delta\omega_n = \omega_n^I - \omega_n$$

$$\delta\omega_e = \omega_e^I - \omega_e$$

$$\delta\lambda = \lambda^I - \lambda$$

$$\delta\ell = \ell^I - \ell$$

The attitude angles α, β, and γ between the indicated and the true geographic axes are already in the form of errors about their nominal values, which are zero. From the constant-speed assumption, it follows that $\dot{\omega}_n = \dot{\omega}_e = 0$. If we assume also that the missile flies a parallel of latitude say $\lambda = \lambda_1$, then $\omega_n = 0$. We can derive the linearized-error-system equations by expanding the right-hand sides of (9.170) through (9.176) in a Taylor series about the nominal trajectory and retaining only the first-order terms (see Appendix 9A). The nominal values of the variables are $\omega_n = 0, \omega_e = V_m \cos\lambda_1$, $\lambda = \lambda_1$, $\alpha_{\text{nom}} = 0$, $\beta_{\text{nom}} = 0$, and $\gamma_{\text{nom}} = 0$.

The resulting linearized error system, written in state-variable form, is

$$\dot{\mathbf{x}} = A\mathbf{x} + \mathbf{f} \tag{9.177}$$

where the state vector $\mathbf{x}$ is defined as

$$\mathbf{x} = [\delta\omega_n \ \delta\omega_e \ \delta\lambda \ \alpha \ \beta \ \gamma \ \delta\ell]' \tag{9.178}$$

and the input vector $\mathbf{f}$ is defined as

$$\mathbf{f} = [f_1 \ f_2 \ f_3 \ f_4 \ f_5 \ f_6 \ f_7]' \tag{9.179}$$

The elements of the 7×7 system matrix A are:

$$\begin{aligned}
a_{12} &= -2(\Omega\sin\lambda + \omega_e\tan\lambda) \\
a_{13} &= -(2\Omega\omega_e\cos\lambda + \omega_e^2\sec^2\lambda) \\
a_{15} &= -(2\Omega\omega_e\cos\lambda + \omega_e^2 - g/R) \\
a_{21} &= 2\Omega\sin\lambda + \omega_e\tan\lambda \\
a_{24} &= -a_{15} \\
a_{26} &= -(2\Omega\omega_e\sin\lambda + \omega_e^2\tan\lambda) \\
a_{31} &= a_{42} = -a_{51} = 1
\end{aligned} \tag{9.180}$$

$$a_{43} = -\Omega \sin \lambda$$

$$a_{45} = -(\Omega \sin \lambda + \omega_e \tan \lambda) = -a_{54}$$

$$a_{56} = \Omega \cos \lambda + \omega_e = -a_{65}$$

$$a_{62} = -\tan \lambda$$

$$a_{63} = -(\Omega \cos \lambda + \omega_e \sec^2 \lambda)$$

$$a_{72} = \sec \lambda$$

$$a_{73} = \omega_e \tan \lambda \sec \lambda$$

The values of λ and ω_e in the foregoing expressions are the nominal values. The remaining coefficients a_{ij} are zero. The inputs f_i depend on the error source. (For example, with gyro drifts U_x, U_y, and U_z, we have $f_4 = U_x, f_5 = U_y, f_6 = U_z$, and $f_1 = f_2 = f_3 = f_7 = 0$. For accelerometer biases a_x and $a_y, f_1 = -a_x/R, f_2 = -a_y/R$, and the other f_i are zero. With component misalignment on the platform (that is, when the sensitive axes of the accelerometers and gyros are mounted slightly askew), the inputs are somewhat more complex. These are derived in Broxmeyer [8].

The crosstrack (x, or north, direction) and downtrack (y, or east, direction) position and velocity errors are obtained from the outputs of the error model (9.177) via the following relationships:

$$\delta x = R\,\delta\lambda \tag{9.181}$$

$$\delta y = R \cos \lambda\, \delta\ell \tag{9.182}$$

$$\delta v_x = R\,\delta\omega_n \tag{9.183}$$

$$\delta v_y = R\,\delta\omega_e \tag{9.184}$$

Initial-attitude errors can be treated by selecting the initial conditions α_0, β_0, and γ_0 in (9.177) accordingly. Similarly, initial-velocity errors δv_{x0} and δv_{y0} imply initial conditions $\delta\omega_{n0} = \delta v_{x0}/R$ and $\delta\omega_{e0} = \delta v_{y0}/R$ in the system equations.

The error-system equations (9.177) are discretized to obtain the form (9.108) to suit the Kalman filter with discrete measurements. The setup is depicted in Figure 9.14. The components of the random-input vector $\mathbf{W}_k$ in this case are gyro drifts and accelerometer biases. The components of the initial-state vector $\mathbf{X}_0$ are errors in initializing the inertial platform, such as initial-attitude errors—that is, $\alpha_0, \beta_0, \gamma_0 \neq 0$. Ground-station measurements of vehicle range, aximuth, and elevation give its position and form the components of the measurement vector $\mathbf{y}$. The errors of the corresponding sensors constitute the components of the measurement-noise vector $\mathbf{V}_k$. The transformation from state vector to measurement is provided by the time-varying matrix C_k.

We note that in this example, unlike many other Kalman filtering problems, the covariance matrices Q_k and R_k for input disturbances and mea-

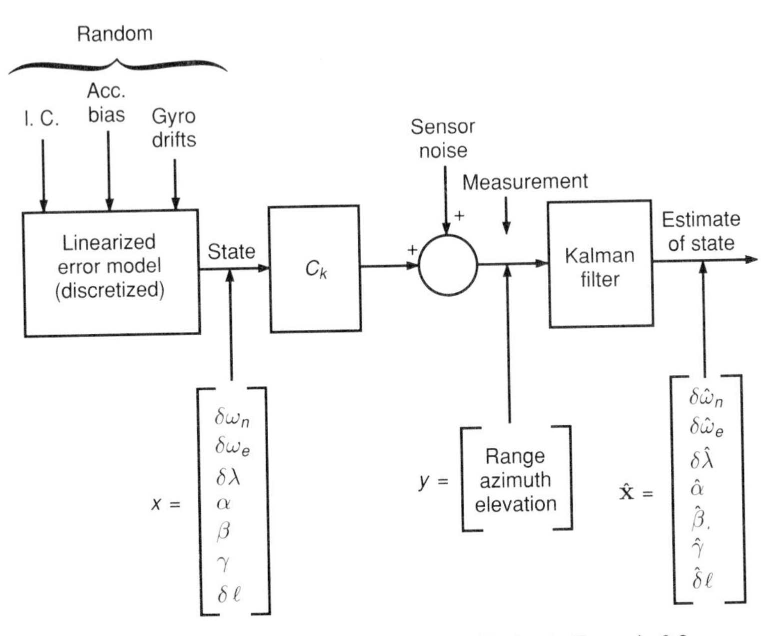

FIGURE 9.14 Schematic for Kalman filtering in Example 9.3.

surement noise, respectively, are well-defined quantities available from the instrument manufacturers.

When the error system was simulated together with the filter algorithms given in Section 9.6, it was possible to determine how the Kalman filter would reduce the navigational error. The amount of improvement depends, in general, on how many position updates, or measurements, $\mathbf{y}_k$ are sent from the ground station. Figure 9.15 indicates how the root mean square (rms) value of the crosstrack error grows until a single measurement is made at $t = 550$ seconds. This reduces the rms error to about 20% of its previous value. Curves for 5 and 10 updates are also shown. These do not seem to reduce the error by much more than one update did. This is because the sensor is measuring position and hence has a direct effect on estimates of position error.

Let us consider how the curves in Figure 9.15 were obtained. Take the uppermost curve ($N = 1$). If there is no update, the covariance of the estimation error propagates in time according to the equation

$$P_{n+1} = \phi_n P_n \phi_n' + \Gamma_n Q_n \Gamma_n' \tag{9.185}$$

as derived in Appendix 9A. At the time $n = k$ at which the update is made, P_n as given by (9.185) has grown to P_k^*, the covariance of the predicted error in estimation. Then, when the update measurement is taken into account in the Kalman filter, the covariance of the updated error in estimation drops sharply from the P_k^* value. We note that the curve is drawn for one element

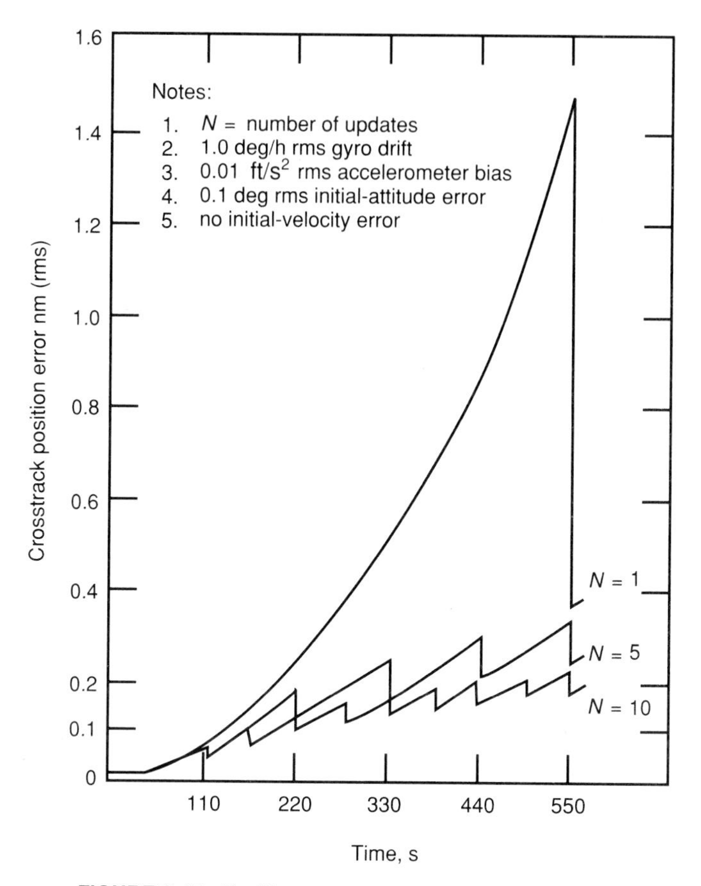

FIGURE 9.15 Position error correction by periodic updates.

of the covariance matrix, the diagonal term corresponding to the variance of the crosstrack error.

Figure 9.16 shows similar correction curves for crosstrack velocity error. Again, most of the improvement is obtained with one measurement to the Kalman filter; additional measurements show diminishing returns in error reduction.

Figure 9.17 indicates that for attitude error correction, taking more measurements substantially reduces the error. This is because the platform attitude is not directly related to the position measurement, as the position and velocity errors were. Therefore, more information (more measurements) is required for estimating attitude error.

The foregoing findings are summarized in Figure 9.18, which indicates how the estimation error depends on the number of updates. Judging by these curves, it would appear that there is not much advantage, in this case, in making more than five or six measurements.

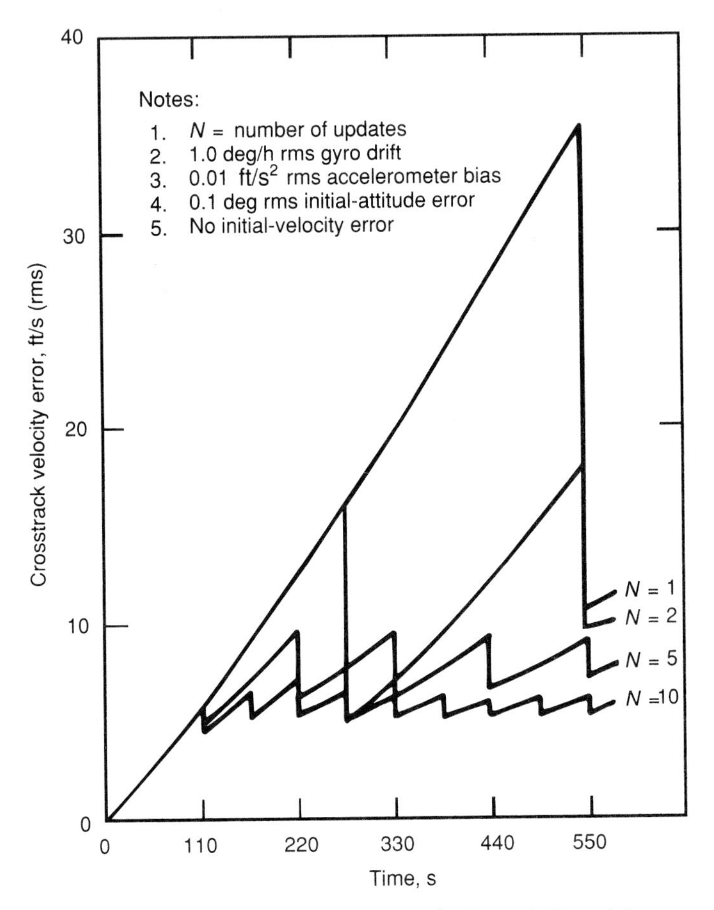

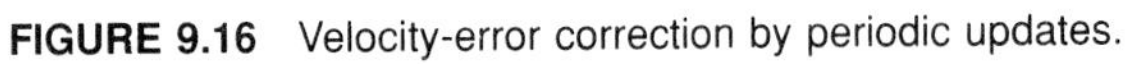

FIGURE 9.16 Velocity-error correction by periodic updates.

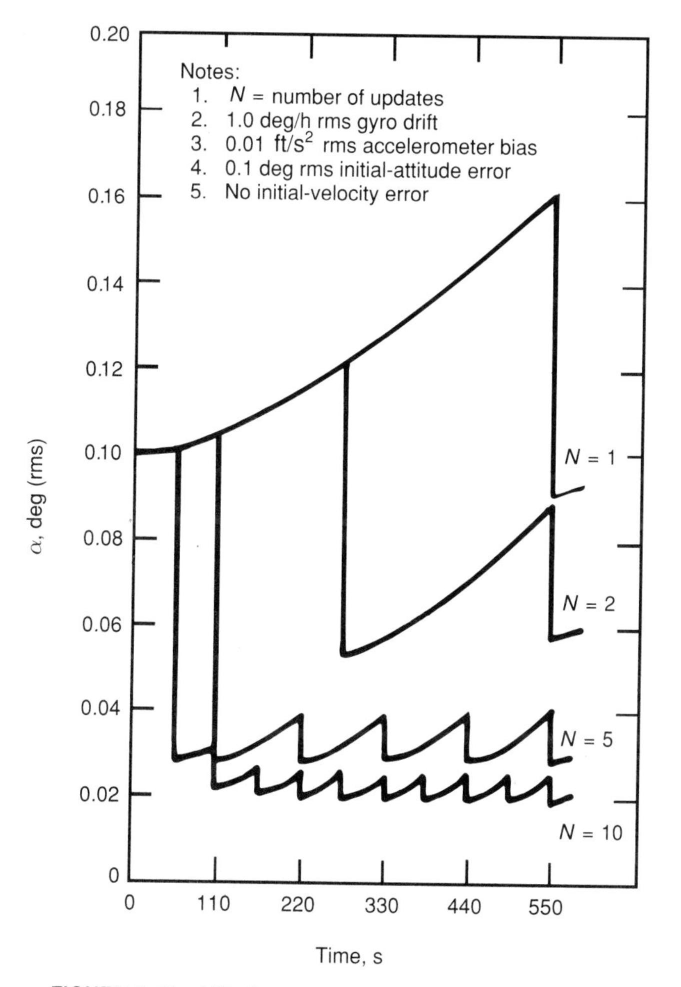

FIGURE 9.17 Attitude-error correction by periodic updates.

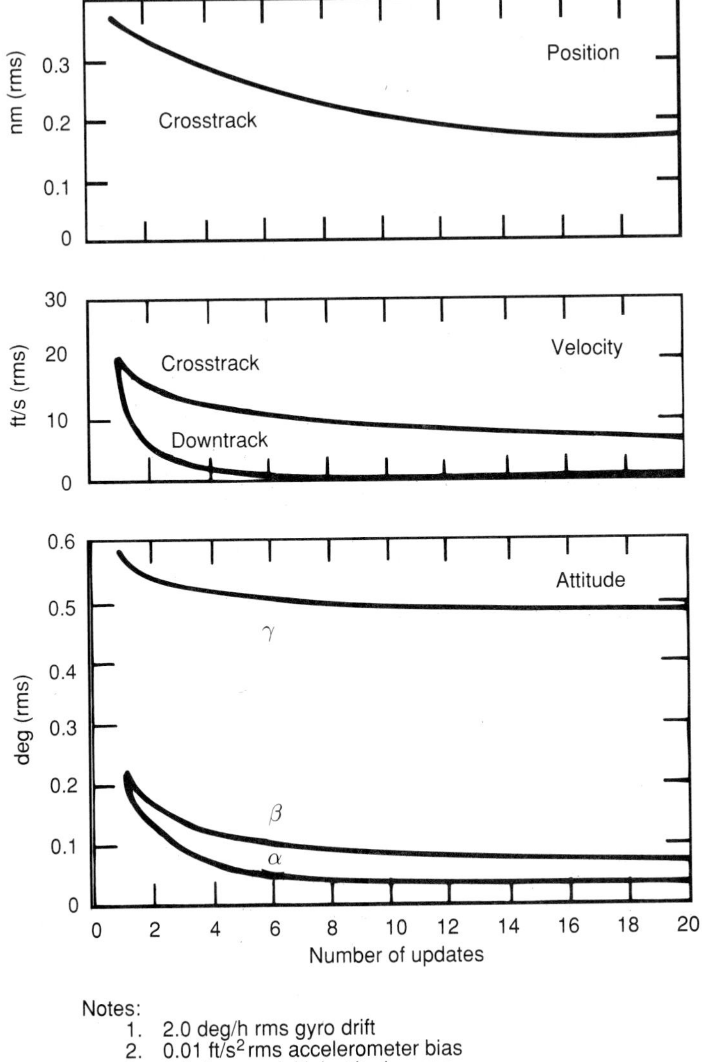

Notes:
1. 2.0 deg/h rms gyro drift
2. 0.01 ft/s^2 rms accelerometer bias
3. 0.5 deg rms initial-attitude error
4. 10.0 ft/s rms initial-velocity error

FIGURE 9.18 Dependence of estimation error on number of updates.

9.9 SUMMARY

Some basic operations in estimation are introduced. A useful application is the alpha–beta tracker, which, because of its constant gains, can be analyzed by means of the z-transform. The discrete Kalman filter is derived from first principles. The resulting set of filter equations can be programmed to estimate the state of any finite-order linear process. Examples of Kalman filtering applications are discussed.

REFERENCES FOR CHAPTER 9

1. A. E Bryson and Y. C. Ho, *Applied Optimal Control* (Waltham, MA: Blaisdell, 1969), Chap. 12.
2. R. E. Kalman, "A New Approach to Linear Filtering and Prediction Problems," *Jour. Basic Eng.*, March 1960: 35–46.
3. T. R. Benedict and G. W. Bordner, "Synthesis of an Optimal Set of Radar Track-While-Scan Smoothing Equations," *IRE Transactions on Automatic Control*, AC-7, no. 4 (July 1962): 27–32.
4. J. A. Cadzow, *Discrete Time Systems* (Englewood Cliffs, NJ: Prentice-Hall, 1973), Chaps. 2 and 8.
5. H. Sorensen, "Kalman Filtering," in *Advances in Control Systems*, vol. 3, ed. C. T. Leondes (New York: Academic, 1966).
6. K. Ogata, *State Space Analysis of Control Systems* (Englewood Cliffs, NJ: Prentice-Hall, 1967).
7. A. Gelb, ed., *Applied Optimal Estimation* (Cambridge, MA: The M.I.T. Press, 1974).
8. C. Broxmeyer, *Inertial Navigation Systems* (New York: McGraw-Hill, 1964).
9. K. R. Britting, *Inertial Navigation Systems Analysis* (New York: Wiley, 1971).

EXERCISES FOR CHAPTER 9

1. The following noisy measurements were made of the speed of a particular body:

$t_i = iT$, s	1	2	3	4	5	6	7	8
speed, ft/s	21	37	58	82	104	119	138	163

a) Use expression (9.14) to find the least-squares estimate $\hat{a}$ of acceleration a (assumed constant).

b) If the measurement noise has mean zero and variance σ_V^2, find the mean and variance of the estimator $\hat{A}$. Is it an unbiased estimator?

2. A particle starts with a speed u, and moves with constant acceleration a, i.e., the position of the particle at time t is

$$s = ut + at^2/2$$

Find the least squares estimate of u and of a based on the following noisy measurements y_i of position:

t_i (sec)	1	2	3	4	5
y_i (ft)	11	30	49	69	101

Do your results imply recursive or nonrecursive operations on y_i?

3. Can you find a recursive form of the least-squares estimate of Exercise 1?

4. Prove Equation (9.62).

5. Use the discrete Kalman filter equations, summarized in Section 9.6, to get a recursive estimate of a constant scalar random variable X (with mean $\bar{x}$ and variance σ_X^2), based on measurements

$$y_i = x + v_i, \quad i = 1, 2, \ldots$$

The measurement noise is white with mean zero and variance $r = \sigma_V^2$. Write the recursive difference equation for the estimate $\hat{x}_k$. (*Hint*: Because x is not time-varying, $x_{k+1} = x_k$. Also, $p_0 = s_0 = \sigma_X^2$).

6. Write a FORTRAN program to give an iterative solution for the second-order Kalman filtering problem in Example 9.2.

7. Write a FORTRAN program based on the equations summarized in Section 9.6 to provide solutions for general Kalman filtering problems.

8. Suppose that we want to estimate the state of a dynamic system

$$\dot{\mathbf{X}} = A\mathbf{X} + B\mathbf{u}(t) + B_w W(t)$$

where the disturbance input $W(t)$, which is scalar, is *correlated* noise. The autocovariance function of W is

$$\gamma_W(\tau) = \sigma^2 e^{-\beta|\tau|}$$

with power spectral density

$$S_W(s) = \frac{2\sigma^2\beta}{-s^2 + \beta^2}$$

Define a new state variable $x_{n+1} = w$ as the output of a first-order system

$$G(s) = \frac{\sigma\sqrt{2\beta}}{s + \beta}$$

excited by white noise. Augment the original state vector $\mathbf{x}$ with this additional component, and write the new matrices A and B_w for the augmented system.

9. Consider a discrete second-order system

$$\mathbf{X}_{k+1} = \phi\mathbf{X}_k + \Gamma W_k + \mathbf{h}_k$$

where $\mathbf{h}_k$ is a deterministic-input vector, Γ is given by (9.134), and W_k is a purely random input scalar with mean zero and variance $q = \sigma_w^2 = 4$. The elements of ϕ are given by (9.153) through (9.156), with $m = 100$, $a = 0.5$, and $T = 0.1$. Suppose that we want to estimate $\mathbf{X}_k$ without making measurements. Compute and plot the elements of the estimation-error covariance matrix P_k as a function of time.

APPENDIX 9A

State Variables and Discretization of Continuous Linear Systems

9A.1 STATE-VARIABLE REPRESENTATION

A continuous linear system described by an nth-order differential equation can be represented by a set of n first-order differential equations. Consider, for example, the system

$$\frac{d^3y}{dt^3} = a_2\frac{d^2y}{dt^2} + a_1\frac{dy}{dt} + a_0y + b_0u(t) \tag{9A.1}$$

Let $x_1 = y$, $x_2 = dy/dt = dx_1/dt$, and $x_3 = d^2y/dt^2 = dx_2/dt$. Then

$$\begin{bmatrix} \frac{dx_1}{dt} \\ \frac{dx_2}{dt} \\ \frac{dx_3}{dt} \end{bmatrix} = \begin{bmatrix} 0 & 1 & 0 \\ 0 & 0 & 1 \\ a_0 & a_1 & a_2 \end{bmatrix} \begin{bmatrix} x_1 \\ x_2 \\ x_3 \end{bmatrix} + \begin{bmatrix} 0 \\ 0 \\ b_0 \end{bmatrix} u(t) \tag{9A.2}$$

or, in state-variable or matrix form,

$$\dot{\mathbf{x}} = A\mathbf{x} + B\mathbf{u}(t) \tag{9A.3}$$

where the state vector $\mathbf{x}$ is defined by

$$\mathbf{x}' = [x_1 \; x_2 \; x_3]$$

and x_1, x_2, and x_3 are the state variables in this case.

Alternatively, suppose that the system transfer function is given by

$$\begin{aligned} \frac{Y(s)}{U(s)} &= \frac{b_ms^m + b_{m-1}s^{m-1} + \cdots + b_1s + b_0}{s^n + a_{n-1}s^{n-1} + \cdots + a_1s + a} \\ &= \frac{b_ms^{m-n} + b_{m-1}s^{m-1-n} + \cdots + b_1s^{1-n} + b_0s^{-n}}{1 + a_{n-1}s^{-1} + \cdots + a_1s^{1-n} + a_0s^{-n}} \end{aligned} \tag{9A.4}$$

Then

$$Y(s) = (b_ms^{m-n} + b_{m-1}s^{m-1-n} + \cdots + b_1s^{1-n} + b_0s^{-n})M(s) \tag{9A.5}$$

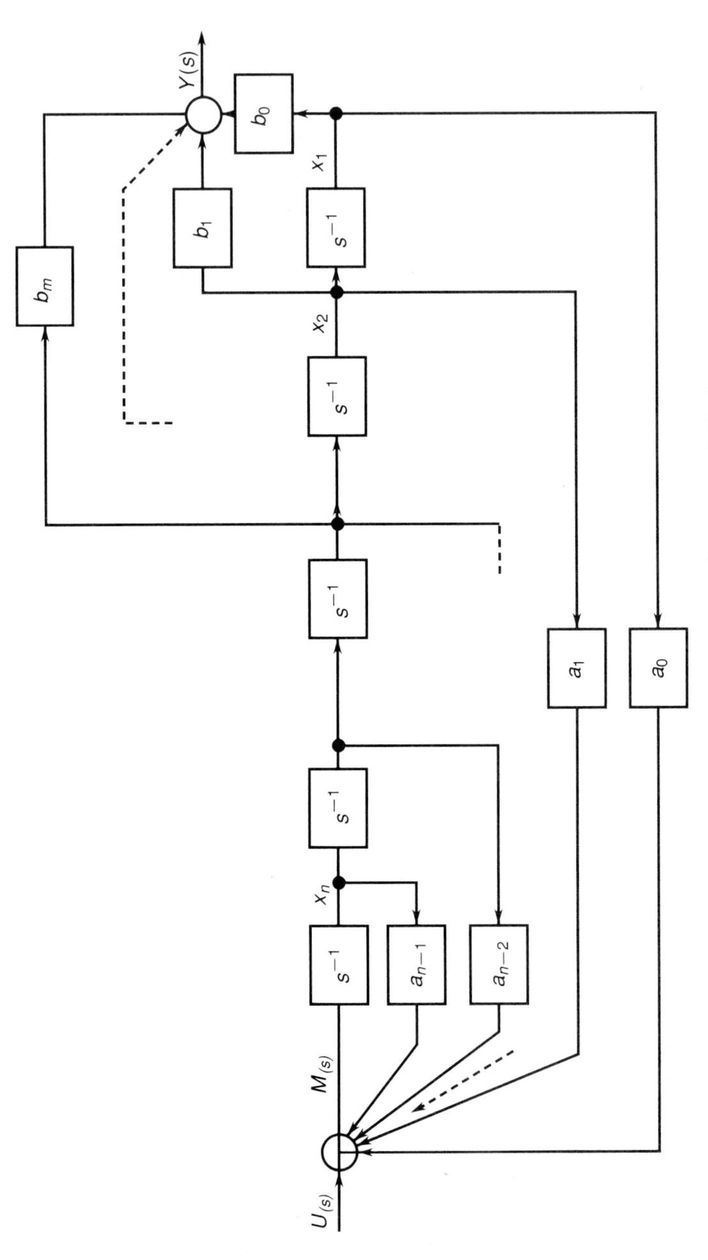

FIGURE 9A.1 System arrangement for state-variable form.

where

$$M(s) = \frac{U(s)}{1 + a_{n-1}s^{-1} + \cdots + a_1 s^{1-n} + a_0 s^{-n}} \tag{9A.6}$$

and $n > m$. The system is represented in block diagram form in Figure 9A.1. Take the output of the n integrators as the n state variables $x_1, x_2, \ldots, x_n$. Then, from the diagram, we can write

$$\begin{bmatrix} \dot{x}_1 \\ \dot{x}_2 \\ \vdots \\ \dot{x}_{n-1} \\ \dot{x}_n \end{bmatrix} = \begin{bmatrix} 0 & 1 & 0 & . & . & . & . & 0 \\ 0 & 0 & 1 & 0 & . & . & . & 0 \\ & & & & & & & \\ 0 & 0 & . & . & . & . & 0 & 1 \\ -a_0 & -a_1 & . & . & . & . & . & -a_{n-1} \end{bmatrix} \begin{bmatrix} x_1 \\ x_2 \\ \vdots \\ x_{n-1} \\ x_n \end{bmatrix} + \begin{bmatrix} 0 \\ 0 \\ \vdots \\ 0 \\ 1 \end{bmatrix} u(t) \tag{9A.7}$$

or

$$\dot{\mathbf{x}} = A\mathbf{x} + B\mathbf{u}(t) \tag{9A.8}$$

Also,

$$\mathbf{y}(t) = b_0 x_1 + b_1 x_2 + \cdots + b_m x_{m+1} \tag{9A.9}$$

or

$$\mathbf{y}(t) = [b_0, b_1, \ldots b_m, 0, \ldots, 0] \begin{bmatrix} x_1 \\ x_2 \\ . \\ . \\ . \\ x_n \end{bmatrix}$$

$$= C\mathbf{x}(t) \tag{9A.10}$$

Even though the input $u(t)$ and output $y(t)$ are scalars in the foregoing example, we could have multiple inputs and multiple outputs, in general. Hence we represent the input and output as vector quantities.

9A.2 SOLUTION OF SYSTEM EQUATIONS IN STATE-VARIABLE FORM

In order to find the solution or output vector $\mathbf{x}(t)$ for a linear dynamic system represented by the differential equation in state-variable form

$$\dot{\mathbf{x}} = A(t)\mathbf{x}(t) + B(t)\mathbf{u}(t) \tag{9A.11}$$

with initial condition $\mathbf{x}(t_0)$, where $\mathbf{x}$ is $n \times 1$ and input vector $\mathbf{u}$ is $r \times 1$, we consider first the homogeneous system

$$\dot{\mathbf{x}} = A(t)\mathbf{x}(t) \tag{9A.12}$$

Let the solution to (9A.12) be

$$\mathbf{x}(t) = \psi(t)\mathbf{c} \tag{9A.13}$$

where ψ is a matrix the columns of which are n linearly independent solutions of (9A.12), and $\mathbf{c}$ is a constant vector to be determined. The matrix ψ is nonsingular and is called the fundmental matrix of the system (9A.12). From (9A.13), it follows that

$$\mathbf{c} = \psi^{-1}(t)\mathbf{x}(t) \tag{9A.14}$$

Differentiate (9A.13) with respect to t, obtaining

$$\dot{\mathbf{x}}(t) = \dot{\psi}(b)\mathbf{c} \tag{9A.15}$$

Substituting for $\mathbf{c}$ from (9A.14) gives

$$\dot{\mathbf{x}}(t) = \dot{\psi}(t)\psi^{-1}(t)\mathbf{x}(t) \tag{9A.16}$$

Comparing (9A.13) and (9A.15) implies that

$$\dot{\psi}(t)\psi^{-1}(t) = A(t) \tag{9A.17}$$

or

$$\dot{\psi}(t) = A(t)\psi(t) \tag{9A.18}$$

which confirms that the fundamental matrix satisfies the homogeneous-system equation.

From (9A.13), the initial state

$$\begin{aligned} \mathbf{x}(t_0) &= \psi(t_0)\,\mathbf{c} \\ &= \psi(t_0)\psi^{-1}(t)\mathbf{x}(t) \end{aligned} \tag{9A.19}$$

Therefore,

$$\begin{aligned} \mathbf{x}(t) &= [\psi(t_0)\psi^{-1}(t)]^{-1}\mathbf{x}(t_0) \\ &= \psi(t)\psi^{-1}(t_0)\mathbf{x}(t_0) \\ &= \phi(t, t_0)\mathbf{x}(t_0) \end{aligned} \tag{9A.20}$$

where $\phi(t, t_0) = \psi(t)\psi^{-1}(t_0)$ is called the state transition matrix because any initial state $\mathbf{x}(t_0)$ can be brought to another state $\mathbf{x}(t)$ by multiplying $\mathbf{x}(t_0)$ by $\phi(t, t_0)$. If we can find $\phi(t, t_0)$ or $\psi(t)$, we can solve the homogeneous system (9A.12). We note that

$$\phi^{-1}(t, t_0) = \phi(t_0, t) \tag{9A.21}$$

$$\phi^{-1}(t_1, t_2) = \phi(t_2, t_1) \tag{9A.22}$$

in general. Also,

$$\phi(t, t) = I \tag{9A.23}$$

the identity matrix.

If $A(t) = A$, a constant matrix, then

$$\phi(t, t_0) = \phi(t - t_0) \tag{9A.24}$$

which is a function of a single variable $r = t - t_0$, the time difference between the two states. Then, from (9A.17),

$$\dot{\phi}(t) = A\phi(t) \tag{9A.25}$$

Take Laplace transforms of both sides of (9A.25) to get

$$s\Phi(s) - \phi(0) = A\Phi(s) \tag{9A.26}$$

or, because $\phi(0) = I$,

$$\Phi(s) = (sI - A)^{-1} \tag{9A.27}$$

Take the inverse Laplace transform to get the state transition matrix for a time-invariant system (A = constant) as

$$\phi(t) = \mathcal{L}^{-1}\{[sI - A]^{-1}\} \tag{9A.28}$$

For example, if

$$A = \begin{bmatrix} 0 & 1 \\ 0 & 0 \end{bmatrix} \tag{9A.29}$$

then, carrying out the operations indicated by (9A.28), we obtain

$$\phi(t) = \begin{bmatrix} 1 & t \\ 0 & 1 \end{bmatrix} \tag{9A.30}$$

and therefore,

$$\phi(t - t_0) = \begin{bmatrix} 1 & (t - t_0) \\ 0 & 1 \end{bmatrix} \tag{9A.31}$$

Alternatively,

$$\begin{aligned} \phi(t) &= e^{At} \\ &= I + At + A^2t^2/2! + \cdots + A^kt^k/k! + \cdots \end{aligned} \tag{9A.32}$$

This can be proved by showing that (9A.32) satisfies the homogeneous equation (9A.12). However, unlike (9A.28), expression (9A.32) does not give a closed-form solution for $\phi(t)$ in general.

We will now consider how to solve the inhomogeneous-system equation (9A.11). Proceed as for the homogeneous system (9A.12), but let the solution be

$$\mathbf{x}(t) = \psi(t)\,\mathbf{c}(t) \tag{9A.33}$$

where $\psi(t)$ is as before, but now $\mathbf{c}(t)$ is a time-varying vector to allow for the effect of the input $\mathbf{u}(t)$. Then

$$\mathbf{c}(t) = \psi^{-1}(t)\mathbf{x}(t) \tag{9A.34}$$

Differentiate (9A.33) to get

$$\dot{\mathbf{x}}(t) = \dot{\psi}(t)\mathbf{c}(t) + \psi(t)\dot{\mathbf{c}}(t) \tag{9A.35}$$

Replace $\mathbf{c}(t)$ by its value from (9A.34). Noting that $\dot{\psi}(t) = A(t)\psi(t)$, we get

$$\begin{aligned}\dot{\mathbf{x}}(t) &= A(t)\psi(t)\psi^{-1}(t)\mathbf{x}(t) + \psi(t)\,\dot{\mathbf{c}}(t) \\ &= A(t)\mathbf{x}(t) + \psi(t)\,\dot{\mathbf{c}}(t) \qquad (9A.36)\end{aligned}$$

Comparing (9A.36) and (9A.11) reveals that

$$\psi(t)\dot{\mathbf{c}}(t) = B(t)\mathbf{u}(t) \qquad (9A.37)$$

Therefore,

$$\dot{\mathbf{c}}(t) = \psi^{-1}(t)B(t)\mathbf{u}(t)$$

and

$$\begin{aligned}\dot{\mathbf{c}}(t) &= \mathbf{c}(t_0) + \int_{t_0}^{t} \psi^{-1}(\sigma)B(\sigma)\mathbf{u}(\sigma)d\sigma \\ &= \psi^{-1}(t_0)\mathbf{x}(t_0) + \int_{t_0}^{t} \psi^{-1}(\sigma)B(\sigma)\mathbf{u}(\sigma)d\sigma \qquad (9A.38)\end{aligned}$$

Then, from (9A.33), it follows that

$$\begin{aligned}\mathbf{x}(t) &= \psi(t)\psi^{-1}(t_0)\mathbf{x}(t_0) + \psi(t)\int_{t_0}^{t} \psi^{-1}(\sigma)B(\sigma)\mathbf{u}(\sigma)d\sigma \\ &= \phi(t, t_0)\mathbf{x}(t_0) + \int_{t_0}^{t} \phi(t, \sigma)B(\sigma)\mathbf{u}(\sigma)d\sigma \qquad (9A.39)\end{aligned}$$

This is the solution of the inhomogeneous equation (9A.11). If state matrix A is constant, then

$$\mathbf{x}(t) = \phi(\tau)\mathbf{x}(t_0) + \int_{t_0}^{t} \phi(t - \sigma)B(\sigma)\mathbf{u}(\sigma)d\sigma \qquad (9A.40)$$

where $\tau = t - t_0$ and the state transition matrix ϕ is evaluated as already described.

9A.3 DISCRETIZING THE CONTINUOUS SYSTEM

We can discretize the system (9A.11) by considering the solution (9A.40) over one sample period, letting $t_0 = t_k = kT$ and $t = t_{k+1} = (k + 1)T$. Then (9A.40) becomes

$$\mathbf{x}_{k+1} = \phi(T)\mathbf{x}_k + \int_{kT}^{(k+1)T} \phi(kT + T - \sigma)B(\sigma)\mathbf{u}(\sigma)d\sigma \qquad (9A.41)$$

If $\mathbf{u}(\sigma)$ is assumed to be constant over the sampling period, as it often is in practice—that is,

$$\mathbf{u}(\sigma) = \mathbf{u}(kT) = \mathbf{u}_k,\, kT \leq \sigma < (k + 1)T \qquad (9A.42)$$

and B is a constant matrix, then (9A.41) can be written

$$\mathbf{x}_{k+1} = \phi(T)\mathbf{x}_k + \int_{kT}^{(k+1)T} \phi(kT + T - \sigma)d\sigma B\mathbf{u}_k \tag{9A.43}$$

Change the variable of integration. Let

$$\eta = kT + T - \sigma$$

Equation (9A.43) becomes

$$\mathbf{x}_{k+1} = \phi(T)\mathbf{x}_k + \int_0^T \phi(\eta)d\eta B\mathbf{u}_k \tag{9A.44}$$

Define

$$\Gamma(T) = \int_0^T \phi(\eta)d\eta B \tag{9A.45}$$

Then (9A.44) can be written

$$\mathbf{x}_{k+1} = \phi(T)\mathbf{x}_k + \Gamma(T)\mathbf{u}_k \tag{9A.46}$$

This is the discretized form of system (9A.11).

The argument T is often omitted because it is understood. When the state transition matrix ϕ has been found for the system, the control matrix Γ can be evaluated from (9A.45).

9A.4 LINEARIZING A NONLINEAR SYSTEM ABOUT A NOMINAL TRAJECTORY

Consider a continuous nonlinear system

$$\dot{\mathbf{x}} = \mathbf{f}(\mathbf{x}, \mathbf{u}(t)) \tag{9A.47}$$

where $\mathbf{x}$ is $n \times 1$, $\mathbf{u}$ is $r \times 1$, and $\mathbf{f}$ is an $n \times 1$ differentiable vector-valued function:

$$\mathbf{f} = [f_1, f_2, \ldots, f_n]'$$

From a given initial state $\mathbf{x}_0$, we can solve (9A.47) by programming it on computer, thereby obtaining a *trajectory* $\mathbf{x}(t)$. In general, however, analysis of such a system is difficult. In many cases, we are interested not in the trajectory $\mathbf{x}(t)$ itself, but in deviations $\Delta\mathbf{x}$ about a known trajectory that are caused by perturbations $\Delta\mathbf{u}$ in the input vector $\mathbf{u}(t)$. Suppose we have a nominal trajectory $\mathbf{x}_0(t)$ resulting from a nominal input vector $\mathbf{u}_0(t)$. Then

$$\dot{\mathbf{x}}_0 = f(\mathbf{x}_0, \mathbf{u}_0(t)) \tag{9A.48}$$

Let $\mathbf{x} = \mathbf{x}_0 + \Delta\mathbf{x}$ and $\mathbf{u} = \mathbf{u}_0 + \Delta\mathbf{u}$. By (9A.47),

$$\frac{d}{dt}(\mathbf{x}_0 + \Delta\mathbf{x}) = f(\mathbf{x}_0 + \Delta\mathbf{x}, \mathbf{u}_0(t) + \Delta\mathbf{u}) \tag{9A.49}$$

Expand the right-hand side in Taylor series about the nominal values, obtaining

$$\dot{\mathbf{x}}_0 + \frac{d}{dt}(\Delta \mathbf{x}) = \mathbf{f}(\mathbf{x}_0, \mathbf{u}_0(t)) + \left.\frac{\partial \mathbf{f}}{\partial \mathbf{x}}\right|_0 \Delta \mathbf{x} + \left.\frac{\partial \mathbf{f}}{\partial \mathbf{u}}\right|_0 \Delta \mathbf{u} + \text{(higher-order terms)} \tag{9A.50}$$

where the derivative terms are evaluated on the nominal trajectory. The first terms on both sides of (9A.50) cancel, by (9A.48). If we assume that the deviations $\Delta \mathbf{x}$ and $\Delta \mathbf{u}$ are small enough for us to neglect the higher-order terms, (9A.50) becomes

$$\frac{d}{dt}(\Delta \mathbf{x}) = \left.\frac{\partial \mathbf{f}}{\partial \mathbf{x}}\right|_0 \Delta \mathbf{x} + \left.\frac{\partial \mathbf{f}}{\partial \mathbf{u}}\right|_0 \Delta \mathbf{u} \tag{9A.51}$$

If we write

$$A(t) = \left.\frac{\partial \mathbf{f}}{\partial \mathbf{x}}\right|_0$$

and

$$B(t) = \left.\frac{\partial \mathbf{f}}{\partial \mathbf{u}}\right|_0$$

where A is an $n \times n$ matrix and B is an $n \times r$ matrix, then (9A.51) is in the form of a linear dynamic system (9A.11) with state vector $\mathbf{x}$ replaced by $\Delta \mathbf{x}$, and $\mathbf{u}$ replaced by $\Delta \mathbf{u}$. In other words, we have linearized the system (9A.47) about the nominal trajectory $\mathbf{x}_0$. This technique is used in Example 9.3.

9A.5 PROPAGATION OF ESTIMATION ERROR COVARIANCE

Suppose we wish to estimate the state of a discrete linear system that is acted on by a random-disturbance input. Let the system be

$$\mathbf{X}_{k-1} = \phi_k \mathbf{X}_k + \Gamma_k \mathbf{W}_k + \mathbf{h}_k \tag{9A.52}$$

where $\mathbf{h}_k$ is a deterministic-input vector and $\mathbf{W}_k$ is a purely random-input vector with mean zero and covariance matrix $Q\delta_{\ell,0}$. The state vector $\mathbf{X}_k$ is a random process because of the random input. If we make measurements linearly related to the state, we can get an estimate of $\mathbf{X}_k$ by means of a Kalman filter, as described in Section 9.5. This estimate is best in the sense of minimizing the mean square error between the true value and the estimate.

What happens if we have no measurement to help us in estimating $\mathbf{X}_k$? Suppose that $\hat{\mathbf{X}}_k$ is the estimate. In the absence of a measurement, the best we can do is to assume that the estimate also satisfies an equation of the form (9A.52),

$$\hat{\mathbf{X}}_{k+1} = \phi \hat{\mathbf{X}}_k + \mathbf{h}_k \tag{9A.53}$$

where we omit the random input $\mathbf{W}_k$ because it is unpredictable. Subtract (9A.53) from (9A.52) to get the equation satisfied by the estimation error $\mathbf{E}_k = \mathbf{X}_k - \hat{\mathbf{X}}_k$ as

$$\mathbf{E}_{k+1} = \phi_k \mathbf{E}_k + \Gamma_k \mathbf{W}_k \tag{9A.54}$$

Take expected values of both sides

$$\begin{aligned} \langle \mathbf{E}_{k+1} \rangle &= \phi_k \langle \mathbf{E}_k \rangle + \Gamma_k \langle \mathbf{W}_k \rangle \\ &= \phi_k \langle \mathbf{E}_k \rangle \end{aligned} \tag{9A.55}$$

because $\langle \mathbf{W}_k \rangle = 0$. Because ϕ_k is arbitrary, (9A.55) will be satisfied only if $\langle \mathbf{E}_k \rangle = 0$. Hence $\langle \mathbf{X}_k - \hat{\mathbf{X}}_k \rangle = 0$, and therefore $\hat{\mathbf{X}}_k$, satisfying (9A.53), is an unbiased estimator of $\mathbf{X}_k$.

Define the error covariance as

$$P_k = \langle \mathbf{E}_k \mathbf{E}'_k \rangle \tag{9A.56}$$

Let us determine how this propagates with time. We know that

$$\begin{aligned} P_{k+1} &= \langle \mathbf{E}_{k+1} \mathbf{E}'_{k+1} \rangle \\ &= \langle (\phi_k \mathbf{E}_k + \Gamma_k W_k)(\phi_k \mathbf{E}_k + \Gamma_k \mathbf{W}_k)' \rangle \end{aligned} \tag{9A.57}$$

from (9A.54). Multiplying the terms, we get

$$\begin{aligned} P_{k+1} &= \langle \phi_k \mathbf{E}_k \mathbf{E}'_k \phi_k + \Gamma_k \mathbf{W}_k \mathbf{E}'_k \phi'_k + \phi_k \mathbf{E}_k \mathbf{W}'_k \Gamma'_k + \Gamma_k \mathbf{W}_k \mathbf{W}'_k \Gamma'_k \rangle \\ &= \phi_k \langle \mathbf{E}_k \mathbf{E}'_k \rangle \phi'_k + \Gamma_k \langle \mathbf{W}_k \mathbf{E}'_k \rangle \phi'_k + \phi_k \langle \mathbf{E}_k \mathbf{W}'_k \rangle \Gamma'_k + \Gamma_k \langle \mathbf{W}_k \mathbf{W}'_k \rangle \Gamma'_k \end{aligned}$$

The second and third terms on the right-hand side are zero, because $\mathbf{E}_k$ is independent of $\mathbf{W}_k$ ($\mathbf{E}_k$ depends on $\mathbf{W}_{k-1}$). Therefore,

$$\begin{aligned} P_{k+1} &= \phi_k \langle \mathbf{E}_k \mathbf{E}'_k \rangle \phi'_k + \Gamma_k \langle \mathbf{W}_k \mathbf{W}'_k \rangle \Gamma'_k \\ &= \phi_k P_k \phi'_k + \Gamma_k Q_k \Gamma'_k \end{aligned} \tag{9A.58}$$

This equation indicates how the error covariance propagates with time when there are no measurements to use to correct the estimates.

REFERENCES FOR APPENDIX 9A

1. K. Ogata, *Modern Control Engineering* (Englewood Cliffs, NJ: Prentice-Hall, 1970).

APPENDIX 9B

Differentiation of a Quadratic Form

9B.1 INTRODUCTION

A quadratic form Q in n variables $x_1, x_2, \ldots, x_n$ is one in which each term contains either the square of a variable or the product of two variables. Thus

$$Q = \sum_{i=1}^{n} \sum_{j=1}^{n} a_{ij} x_i x_j$$

$$= [x_1 \ldots x_n] \begin{bmatrix} a_{11} & \cdots & a_{1n} \\ \vdots & & \vdots \\ a_{n1} & \cdots & a_{nn} \end{bmatrix} \begin{bmatrix} x_i \\ \vdots \\ x_n \end{bmatrix}$$

$$= \mathbf{x}' A \mathbf{x} \tag{9B.1}$$

where $\mathbf{x}' = [x_1, \ldots, x_n]$ is the row vector that is the transpose of the column vector $\mathbf{x}$. From (9B.1), it is evident that Q is a scalar quantity (a 1×1 matrix).

Often, arriving at the optimal design of a filter or control system involves differentiating a performance index in a quadratic form in order to minimize or maximize the index with respect to some variable. If the quadratic form is expressed in matrix form, as in (9B.1), it is often desirable to differentiate it with respect to a vector (column matrix) or to a matrix. Differentiating with respect to a vector or matrix means differentiating with respect to the elements of the vector or matrix. The differentiation is reviewed briefly below. Further details can be found in the literature (for example, [1]).

9B.2 DIFFERENTIATION WITH RESPECT TO A SCALAR

Suppose we want to differentiate Q as defiined in (9B.1) with respect to one element of vector $\mathbf{x}$, namely x_k. Then

$$\frac{\partial Q}{\partial x_k} = \frac{\partial}{\partial x_k}\left[a_{kk}x_k^2 + x_k\left(\sum_{\substack{j=1\\j\neq k}}^{n} a_{kj}x_j + \sum_{\substack{i=1\\i\neq k}}^{n} a_{ik}x_i\right)\right]$$

$$= 2a_{kk}x_k + \sum_{\substack{j=1\\j\neq k}}^{n} a_{kj}x_j + \sum_{\substack{i=1\\i\neq k}}^{n} a_{ik}x_i$$

$$= \sum_{j=1}^{n} a_{kj}x_j + \sum_{i=1}^{n} a_{ik}x_i$$

$$= (Ax)_k + (A'\mathbf{x})_k, \quad k = 1, 2, \ldots, n \tag{9B.2}$$

where $(A\mathbf{x})_k$ and $(A'\mathbf{x})_k$ are the kth rows of matrices $A\mathbf{x}$ and $A'\mathbf{x}$, respectively.

9B.3 DIFFERENTIATION WITH RESPECT TO A VECTOR

Define the vector differential operator

$$\nabla_{\mathbf{x}} = \begin{bmatrix} \frac{\partial}{\partial x_1} \\ \vdots \\ \frac{\partial}{\partial x_n} \end{bmatrix} \tag{9B.3}$$

which is called del $\mathbf{x}$. Then, from (9B.2),

$$\nabla_{\mathbf{x}} Q = \begin{bmatrix} (Ax)_1 + (A'x)_1 \\ \vdots \\ (Ax)_n + (A'x)_n \end{bmatrix} = A\mathbf{x} + A'\mathbf{x} \tag{9B.4}$$

If A is symmetric (that is, $A' = A$), then

$$\nabla_{\mathbf{x}} Q = 2A\mathbf{x} \tag{9B.5}$$

Often it is required to differentiate an unsymmetric form such as

$$R = \mathbf{x}'F\mathbf{y}$$

$$= [x_1 \ldots x_n] \begin{bmatrix} f_{11} & \cdots & f_{1m} \\ & \vdots & \\ f_{n1} & \cdots & f_{nm} \end{bmatrix} \begin{bmatrix} y_1 \\ \vdots \\ y_m \end{bmatrix} \tag{9B.6}$$

with respect to vector $\mathbf{x}$. If we write (9B.6) in the form

$$R = x_1 \sum_{j=1}^{m} f_{1j}y_j + \cdots + x_k \sum_{j=1}^{m} f_{kj}y_j + \cdots + x_n \sum_{j=1}^{m} f_{nj}y_j \tag{9B.7}$$

it is evident that

$$\frac{\partial R}{\partial x_k} = \sum_{j=1}^{n} f_{kj} y_j = k^{th} \text{ row of } F\mathbf{y} \tag{9B.8}$$

Therefore,

$$\nabla_{\mathbf{x}}(\mathbf{x}'\mathbf{F}\mathbf{y}) = \begin{bmatrix} \text{1st row of } F\mathbf{y} \\ \vdots \\ n\text{th row of } F\mathbf{y} \end{bmatrix} = F\mathbf{y} \tag{9B.9}$$

Similarly,

$$(F\mathbf{y})'\mathbf{x} = \left(\sum_{j=1}^{m} f_{1j} y_j\right) x_1 + \cdots + \left(\sum_{j=1}^{m} f_{kj} y_j\right) x_k + \cdots + \left(\sum_{j=1}^{m} f_{nk} y_j\right) x_n$$

Hence

$$\begin{aligned} \frac{\partial[(F\mathbf{y})'\mathbf{x}]}{\partial x_k} &= \sum_{j=1}^{m} f_{kj} y_j \\ &= k\text{th row of } F\mathbf{y} \end{aligned} \tag{9B.10}$$

Therefore,

$$\nabla_{\mathbf{x}}(F\mathbf{y})'\mathbf{x} = F\mathbf{y} \tag{9B.11}$$

9B.4 DIFFERENTIATION OF THE QUADRATIC FORM $y'F'Fy$ WITH RESPECT TO MATRIX F

We have

$$\begin{aligned} \mathbf{y}'F'F\mathbf{y} &= [y_1 \dots y_m] \begin{bmatrix} f_{11} & \cdots & f_{n1} \\ \vdots & & \vdots \\ f_{1m} & \cdots & f_{nm} \end{bmatrix} \begin{bmatrix} f_{11} & \cdots & f_{1m} \\ \vdots & & \vdots \\ f_{n1} & \cdots & f_{nm} \end{bmatrix} \begin{bmatrix} y_1 \\ \vdots \\ y_m \end{bmatrix} \\ &= \left(\sum_{i=1}^{m} f_{1i} y_i, \dots, \sum_{i=1}^{m} f_{ni} y_i\right) \begin{bmatrix} \sum_{j=1}^{m} f_{1j} y_j \\ \vdots \\ \sum_{j=1}^{m} f_{nj} y_j \end{bmatrix} \\ &= \sum_{i=1}^{m} f_{1i} y_i \sum_{j=1}^{m} f_{1j} y_j + \cdots + \sum_{i=1}^{m} f_{ni} y_i \sum_{j=1}^{m} f_{nj} y_j \end{aligned} \tag{9B.12}$$

General term:

$$\sum_{i=1}^{m} f_{ki} y_i \sum_{j=1}^{m} f_{kj} y_j$$

$$\frac{\partial}{\partial f_{k\ell}}(\mathbf{y}' F' F \mathbf{y}) = y_\ell \sum_{j=1}^{m} f_{kj} y_j + \sum_{i=1}^{m} f_{ki} y_i y_\ell$$

$$= 2 y_\ell \sum_{j=1}^{m} f_{kj} y_j, \quad \ell = 1, \ldots, m \tag{9B.13}$$

Let $S = \mathbf{y}' F' F \mathbf{y}$.

$$\nabla_F S \triangleq \begin{bmatrix} \frac{\partial S}{\partial f_{11}} & \cdots & \frac{\partial S}{\partial f_{1m}} \\ \vdots & & \vdots \\ \frac{\partial S}{\partial f_{n1}} & \cdots & \frac{\partial S}{\partial f_{nm}} \end{bmatrix}$$

$$= 2 \begin{bmatrix} y_1 \sum_{j=1}^{m} f_{1j} y_j & \cdots & y_m \sum_{j=1}^{m} f_{1j} y_j \\ \vdots & & \vdots \\ y_1 \sum_{j=1}^{m} f_{nj} y_j & \cdots & y_m \sum_{j=1}^{m} f_{nj} y_j \end{bmatrix}$$

$$= 2 \begin{bmatrix} f_{11} & \cdots & f_{1m} \\ \vdots & & \vdots \\ f_{n1} & \cdots & f_{nm} \end{bmatrix} \begin{bmatrix} y_1 \\ \vdots \\ y_m \end{bmatrix} [y_1 \ldots y_m] \tag{9B.14}$$

Therefore,

$$\nabla_F(\mathbf{y}' F' F \mathbf{y}) = 2 F \mathbf{y} \mathbf{y}' \tag{9B.15}$$

9B.5 DIFFERENTIATION OF $x'Fy$ AND $y'F'x$ WITH RESPECT TO MATRIX F

We can write

$$\mathbf{x}' F \mathbf{y} = (x_1 \cdots x_n) \begin{bmatrix} f_{11} & \cdots & f_{1m} \\ \vdots & & \vdots \\ f_{n1} & \cdots & f_{nm} \end{bmatrix} \begin{bmatrix} y_1 \\ \vdots \\ y_m \end{bmatrix}$$

$$= x_1 \sum_{j=1}^{m} f_{1j} y_j + \cdots + x_n \sum_{j=1}^{m} f_{nj} y_j \tag{9B.16}$$

Therefore,

$$\frac{\partial \mathbf{x}' F \mathbf{y}}{\partial f_{k\ell}} = x_k y_\ell \tag{9B.17}$$

Hence

$$\nabla_F(\mathbf{x}' F \mathbf{y}) = \begin{bmatrix} x_1 y_1 & \cdots & x_1 y_m \\ \vdots & & \vdots \\ x_n y_1 & \cdots & x_n y_m \end{bmatrix} = \mathbf{x}\mathbf{y}' \tag{9B.18}$$

Similarly,

$$\mathbf{y}' F' \mathbf{x} = [y_1 \cdots y_m] \begin{bmatrix} f_{11} & \cdots & f_{n1} \\ \vdots & & \vdots \\ f_{1m} & \cdots & f_{nm} \end{bmatrix} \begin{bmatrix} x_1 \\ \vdots \\ x_n \end{bmatrix}$$

$$= y_1 \sum_{j=1}^{n} f_{j1} x_j + \cdots + y_m \sum_{j=1}^{n} f_{jm} x_j \tag{9B.19}$$

Therefore,

$$\frac{\partial (\mathbf{y}' F' \mathbf{x})}{\partial f_{k\ell}} = y_\ell x_k \tag{9B.20}$$

Hence

$$\nabla_F(\mathbf{y}' F' \mathbf{x}) = \mathbf{x}\mathbf{y}' \tag{9B.21}$$

9B.6 SUMMARY

The main results of this appendix can be summarized as follows:

1.
$$\nabla_{\mathbf{x}}(\mathbf{x}' A \mathbf{x}) = A\mathbf{x} + A'\mathbf{x} \tag{9B.22}$$
$$= 2A\mathbf{x} \tag{9B.23}$$

if A is symmetric.

2.
$$\nabla_{\mathbf{x}}(\mathbf{x}' F \mathbf{y}) = F\mathbf{y} \tag{9B.24}$$

3.
$$\nabla_{\mathbf{x}}(\mathbf{y}' F' \mathbf{x}) = \nabla_{\mathbf{x}}[(F\mathbf{y})'\mathbf{x}] = F\mathbf{y} \tag{9B.25}$$

4.
$$\nabla_F(\mathbf{y}' F' F \mathbf{y}) = 2F\mathbf{y}\mathbf{y}' \tag{9B.26}$$

5. $$\nabla_F(\mathbf{x}'F\mathbf{y}) = \mathbf{x}\mathbf{y}' \tag{9B.27}$$

6. $$\nabla_F(\mathbf{y}'F'\mathbf{x}) = \mathbf{x}\mathbf{y}' \tag{9B.28}$$

REFERENCES FOR APPENDIX 9B

1 Tou, J. T., *Modern Control Theory* (New York: McGraw-Hill, 1964), Chap. 2.

APPENDIX 9C

Diskette Program Example

The FORTRAN program TRACK implements an alpha–beta tracker used to estimate range and range rate from noisy measurements of range. It is arranged for interactive operation.

The user is prompted for alpha, beta, and pulse period values. He or she may input a particular range measurement or use the default option of a constant-speed target with range measurements corrupted by Gaussian noise. Estimates of range and range rate are stored in a file TRACK.DAT created by the program. The default option is used in the example.

```
ENTER VALUE FOR ALPHA:
.2
ENTER VALUE FOR BETA:
.2
ENTER VALUE FOR T, TIME BETWEEN RADAR PULSES:
1.
THE USER MAY INPUT A NOISY RANGE MEASUREMENT Y(K) FOR
FILTERING, OR USE THE DEFAULT OPTION OF A CONSTANT SPEED TARGET
WITH RANGE MEASUREMENTS CORRUPTED BY GAUSSIAN
NOISE
DO YOU WANT TO PROCESS AN ARBITRARY INPUT SEQUENCE,
(IS), OR THE EXAMPLE (EX) (ENTER IS/EX):
EX
INPUT AND OUTPUT DATA IN FILE TRACK.DAT
Stop - Program terminated.
```

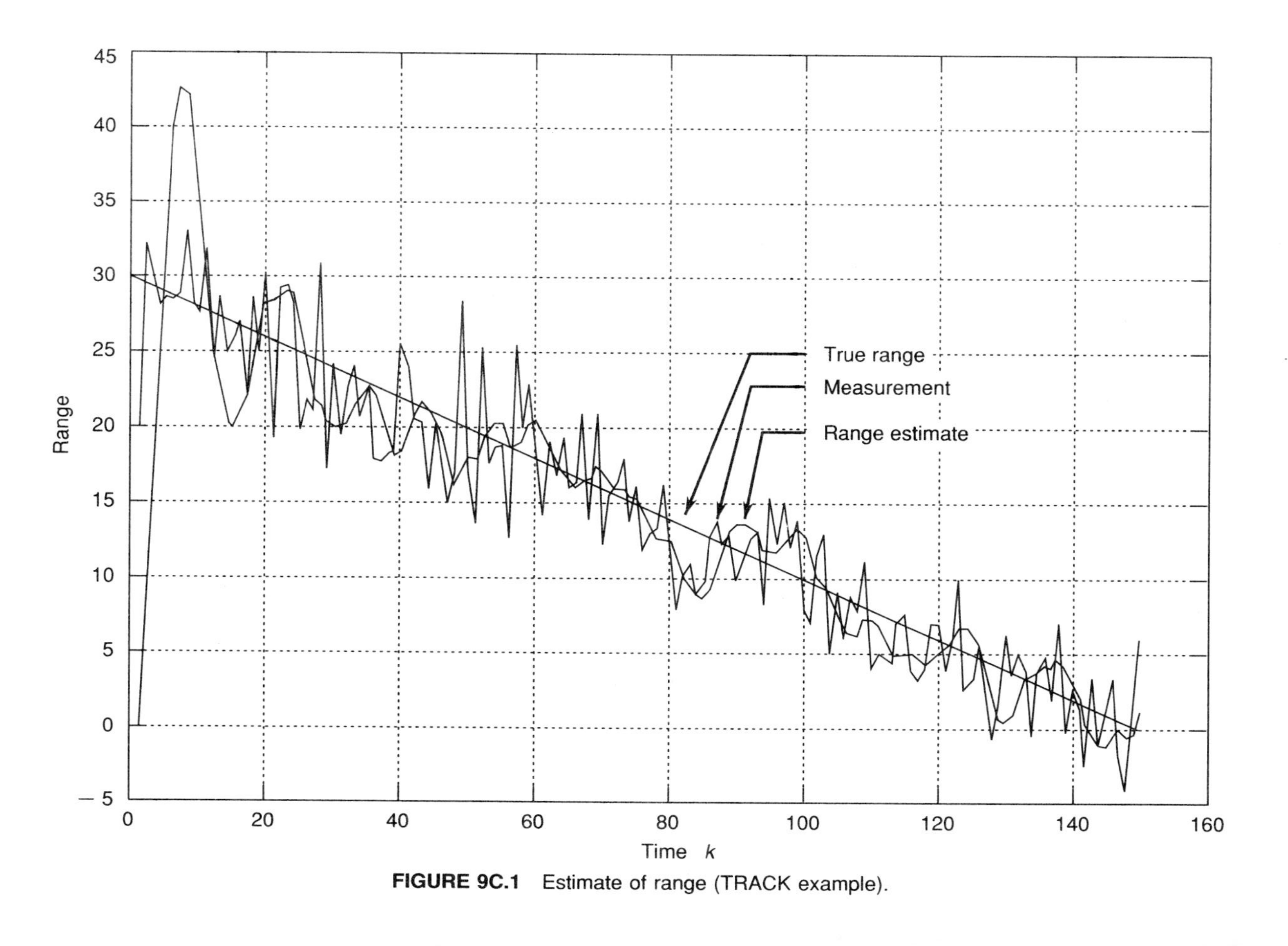

FIGURE 9C.1 Estimate of range (TRACK example).

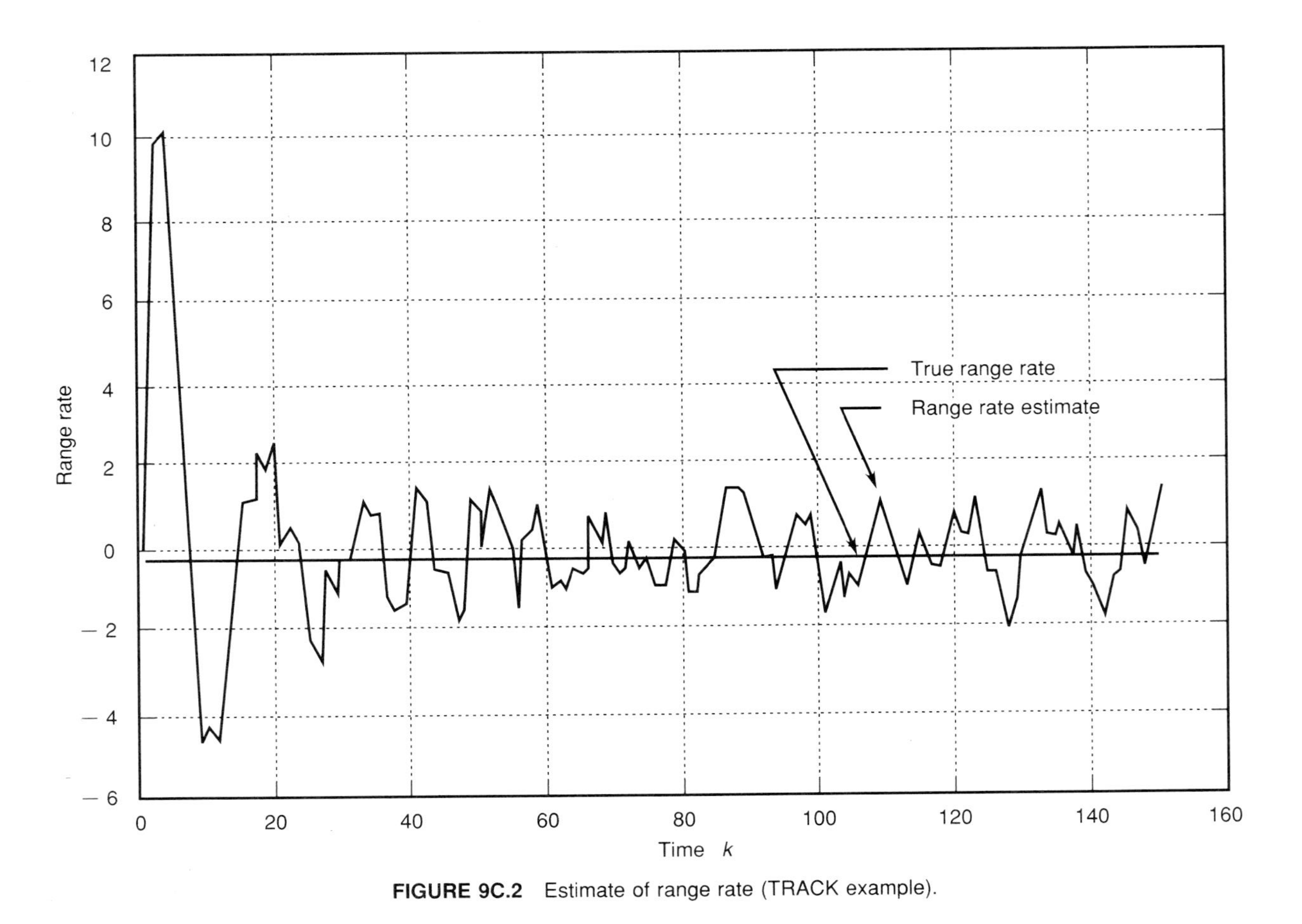

FIGURE 9C.2 Estimate of range rate (TRACK example).

Index